Lecture Notes in Computer Science 9262

Commenced Publication in 1973
Founding and Former Series Editors:
Gerhard Goos, Juris Hartmanis, and Jan van Leeuwen

More information about this series at http://www.springer.com/series/7409

Qiming Chen · Abdelkader Hameurlain
Farouk Toumani · Roland Wagner
Hendrik Decker (Eds.)

Database and Expert Systems Applications

26th International Conference, DEXA 2015
Valencia, Spain, September 1–4, 2015
Proceedings, Part II

 Springer

Editors

Qiming Chen
Hewlett-Packard Enterprise
Sunnyvale, CA
USA

Roland Wagner
University of Linz
Linz
Austria

Abdelkader Hameurlain
Paul Sabatier University
Toulouse
France

Hendrik Decker
Universidad Politécnica de Valencia
Valencia
Spain

Farouk Toumani
Blaise Pascal University
Aubiere
France

ISSN 0302-9743   ISSN 1611-3349 (electronic)
Lecture Notes in Computer Science
ISBN 978-3-319-22851-8   ISBN 978-3-319-22852-5   (eBook)
DOI 10.1007/978-3-319-22852-5

Library of Congress Control Number: 2015946079

LNCS Sublibrary: SL3 – Information Systems and Applications, incl. Internet/Web, and HCI

Springer International Publishing AG Switzerland is part of Springer Science+Business Media
(www.springer.com)

Preface

Since 1990, DEXA has established itself as a premier forum in the intersection of data management, knowledge engineering, and artificial intelligence, providing a unique opportunity for the researchers, users, practitioners, and developers from multiple disciplines to present the state-of-the-art in these fields, exchange research ideas, share industry experiences, and explore future directions, which will help proliferate inter-disciplinary discovery, drive innovation, and provide commercial opportunity. The 26th International Conference on Database and Expert Systems Applications took place in Valencia, Spain, during September 1–4, 2015. We are proud to present its program in these proceedings.

Each of the main days of the conference started out with a keynote by a distin-guished scientist: Shahram Ghandeharizadeh from USC Database Laboratory of Uni-versity of Southern California, USA; Juan Carlos Perez Cortes from Universidad Politecnica de Valencia (UPV), Spain; and Roland Traunmüller from Johannes Kepler Universität Linz, Austria. Accompanying the main conference were eight related conferences and workshops.

We received 125 submissions for the research track and industrial tracks, which is a testament to our community's high regard for DEXA, and we thank all authors for submitting their innovative work to the conference.

The Program Committee (PC) comprised highly reputational scientists from all over the word. The paper selection process consisted of three distinct phases: the initial reviews by the PC members with each paper being evaluated by four reviewers on average, the early opinion exchanges, and the further discussion for fine-tuning of the review results. This year many submissions report competitive scientific and engi-neering breakthroughs; however, because of space limitations we had to reject some high-quality papers. The final program featured 40 full papers and 32 short papers.

In these volumes, we also include, as a special section, the papers accepted for the 8th International Conference on Data Management in Cloud, Grid and P2P Systems (Globe 2015), which took place in Valencia, Spain, on September 1, 2015. The Globe conference provides opportunities for academics and industry researchers to present, exchange, and discuss the latest data management research and applications in cloud, grid, and peer-to-peer systems. Globe 2015 received 13 submissions for the research track. Each paper was evaluated by three reviewers on average. The reviewing process led to the acceptance of eight full papers for presentation at the conference and inclusion in this LNCS volume. The selected papers focus mainly on MapReduce framework (e.g., load balancing, optimization), security, data privacy, query rewriting, and streaming.

As is the tradition of DEXA conferences, all accepted papers are published by Springer. Authors of selected papers presented at the conference will be invited to submit extended versions of their papers for publication in the Springer journal *Transactions on Large-Scale Data- and Knowledge-Centered Systems (TLDKS)*.

The success of DEXA 2015 is a result of collegial teamwork from many individuals. We thank Isidro Ramos, Technical University of Valencia, Spain, the honorary chair of DEXA 2015. We thank Vladimir Marik, Czech Technical University, Czech Republic, the publication chair, as well as Marcus Spies, Ludwig-Maximilians-Universität München, Germany, A. Min Tjoa, Technical University of Vienna, Austria, and Roland R. Wagner, FAW, University of Linz, Austria, the workshop chairs. Our appreciations also go to all the PC members, the external reviewers, and the session chairs. Furthermore, we especially express our deep appreciation to Gabriela Wagner, the DEXA secretary, and to the local organization team, for their outstanding work put in over many months.

Finally, we once again thank all the authors, presenters, and participants of the conference. We hope that you enjoy the proceedings.

June 2015

Abdelkader Hameurlain
Qiming Chen
Farouk Toumani

Organization

General Chair

Hendrik Decker   Instituto Tecnológico de Informática, Valencia, Spain
Roland R. Wagner   Johannes Kepler University Linz, Austria

Program Committee Co-chairs

Abdelkader Hameurlain   IRIT, Paul Sabatier University, Toulouse, France
Qiming Chen   HP Labs Palo Alto, CA, USA

Honorary Chair

Isidro Ramos   Technical University of Valencia, Spain

Program Committee

Abdennadher Slim   German University, Cairo, Egypt
Abramowicz Witold   The Poznan University of Economics, Poland
Afsarmanesh Hamideh   University of Amsterdam, The Netherlands
Albertoni Riccardo   Institute of Applied Mathematics and Information
                          Technologies - Italian National Council of Research,
                          Italy
Alfaro Eva   UPV, Spain
Anane Rachid   Coventry University, UK
Appice Annalisa   Università degli Studi di Bari, Italy
Atay Mustafa   Winston-Salem State University, USA
Bakiras Spiridon   City University of New York, USA
Bao Jie   Microsoft Research Asia, China
Bao Zhifeng   National University of Singapore, Singapore
Bellatreche Ladjel   ENSMA, France
Bennani Nadia   INSA Lyon, France
Benyoucef Morad   University of Ottawa, Canada
Berrut Catherine   Grenoble University, France
Biswas Debmalya   Swisscom, Switzerland
Bouguettaya Athman   RMIT, Australia
Boussaid Omar   University of Lyon, France
Bressan Stephane   National University of Singapore, Singapore
Bu Yingyi   University of California, Irvine, USA
Camarinha-Matos Luis M.   Universidade Nova de Lisboa + Uninova, Portugal
Catania Barbara   DISI, University of Genoa, Italy

Ceci Michelangelo	University of Bari, Italy
Chen Cindy	University of Massachusetts Lowell, USA
Chen Phoebe	La Trobe University, Australia
Chen Qiming	HP Labs Palo Alto, CA, USA
Chen Shu-Ching	Florida International University, USA
Cheng Hao	Yahoo, USA
Chevalier Max	IRIT - SIG, Université de Toulouse, France
Choi Byron	Hong Kong Baptist University, Hong Kong, SAR China
Christiansen Henning	Roskilde University, Denmark
Chun Soon Ae	City University of New York, USA
Dahl Deborah	Conversational Technologies, USA
Darmont Jérôme	Université de Lyon, ERIC Lyon 2, France
de Carvalho Andre	University of Sao Paulo, Brazil
De Virgilio Roberto	Università Roma Tre, Italy
Decker Hendrik	Instituto Tecnológico de Informática, Valencia, Spain, Spain
Deng Zhi-Hong	Peking University, China
Deufemia Vincenzo	Università degli Studi di Salerno, Italy
Dibie-Barthélemy Juliette	AgroParisTech, France
Ding Ying	Indiana University, USA
Dobbie Gill	University of Auckland, New Zealand
Dou Dejing	University of Oregon, USA
du Mouza Cedric	CNAM, France
Eder Johann	University of Klagenfurt, Austria
El-Beltagy Samhaa	Nile University, Cairo, Egypt
Embury Suzanne	The University of Manchester, UK
Endres Markus	University of Augsburg, Germany
Fazzinga Bettina	ICAR-CNR, Italy
Fegaras Leonidas	The University of Texas at Arlington, USA
Felea Victor	"Al. I. Cuza" University of Iasi, Romania
Ferilli Stefano	University of Bari, Italy
Ferrarotti Flavio	Software Competence Center Hagenberg, Austria
Ferrucci Filomena	Università di Salerno, Italy
Fomichov Vladimir	School of Business Informatics, National Research University Higher School of Economics, Moscow, Russian Federation
Frasincar Flavius	Erasmus University Rotterdam, The Netherlands
Freudenthaler Bernhard	Software Competence Center Hagenberg GmbH, Austria
Fukuda Hiroaki	Shibaura Institute of Technology, Japan
Furnell Steven	Plymouth University, UK
Gangopadhyay Aryya	University of Maryland Baltimore County, USA
Gao Yunjun	Zhejiang University, China
Garfield Joy	University of Worcester, UK
Gergatsoulis Manolis	Ionian University, Greece

Grabot Bernard	LGP-ENIT, France
Grandi Fabio	University of Bologna, Italy
Gravino Carmine	University of Salerno, Italy
Groppe Sven	Lübeck University, Germany
Grosky William	University of Michigan, USA
Grzymala-Busse Jerzy	University of Kansas, USA
Guerra Francesco	Università degli Studi Di Modena e Reggio Emilia, Italy
Guerrini Giovanna	University of Genova, Italy
Guzzo Antonella	University of Calabria, Italy
Hameurlain Abdelkader	Paul Sabatier University, France
Hamidah Ibrahim	Universiti Putra Malaysia, Malaysia
Hara Takahiro	Osaka University, Japan
Hsu Wynne	National University of Singapore, Singapore
Hua Yu	Huazhong University of Science and Technology, China
Huang Jimmy	York University, Canada, Canada
Huang Xiao-Yu	South China University of Technology, China
Huptych Michal	Czech Technical University in Prague, Czech Republic
Hwang San-Yih	National Sun Yat-Sen University, Taiwan
Härder Theo	TU Kaiserslautern, Germany
Iacob Ionut Emil	Georgia Southern University, USA
Ilarri Sergio	University of Zaragoza, Spain
Imine Abdessamad	Inria Grand Nancy, France
Ishihara Yasunori	Osaka University, Japan
Jin Peiquan	University of Science and Technology of China, China
Kao Anne	Boeing, USA
Karagiannis Dimitris	University of Vienna, Austria
Katzenbeisser Stefan	Technische Universität Darmstadt, Germany
Kim Sang-Wook	Hanyang University, Republic of Korea
Kitagawa Hiroyuki	University of Tsukuba, Japan
Kleiner Carsten	University of Applied Sciences & Arts Hannover, Germany
Koehler Henning	Massey University, New Zealand
Korpeoglu Ibrahim	Bilkent University, Turkey
Kosch Harald	University of Passau, Germany
Krátký Michal	Technical University of Ostrava, Czech Republic
Kremen Petr	Czech Technical University in Prague, Czech Republic
Küng Josef	University of Linz, Austria
Lammari Nadira	CNAM, France
Lamperti Gianfranco	University of Brescia, Italy
Laurent Anne	LIRMM, University of Montpellier 2, France
Léger Alain	FT R&D Orange Labs Rennes, France
Lhotska Lenka	Czech Technical University, Czech Republic
Liang Wenxin	Dalian University of Technology, China
Ling Tok Wang	National University of Singapore, Singapore

Link Sebastian	The University of Auckland, New Zealand
Liu, Chuan-Ming	National Taipei University of Technology, Taiwan
Liu Hong-Cheu	University of South Australia, Australia
Liu Rui	HP, USA
Lloret Gazo Jorge	University of Zaragoza, Spain
Loucopoulos Peri	Harokopio University of Athens, Greece
Lu Jianguo	University of Windsor, Canada
Lumini Alessandra	University of Bologna, Italy
Ma Hui	Victoria University of Wellington, New Zealand
Ma Qiang	Kyoto University, Japan
Maag Stephane	TELECOM SudParis, France
Masciari Elio	ICAR-CNR, Università della Calabria, Italy
Medjahed Brahim	University of Michigan - Dearborn, USA
Mishra Alok	Atilim University, Ankara, Turkey
Mishra Harekrishna	Institute of Rural Management Anand, India
Misra Sanjay	University of Technology, Minna, Nigeria
Mocito Jose	MakeWise, Portugal
Moench Lars	University of Hagen, Germany
Mokadem Riad	IRIT, Paul Sabatier University, France
Moon Yang-Sae	Kangwon National University, Republic of Korea
Morvan Franck	IRIT, Paul Sabatier University, France
Munoz-Escoi Francesc	Universitat Politecnica de Valencia, Spain
Navas-Delgado Ismael	University of Málaga, Spain
Ng Wilfred	Hong Kong University of Science & Technology, Hong Kong, SAR China
Nieves Acedo Javier	University of Deusto, Spain
Oussalah Mourad	University of Nantes, France
Ozsoyoglu Gultekin	Case Western Reserve University, USA
Pallis George	University of Cyprus, Cyprus
Paprzycki Marcin	Polish Academy of Sciences, Warsaw Management Academy, Poland
Pastor Lopez Oscar	Universidad Politecnica de Valencia, Spain
Patel Dhaval	Indian Institute of Technology Roorkee, India
Pivert Olivier	Ecole Nationale Supérieure des Sciences Appliquées et de Technologie, France
Pizzuti Clara	Institute for High Performance Computing and Networking (ICAR)-National Research Council (CNR), Italy
Poncelet Pascal	LIRMM, France
Pourabbas Elaheh	National Research Council, Italy
Qin Jianbin	University of New South Wales, Australia
Rabitti Fausto	ISTI, CNR Pisa, Italy
Raibulet Claudia	Università degli Studi di Milano-Bicocca, Italy
Ramos Isidro	Technical University of Valencia, Spain
Rao Praveen	University of Missouri-Kansas City, USA
Rege Manjeet	University of St. Thomas, USA

Resende Rodolfo F.	Federal University of Minas Gerais, Brazil
Roncancio Claudia	Grenoble University/LIG, France
Ruckhaus Edna	Universidad Simon Bolivar, Venezuela
Ruffolo Massimo	ICAR-CNR, Italy
Sacco Giovanni Maria	University of Turin, Italy
Saltenis Simonas	Aalborg University, Denmark
Sansone Carlo	Università di Napoli "Federico II", Italy
Santos Grueiro Igor	Deusto University, Spain
Sanz Ismael	Universitat Jaume I, Spain
Sarda N.L.	I.I.T. Bombay, India
Savonnet Marinette	University of Burgundy, France
Scheuermann Peter	Northwestern University, USA
Schewe Klaus-Dieter	Software Competence Centre Hagenberg, Austria
Schweighofer Erich	University of Vienna, Austria
Sedes Florence	IRIT, Paul Sabatier University, Toulouse, France
Selmaoui Nazha	University of New Caledonia, New Caledonia
Siarry Patrick	Université Paris 12 (LiSSi), France
Silaghi Gheorghe Cosmin	Babes-Bolyai University of Cluj-Napoca, Romania
Skaf-Molli Hala	Nantes University, France
Sokolinsky Leonid	South Ural State University, Russian Federation
Srinivasan Bala	Monash University, Australia
Straccia Umberto	ISTI - CNR, Italy
Sunderraman Raj	Georgia State University, USA
Taniar David	Monash University, Australia
Teisseire Maguelonne	Irstea - TETIS, France
Tessaris Sergio	Free University of Bozen-Bolzano, Italy
Teste Olivier	IRIT, University of Toulouse, France
Teufel Stephanie	University of Fribourg, Switzerland
Teuhola Jukka	University of Turku, Finland
Thalheim Bernhard	Christian-Albrechts-Universität zu Kiel, Germany
Thevenin Jean-Marc	University of Toulouse 1 Capitole, France
Thoma Helmut	Thoma SW-Engineering, Basel, Switzerland
Tjoa A. Min	Vienna University of Technology, Austria
Torra Vicenc	University of Skövde, Sweden
Truta Traian Marius	Northern Kentucky University, USA
Tzouramanis Theodoros	University of the Aegean, Greece
Vaira Lucia	University of Salento, Italy
Vidyasankar Krishnamurthy	Memorial University of Newfoundland, Canada
Vieira Marco	University of Coimbra, Portugal
Wang Junhu	Griffith University, Brisbane, Australia
Wang Qing	The Australian National University, Australia
Wang Wendy Hui	Stevens Institute of Technology, USA
Weber Gerald	The University of Auckland, New Zealand
Wijsen Jef	Université de Mons, Belgium
Wu Huayu	Institute for Infocomm Research, A*STAR, Singapore
Xu Lai	Bournemouth University, UK

Yang Ming Hour	Chung Yuan Christian University, Taiwan
Yang Xiaochun	Northeastern University, China
Yao Junjie	ECNU, China
Yin Hongzhi	The University of Queensland, Australia
Yokota Haruo	Tokyo Institute of Technology, Japan
Zeng Zhigang	Huazhong University of Science and Technology, China
Zhang Xiuzhen (Jenny)	RMIT University Australia, Australia
Zhao Yanchang	RDataMining.com, Australia
Zhu Qiang	The University of Michigan, USA
Zhu Yan	Southwest Jiaotong University, China

External Reviewers

Yimin Yang	Florida International University, USA
Hsin-Yu Ha	Florida International University, USA
Samira Pouyanfar	Florida International University, USA
Anas Katib	University of Missouri-Kansas City, USA
Ermelinda Oro	ICAR-CNR - Italy
Ángel Luis Garrido	University of Zaragoza, Spain
Pasquale Salza	University of Salerno, Italy
Erald Troja	City University of New York, USA
Konstantinos Nikolopoulos	City University of New York, USA
Bin Mu	City University of New York, USA
Gang Qian	University of Central Oklahoma, USA
Alok Watve	Google, USA
Satya Motheramgari	Harman International, USA
Adrian Caciula	Georgia Southern University, USA
Weiqing Wang	The University of Queensland, Australia
Hao Wang	University of Oregon, USA
Fernando Gutierrez	University of Oregon, USA
Sabin Kafle	University of Oregon, USA
Yun Peng	Qilu University of Technology, China
Kiki Maulana	Monash University, Australia
Qiong Fang	HKUST, Hong Kong, SAR China
Valentina Indelli Pisano	University of Salerno, Italy
Meriem Laifa	University Bordj Bou Arreridj, Algeria
Jiyi Li	Kyoto University, Japan
Chenyi Zhuang	Kyoto University, Japan
Matthew Damigos	NTUA, Greece
Lefteris Kalogeros	Ionian University, Greece
Franca Debole	ISTI-CNR, Italy
Claudio Gennaro	ISTI-CNR, Italy
Fabrizio Falchi	ISTI-CNR, Italy
Sajib Mistry	RMIT, Australia
Hai Dong	RMIT, Australia

Azadeh Ghari-Neiat	RMIT, Australia
Xu Zhuang	Southwest Jiaotong University, China
Paul de Vrieze	Bournemouth University, UK
Amine Abdaoui	LIRMM, France
Mike Donald Tapi Nzali	LIRMM, France
Florence Wang	LIRMM, France
Wenxin Liang	Dalian University of Technology, China
Yosuke Watanabe	Nagoya University, Japan
Xi Fang	Tokyo Institute of Technology, Japan
Gianvito Pio	University of Bari, Italy
Cyril Labbé	University of Grenoble, France
Patrick Roocks	University of Augsburg, Germany
Jorge Bernardino	Polytechnic Institute of Coimbra, Portugal
Bruno Cabral	University of Coimbra, Portugal
Ivano Elia	University of Coimbra, Portugal
Nuno Antunes	University of Coimbra, Portugal
Souad Boukhadouma	University of Nantes, France
Nouredine Gasmallah	University of Nantes, France
Quentin Grossetti	CNAM, France
Camelia Constantin	UPMC, France
Frederic Flouvat	PPME, University of New Caledonia
Jia-Ning Luo	Ming Chuan University, Taiwan
Christos Kalyvas	University of the Aegean, Greece
Athanasios Kokkos	University of the Aegean, Greece
Eirini Molla	University of the Aegean, Greece
Stéphane Jean	LIAS/ISAE-ENSMA, France
Selma Khouri	LIAS/ISAE-ENSMA, France
Selma Bouarar	LIAS/ISAE-ENSMA, France
Julius Köpke	University of Klagenfurt, Austria
Frédéric Flouvat	PPME University of New Caledonia, New Caledonia

Organization of the Special Section Globe 2015 (8th International Conference on Data Management in Cloud, Grid and P2P Systems)

Conference Program Chairs

Abdelkader Hameurlain	IRIT, Paul Sabatier University, Toulouse, France
Farouk Toumani	LIMOS, Blaise Pascal University, Clermont-Ferrand, France

Program Committee

Fabricio B. Alves	FIOCRUZ - Fundação Oswaldo Cruz - Rio de Janeiro, Brazil
Djamal Benslimane	LIRIS, University of Lyon, France
Qiming Chen	HP Labs Palo Alto, CA, USA
Thomas Cerqueus	University College Dublin, Ireland
Frédéric Cuppens	Telecom, Bretagne, France
Tran Khanh Dang	HCMC University of Technology, Ho Chi Minh City, Vietnam
Bruno Defude	Telecom INT, Evry, France
Tasos Gounaris	Aristotle University of Thessaloniki, Greece
Maria Indrawan-Santiago	Faculty of Information Technology Monash University, Melbourne, Australia
Sergio Ilarri	University of Zaragoza, Spain
Rui Liu	HP Labs, Palo Alto, CA, USA
Gildas Menier	IRISA-UBS, University of South Bretagne, France
Anirban Mondal	Xerox Research Lab, Bangladore, India
Riad Mokadem	IRIT, Paul Sabatier University, Toulouse, France
Franck Morvan	IRIT, Paul Sabatier University, Toulouse, France
Kjetil Nørvåg	Norwegian University of Science and Technology, Trondheim, Norway
Jean-Marc Pierson	IRIT, Paul Sabatier University, Toulouse, France
Claudia Roncancio	LIG, Grenoble University, France
Soror Sahri	LIPADE, Descartes Paris University, France
Florence Sedes	IRIT, Paul Sabatier University, Toulouse, France
Mário J. Gaspar da Silva	ST/INESC-ID, Lisbon, Portugal
Hala Skaf-Molli	LINA, Nantes University, France
Wolfram Wöß	FAW, University of Linz, Austria
Shaoyi Yin	IRIT, Paul Sabatier University, Toulouse, France

Contents – Part II

XVIII   Contents – Part II

Data Partitioning, Indexing

Data Mining IV, Applications

WWW and Databases

Data Management Algorithms

Special Section Globe 2015 – MapReduce Framework: Load Balancing, Optimization and Classification

Special Section Globe 2015 – Security, Data Privacy and Consistency

Special Section Globe 2015 – Query Rewriting and Streaming

Contents – Part I

NoSQL, NewSQL, Data Integration

Uncertain Data and Inconsistency Tolerance

Database System Architecture

Data Mining I

Data Mining III

Modeling, Extraction, Social Networks

Knowledge Management
and Consistency

A Logic Based Approach for Restoring Consistency in P2P Deductive Databases

Luciano Caroprese[✉] and Ester Zumpano

DEIS, University Della Calabria, 87030 Rende, Italy
{lcaroprese,zumpano}@deis.unical.it

Abstract. This paper stems from the work in [10] in which the declarative semantics of a P2P system is defined in terms of minimal weak models. Under this semantics each peer uses its mapping rules to import minimal sets of mapping atoms allowing to satisfy its local integrity constraints. This behavior results to be useful in real world P2P systems in which peers often use the available import mechanisms to extract knowledge from the rest of the system only if this knowledge is strictly needed to repair an inconsistent local database. Then, an inconsistent peer, in the interaction with different peers, just imports the information allowing to restore consistency, that is *minimal sets of atoms allowing the peer to enrich its knowledge so that restoring inconsistency anomalies*. The paper extends previous work by proposing a rewriting technique that allows modeling a P2P system, $\mathcal{PS}$, as a unique logic program whose minimal models correspond to the minimal weak models of $\mathcal{PS}$.

1 Introduction

Recently, there have been several proposals which consider the integration of information and the computation of queries in an open ended network of distributed peers [1,4,12] as well as the problem of schema mediation and query optimization in P2P environments [13–15,17].

However, many serious theoretical and practical challenges need an answer. Previously proposed approaches investigate the data integration problem in a P2P system by considering each peer as initially consistent, therefore the introduction of inconsistency is just relied to the operation of importing data from other peers. These approaches assume that for each peer it is preferable to import as much knowledge as possible. Our previous works in [5–9] follow this same direction. The interaction among deductive databases in a P2P system has been modeled by importing, by means of mapping rules, maximal sets of atoms not violating integrity constraints, that is *maximal sets of atoms that allow the peer to enrich its knowledge while preventing inconsistency anomalies*. This paper has a different perspective and extends the work in [10]. The declarative semantics in [10], reviewed in this paper, stems from the observations that in real world P2P systems peers often use the available import mechanisms to extract knowledge from the rest of the system only if this knowledge is strictly needed to repair an inconsistent local database.

© Springer International Publishing Switzerland 2015
Q. Chen et al. (Eds.): DEXA 2015, Part II, LNCS 9262, pp. 3–12, 2015.
DOI: 10.1007/978-3-319-22852-5_1

A peer can be initially inconsistent. In this case, the P2P system it belongs to has to provide support to restore consistency. The basic idea, yet very simple, is the following: in the case of an inconsistent database the information provided by the neighbors can be used in order to restore consistency, that is to only integrate the missing portion of a correct, but incomplete database. Then, an inconsistent peer, in the interaction with different peers, just imports the information allowing to restore consistency, that is *minimal sets of atoms allowing the peer to enrich its knowledge so that restoring inconsistency anomalies*. The following example will intuitively clarify our perspective.

Example 1. Consider the P2P system consisting of the following two peers[1]

- The peer $\mathcal{P}_2$ stores information about vendors of devices and contains the facts *vendor*(*dan*, *laptop*), whose meaning is *'Dan is a vendor of laptops'*, and *vendor*(*bob*, *laptop*), whose meaning is *'Bob is a vendor of laptops'*.
- The peer $\mathcal{P}_1$ contains the fact *order*(*laptop*), stating that there exists the order of a laptop, the standard rule *available*(*Y*) ← *supplier*(*X*, *Y*), stating that a device *Y* is available if there is a supplier *X* of *Y*, and the constraint ← *order*(*X*), *not available*(*X*), stating that there cannot exist the order of a device which is not available. Moreover, it also exhibits the mapping rule *supplier*(*X*, *Y*) ← *vendor*(*X*, *Y*), used to import tuples from the relation *vendor* of $\mathcal{P}_2$ into the relation *supplier* of $\mathcal{P}_1$.

The local database of $\mathcal{P}_1$ is inconsistent because the ordered device *laptop* is not available (there is no supplier of laptops). $\mathcal{P}_1$ 'needs' to import the minimal set of atoms in order to restore consistency from its neighbors. □

A pragmatic solution for assigning semantics to a P2P system, that can be effectively implemented in current P2P systems, consists in a rewriting technique that allows modeling a P2P system, $\mathcal{PS}$, as a unique disjunctive logic program whose models capture the P2P system semantics.

2 Background

It is assumed there are finite sets of *predicate symbols*, *constants* and *variables*. A *term* is either a constant or a variable. An *atom* is of the form $p(t_1, \ldots, t_n)$ where p is a predicate symbol and $t_1, \ldots, t_n$ are terms. A *literal* is either an atom A or its negation *not A*. A (*disjunctive*) *rule* is of the form $\mathcal{H} \leftarrow \mathcal{B}$, where $\mathcal{H}$ is a disjunction of atoms (*head* of the rule) and $\mathcal{B}$ is a conjunction of literals (*body* of the rule). A rule is *normal* if just an atom appears in its head. It is assumed that each rule is *safe*, i.e. variables appearing in the head or in negated body literals also appear in some positive body literal. A (*disjunctive*) *program* is a finite set of rules. A program $\mathcal{P}$ is *normal* if each rule in $\mathcal{P}$ is normal; it is *positive* if the body of each rule in $\mathcal{P}$ is negation-free. A term (resp. literal, rule, program) is

---

[1] The syntax of facts, standard rules and mapping rules, here presented in an informal way, will be formally defined in the following section.

ground if no variable appears in it. The set of all ground atoms obtained from predicate symbols and constants occurring in a program $\mathcal{P}$, is called *Herbrand base* of $\mathcal{P}$ and is denoted as $\mathcal{HB}(\mathcal{P})$. Given a rule r, $ground(r)$ denotes the set of all ground rules obtained from r by replacing variables with constants in all possible ways. Given a program $\mathcal{P}$, $ground(\mathcal{P}) = \bigcup_{r \in \mathcal{P}} ground(r)$. A rule with empty head is a *constraint*. A normal ground rule with empty body is a *fact*. In this case the implication symbol ($\leftarrow$) can be omitted. An *interpretation* is a set of ground atoms. The *truth value* of a ground atom, a literal and a rule with respect to an interpretation M is as follows[2]: $val_M(A) = A \in M$, $val_M(not\ A) = not\ val_M(A)$, $val_M(L_1, \ldots, L_n) = min\{val_M(L_1), \ldots, val_M(L_n)\}$, $val_M(L_1 \vee \cdots \vee L_n) = max\{val_M(L_1), \ldots, val_M(L_n)\}$ and $val_M(\mathcal{H} \leftarrow \mathcal{B}) = val_M(\mathcal{H}) \geq val_M(\mathcal{B})$, where A is an atom, $L_1, \ldots, L_n$ are literals, $H \leftarrow \mathcal{B}$ is a rule and $false < true$. An interpretation M is a *model* for a program $\mathcal{P}$ (or $M \models \mathcal{P}$), if all rules in $ground(\mathcal{P})$ are *true* w.r.t. M (i.e. $val_M(r) = true$ for each $r \in ground(\mathcal{P})$). A model M is said to be *minimal* if there is no model N such that $N \subset M$. The set of minimal models of a program $\mathcal{P}$ is denoted as $\mathcal{MM}(\mathcal{P})$.

An interpretation M is a *stable model* (or *answer set*) of $\mathcal{P}$ if M is the unique minimal model of the positive program $\mathcal{P}^M$, where $\mathcal{P}^M$ is obtained from $ground(\mathcal{P})$ by (i) removing all rules r such that there exists a negative literal $not\ A$ in the body of r and A is in M and (ii) removing all negative literals from the remaining rules. It is well known that stable models are minimal models (i.e. $\mathcal{SM}(\mathcal{P}) \subseteq \mathcal{MM}(\mathcal{P})$) and that for negation free programs, minimal and stable model semantics coincide (i.e. $\mathcal{SM}(\mathcal{P}) = \mathcal{MM}(\mathcal{P})$).

Prioritized Logic Programs

Several works have investigated various forms of priorities into logic languages [2,3,11,16]. In this paper we refer to the extension proposed in [16]. A *(partial) preference relation* $\succeq$ among atoms is defined as follows. Given two atoms e_1 and e_2, the statement $e_2 \succeq e_1$ is a *priority* stating that for each a_2 instance of e_2 and for each a_1 instance of e_1, a_2 has higher priority than a_1. If $e_2 \succeq e_1$ and $e_1 \not\succeq e_2$ we write $e_2 \succ e_1$. If $e_2 \succ e_1$ the sets of ground instantiations of e_1 and e_2 have empty intersection. This property is evident. Indeed, assuming that there is a ground atom a which is an instance of e_1 and e_2, the statements $a \succeq a$ and $a \not\succeq a$ would hold at the same time (a contradiction). The relation $\succeq$ is transitive and reflexive. A *prioritized logic program* (PLP) is a pair $(\mathcal{P}, \Phi)$ where $\mathcal{P}$ is a program and Φ is a set of priorities. Φ^* denotes the set of priorities which can be reflexively or transitively derived from Φ. Given a prioritized logic program $(\mathcal{P}, \Phi)$, the relation $\sqsupseteq$ is defined over the minimal models of $\mathcal{P}$ as follows. For any minimal models M_1, M_2 and M_3 of $\mathcal{P}$: (i) $M_1 \sqsupseteq M_1$; (ii) $M_2 \sqsupseteq M_1$ if (a) $\exists\ e_2 \in M_2 - M_1$, $\exists\ e_1 \in M_1 - M_2$ such that $(e_2 \succeq e_1) \in \Phi^*$ and (b) $\nexists\ e_3 \in M_1 - M_2$ such that $(e_3 \succ e_2) \in \Phi^*$; (iii) if $M_2 \sqsupseteq M_1$ and $M_1 \sqsupseteq M_0$, then $M_2 \sqsupseteq M_0$.

If $M_2 \sqsupseteq M_1$ holds, then we say that M_2 is *preferable* to M_1 w.r.t. Φ. Moreover, we write $M_2 \sqsupset M_1$ if $M_2 \sqsupseteq M_1$ and $M_1 \not\sqsupseteq M_2$. An interpretation M is a *preferred*

---

[2] Assuming that $max(\emptyset) = false$ and $min(\emptyset) = true$.

minimal model of $(\mathcal{P}, \Phi)$ if M is a minimal model of $\mathcal{P}$ and there is no minimal model N such that $N \sqsupset M$. The set of preferred minimal models of $(\mathcal{P}, \Phi)$ will be denoted by $\mathcal{PMM}(\mathcal{P}, \Phi)$.

3   P2P Systems: Syntax and Semantics

3.1   Syntax

A *(peer) predicate symbol* is a pair $i : p$, where i is a *peer identifier* and p is a predicate symbol. A *(peer) atom* is of the form $i : A$, where i is a *peer identifier* and A is a standard atom. A *(peer) literal* is a peer atom $i : A$ or its negation *not* $i : A$. A conjunction $i : A_1, \ldots, i : A_m, not\ i : A_{m+1}, \ldots, not\ i : A_n, \phi$, where ϕ is a conjunction of built-in atoms[3], will be also denoted as $i : \mathcal{B}$, with $\mathcal{B}$ equals to $A_1, \ldots, A_m, not\ A_{m+1}, \ldots, not\ A_n, \phi$.

A *(peer) rule* can be of one of the following three types:

1. STANDARD RULE. It is of the form $i : H \leftarrow i : \mathcal{B}$, where $i : H$ is an atom and $i : \mathcal{B}$ is a conjunction of atoms and built-in atoms.
2. INTEGRITY CONSTRAINT. It is of the form $\leftarrow i : \mathcal{B}$, where $i : \mathcal{B}$ is a conjunction of literals and built-in atoms.
3. MAPPING RULE. It is of the form $i : H \leftarrowtail j : \mathcal{B}$, where $i : H$ is an atom, $j : \mathcal{B}$ is a conjunction of atoms and built-in atoms and $i \neq j$.

In the previous rules $i : H$ is called *head*, while $i : \mathcal{B}$ (resp. $j : \mathcal{B}$) is called *body*. Negation is allowed just in the body of integrity constraints. The concepts of *ground rule* and *fact* are similar to those reported in Sect. 2. The definition of a predicate $i : p$ consists of the set of rules in whose head the predicate symbol $i : p$ occurs. A predicate can be of three different kinds: *base predicate*, *derived predicate* and *mapping predicate*. A base predicate is defined by a set of ground facts; a derived predicate is defined by a set of standard rules and a mapping predicate is defined by a set of mapping rules.

An atom $i : p(X)$ is a *base atom* (resp. *derived atom*, *mapping atom*) if $i : p$ is a base predicate (resp. standard predicate, mapping predicate). Given an interpretation M, $M[\mathcal{D}]$ (resp. $M[\mathcal{LP}]$, $M[\mathcal{MP}]$) denotes the subset of base atoms (resp. derived atoms, mapping atoms) in M.

Definition 1. P2P SYSTEM. *A peer* $\mathcal{P}_i$ *is a tuple* $\langle \mathcal{D}_i, \mathcal{LP}_i, \mathcal{MP}_i, \mathcal{IC}_i \rangle$, *where* (i) $\mathcal{D}_i$ *is a set of facts (local database); (ii)* $\mathcal{LP}_i$ *is a set of standard rules; (iii)* $\mathcal{MP}_i$ *is a set of mapping rules and (iv)* $\mathcal{IC}_i$ *is a set of constraints over predicates defined by* $\mathcal{D}_i$, $\mathcal{LP}_i$ *and* $\mathcal{MP}_i$. *A P2P system* $\mathcal{PS}$ *is a set of peers* $\{\mathcal{P}_1, \ldots, \mathcal{P}_n\}$. □

Without loss of generality, we assume that every mapping predicate is defined by only one mapping rule of the form $i : p(X) \leftarrowtail j : q(X)$. The definition of a mapping

---

[3] A *built-in atom* is of the form $X \theta Y$, where X and Y are terms and θ is a comparison predicate.

predicate $i : p$ consisting of n rules of the form $i : p(X) \leftharpoonup \mathcal{B}_k$, with $k \in [1..n]$, can be rewritten into $2 * n$ rules of the form $i : p_k(X) \leftharpoonup \mathcal{B}_k$ and $i : p(X) \leftarrow i : p_k(X)$, with $k \in [1..n]$. Given a P2P system $\mathcal{PS} = \{\mathcal{P}_1, \ldots, \mathcal{P}_n\}$, where $\mathcal{P}_i = \langle \mathcal{D}_i, \mathcal{LP}_i, \mathcal{MP}_i, \mathcal{IC}_i \rangle$, $\mathcal{D}, \mathcal{LP}, \mathcal{MP}$ and $\mathcal{IC}$ denote, respectively, the global sets of ground facts, standard rules, mapping rules and integrity constraints, i.e. $\mathcal{D} = \bigcup_{i \in [1..n]} \mathcal{D}_i$, $\mathcal{LP} = \bigcup_{i \in [1..n]} \mathcal{LP}_i$, $\mathcal{MP} = \bigcup_{i \in [1..n]} \mathcal{MP}_i$ and $\mathcal{IC} = \bigcup_{i \in [1..n]} \mathcal{IC}_i$. In the rest of this paper, with a little abuse of notation, $\mathcal{PS}$ will be also denoted both with the tuple $\langle \mathcal{D}, \mathcal{LP}, \mathcal{MP}, \mathcal{IC} \rangle$ and the set $\mathcal{D} \cup \mathcal{LP} \cup \mathcal{MP} \cup \mathcal{IC}$; moreover whenever the peer is understood, the peer identifier will be omitted.

3.2 The Minimal Weak Model Semantics

This section reviews the *Minimal Weak Model* semantics for P2P systems [10] which is based on a special interpretation of mapping rules. The semantics presented in this paper stems from the observations that in real world P2P systems often the peers use the available import mechanisms to extract knowledge from the rest of the system only if this knowledge is strictly needed to repair an inconsistent local database. In more formal terms, each peer uses its mapping rules to import minimal sets of mapping atoms allowing to satisfy local integrity constraints.

In this paper we refer to a particular interpretation of mapping rules. Intuitively, a mapping rule $H \leftharpoonup \mathcal{B}$ states that if the body conjunction $\mathcal{B}$ is *true* in the source peer the atom H can be imported in the target peer, that is H is *true* in the target peer only if it implies (directly or even indirectly) the satisfaction of some constraints that otherwise would be violated. The following example should make the meaning of mapping rules crystal clear.

Example 2. Consider the P2P system presented in Example 1. As we observed, the local database of $\mathcal{P}_1$ is inconsistent because the ordered device *laptop* is not available. The peer $\mathcal{P}_1$ has to import some supplier of laptops in order to make its database consistent. Then, the mapping rule $supplier(X, Y) \leftharpoonup vendor(X, Y)$ will be used to import one supplier from the corresponding facts of $\mathcal{P}_2$: $supplier(dan, laptop)$ or $supplier(bob, laptop)$. $\mathcal{P}_1$ will not import both facts because just one of them is sufficient to satisfy the local integrity constraint $\leftarrow order(X), not\ available(X)$.

We observe that if $\mathcal{P}_1$ does not contain any fact its database is consistent and no fact is imported from $\mathcal{P}_2$. □

Before formally presenting the minimal weak model semantics we introduce some notation. Given a mapping rule $r = A \leftharpoonup \mathcal{B}$, with $St(r)$ we denote the corresponding logic rule $A \leftarrow \mathcal{B}$. Analogously, given a set of mapping rules $\mathcal{MP}$, $St(\mathcal{MP}) = \{St(r) \mid r \in \mathcal{MP}\}$ and given a P2P system $\mathcal{PS} = \mathcal{D} \cup \mathcal{LP} \cup \mathcal{MP} \cup \mathcal{IC}$, $St(\mathcal{PS}) = \mathcal{D} \cup \mathcal{LP} \cup St(\mathcal{MP}) \cup \mathcal{IC}$.

Informally, the idea is that for a ground mapping rule $A \leftharpoonup \mathcal{B}$, the atom A *could be inferred* only if the body $\mathcal{B}$ is *true*. Formally, given an interpretation M, a ground standard rule $C \leftarrow \mathcal{D}$ and a ground mapping rule $A \leftharpoonup \mathcal{B}$, $val_M(C \leftarrow \mathcal{D}) = val_M(C) \geq val_M(\mathcal{D})$, whereas $val_M(A \leftharpoonup \mathcal{B}) = val_M(A) \leq val_M(\mathcal{B})$.

Definition 2. WEAK MODEL. Given a P2P system $\mathcal{PS} = \mathcal{D} \cup \mathcal{LP} \cup \mathcal{MP} \cup \mathcal{IC}$, an interpretation M is a *weak model* for $\mathcal{PS}$ if $\{M\} = \mathcal{MM}(St(\mathcal{PS}^M))$, where $\mathcal{PS}^M$ is the program obtained from $ground(\mathcal{PS})$ by removing all mapping rules whose head is *false* w.r.t. M. □

We shall denote with $M[\mathcal{D}]$ (resp. $M[\mathcal{LP}]$, $M[\mathcal{MP}]$) the set of ground atoms of M which are defined in $\mathcal{D}$ (resp. $\mathcal{LP}$, $\mathcal{MP}$).

Definition 3. MINIMAL WEAK MODEL. Given two weak models M and N, we say that M is *preferable* to N, and we write $M \sqsupseteq N$, if $M[\mathcal{MP}] \subseteq N[\mathcal{MP}]$. Moreover, if $M \sqsupseteq N$ and $N \not\sqsupseteq M$ we write $M \cdot \sqsupseteq N$. A weak model M is said to be *minimal* if there is no weak model N such that $N \sqsubset M$. □

The set of weak models for a P2P system, $\mathcal{PS}$, will be denoted by $\mathcal{WM}(\mathcal{PS})$, whereas the set of minimal weak models will be denoted by $\mathcal{MWM}(\mathcal{PS})$. We say that a P2P system $\mathcal{PS}$ is *consistent* if $\mathcal{MWM}(\mathcal{PS}) \neq \emptyset$; otherwise it is *inconsistent*.

We observe that, if each peer of a P2P system is locally consistent then no mapping atom is inferred. Clearly not always a minimal weak model exists. This happens when there is at least a peer which is locally inconsistent and there is no way to import mapping atoms that could repair its local database so that its consistency can be restored.

Example 3. Consider the P2P system $\mathcal{PS}$ presented in Example 2. The weak models of the system are:

$$M_1 = \{vendor(dan, laptop), vendor(bob, laptop), order(laptop),$$
$$\quad supplier(dan, laptop), available(laptop)\},$$
$$M_2 = \{vendor(dan, laptop), vendor(bob, laptop), order(laptop),$$
$$\quad supplier(bob, laptop), available(laptop)\}, \text{ and}$$
$$M_3 = \{vendor(dan, laptop), vendor(bob, laptop), order(laptop),$$
$$\quad supplier(dan, laptop), supplier(bob, laptop), available(laptop)\},$$

whereas the minimal weak models are M_1 and M_2 because they contain minimal subsets of mapping atoms (resp. $\{supplier(dan, laptop)\}$ and $\{supplier(bob, laptop)\}$). □

3.3   Prioritized Programs and Preferred Minimal Models

Now we present an alternative characterization of the minimal weak model semantics based on the rewriting of mapping rules into prioritized rules [3,16]. For the sake of notation, we consider exclusive disjunctive rules of the form $A \oplus A' \leftarrow B$ whose meaning is that if B is *true* then exactly one of A or A' must be *true*. Note that, the rule $A \oplus A' \leftarrow B$ is just a shorthand for the rules $A \leftarrow B, not\ A'$ and $A' \leftarrow B, not\ A$ and the integrity constraint $\leftarrow A, A'$.

Given a pair $P = (A, B)$, where A and B are generic objects, $P[1]$ (resp. $P[2]$) denotes the object A (resp. B).

Definition 4. Given a P2P system $\mathcal{PS} = \mathcal{D} \cup \mathcal{LP} \cup \mathcal{MP} \cup \mathcal{IC}$ and a mapping rule $r = i : p(x) \leftarrow \mathcal{B}$, then

- $Rew(r)$ denotes the pair $(i{:}p(x) \oplus i{:}p'(x) \leftarrow \mathcal{B},\ i{:}p'(x) \succeq i{:}p(x))$, consisting of a disjunctive mapping rule and a priority statement,
- $Rew(\mathcal{MP}) = (\{Rew(r)[1]|\ r \in \mathcal{MP}\}, \{Rew(r)[2]|\ r \in \mathcal{MP}\})$ and
- $Rew(\mathcal{PS}) = (\mathcal{D} \cup \mathcal{LP} \cup Rew(\mathcal{MP})[1] \cup \mathcal{IC},\ Rew(\mathcal{MP})[2])$. □

In the above definition the atom $i : p(x)$ (resp. $i : p'(x)$) means that the fact $p(x)$ is imported (resp. not imported) in the peer $\mathcal{P}_i$. For a given mapping rule r, $Rew(r)[1]$ (resp. $Rew(r)[2]$) denotes the first (resp. second) component of $Rew(r)$.

Intuitively, the rewriting $Rew(r) = (A \oplus A' \leftarrow \mathcal{B}, A' \succeq A)$ of a mapping rule $r = A \leftarrow \mathcal{B}$ means that if $\mathcal{B}$ is *true* in the *source peer* then two alternative actions can be performed in the *target peer*: A can be either imported or not imported; but the action of *not importing* A is preferable over the action of *importing* A.

Example 4. Consider again the system analyzed in Example 3. The rewriting of the system is:
$Rew(\mathcal{PS}) = \{vendor(dan, laptop), vendor(bob, laptop), order(laptop),$
$\qquad\qquad supplier(X, Y) \oplus supplier'(X, Y) \leftarrow vendor(X, Y),$
$\qquad\qquad available(Y) \leftarrow supplier(X, Y),$
$\qquad\qquad \leftarrow order(X), not\ available(X)\},$
$\qquad\qquad \{supplier'(X, Y) \succeq supplier(X, Y)\}).$

$Rew(\mathcal{PS})[1]$ has three minimal models:

$M_1 = \{vendor(dan, laptop), vendor(bob, laptop), order(laptop),$
$\qquad supplier(dan, laptop), supplier'(bob, laptop), available(laptop)\},$
$M_2 = \{vendor(dan, laptop), vendor(bob, laptop), order(laptop),$
$\qquad supplier'(dan, laptop), supplier(bob, laptop), available(laptop)\},$
$M_3 = \{vendor(dan, laptop), vendor(bob, laptop), order(laptop),$
$\qquad supplier(dan, laptop), supplier(bob, laptop), available(laptop)\},$

The preferred minimal models are M_1 and M_2. □

Given a P2P system $\mathcal{PS}$ and a preferred minimal model M for $Rew(\mathcal{PS})$ we denote with $St(M)$ the subset of non-primed atoms of M and we say that $St(M)$ is a preferred minimal model of $\mathcal{PS}$. We denote the set of preferred minimal models of $\mathcal{PS}$ as $\mathcal{PMM}(\mathcal{PS})$. The following theorem shows the equivalence of preferred minimal models and minimal weak models.

Theorem 1. *For every P2P system $\mathcal{PS}$, $\mathcal{PMM}(\mathcal{PS}) = \mathcal{MWM}(\mathcal{PS})$.* □

Additionally the particular structure of $Rew(\mathcal{PS})[1]$ ensures that the set of stable models of $Rew(\mathcal{PS})[1]$ coincides with the set of its minimal models. Therefore, the following result holds.

Theorem 2. *For every P2P system $\mathcal{PS}$, $\mathcal{PMM}(\mathcal{PS}) = \mathcal{PSM}(\mathcal{PS})$.* □

4   Computing the Minimal Weak Model Semantics

The previous section has reviewed the semantics of a P2P system in terms of minimal weak models. This section presents an alternative characterization of the minimal weak model semantics allowing to model a P2P system $\mathcal{PS}$ with a single logic program $Rew_m(\mathcal{PS})$. The logic program $Rew_m(\mathcal{PS})$ is then used as a computational vehicle to calculate the semantics of the P2P system as its minimal models correspond to the minimal weak models of $\mathcal{PS}$. With this technique the computation of the minimal weak models is performed in a *centralized way*, however the program $Rew_m(\mathcal{PS})$ can be used as a starting point for a *distributed technique*.

Definition 5. Given a P2P system $\mathcal{PS} = \mathcal{D} \cup \mathcal{LP} \cup \mathcal{MP} \cup \mathcal{IC}$ and a mapping rule $r = i : p(x) \leftarrow \mathcal{B}$, then[4]

- $Rew_m(r)$ denotes the sets of rules $\{i{:}p(x) \vee i{:}p_0(x) \leftarrow \mathcal{B},\ i{:}p_0(x) \leftarrow i{:}p(x)\}$,
- $Rew_m(\mathcal{MP}) = \bigcup_{r \in \mathcal{MP}} Rew_m(r)$
- $Rew_m(\mathcal{PS}) = \mathcal{D} \cup \mathcal{LP} \cup Rew_m(\mathcal{MP}) \cup \mathcal{IC}.$ □

The rewriting of the mapping rule allows to capture the mechanism at the basis of our framework, i.e. import only if necessary in order to restore consistency.

The import of the atom $i : p(x)$ occurs if the couple $\{i : p_0(x), i : p(x)\}$ is derived, whereas the only presence of $i{:}p_0(x)$ means that $i{:}p(x)$ is not imported. The rule $i : p(x) \vee i : p_0(x) \leftarrow \mathcal{B}$ allows to derive (non exclusively) $i : p_0(x)$ or $i{:}p(x)$. Anyhow, the presence of the rule $i{:}p_0(x) \leftarrow i{:}p(x)$ ensures the presence of $i{:}p_0(x)$ whenever $i{:}p(x)$ is derived.

Example 5. Consider again the system $\mathcal{PS}$ analyzed in Example 3.
The rewriting $Rew_m(\mathcal{PS})$ is the following disjunctive logic program:

$$Rew_m(\mathcal{PS}) = \{vendor(dan, laptop), vendor(bob, laptop), order(laptop),$$
$$supplier(X, Y) \vee supplier_0(X, Y) \leftarrow vendor(X, Y),$$
$$supplier_0(X, Y) \leftarrow supplier(X, Y),$$
$$available(Y) \leftarrow supplier(X, Y),$$
$$\leftarrow order(X), not\ available(X)\}$$

$Rew_m(\mathcal{PS})$ has two minimal models:

$$M_1 = \{vendor(dan, laptop), vendor(bob, laptop), order(laptop),$$
$$supplier(dan, laptop), supplier_0(dan, laptop), supplier_0(bob, laptop),$$
$$available(laptop)\}\ \text{and}$$

---

[4] The symbol $\vee$ denotes inclusive disjunction and is different from $\oplus$ as the latter denotes exclusive disjunction. It should be recalled that inclusive disjunction allows more than one atom to be true while exclusive disjunction allows only one atom to be true.

$M_2 = \{vendor(dan, laptop), vendor(bob, laptop), order(laptop),$
$\qquad supplier_0(dan, laptop), supplier(bob, laptop), supplier_0(bob, laptop),$
$\qquad available(laptop)\}$

Observe that the model:

$M_3 = \{vendor(dan, laptop), vendor(bob, laptop), order(laptop),$
$\qquad supplier(dan, laptop), supplier_0(dan, laptop), supplier(bob, laptop),$
$\qquad supplier_0(bob, laptop), available(laptop)\}$

is not minimal.   □

Note that, the selection mechanism of minimal model ensures the presence in a model of both the mapping atom and its auxiliary atom, only if the presence of just the auxiliary atom (stating for a non import action) violates some integrity constraint.

Given a P2P system $\mathcal{PS}$ and a minimal model M for $Rew_m(\mathcal{PS})$ we denote with $St_0(M)$ the subset of M not containing auxiliary atoms of the form $i : p_0(t)$ and we say that $St_0(M)$ is a minimal model of $\mathcal{PS}$. We denote the set of minimal models of $\mathcal{PS}$ as $\mathcal{MM}(\mathcal{PS})$. The following theorem shows the equivalence of minimal models and minimal weak models.

Theorem 3. *For every P2P system $\mathcal{PS}$, $\mathcal{MM}(\mathcal{PS}) = \mathcal{MWM}(\mathcal{PS})$.*   □

5   Complexity Results

We consider now the computational complexity of calculating minimal weak models and answers to queries.

Proposition 1. *Given a P2P system $\mathcal{PS}$, checking if there exists a minimal weak model for $\mathcal{PS}$ is a NP-complete problem.*   □

As a P2P system may admit more than one preferred weak model, the answer to a query is given by considering *brave* or *cautious* reasoning (also known as *possible* and *certain* semantics).

Definition 6. Given a P2P system $\mathcal{PS} = \{\mathcal{P}_1, \dots, \mathcal{P}_n\}$ and a ground peer atom A, then A is *true* under

– brave reasoning if $A \in \bigcup_{M \in \mathcal{MWM}(\mathcal{PS})} M$,
– cautious reasoning if $A \in \bigcap_{M \in \mathcal{MWM}(\mathcal{PS})} M$.   □

We assume here a simplified framework not considering the distributed complexity as we suppose that the complexity of communications depends on the number of computed atoms which are the only elements exported by peers.

The upper bound results can be immediately found by considering analogous results on stable model semantics for prioritized logic programs. For disjunction-free ($\lor - free$) prioritized programs deciding whether an atom is *true* in some preferred model is Σ_2^p-complete, whereas deciding whether an atom is *true* in every preferred model is Π_2^p-complete [16].

Theorem 4. *Let* $\mathcal{PS}$ *be a consistent P2P system, then*

1. *Deciding whether an atom* A *is* true *in some preferred weak model of* $\mathcal{PS}$ *is in* Σ_2^p.
2. *Deciding whether an atom* A *is* true *in every preferred weak model of* $\mathcal{PS}$ *is in* Π_2^p *and* co$\mathcal{NP}$-hard. $\qquad\square$

References

1. Bernstein, P.A., Giunchiglia, F., Kementsietsidis, A., Mylopulos, J., Serafini, L., Zaihrayen, I.: Data management for peer-to-peer computing: a vision. WebDB **2002**, 89–94 (2002)
2. Brewka, G., Eiter, T.: Preferred answer sets for extended logic programs. Artif. Intell. **109**(1–2), 297–356 (1999)
3. Brewka, G., Niemela, I., Truszczynski, M.: Answer set optimization. In: International Joint Conference on Artificial Intelligence, pp. 867–872 (2003)
4. Calvanese, D., De Giacomo, G., Lenzerini, M., Rosati, R.: Logical foundations of peer-to-peer data integration. In: PODS Conference, pp. 241–251 (2004)
5. Caroprese, L., Greco, S., Zumpano, E.: A logic programming approach to querying and integrating P2P deductive databases. In: FLAIRS, pp. 31–36 (2006)
6. Caroprese, L., Molinaro, C., Zumpano, E.: Integrating and querying P2P deductive databases. In: IDEAS, pp. 285–290 (2006)
7. Caroprese, L., Zumpano, E.: Consistent data integration in P2P deductive databases. In: Prade, H., Subrahmanian, V.S. (eds.) SUM 2007. LNCS (LNAI), vol. 4772, pp. 230–243. Springer, Heidelberg (2007)
8. Caroprese, L., Zumpano, E.: Modeling cooperation in P2P data management systems. In: ISMIS, pp. 225-235 (2008)
9. Caroprese, L., Zumpano, E.: Handling preferences in P2P systems. In: FOIKS, pp. 91–106 (2012)
10. Caroprese, L., Zumpano, E.: Restoring consistency in P2P deductive databases. In: Hüllermeier, E., Link, S., Fober, T., Seeger, B. (eds.) SUM 2012. LNCS, vol. 7520, pp. 168–179. Springer, Heidelberg (2012)
11. Delgrande, J.P., Schaub, T., Tompits, H.: A framework for compiling preferences in logic programs. TPLP **3**(2), 129–187 (2003)
12. Franconi, E., Kuper, G.M., Lopatenko, A., Zaihrayeu, I.: A robust logical and computational characterisation of peer-to-peer database systems. In: DBISP2P, pp. 64–76 (2003)
13. Gribble, S., Halevy, A., Ives, Z., Rodrig, M., Suciu, D.: What can databases do for peer-to-peer? In: WebDB, pp. 31–36 (2006)
14. Halevy, A., Ives, Z., Suciu, D., Tatarinov, I.: Schema mediation in peer data management systems. In: International Conference on Database Theory, pp. 505–516 (2003)
15. Madhavan, J., Halevy, A.Y.: Composing mappings among data sources. In: International Conference on Very Large Data Bases, pp. 572–583 (2003)
16. Sakama, C., Inoue, K.: Prioritized logic programming and its application to commonsense reasoning. Artif. Intell. **123**, 185–222 (2000)
17. Tatarinov, I., Halevy., A.: Efficient query reformulation in peer data management systems. In: SIGMOD, pp. 539–550 (2004)

Expert System with Web Interface Based on Logic of Plausible Reasoning

Grzegorz Legień[1]([✉]), Bartłomiej Śnieżyński[1], Dorota Wilk-Kołodziejczyk[1,2],
Stanisława Kluska-Nawarecka[2], Edward Nawarecki[1], and Krzysztof Jaśkowiec[2]

[1] AGH University of Science and Technology, Al. Mickiewicza 30,
30-059 Krakow, Poland
{glegien,bartlomiej.sniezynski}@agh.edu.pl
[2] Foundry Research Institute in Krakow, Zakopianska Street 73, Krakow, Poland

Abstract. The paper presents an expert system based on Logic of Plausible Reasoning (LPR). This formalism reflects human ways of knowledge representation and reasoning. The knowledge is modeled using several kinds of formulas representing statements, hierarchies, similarities, dependencies and implications. Several types of inference patterns are defined. Knowledge uncertainty can be modeled. The paper is structured as follows. Research related to LPR is presented. Next, the formalism is introduced and a Web-based application, which was developed for this research, is described. Finally, a case study is presented – a prototype expert system which recommends a material and a technology for a casting process.

Keywords: Web-based expert system · Logic of plausible reasoning · Knowledge representation and processing

1 Introduction

There are many knowledge modeling and reasoning techniques used in Expert Systems [12]. Some of them allow to represent certain knowledge only (like very popular rule-based systems based on classical logic [13,17]). There are also many approaches that take into account uncertainty of knowledge (e.g. Bayesian networks [15], fuzzy logic [26], certainty factors [19], Dempster-Shafer theory [18], and rough sets [16]). These techniques are based mostly on logic and probability theory. The common feature is that all of them are strong simplifications of human reasoning methods. As a result, application of these techniques may be unnatural to users. For example, statement generalization or similarity between entities can not be represented (e.g. in Bayesian networks or fuzzy logic) or should be defined using rules (implications) instead of relations (in rule-based systems).

Logic of plausible reasoning (LPR) applied in this research has a different origin. Collins analyzed scripts describing human ways of answering common life questions [7] and extracted frequently repeating patterns, which were next formalized as LPR in cooperation with Michalski [8]. As a consequence, LPR

© Springer International Publishing Switzerland 2015
Q. Chen et al. (Eds.): DEXA 2015, Part II, LNCS 9262, pp. 13–20, 2015.
DOI: 10.1007/978-3-319-22852-5_2

provides several inference patterns and many parameters to represent uncertainty. Application of this method allows to model and process the knowledge in a natural way.

In the following sections a related research is discussed and the LPR language is presented. Next, web-based software developed is described and a case study is discussed.

2   Related Research

The experimental results confirming that the methods of reasoning used by humans can be represented in the LPR are presented in subsequent papers [4,5]. The objective set by the creators has caused that LPR is significantly different from other known knowledge representation methods mentioned in the Introduction. Firstly, there are many inference rules in LPR, which are not present in the formalisms mentioned above. Secondly, many parameters are specified for representing the uncertainty of knowledge.

Studies described in [25] present RESCUER, a UNIX shell support system. By tracing changes in the file system and knowledge of the interpreter commands, the system is able to recognize the wrong commands and suggest appropriate substitutes. In this work only small part of LPR is used. In [24] LPR-based tutoring tool is presented. Knowledge for teaching is represented in LPR and the system is able to infer what the student should know after learning.

Another field of formalism application is presented in the work done by Cawsey [6]. It discloses a system generating a description of the concepts based on the recipient's model, taking into account his/her current knowledge.

Hierarchies' concept is also core element of ScubAA system [1], recommending best services in analyzed context, e.g. most accurate Internet search engines. It stores system knowledge in a tree, automatically updated during reasoning process and according to users' feedback. Moreover, in comparison to LIIS, solution uses only three transformations (generalization, specialization and similarity) and limits statements to hierarchy-related.

Research on LPR applications has been also performed at the AGH University. It concerned, in particular, diagnostics, knowledge representation and machine learning [11,22].

Important factor during design and implementation of expert system with web interface is creation of intuitive and user-friendly GUI. Verification of this assumption was important part of works on eXtraSpec [2]. When problem complexity is affecting user interface, like query specification in mentioned system, application should provide supporting tools. For example, when filling a form in eXtraSpec, system suggests correct values in the current edit box.

In CoMES system [3] authors attempted to join many popular techniques from Artificial Intelligence and Software Engineering. Machine learning is used for updating the knowledge base, which can be accessed by few algorithms in parallel. The system use agent architecture to integrate knowledge from human experts and other expert systems.

3 Outline of the Logic of Plausible Reasoning

LPR may be formalized as a labeled deductive system (LDS) [10]. The *language* consists of a finite set of constant symbols C, variables X (represented by capital letters), seven relational symbols and logical connectives: $\rightarrow$, $\wedge$. The relational symbols are: V, H, B, S, E, P, N. They are used to represent: statements (V), hierarchy (H, B), similarity (S), dependency (E), precedence (P) and negation (N).

Statements are represented as object-attribute-value triples: $V(o, a, v)$, where $o, v \in C \cup X, c \in C$. It is a representation of the fact that object o has an attribute a that equals v. If object o has several values of a, there should be several appropriate statements in a knowledge base. To represent vagueness of knowledge it is possible to extend this definition and allow to use composite value $[v_1, v_2, \ldots, v_n]$, list of elements of $C \cup X$. It can be interpreted that object o has an attribute a equals v_1 or v_2, ..., or v_n. If $n = 1$ instead of $V(o, a, [v_1])$ notation $V(o, a, v_1)$ is used. Relation $H(o_1, o, c)$, where $o_1, o \in C \cup X, c \in C$, means that o_1 is o in a context c. Context is used for specification of the range of inheritance. o_1 and o have the same value for all attributes which depend on attribute c of object o. To show that one object is below the other in any hierarchy, relation $B(o_1, o)$, where $o_1, o \in C$, should be used. Relation $S(o_1, o_2, c)$ represents a fact that o_1 is similar to o_2; $o_1, o_2, \in C \cup X, c \in C$. Context, as above, specifies the range of similarity. Only these attributes of o_1 and o_2 have the same values which depend on c. Dependency relation $E(o_1, a_1, o_2, a_2)$, where $o_1, o_2 \in C \cup X$, $a_1, a_2 \in C$, means that values of attribute a_1 of object o_1 depend on attribute a_2 of the second object (o_2). Precedence relation $P(o_1, o_2)$, where $o_1, o_2 \in C \cup X$, says that object o_2 follows o_1 in some order. Negation relation is represented with $N(o_1, o_2)$, where object $o_1, o_2 \in C \cup X$. It represents situation where o_1 and o_2 can never have equal value. This relation do not exist independently, always as a premise of some implication.

In object-attribute-value triples, value should be placed below an attribute in a hierarchy: if $V(o, a, [v_1, v_2, \ldots, v_n])$ is in a knowledge base, there should be also $H(v_i, a, c)$ for any $1 \leq i \leq n, c \in C$.

Using relational symbols, *formulas* of LPR can be defined. If $o, o_1, ..., o_n, a$, $a_1, ..., a_n, v, c \in C$, $v_1, ..., v_n$ are lists of elements of C, then $V(o, a, v)$, $H(o_1, o, c)$, $B(o_1, o)$, $S(o_1, o_2, o, a)$, $E(o_1, a_1, o_2, a_2)$, $P(o_1, o_2)$ and $\alpha_1 \wedge ... \wedge \alpha_n \rightarrow V(o, a, v)$, where α_i is $V(o_i^\alpha, a_i^\alpha, v_i^\alpha)$, $P(v_i^\alpha, w_i^\alpha)$ or $N(v_i^\alpha, w_i^\alpha)$, are LPR formulas (variables can be used in implications).

To manage uncertainty, the following *label algebra* is used: $\mathcal{A} = (A, \{f_{r_i}\})$. A is a set of labels which estimate uncertainty of formulas. *Labeled formula* is a pair $f : l$ where f is a formula and $l \in A$ is a label. A set of labeled formulas can be considered as a knowledge base.

LPR inference patterns are defined as *proof rules*. Every proof rule r_i has a sequence of premises α_i (of length n_i) and a conclusion α:

$$\frac{\alpha_1 : l_1, \alpha_2 : l_2, \ldots, \alpha_n : l_n}{\alpha : l} \tag{1}$$

$\{f_{r_i}\}$ is a set of functions which are used in proof rules to generate a label of a conclusion: for every proof rule r_i an appropriate function $f_{r_i} : A^{n_i} \to A$ should be defined. For rule r_i the plausible label of its conclusion is calculated using $f_{r_i}(l_1, ..., l_n)$. Examples of definitions of plausible algebras can be found in [20].

There are five main types of proof rules: GEN, $SPEC$, SIM, $TRAN$ and MP. They correspond to the following inference patterns: generalization, specialization, similarity transformation, transitivity of relations and modus ponens. Some transformations can be applied to different types of formulas, therefore indexes are used to distinguish different versions of rules. Formal definitions of all these rules can be found in [8,21]. GEN_o and $SPEC_o$ change the scope of objects in statements. GEN_v and $SPEC_v$ change the value in statements decreasing or increasing the description detail level. SIM_o allows to reason by analogy by changing the object, while SIM_v changes the value. MP is a classical Modus Ponens inference rule representing deductive reasoning pattern.

Proof can be defined as a tree. A tree P is a proof of labeled formula $\varphi : l$ from a set of labeled formulas KB if a root node of P is equal $\varphi : l$ and for every node $\psi : l_\psi$:

- if $\psi : l_\psi$ is a leaf, then $\psi : l_\psi \in KB$,
- else, there are nodes $(\psi_1 : l_{\psi_1}, ..., \psi_k : l_{\psi_k})$, connected to $\psi : l_\psi$ and a proof rule r_i such, that $\psi : l_\psi$ is a consequence of r_i and $(\psi_1 : l_{\psi_1}, ..., \psi_k : l_{\psi_k})$ are its premises (label of ψ is calculated using f_{r_i}).

We say, that a labeled formula ψ is a syntactic consequence of a set of labeled formulas KB $(KB \vdash \psi : l)$ if there exist a proof of $\psi : l$ from KB.

To represent reasoning complexity, every proof rule has its cost assigned. Cost of a proof is equal to sum of costs of its proof rules.

4   LIIS System

LPR Intelligent Information System (abbr. LIIS) is a web-based application written in Java. It is created with Google Web Toolkit, supporting browser-based application development. The toolkit provides Java API and widgets, which can be compiled to JavaScript frontend. LIIS uses MySQL database for storing knowledge bases (formulas) and system data (e.g. user profiles, privileges). Object-relational mapping between Java objects and database records is provided by Hibernate framework. System builds with Maven, project dependency management tool. LIIS architecture is presented in Fig. 1.

A core element of LIIS is its reasoning engine, called LPR-Library. It implements the proof searching algorithm discussed above. In the current version of library E-formulas are omitted. Therefore context in hierarchies and similarities specifies a single attribute that may be inferred.

Reasoning process is performed with the LPA algorithm [21] based on the $AUTOLOGIC$ system developed by Morgan [14]. Around hypothesis, taken as input, it constructs a reasoning tree, which nodes are lists of LPR formulas.

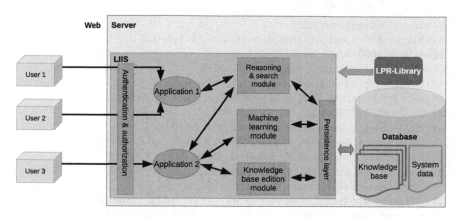

Fig. 1. LIIS architecture

Label algebra is defined as follows. Formulas are labeled by real numbers from range $(0, 1]$. Label of the consequence is equal to the product of premises' certainty parameters. The only exception is generalization of value, that copies statement's label from the first premise.

5 Experimental Results

In this section a case study is discussed. LPR is used for material and technology recommendation in a casting process. The problem is that for the same purpose many materials may be applied. The goal of the system is to chose the material for fitting the requirements provided by the user (application of the product, maximal acceptable production cost, mass of the product etc.). Below we present fragments of a prototype knowledge base, which was built in cooperation with casting technicians, and three scenarios showing how this knowledge is used for recommendation.

The knowledge base starts with statements describing possible applications and properties of the materials that can be used. Hierarchies present facts that ADI is a type of cast iron (in the context of attributes related to cost and production volume) and there are five subtypes of ADI (ADI_GJS-1000-5, ADI_GJS-1200-2, ADI_GJS-1400-1, ADI_GJS-800-8). Similarity formula represents the fact that ADI is similar tu carburized steel in the context of application.

There are two groups of implications (rules). Implications with conclusion V(casting, material_alternative, X) allow to find the recommended material X. The rule with more tests in premise part (e.g. rule no. 30) is more certain then one with less tests (e.g. no. 32). Implications with conclusion V(M,cost,C) define the production cost C of the material M, which depends on the weight of the product and the production volume.

```
1. V(adi, application, rake):0.9
2. V(steel_carburized,application,rack):0.9
3. V(adi_gjs-1000-5,minimum_elongation_A,5):1.0
```

```
11. H(adi,cast iron,cost):1.0:0.0
24. H(adi_gjs-800-8,adi,minimum_elongation_A):1.0:0.0
25. H(adi_gjs-1000-5,adi,cost):1.0:1.0
29. S(adi,steel_carburized,application):0.8
30. [V(casting,application_required,A), V(X,application,A),
V(casting,cost_required,COST_MAX), V(X,cost,COST_CALCULATED),
P(COST_CALCULATED,COST_MAX),
V(casting, strength_tensile_Rm_required,STRENGTH_MIN),
V(X, strength_tensile_Rm,C), P(STRENGTH_MIN,C),
V(casting,minimum_elongation_A_required,ELONG_MIN),
V(X,minimum_elongation_A,E), P(ELONG_MIN,E)]
-> V(casting,material_alternative,X):1.0
32. [V(casting, application_required,A), V(X, application,A),
V(casting,cost_required,COST_MAX), V(X,cost,COST_CALCULATED),
P(COST_CALCULATED,COST_MAX)] -> V(casting ,material_alternative,X):0.5
39. [V(casting, weight, medium), V(casting, volume_production,large)]
-> V(adi,cost,14):1.0
40. [V(casting , weight,heavy), V(casting, volume_production,small)]
-> V(adi,cost,16):0.8
42. [V(casting, weight, heavy), V(casting, volume_production,large)]
-> V(adi,cost,12):1.0
```

Scenario 1. The first scenario is the simplest. Application of the material is a rake[1], the maximum cost required is equal to 15, product weight is heavy, and the batch size is large. System recommends ADI with the confidence 0.45. The recommendation is general (no specific subtype is chosen) because there is not enough information. To infer the conclusion, Modus Ponens (MP) rule is applied twice to implications no. 32 and 42.

Scenario 2. In this scenario more requirements are provided by the user: application, cost, weight and batch size as above, minimal tensile strength Rm is equal to 1100, and minimal elongation A is equal to 2. As a result, the system recommends ADI_GJS-1200-2 with confidence 0.9. The result is more specific and the confidence is larger. The proof was obtained by double application of the MP rule (to implications no. 30 and 40) and double object specialization (SPECo) rule. Certainty is higher because more information is provided by the user and implication 30 instead of 32 may be used.

Scenario 3. In this scenario other application is required by a user – a rack. The maximum cost is set to 15, casting weight is medium, batch size is large. The answer of the system is again ADI with confidence 0.45 This proof will be discussed in a more detailed way. It was obtained by double application of the MP rule and object similarity (SIMo) rule. In the first step, the MP rule was applied to implication no. 32, which means that if the required application of casting under consideration is equal to A (premise 1) and is the same as the application allowed for an alternative material in the rule marked by variable L (premise 2), the required maximum cost is equal to COST_MAX (premise 3), and the cost calculated for an alternative material is equal to

---

[1] A rake is a tool used in sewage-treatment plants. Its main task is to mix organic materials such as straw, grass, hay, etc. with semi-liquid material.

COST_CALCULATED (premise 4), and is lower than the maximum cost (premise 5), then the alternative material (L) should be used with confidence 0.5. Premises 1, 3 and 5 can be unified with the knowledge base elements or answers to questions provided by the user. Premise 2 (application acceptable for ADI) was inferred using object similarity rule (SIMo) for a similarity between ADI and carburized steel, the premises of which are included in the knowledge base. Premise 4 (ADI cost at a given weight and size of the batch) was inferred using MP rule and implication no. 39 representing the production cost. Thus the value of COST_CALCULATED equal to 14 was obtained.

6   Conclusions and Further Works

The presented LIIS system based on LPR allowed to create a web-based expert system for material and technology recommendation in a casting process. The system was tested in many scenarios, three examples of which are described above. Technologists confirmed that the answers are right and the proofs are valid and easy to follow.

The knowledge base created consists of various types of formulas: statements, hierarchies, similarities and implications. Therefore, in the knowledge processing various types of inference patterns are applied (deductive reasoning, generalization, and similarity). As a result, the knowledge and reasoning reflect human way of thinking, what makes the creation of the knowledge base more natural.

Further works will concern adding learning capabilities to the system. Learning module is already implemented (see Sect. 4). However, appropriate knowledge base and use cases are still under construction. Other works concern application in a system in other domains. Knowledge-based systems for telemetry-oriented applications [23] and money laundering detection [9] are under investigation. Also consistency check of the Knowledge Base should be added.

Acknowledgments. The research reported in the paper was supported by the grant of The National Centre for Research and Development (LIDER/028/593/L-4/12/NCBR/ 2013) and by the Polish Ministry of Science and Higher Education under AGH University of Science and Technology Grant 11.11.230.124.

References

1. Abedinzadeh, S., Sadaoui, S.: A trust-based service suggestion system using human plausible reasoning. Appl. Intell. **41**(1), 55–75 (2014)
2. Abramowicz, W., Bukowska, E., Dzikowski, J., Filipowska, A., Kaczmarek, M.: Semantically enabled experts finding system ontologies, reasoning approach and web interface design. In: 15th ADBIS Conference (2011)
3. Althoff, K., et al.: Collaborative multi-expert-systems realizing knowledge-lines with case factories and distributed learning systems. In: Proceedings of the 3rd KESE Workshop (2007)
4. Boehm-Davis, D., Dontas, K., Michalski, R.S.: Plausible reasoning: an outline of theory and the validation of its structural properties. In: Intelligent Systems: State of the Art and Future Directions, North Holland (1990)
5. Boehm-Davis, D., Dontas, K., Michalski, R.S.: A validation and exploration of the Collins-Michalski theory of plausible reasoning. Technical report, George Mason University (1990)

6. Cawsey, A.: Using plausible inference rules in description planning. In: Proceedings of the 5th EACL Conference, Congress Hall, Berlin (1991)
7. Collins, A.: Human plausible reasoning. Technical Report 3810, Bolt Beranek and Newman Inc. (1978)
8. Collins, A., Michalski, R.S.: The logic of plausible reasoning: a core theory. Cogn. Sci. **13**, 1–49 (1989)
9. Drezewski, R., Sepielak, J., Filipkowski, W.: The application of social network analysis algorithms in a system supporting money laundering detection. Inf. Sci. **295**, 18–32 (2015)
10. Gabbay, D.M.: LDS - Labeled Deductive Systems. Oxford University Press, Oxford (1991)
11. Kluska-Nawarecka, S., Nawarecki, E., Śnieżyński, B., Wilk-Kołodziejczyk, D.: The recommendation system knowledge representation and reasoning procedures under uncertainty for metal casting. Metalurgija **54**(1), 263–266 (2015)
12. Liao, S.H.: Expert system methodologies and applications a decade review from 1995 to 2004. Expert Syst. Appl. **28**(1), 93–103 (2005)
13. Ligeza, A.: Logical Foundations for Rule-Based Systems. Springer, Heidelberg (2006)
14. Morgan, C.G.: Autologic. Logique et Analyse **28**(110–111), 257–282 (1985)
15. Neapolitan, R.E.: Probabilistic Reasoning In Expert Systems: Theory and Algorithms. CreateSpace Independent Publishing Platform, USA (2012)
16. Pawlak, Z.: Rough sets. Int. J. Comp. Inf. Sci. **11**, 344–356 (1982)
17. Riley, G.: Clips - an expert system building tool. In: Proceedings of Technology 2001 Conference, San Jose, CA (1991)
18. Shafer, G.: A Mathematical Theory of Evidence. Princeton University Press, Princeton (1976)
19. Shortliffe, H., Buchanan, B.G.: A model of inexact reasoning in medicine. Math. Biosci. **23**, 351–379 (1975)
20. Śnieżyński, B.: Probabilistic label algebra for the logic of plausible reasoning. In: Kłopotek, M., et al. (eds.) Intelligent Information Systems 2002. Advances in Soft Computing, vol. 17, pp. 267–277. Springer, Heidelberg (2002)
21. Śnieżyński, B.: Proof searching algorithm for the logic of plausible reasoning. In: Kłopotek, M., et al. (eds.) Intelligent Information Processing and Web Mining. Advances in Soft Computing, vol. 22, pp. 393–398. Springer, Heidelberg (2003)
22. Sniezynski, B., Kluska-Nawarecka, S., Nawarecki, E., Wilk-Kołodziejczyk, D.: Intelligent information system based on logic of plausible reasoning. In: Hippe, Z.S., Kulikowski, J.L., Mroczek, T., Wtorek, J. (eds.) Issues and Challenges in Artificial Intelligence. Studies in Computational Intelligence, vol. 559, pp. 57–74. Springer, Switzerland (2014)
23. Szydlo, T., Nawrocki, P., Brzoza-Woch, R., Zielinski, K.: Power aware MOM for telemetry-oriented applications using gprs-enabled embedded devices - levee monitoring use case. In: Proceedings of the FedCSIS 2014, pp. 1059–1064, IEEE, Sept 2014
24. Virvou, M.: A cognitive theory in an authoring tool for intelligent tutoring systems. In: Kamel, A.E., Mellouli, K., Borne, P. (eds.) Proceedings of the 2002 IEEE International Conference on Systems, Man and Cybernetics (2002)
25. Virvou, M., Boulay, B.D.: Human plausible reasoning for intelligent help. User Model. User-Adap. Inter. **9**, 321–375 (1999)
26. Zadeh, L.A.: Fuzzy sets. Inf. Control **8**, 338–353 (1965)

Extending Knowledge-Based Profile Matching in the Human Resources Domain

Alejandra Lorena Paoletti[1]([✉]), Jorge Martinez-Gil[1],
and Klaus-Dieter Schewe[1,2]

[1] Software Competence Center Hagenberg, Hagenberg, Austria
{Lorena.Paoletti,Jorge.Martinez-Gil,kd.schewe}@scch.at
[2] Research Institute for Applied Knowledge Processing,
Johannes-Kepler-University, Linz, Austria
kd.schewe@cdcc.faw.jku.at

Abstract. In the Human Resources domain the accurate matching between job positions and job applicants profiles is crucial for job seekers and recruiters. The use of recruitment taxonomies has proven to be of significant advantage in the area by enabling semantic matching and reasoning. Hence, the development of Knowledge Bases (KB) where curricula vitae and job offers can be uploaded and queried in order to obtain the best matches by both, applicants and recruiters is highly important. We introduce an approach to improve matching of profiles, starting by expressing jobs and applicants profiles by filters representing skills and competencies. Filters are used to calculate the similarity between concepts in the subsumption hierarchy of a KB. This is enhanced by adding weights and aggregates on filters. Moreover, we present an approach to evaluate over-qualification and introduce blow-up operators that transform certain role relations such that matching of filters can be applied.

1 Introduction

In the Human Resources (HR) domain the accurate matching of job applicants to position descriptions and vice versa is of central importance for employers and job seekers. Therefore, the development of data or knowledge bases to which job descriptions and curricula vitae (CV) can be uploaded and which can be queried effectively and efficiently by both, employers and job seekers to find best matching candidates for a given job profile and best suitable job offers matching a given applicant skill set, respectively, is of high importance.

It seems appropriate to consider knowledge bases for the representation and thus the storage of the (job and CV) profiles, which in addition to pure storage would support reasoning about profiles and their classification by exploiting the underlying lattice structure of knowledge bases, i.e., the partial order on concepts

The research reported in this paper was supported by the Austrian Forschungs-förderungsgesellschaft (FFG) for the Bridge project "Accurate and Efficient Profile Matching in Knowledge Bases" (ACEPROM) under contract **841284**.

© Springer International Publishing Switzerland 2015
Q. Chen et al. (Eds.): DEXA 2015, Part II, LNCS 9262, pp. 21–35, 2015.
DOI: 10.1007/978-3-319-22852-5_3

representing skills. For instance, a skill such as "knowledge of C" is more detailed than "programming knowledge". Thus, defining profiles by filters, i.e., upward-closed sets of skills (e.g., if "knowledge of C" is in the profile, then "programming knowledge" is in there as well) and using measures on such filters as the basis for the matching seems adequate.

Concerning automatic matching of candidate profiles and job profiles, the commercial practice is largely dominated by Boolean matching, i.e. for a requested profile it is merely checked how many of the requested terms are in the candidate profile [11,12] which amounts to simply counting the number of elements in different sets. This largely ignores similarity between skills, e.g. programming skills in C++ or Java would be rated similar by a human expert.

Improving this primitive form of matching requires at least taking hierarchical dependencies between skill terms into account. Various taxonomies have already been developed for this purpose: DISCO competences [1], ISCO [3] and ISCED [2]. Taxonomies can then be refined by using knowledge bases (ontologies) based on common description logics, which have been studied in depth for more than 20 years [4]. However, sophisticated knowledge bases in the HR domain are still rare, as building up a good, large KB is a complex and time-consuming task, though in principle this can be done as proven by experiences in many other application domains [9].

Ontologies and more precisely description logics have been used as the main means for knowledge representation for a long time [8]. The approach is basically to take a fraction of first-order logic for which implication is decidable. The common form adopted in description logics is to concentrate on unary and binary predicates known as concepts and roles, and to permit a limited set of constructors for concepts and roles. Then the terminological layer (TBox) is defined by axioms usually expressing implication between concepts. In addition, an assertional layer (ABox) is defined by instances of the TBox satisfying the axioms. The various description logics differ mainly by their expressiveness. A prominent representative of the family of description logics is $\mathcal{SROIQ\text{-}D}$, which forms the formal basis of the web ontology language OWL-2 [7], which is one of the more expressive description logics. As the aim of this work is not focused on developing novel ideas for knowledge representation, but merely intends to use knowledge representation as grounding technology for the semantic representation of job offers and candidate CVs, it appears appropriate to fix $\mathcal{SROIQ\text{-}D}$ as the description logics to be used in this work.

The lattice-like structure of concepts within a KB provides basic characteristics to determine the semantic similarity between concepts included in both, job descriptions and curricula vitae. The matching algorithms implemented to determine the semantic similarity between concepts should allow to compare job descriptions and applicants profiles based on their semantics. By comparing the concepts contained within a particular job description against the applicants profile through different categories, (i.e., competencies, education, skills) it is possible to rank the candidates and select the best matches for the job.

The two profiles (job descriptions and applicants) are defined by means of *filters*. If $\leq$ denotes the partial order of the lattice in the TBox, then a filter on the TBox is an *upward-closed*, non-empty set of concepts. Filter-based matching on grounds of partially ordered sets are the starting point of this work, this has been investigated previously [13]. The simple idea is that, for two filters $\mathcal{F}_1$ and $\mathcal{F}_2$ a matching value $m(\mathcal{F}_1, \mathcal{F}_2)$ is computed as $\#(\mathcal{F}_1, \mathcal{F}_2)/ \#\mathcal{F}_2$, i.e. by counting numbers of elements in filters. Experiments based on DISCO already showed that this simple filter-based measure significantly improves the matching accuracy [10].

The goal of our research is to provide solid techniques to improve matching between job and CVs profiles within the HR domain. We will show how adding weights on filters and categories can significantly improve the quality of the matching results based on filter-based matching on grounds of partially ordered sets. As part of the matching process, we also address the problem of over-qualification that cannot be captured solely by means of filters. Finally, we introduce the novel concept of "blow-up" operators in order to extend the matching by integrating roles in the TBox. The idea is to expand the TBox by using roles in order to define arbitrarily many sub-concepts so that the original matching measures could again be applied.

The paper is organized as follows. A subset of the description logic $\mathcal{SROIQ\text{-}D}$ is introduced in Sect. 2. An example of a TBox and how to manipulate concepts in order to perform reasoning is presented in Sect. 3. We define the filter-based matching in Sect. 4. Weights on filters and weighted aggregates on categories are presented in Sect. 4.1 and Sect. 4.2 respectively. In Sect. 4.3 the problem of over-qualification is addressed. And finally, "blow-up" operators is introduced in Sect. 4.4.

2 Profile Matching in Description Logic

The representation of knowledge within taxonomies is used to represent the conceptual terminology of a problem domain in a structured way in order to perform reasoning about it. In this section, we introduce the syntax and the semantics of the language we use to represent the conceptual knowledge of the HR domain within this work, a subset of the description logic $\mathcal{SROIQ\text{-}D}$.

The most elementary components of the logic are *atomic concepts* and *atomic roles*, denoted by the letters C and R respectively. Atomic concepts denote sets of objects and atomic roles denote binary relationships between atomic concepts. Note that the terms "concepts" and "sets" are not synonyms. While a set is a collection of arbitrary elements of the universe, a concept is an expression of the formal language of the description logic. *Atoms* or *nominal* are singleton sets containing one element of the domain, denoting individuals in the description language. Concept descriptions can be build using concept constructors as well as role descriptions can be build from role names as defined below.

Definition 1 *(Syntax of Concept Descriptions).* *Concept description are defined by the following syntax rules:*

$C_1, C_2 \longrightarrow$	A		
	$\top \mid \bot$		*top and bottom*
	$\{a\}$		*atoms or nominal*
	$\neg C_1$		*negation of a concept C_1 (or, complement of C_1)*
	$C_1 \sqcup C_2$		*union*
	$C_1 \sqcap C_2$		*intersection*
	$\exists R.C_1$		*existential restriction*
	$\forall R.C_1$		*value restriction*
	$\leq nR.C_1$		*cardinality restriction $\leq$*
	$\geq nR.C_1$		*cardinality restriction $\geq$*
	$= nR.C_1$		*cardinality restriction $=$*

where A denotes atomic concepts (also known as concept name), $\top$ and $\bot$ denote the two reserved atomic concepts top and bottom that represent the universe and empty set, respectively, a denotes an atom, R denotes an atomic role (also known as role name), C_1 and C_2 denote concept descriptions and $n \in \mathbb{N}$.

Definition 2 *(Syntax of Role Descriptions).* *Given two role names R_1, R_2 and an atom a, the* inverse role R_1^-, *the roles involving atoms $\exists R_1.\{a\}$, and the role chain $R_1 \circ R_2$ are role descriptions.*

A role involving atoms of the form $\exists R.\{a\}$ denotes the set of all objects that have a as a "filler" of the role R. For example, $\exists$SpokenLanguage.{Russian} denotes that Russian is a spoken language. Inverse roles R_1^- are used to describe passive constructions, i.e., *a person owns something* (Owns.Person) can be expressed as *something is owned by a person* (Owns$^-$.Thing). Two binary relations can be composed to create a third relation. For instance, having a role R_1 that relates the element a_1 to element a_2 and role R_2 that relates a_2 with a_3, we can relate a_1 with a_3 by using role chain, this is $R_1 \circ R_2$. For example: by building a composition of the role hasSkill, that relates elements of concept Person with elements of a given Competency, with the role hasProficiencyLevel, that relates Competences with ProficiencyLevel, we have:

$$\text{hasSkill} \circ \text{hasProficiencyLevel}$$

that produces the proficiency level of individuals with experience in a particular competency. We can also define a role hasSkillExperience and express it as:

$$\text{hasSkill} \circ \text{hasProficiencyLevel} \sqsubseteq \text{hasSkillExperience}$$

In general terms, n roles can be chained to form a new role $R_1 \circ \cdots \circ R_n$.

We introduce the concept of an *interpretation* in order to define the formal semantic of the language. Concrete situations are modeled in logic through interpretations that associate specific concept names to individuals of the universe.

An interpretation $\mathcal{I}$ is a non-empty set $\Delta^{\mathcal{I}}$ called the domain of the interpretation $\mathcal{I}$. We sometimes use also $\mathcal{D}$ to denote $\Delta^{\mathcal{I}}$. The interpretation function assigns, to every atomic concept C a set $\Delta(C) \subseteq \mathcal{D}$ and, to every role R a binary relation $\Delta(R) \subseteq \mathcal{D} \times \mathcal{D}$.

Definition 3 *(Semantic of the Language).* *Given an interpretation $\mathcal{I}$, the atomic concepts top and bottom are interpreted as $\Delta(\top) = \mathcal{D}$ and $\Delta(\bot) = \varnothing$ and, the interpretation function can be extended to arbitrary concept and role descriptions as follows:*

$$
\begin{aligned}
\Delta(\{a\}) &= \{\bar{a}\}, a \in \mathcal{D}, \\
\Delta(C_1 \sqcap C_2) &= \Delta(C_1) \cap \Delta(C_2), \\
\Delta(C_1 \sqcup C2) &= \Delta(C_1) \cup \Delta(C_2), \\
\Delta(\neg C) &= \mathcal{D} \backslash \Delta(C), \\
\Delta(\forall R.C) &= \{a \in \mathcal{D} | \forall b.(a,b) \in \Delta(R) \to b \in \Delta(C)\}, \\
\Delta(\exists R.C) &= \{a \in \mathcal{D} | \exists b.(a,b) \in \Delta(R)\}, \\
\Delta(\leq nR.C) &= \{a \in \mathcal{D} | \#\{b \in \Delta(C) | (a,b) \in \Delta(R)\} \leq n\}, \\
\Delta(\geq nR.C) &= \{a \in \mathcal{D} | \#\{b \in \Delta(C) | (a,b) \in \Delta(R)\} \geq n\}, \\
\Delta(= nR.C) &= \{a \in \mathcal{D} | \#\{b \in \Delta(C) | (a,b) \in \Delta(R)\} = n\}, \\
\Delta(R.\{a\}) &= \{b \in \mathcal{D} | (b,a) \in \Delta(R)\}, \\
\Delta(R^-) = \Delta(R)^{-1} &= \{(b,a) \in \mathcal{D}^2 | (a,b) \in \Delta(R)\}, \\
\Delta(R_1 \circ \cdots \circ R_n) \sqsubseteq \Delta(S) &\equiv \{(a_0, a_1) \in \Delta(R_1), \dots, (a_{n-1}, a_n) \in \Delta(R_n) | \\
&\quad (a_0, a_n) \in \Delta(S)\}.
\end{aligned}
$$

The *number restrictions*, $\leq nR.C$, $\geq nR.C$ and, $= nR.C$ denote all elements that are related through the role R to at least n, at most n or, exactly n elements of the universe, respectively, where $n \in \mathbb{N}$ and $\#$ denotes the cardinality of the set.

Subsumption, written as $C_1 \sqsubseteq C_2$ denotes that C_1 is a subset of C_2. It is considered the basic reasoning service of the KB. When using subsumption, it is important to determine whether concept C_2 is more general than concept C_1. For example, to express that the programming language C is a subset of Programming Languages is written as C $\sqsubseteq$ Programming Languages. Expressions of this sort are statements because they may be true or false depending on the circumstances. The truth conditions are: if C_1 and C_2 are concepts, the expression $C_1 \sqsubseteq C_2$ is true under interpretation I if and only if, the elements of C_1 in I are a subset of the elements of C_2 in I. This is, $\Delta(C_1 \sqsubseteq C2) = \Delta(C_1) \subseteq \Delta(C_2)$.

New concepts can be introduced from previously defined concepts by using logical equivalence $C_1 \equiv C_2$. For instance, FunctionalProgrammer $\equiv$ Lisp$\sqcup$Haskell introduces the concept FunctionalProgrammer denoting all individuals that have experience programming in Lisp or Haskell, or both. In this context, a concept name occurring in the left hand side of a concept definition of the form $C_1 \equiv C_2$ is called a *defined concept*.

We have introduced in this section a subset of $\mathcal{SROIQ\text{-}D}$ that is sufficient for this work. Although, for a comprehensive detail of description logics we recommend [5].

3   Representation of Profile Knowledge

Knowledge representation based on description logic is comprised by two main components, the Terminological layer or *TBox* for short, and the Assertional layer, or *ABox*. The TBox contains the terminology of the domain. This is the general knowledge description about the problem domain. The ABox contains knowledge in extensional form, describing characteristics of a particular domain by specifying it through individuals.

Within the TBox, it is possible to describe inclusion relation between concepts by using subsumption. Hence, we can specify, for instance that, Computing is part of Competences and, Programming is part of Computing and, different Programming Languages are included within Programming such that:

LISP ⊑ Programming Languages ⊑ Programming ⊑ Computing ⊑ Competences

Java ⊑ Programming Languages ⊑ Programming ⊑ Computing ⊑ Competences

this gives rise to a partial order on the elements of the KB. Given the nature of subsumption of concepts within Knowledge Bases, TBoxes are lattice-like structures. This is purely determined by the subsumption relationship between the concepts that determine a partially ordered set of elements. In this partially ordered set, the existence of the greatest lower bound (LISP, Java) is trivial which also implies the existence of the least upper bound (Competences).

In ABoxes, we specify properties about individuals characterized under a specific situation in terms of concepts and roles. Some of the concept and role atoms in the ABox may be defined names of the TBox. Thus, within an ABox, we introduce individuals by giving them names $(a_1, a_2, \dots)$, and we assert their properties trough concepts C and roles R. This is, *concept assertions* $C(a_1)$, denote that a_1 belongs to the interpretation of C and, *role assertions* $R(a_1, a_2)$, denote that a_1 is a *filler* of the role R for a_2.

As an example, we consider the TBox in Fig. 1 corresponding to the Competences sub-lattice in Fig. 2 that represents a small set of Programming Languages. Note that, we have refined the relation between the concepts in order to reflect the conceptual influence between the different programming languages. Note also that Programming Languages (PL) is not the least upper bound in Fig. 2. For convenience, we have suppressed the upper part of the subsumption structure of the sub-lattice (Programming Languages ⊑ Programming ⊑ Computing ⊑ Competences).

Note that atomic concepts are not defined as such in Fig. 1 but, they are used in concept descriptions and defined concepts. Concept descriptions describe mainly the subsumption structure of the atomic concepts while defined concepts describe the following characteristics of programming languages. The set

Concept Description	
Imperative ⊔ Object Oriented ⊔ Unix Shell ⊔ Functional ⊑ Programming Languages	
C# ⊑ C++ ⊑ C ⊑ Imperative	
C++ ⊑ Object Oriented	
C++ ⊑ FORTRAN ⊑ Imperative	
Defined Concepts	**Roles**
C-Family ≡ C# ⊔ C++ ⊔ C ⊔ Java ⊔ Perl	hasSkill
NoJava ≡ ∀hasSkill.¬Java	
Programmer ≡ ∃hasSkill.C_i	
Polyglot ≡ ⩾2∃hasSkill.C_i	

Fig. 1. Programming languages TBox

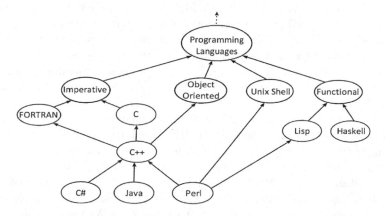

Fig. 2. Programming languages sub-lattice

of programming languages with a C-like structure, this is C-Family; the set of all programming languages but Java, NoJava; Programmer defines every individual that has experience programming with at least one programming language and Polyglot describes all individuals that have experience in programming in two or more programming languages. There is only one role here, hasSkill denoting all objects having some experience in certain domain. Under a given interpretation $\mathcal{I}$ with individuals $a_1, a_2 \in \mathcal{D}$, we can for instance express the queries C_0 and C_1 below. C_0 expresses that the individual a_1 has some experience in programming in Haskell while C_1 states that a_1 is a programmer in at least one of the C-Family languages but Java:

$$C_0 := \{(a_1, a_2) \in \Delta(\text{hasSkill}) \wedge a_2 \in \Delta(\text{Haskell})\}$$
$$C_1 := \{(a_1, a_2) \in \Delta(\text{hasSkill}) \wedge a_2 \in \Delta(\exists\text{C-Family}) \wedge a_2 \in \Delta(\text{NoJava})\}$$

If a_1 satisfies $C_0 \sqcup C_1$ and given that Δ(C-Family) is the set composed by {C#, C++, C, Java, Perl}, we can deduce other characteristics of a_1 in this ABox:

$a_1 \in \Delta(\text{Programmer})$ a_1 is a programmer
$a_1 \in \Delta(\text{Polyglot})$ a_1 is a polyglot programmer
$a_1 \in \Delta(\text{Imperative})$ a_1 has knowledge in Imperative Paradigm
$a_1 \in \Delta(\text{Functional})$ a_1 has knowledge in Functional Paradigm
$a_1 \in \Delta(\text{Objec Oriented})$ a_1 has knowledge in Object Oriented Paradigm

4 Matching Theory

In the HR sector, the data exchange between employers and job applicants is based on a set of shared vocabularies or taxonomies describing relevant terms within the domain, i.e.: competencies, education, skills, etc. Knowledge bases act as repository-like structures for the domain specific knowledge. The lattice-like structure of concepts within a KB provides basic characteristics to determine the semantic similarity between concepts included within the two profiles: job descriptions and CV. In the HR sector, the data exchange between employers and job applicants is based on a set of shared vocabularies or taxonomies describing relevant terms within the domain, i.e.: competencies, education, skills, etc. knowledge bases act as repository-like structures for the domain specific knowledge. The lattice-like structure of concepts within a KB provides basic characteristics to determine the semantic similarity between concepts included within the two profiles: job descriptions and CV. We distinguish the two profiles involved by identifying them as, the *required competencies* to all characteristics included in a job description and, the *given competencies* to all characteristics of an applicant skill sets contained in a CV. The two profiles are defined by means of *filters*. If $\leq$ denotes the partial order of the lattice in the TBox, then a filter on the TBox is an *upward-closed*, non-empty set of concepts. More precisely, we can assume that each profile in the KB representing either a candidate CV or a job offer, is defined by a set of (given or required) skills, each modelled as subconcepts of a concept "skill". Thus, it is possible to concentrate on filters on the sub-lattice of sub-concepts of "skill".

An example of filters taken from Fig. 2 could be for instance, "someone with experience programming in C#". In this example, the upward-closed set of concepts is defined as:

C# $\sqsubseteq$ C++ $\sqsubseteq$ Object Oriented $\sqsubseteq$ PL $\sqsubseteq$ Programming $\sqsubseteq$ Computing $\sqsubseteq$ Competences

For a given job position (and applicant profile) it is expected to find many different filters that represent subsets of the applicant profiles and the job description. Note that, every job offer (and also applicants profiles) is comprised by a number of categories (Competences, Languages, Education, Skills, Social Skills, etc.). In turns, every category is expected to consist of at least one filter. For instance, for a given job advert it could be requested that candidates comply with $\mathcal{F}_j =$ knowledge of Java, $\mathcal{F}_l =$ knowledge of Linux, $\mathcal{F}_{db} =$ knowledge of database programming, etc. within the Competency category.

The filtered-based matching on partially ordered sets has been investigated in [13]. The basic idea is defined as follows:

Definition 4. *Let $\mathcal{F}_1$ and $\mathcal{F}_2$ be filters in the given profile and in the required profile, respectively. The matching value $m(\mathcal{F}_1, \mathcal{F}_2)$ for $\mathcal{F}_1$ and $\mathcal{F}_2$ is computed as:*

$$m(\mathcal{F}_1, \mathcal{F}_2) = \frac{\#(\mathcal{F}_1 \cap \mathcal{F}_2)}{\#\mathcal{F}_2}$$

where $\#\mathcal{F}_2$ and $\#(\mathcal{F}_1 \cap \mathcal{F}_2)$ denote the cardinality of $\mathcal{F}_2$ and $\mathcal{F}_1 \cap \mathcal{F}_2$, respectively.

Note that the matching values are normalized in the range of $[0, 1]$ and satisfy the Bayesian-type rule $m(\mathcal{F}_1, \mathcal{F}_2) \cdot \#\mathcal{F}_2 = m(\mathcal{F}_2, \mathcal{F}_1) \cdot \#\mathcal{F}_1$.

An example taken from Fig. 2 could be a particular job description looking for applicants with experience programming in C# and, a particular applicant profile having some experience programming in Java. The two filters are:

$\mathcal{F}_1 = $ experience in Java $\mathcal{F}_1 := \{(a_1, b_1) \in \Delta(\text{hasSkill}) \wedge b_1 \in \Delta(\text{Java})\}$
$\mathcal{F}_2 = $ experience in C# $\mathcal{F}_2 := \{(a_2, b_2) \in \Delta(\text{hasSkill}) \wedge b_2 \in \Delta(\text{C\#})\}$

The simplest algorithm would take the shortest distance between the two concepts from the least upper concept in the sub-lattice and calculate the distance between the two concepts (Java and C++) by counting cardinality of concepts.

$\mathcal{F}_1 = $ Java $\sqsubseteq$ C++ $\sqsubseteq$ ObjectO $\sqsubseteq$ PL $\sqsubseteq$ Programming $\sqsubseteq$ Computing $\sqsubseteq$ Competences
$\mathcal{F}_2 = $ C# $\sqsubseteq$ C++ $\sqsubseteq$ ObjectO $\sqsubseteq$ PL $\sqsubseteq$ Programming $\sqsubseteq$ Computing $\sqsubseteq$ Competences

In this particular example, there is a measure of 7 for $\mathcal{F}_1$ and a measure of 7 for $\mathcal{F}_2$ as well, giving that the two elements (Java and C#) are siblings. Although, it is the elements in common between the two filters that counts in here. Therefore, the matchability measurement of the two filters is $0, 86$ calculated: $m(\mathcal{F}_1, \mathcal{F}_2) = \frac{6}{7}$ where, 6 is the number of common elements between $\mathcal{F}_1$ and $\mathcal{F}_2$, and 7 is the total number of elements in $\mathcal{F}_2$. In the context of the TBox in Fig. 2 and given the fact that matching on filters ranges between $[0,1]$, we can say that having some experience in Java results in a relatively high score for the required experience in C#.

We introduce in the following sub-sections the main contribution of our research in this work. The main goal of this research is to provide an improvement on the matching process of job and applicants profiles within the HR domain. We will show how including weights can significantly improve the quality of the matching results based on filter-based matching on grounds of partially ordered sets. The introduction of a measure that improves matching on filters is detailed in Sect. 4.1. And aggregates on categories of profiles is introduced in Sect. 4.2. We have also researched how to address over-qualification, as part of the matching process that, clearly cannot be captured solely by means of filters. This is introduced in Sect. 4.3. Finally, in Sect. 4.4 we introduce the novel concept of "blow-up" operators that allow to extend the matching by integrating roles in the TBox. The idea is to expand the TBox by using roles to define arbitrarily many sub-concepts so that the original matching measures can again be applied.

4.1　Aggregates on Filters

It has already been shown in [13] that the idea of filter-based matching, as described in Sect. 4, significantly improves accuracy in comparison to simply taking differences of skill sets. A new matching measurement is introduced here, achieved by adding weights to the elements of the sub-lattice.

Definition 5. *Let $\mathcal{F}_1$ and $\mathcal{F}_2$ be filters in the given profile and in the required profile, respectively. Let $\mathcal{F}_1 \cap \mathcal{F}_2$ denote the set of concepts that appear in both $\mathcal{F}_1$ and $\mathcal{F}_2$. Let w_i and w_j be the weight associated to every concept C_i in $\mathcal{F}_1 \cap \mathcal{F}_2$ and C_j in $\mathcal{F}_2$, respectively with $i, j \in \mathbb{N}$. Then the aggregate on filters $m_w(\mathcal{F}_1, \mathcal{F}_2)$ is defined:*

$$m_w(\mathcal{F}_1, \mathcal{F}_2) = \frac{\sum_{C_i \in \mathcal{F}_1 \cap \mathcal{F}_2} w_i}{\sum_{C_j \in \mathcal{F}_2} w_j}$$

By adding weights to the concepts of the sub-lattice structure, we are not only improving the matching but also, providing the possibility of adding a ranking of importance to every element in the underlying sub-lattice for a required aspect within a job profile. In this way, one could emphasize the search on the generic areas of required competencies. For instance, if searching for someone with experience in Object Oriented (OO) is more important than someone with experience in a specific Programming Language, then aggregates could be distributed as follows:

C++$_{[0,1]}$, OO$_{[0,6]}$, PL$_{[0,1]}$, Programming$_{[0,1]}$, Computing$_{[0,05]}$, Competences$_{[0,05]}$

Or, emphasize the search in more specific required competencies. For instance, expertise in C++ is absolutely relevant to the position,

C++$_{[0,6]}$, OO$_{[0,1]}$, PL$_{[0,1]}$, Programming$_{[0,1]}$, Computing$_{[0,05]}$, Competences$_{[0,05]}$

Note that, the distribution of aggregates through the elements of the underlying sub-lattice is normalized to be within the range [0,1].

4.2　Aggregates on Categories

Ranking the top candidates for a particular job position can be challenging in situations where, for example, a number of the top candidates result with the same matching measurement. The Fitness column in Table 1 shows an example where three different candidates have a final score of 0,5 as a result of the 4 analyzed categories (Competences, Languages, Education, Skills). To overcome these situations, adding aggregates on the categories of the evaluated job profile has been considered. This normalized adjustable measure defined within $[0, 1]$, intends to provide a more granular set of results.

Consider the aggregates on filters $m_w(\mathcal{F}_1, \mathcal{F}_2)$ as in Definition 5 and, consider as well the set of filters a given category is composed by, i.e., *Category* = $\{\mathcal{F}_1, \mathcal{F}_2, \dots, \mathcal{F}_x\}$. The score of a given category within a given profile, is calculated by $\bar{m}_w = \sum_{l=1}^{x} \frac{m_{w_l}(\mathcal{F}_1, \mathcal{F}_2)}{x}$. Then, the aggregates on categories is defined as follows.

Table 1. Matching measures with same fitness

	Competences	Languages	Education	Skills	Fitness
Candidate 1	1	0	1	0	0,5
Candidate 2	0,5	0,5	0,5	0,5	0,5
Candidate 3	0,8	0,2	0,2	0,8	0,5

Table 2. Ranking candidates with aggregates on categories

	Competences (50 %)	Languages (20 %)	Education (10 %)	Skills (20 %)	Fitness
Candidate 1	1	0	1	0	0,6
Candidate 2	0,5	0,5	0,5	0,5	0,5
Candidate 3	0,8	0,2	0,2	0,8	0,62

Definition 6. *Let $\bar{m}_{wj}$ be the score per category j and be w_j the weights associated to every category within a profile. We define the aggregates on categories m_c as follows:*

$$m_c = \sum_{j=1}^{m} \bar{m}_{wj} * w_j$$

where $m \in \mathbb{N}$ denotes the total number of categories.

As an example, we apply different weights to the categories in Table 1, for instance: 0,5 to Competences, 0,2 to Languages, 0,1 to Education and 0,2 to Skills. This results in a completely different set of scores, that is clearly less problematic to evaluate during the ranking of the candidates, as shown in Table 2. This approach provides the experts in the HR domain, with the option of ranking the different categories in order of importance. One may guess that Skills and Competences could almost always be the most important categories although, this is subject to every particular case where advise from the experts in the area is required.

4.3 Over-Qualification in Profile Matching

An over-qualified applicant for a given job offer is defined as a person that is skilled or educated beyond what it is required to conform to a given advertised position. Our approach is to evaluate over-qualification by the dual notion of [fitness, over-qualification] as part of a person skill set such that, fitness is calculated by means of aggregates on filters, as in Definition 5 and over-qualification is calculated by means of cardinality of filters as in Definition 4. Therefore, we look for the maximality of the fitness scores and the minimality of the over-qualification scores of candidates during the ranking process.

It seems obvious to note that over-qualification is only evaluated in cases where fitness > 0, this is, if $m(\mathcal{F}_1, \mathcal{F}_2) > 0$.

Definition 7. *Given two filters $\mathcal{F}_1$ representing a candidate profile and $\mathcal{F}_2$, representing a job profile, over-qualification is calculated as follows:*

$$O(\mathcal{F}_2, \mathcal{F}_1) = \frac{\#(\mathcal{F}_2 \cap \mathcal{F}_1)}{\#\mathcal{F}_1}$$

where $O(\mathcal{F}_2, \mathcal{F}_1) \in [0, 1]$.

Note that, over-qualification is calculated by counting the number of elements on filters as in Definition 4, it measures the characteristics of the job profile against the characteristics of the applicants profile. This is, skills, academic records, competencies, etc. present in the applicants profile that are not required in the job description. Consider the following example where:

$\mathcal{F}_1 := \{(a_1, b_1) \in \Delta(\mathsf{hasSkill}) \wedge b_1 \in \Delta(\mathsf{Java}) \wedge (a_1, c_1) \in \Delta(\mathsf{hasSkill}) \wedge c_1 \in \Delta(\mathsf{Haskell})\}$

$\mathcal{F}_2 := \{(a_2, b_2) \in \Delta(\mathsf{hasSkill}) \wedge b_2 \in \Delta(\mathsf{C\#})\}$

We calculated the matching measure between Java and C# already in Sect. 4 this is, $m(\mathcal{F}_1, \mathcal{F}_2) = 0,86$. Given that Haskell is clearly not part of the job profile filter, we calculate over-qualification of Haskell in terms of C#. This results in a measure of $0,67$ given that $O(\mathcal{F}_2, \mathcal{F}_1) = \frac{4}{6}$. Then, the measurement [fitness, over-qualification] is [0,86; 0,67]. We could say that this individual is highly competent and over-qualified for the position as a programmer in C#.

4.4 Blow-Up Operators

In order to take full benefit of the knowledge-based representation of profiles, we define TBox transformations by means of operators that remove roles and enlarge the KB by many new sub-concepts. We call these *blow-up operators*, because the removal of a single rule may lead to many new concepts, even infinitely many. However, for the sake of computability, the filters should remain finite. The idea is to extend the approach of matching by integrating new roles in the TBox. This is basically to define operators that expand the TBox by using roles in order to define arbitrarily many new sub-concepts so that the original measures of matching, as defined in Sect. 4, could again be applied. We define a blow-up operator as follows:

Definition 8. *Let C and C' be two concept names and let R be a role name that relates elements of C with elements of C'. Let I be an interpretation such that for a given two atoms a and b, $b \in \Delta(C')$ and $\{a \in \Delta(C)|(a, b) \in \Delta(R)\}$. Then, Blow-up operator is defined as:*

$$blow\text{-}up(C, \{R, C'\}) \equiv C \sqcap \exists R.C'$$

where an instance b of C' defines sub-concepts of C such that, the R-values are restricted to b, then $R(a, b) \in \Delta(R)$.

It is easier to note the new created concepts, denoted $\bar{C}_i$, by extending the definition

$$\bar{C}_1 := \text{blow-up}(C, \{R, C_1'\})$$
$$\bar{C}_2 := \text{blow-up}(C, \{R, C_2'\})$$
$$\vdots$$
$$\bar{C}_n := \text{blow-up}(C, \{R, C_n'\})$$

Note that, an order on the instances of C' defines a subsumption on the elements $\bar{C}_i$. This is, for every $i, j \leq n$ it is $\bar{C}_i \sqsubseteq \bar{C}_j$ if and only if $i \leq j$. The generic definition is:

$$\text{Blow-up}_n(C, \{(R, C_1), \ldots, (R, C_n)\}) \equiv \bigcap_{1 \leq i \leq n} \text{blow-up}(C, \{R, C_i\})$$

Consider an example where people have certain years of experience working with programming languages in general. This can be thought as a relation between two concepts: Years of Experience and Programming Languages related through the role hasExperience as in Fig. 3. By applying blow-up operators, the role hasExperience is removed and instead, creates many sub-concepts of Programming Languages with every instance of Years of Experience. It is important to note that this applies only to every instance of Programming Languages where the value in hasExperience is restricted to Years of Experience. Thereby, after applying blow-up operators, the role in Fig. 3 is expanded, as shown in Fig. 4, where C++ is an instance of Programming Languages, and $\{1\text{ye}, 2\text{ye}, \ldots\}$ are instances of Years of Experience. Where 1ye expresses "1 year of experience", 2ye = "2 years of experience", etc. Note that, an order on the instances of Years of Experience gives rise to subsumption between the so-defined sub-concepts of Programming Languages, i.e., "C++ with n year of experience" $\sqsubseteq$ "C++ with n-1 years of experience" $\sqsubseteq \cdots \sqsubseteq$ "C++ with 2 years of experience" $\sqsubseteq$ "C++ with 1 year of experience". In consequence, integrating matching measurements as explained in Sect. 4 with blow-up operators seems plausible. Consider an example taken from Fig. 2 where, $\mathcal{F}_1$ = a given competency in C++ with 2ye and $\mathcal{F}_2$ = a required competency in C++ with 3ye expressed as follows:

$$\mathcal{F}_1 := \{(a_1, b_1) \in \Delta(\text{hasSkill}) \wedge b_1 \in \Delta(\text{C++}) \wedge (b_1, c_1) \in \Delta(\text{hasExperience})$$
$$\wedge\, c_1 \in \Delta(2\text{ye})\}$$
$$\mathcal{F}_2 := \{(a_2, b_2) \in \Delta(\text{hasSkill}) \wedge b_2 \in \Delta(\text{C++}) \wedge (b_2, c_2) \in \Delta(\text{hasExperience})$$
$$\wedge\, c_2 \in \Delta(3\text{ye})\}$$

The basic matching measurement is $m(\mathcal{F}_1, \mathcal{F}_2) = 1$ given that $m(\mathcal{F}_1, \mathcal{F}_2) = \frac{6}{6}$. While, the basic matching measurement applied to the blow-up of years of experience results as follows:

$$\frac{\#(\text{blow-up}(\text{C++}, \{\text{hasExperience}, 2\text{ye}\}) \cap \text{blow-up}(\text{C++}, \{\text{hasExperience}, 3\text{ye}\}))}{\#\text{blow-up}(\text{C++}, \{\text{hasExperience}, 3\text{ye}\})} = \frac{4}{5}$$

Fig. 3. Role hasExperience between programming languages and years of experience

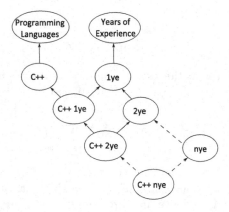

Fig. 4. Role hasExperience blowed-up with C++ and years of experience

Note that a blow-up is performed up-to the level of the required profile. Therefore blow-up operator is defined within the range $[0,1]$.

It is clear to note how applying blow-up operators to the role hasExperience in Fig. 3 influences the result of the original matching measure of 1. We believe blow-up operators is a relevant improvement on how matching of profiles is performed nowadays in relation to complex roles such as "years of experience" on skills and competencies. Blow-up operators allow to expand the KB, and more precisely the TBox, in such a way that filter-based matching measures remain unchanged. Although, blow-up operators are not intended to be part of the TBox itself. Nevertheless, they extend concepts within the TBox in a lattice-like structure by blowing-up the role R. Blow-up operators are intended to be applied to a wider range of circumstances, i.e., years of experience in competences, proficiency level of skills, versioning, etc. Therefore, blowing-up all these possibilities and possibly combining them all together within a given TBox would be rather slow. Then, calculating blow-up operators out of the TBox is a better and a more efficient approach.

5   Conclusion and Further Work

We showed in this paper a number of improvements to the matching process of profiles between job offers and applicants profiles within the HR domain. Starting from a previous work (Popov-Jebelean, 2013), we have introduced an improvement to the matching of filters by adding aggregates on filters and categories of profiles. We have also addressed the problem of over-qualification with the dual measure of [fitness, over-qualification] and introduced the novel idea of blow-up operators by integrating roles in the TBox.

Further research will be focused on the investigation of methods for the learning of matching relations from large ABoxes. With respect to the enrichment of KB, several approaches for ontology learning have been investigated [6,14] but also exploiting Formal Concept Analysis [10].

References

1. European dictionary of skills and competences. http://www.disco-tools.eu
2. International standard classification of education. http://www.uis.unesco.org/Education/Pages/international-standardclassification-of-education.aspx
3. International standard classification of occupations. http://www.ilo.org/public/english/bureau/stat/isco/isco08/index.htm
4. Baader, F., Bürckert, H.-J., Heinsohn, J., Hollunder, B., Müller, J., Nebel, B., Nutt, W., Profitlich, H.-J.: Terminological knowledge representation: a proposal for a terminological logic. In: Description Logics, pp. 120–128 (1991)
5. Baader, F., Calvanese, D., McGuinness, D.L., Nardi, D., Patel-Schneider, P.F. (eds.): The Description Logic Handbook: Theory, Implementation, and Applications. Cambridge University Press, Cambridge (2003)
6. Cimiano, P., Hotho, A., Stumme, G., Tane, J.: Conceptual knowledge processing with formal concept analysis and ontologies. In: Eklund, P. (ed.) ICFCA 2004. LNCS (LNAI), vol. 2961, pp. 189–207. Springer, Heidelberg (2004)
7. Grau, B.C., Horrocks, I., Motik, B., Parsia, B., Patel-Schneider, P.F., Sattler, U.: OWL 2: The next step for OWL. J. Web Sem. **6**(4), 309–322 (2008)
8. Gruber, T.R.: Toward principles for the design of ontologies used for knowledge sharing? Int. J. Hum. Comput. Stud. **43**(5–6), 907–928 (1995)
9. Klein, M.C.A., Broekstra, J., Fensel, D., van Harmelen, F., Horrocks, I.: Ontologies and schema languages on the web. In: Spinning the Semantic Web: Bringing the World Wide Web to its Full Potential [outcome of a Dagstuhl seminar], pp. 95–139 (2003)
10. Looser, D., Ma, H., Schewe, K.-D.: Using formal concept analysis for ontology maintenance in human resource recruitment. In: Ninth Asia-Pacific Conference on Conceptual Modelling, APCCM 2013, Adelaide, Australia, January 29–Feburary 1 2013, pp. 61–68 (2013)
11. Mochol, M., Nixon, L.J.B., Wache, H.: Improving the recruitment process through ontology-based querying. In: Proceedings of the First International Workshop on Applications and Business Aspects of the Semantic Web (SEBIZ 2006) Collocated with the 5th International Semantic Web Conference (ISWC-2006), Athens, Georgia, USA, 6 November 2006 (2006)
12. Mochol, M., Wache, H., Nixon, L.J.B.: Improving the accuracy of job search with semantic techniques. In: Abramowicz, W. (ed.) BIS 2007. LNCS, vol. 4439, pp. 301–313. Springer, Heidelberg (2007)
13. Popov, N., Jebelean, T.: Semantic matching for job search engines: a logical approach. Technical report 13–02, RISC Report Series. University of Linz, Austria (2013)
14. Zavitsanos, E., Paliouras, G., Vouros, G.A.: Gold standard evaluation of ontology learning methods through ontology transformation and alignment. IEEE Trans. Knowl. Data Eng. **23**(11), 1635–1648 (2011)

Sensitive Business Process Modeling for Knowledge Management

Mariam Ben Hassen[✉], Mohamed Turki, and Faïez Gargouri

ISIMS, MIRACL Laboratory, University of Sfax, P.O. Box 242
3021 Sfax, Tunisia
mariembenhassen@yahoo.fr,
mohamed.turki@isetsf.rnu.tn,
faiez.gargouri@isims.rnu.tn

Abstract. Currently, modern organizations are characterized by collaborative, highly dynamic, complex and highly intensive knowledge processes. They are all the more aware of the need to effectively identify, preserve, share and use the knowledge mobilized by their business processes. Thus, in order to improve their performance, business process modeling has become a primary concern for any organization in order to improve the management of its individual and collective knowledge assets. This paper proposes a new meta-model of sensitive business processes modeling for knowledge management, called BPM4KI (Business Process Meta-Model for Knowledge Identification) based on COOP, a core ontology of organization's processes. The aim of this meta-model is to help identify and localize the crucial knowledge that is mobilized and created by these processes. Moreover, it has been illustrated through applying it to a medical process in the context of the organization of protection of the motor disabled people of Sfax-Tunisia (ASHMS).

Keywords: Knowledge management · Sensitive business process · Core ontology of organization's processes · Business process modeling

1 Introduction

Nowadays, organizations are all the more aware of the importance of the immaterial capital owned by their members which corresponds to their experience and accumulated knowledge about the firm activities. Thus, to improve their performance, such organizations have become conscious of the necessity to formalize the organizational knowledge mobilized and created by their business processes (BPs).

The integration of knowledge management (KM) into BPs has been considered one of the most important research approaches for the KM. According to the literature review, several attempts have already been elaborated to integrate KM and BP orientation. We quote process oriented KM approaches [1–3] as well as knowledge oriented business process modeling (BPM) approaches [4–7]. However, the integration of the BPM-KM domain has not yet received sufficient attention. In fact, the knowledge dimension needed for performing BP (i.e., the organizational and individual knowledge mobilized or generated by organizational activities, knowledge sources, explicit and tacit knowledge, individual and collective dimension of knowledge/activities,

© Springer International Publishing Switzerland 2015
Q. Chen et al. (Eds.): DEXA 2015, Part II, LNCS 9262, pp. 36–46, 2015.
DOI: 10.1007/978-3-319-22852-5_4

knowledge flows between sources and activities, the knowledge conversion, etc.) are not explicitly represented, integrated and implemented in BP meta-models.

This research addresses the gap between KM and BPM as well as an important issue that is not often raised in KM methodology: the problem of identification and localization of crucial knowledge that is mobilized by the sensitive business processes (SBPs) [8–10]. In fact, the more BPs are sensitive, the more they can mobilize crucial knowledge (i.e., specific knowledge on which it is necessary to capitalize).

Accordingly, in this paper, we propose a new multi-perspective meta-model of the BPs for KM, called BPM4KI (Business Process Meta-Model for Knowledge Identification). BPM4KI highlights the concepts needed to completely and adequately address all SBP essential characteristics from several perspectives. It covers all aspects of BPM and KM, namely: functional, organizational, behavioral, intentional, informational and knowledge perspectives. BPM4KI is well founded on COOP, a core ontology of organization's processes proposed by [11]. This meta-model aims to enrich the graphical representation of SBPs and improve the localization and identification of crucial knowledge mobilized and created by these processes. Furthermore, we intend to integrate and implement the BPM4KI meta-model in the most suitable notations for BPM, namely, the Business Process Modeling Notation (BPMN 2.0). This extension of BPMN 2.0 will enable us to have a rich and expressive representation of SBPs which are likely to mobilize crucial knowledge.

The remainder of this paper is structured as follows: Sect. 2 briefly presents related works relevant to the research problem. Section 3 presents fundamental characteristics regarding SBP and related work about modeling SBP. Section 4 describes the proposed BPM4KI meta-model. Section 5 presents a practical example to illustrate the application of BPM4KI. Section 6 concludes the paper and highlights future perspectives of this research.

2 Related Works

In this section, we present some methodologies relevant to the BPM-KM domain, addressing the analysis and modeling of BPs for knowledge identification [8–10]. The Global Analysis Methodology (GAMETH) proposed by [8] comprises three main phases gathering the following steps: (i) « Identifying the sensitive processes[1]. » determines the processes targeted to be deeply analyzed, (ii) « Identifying the determining problems » aims at distinguishing the problems which weaken the critical activities (i.e. the activities that could endanger the SBP due to dysfunctions and constraints which affect it), and (iii) « Identifying the crucial knowledge » is intended to define, localize and characterize the knowledge to be capitalized. The methodology for identifying the crucial knowledge proposed by [9] is based on the GAMETH framework. It aims at capitalizing the knowledge mobilized and created in the course of a

---

[1] According to [8] "A sensitive process is a process, which represents the important issues which are collectively acknowledged: weakness of the process which risks not attaining its objectives, obstacles to overcome; difficult challenge to take in charge; produced goods or services which are strategic in regard to the organization's orientations"

project. It is composed of three phases: (i) Determining « reference knowledge » , (ii) Constructing preference model (decision rules), and (iii) Classifying « potential crucial knowledge » . The authors [10] have dealt in depth with the issue of identifying « Sensitive organization's processes » . They have proposed a new multi-criteria methodology entitled SOPIM (Sensitive Organization's Process Identification Methodology) to analyze and identify the SBPs. It is composed of two main phases: (i) Construction of the preference model, and (ii) Exploitation of the preference model to classify the « Potential Sensitive organization's processes » .

The above-mentioned approaches intend to identify, model and analyze the SBPs, in order to localize the crucial knowledge. However, the BPM phase has not been studied in depth. In particular, we have noted the lack of expressiveness of BPM formalisms and BP models with a knowledge dimension and other BPM aspects. Overcoming this lack, this paper aims to extend and consolidate previous work [9, 10] to bridge the gap between BPM and KM and address an important issue that is not often dealt with in KM methodologies. Precisely, our mission aims to enrich the operation of « modeling and representation of identified SBPs » in order to increase the probability of localizing and identifying the crucial knowledge. This reduces the cost of the operation of capitalizing on knowledge. So, the specification of a precise conceptualization, together with a subjacent representation notation, that precisely describes all SBP essential characteristics, is indispensable to reach this objective.

3 Sensitive Business Process

3.1 SBP Fundamentals

An SBP has its own characteristics that distinguish it from classical BPs [11]. A BP is described as « sensitive » [8–11], if at least one of the following requirements is fulfilled:

– It mobilizes « crucial knowledge » , i.e. the risk of their loss and the cost of their (re) creation is considered to be important. Thus, their contribution to reach the firm objectives is very important and their use duration is long [12].
– It contains activities that valorize the acquisition, storage, dissemination, sharing, and creation and (re) use of organizational knowledge.
– It is dependent on the tacit knowledge embedded in the stakeholders' minds.
– It includes a high number of critical activities [8–10] mobilizing crucial knowledge. In our context, a critical activity mobilizes different types of knowledge: (i) imperfect individual and collective knowledge (tacit and/or explicit) (i.e. missing, poorly mastered, incomplete, uncertain, etc.) which are necessary for solving critical determining problems; (ii) heterogeneous knowledge recorded on multiple knowledge sources (dispersed and sometimes lacking accessibility); (iii) rare knowledge held by a very small number of experts (who carry out actions with high levels of experience, expertise, creativity and innovation); (iv) very important tacit knowledge (like competences, abilities and practical experiences of experts).
– It is very complex and dynamic and has a high number of activities which are flexible. It can be semi-structured, structured and unstructured.

- It mobilizes a large number of business domains/skills (in terms of internal and external organization unit operating in the process). Its execution involves many participants and many experts, with different experience and expertise levels.
- It involves a large number of external agents who are not affiliated to the organization. It is then known as a collaborative inter-organizational process.
- It has a high number of collaborative activities that mobilize, share and generate new organizational knowledge (tacit and explicit) created during the interaction among agents. So, it focuses on the dynamic conversion of knowledge [13].
- It possesses a high degree of dynamism in the objectives' change associated to it, essentially, in decision making context. The change of organizational objective leads to a new organizational distal intention which is necessary to control the SBP.
- Its contribution degree to reach strategic objectives of the organization is very important. Moreover, their realization duration and cost are important.

Due to those characteristics, building a SBP model is not an easy task. Several BPM approaches have been proposed in BP engineering as likely to represent SBP.

3.2 Representation Approaches for SBP

Numerous conventional workflows/BPM approaches, such as Event Driven Process Chain (EPC) [14], UML 2.0 Activity Diagrams (AD) [15], Process Specification Language (PSL) [16] and Business Process Modeling Notation (BPMN 2.0) [17] have been adapted to allow the representation of the intrinsic elements of knowledge within BPs. Although these approaches are suitable for process perspective representation, they have many shortcomings concerning flexibility, collaboration, the dynamics and knowledge aspects (such as distinction between data, information and knowledge needed for performing the BP, individual and collective dimension of knowledge/actions, etc.). Besides, other approaches addressing the representation of knowledge-intensive processes [1] (also called Process-oriented knowledge modeling approaches), such as the Business Process Knowledge Method (BPKM) [4], Knowledge Modeling Description Language (KMDL 2.2) [1] and Notation for Knowledge-Intensive Processes (NKIP) [7] have not been widely adopted by organizations and are very incipient. These proposals have poor capabilities to explicitly model the process and knowledge dimensions as a whole.

Although it mobilizes crucial knowledge within an organization and their key role for organizational KM, the current BPM approaches have shortcomings in their ability to adequately address all SBPs essential characteristics as well as relevant issues at the intersection of KM and BPM. This paper proposes a semantically rich conceptualization of a SBP organized in a meta-model, the Business Process Meta-model for Knowledge Identification. BPM4KI explicit and organize the key concepts and relationships to completely characterize a SBP from different perspectives to enable an expressive representation of SBPs, integrating all aspects of BPM and KM.

4   Proposition of a Business Process Meta-model for Knowledge Identification

To improve the SBP representation and localize in depth the crucial knowledge, we propose a novel Business Process Meta-model for Knowledge Identification, entitled BPM4KI addressing all the requirements of the SBP seen before. We structure our meta-model into several aspects according to the framework developed by [18] which consists of five perspectives, namely: functional, organizational, behavioral, informational, and intentional. Besides, we extend these BPM perspectives with the « knowledge perspective » to fully address all relevant KM issues. Moreover, BPM4KI is based on the core ontology COOP proposed by [11] to characterize in depth the concepts useful for the modeling and analysis of SBPs. In fact, COOP provides the taxonomy of semantically rich concepts inherent to the BP domain which are defined in a rigorous and consensual way (such as Action of Organization, Individual Action, Action of Collective, Collective, Distal Intention, Deliberate Action, Critical Activity, etc.). Figure 1 depicts the BPM4KI meta-model in a UML class diagram, in terms of the main concepts related to different aspects of BPM and KM. The defined concepts that are in line with the COOP ontology are marked in grey. In the following, we describe the six perspectives that categorize the BPM4KI concepts.

■ *Functional Perspective*: *What BP elements are being performed?(Collective Action, Organizational Action, Organizational Sub Process, Critical Organizational Activity)* The BPM4KI main concept that reflects the functional perspective is Action.[2] An Action can be individual or collective. An Individual Action *is carried out* by (*hasForAgent*) a Human. While a Collective Action (i.e. a group of several individual actions combining their effects [19]) is *carried out by* a Collective. A Business Process is an Actionof Organization which *hasForproper-Part* a set of Organizational Actions/Activities coordinated and undertaken according to an intentionally defined objective.

■ *Organizational Perspective*: *Where and by whom process elements are performed?* (*Collective, Human, Organization, Organization Unit*) The organizational aspect of BPM4KI includes the basic concept Agentive Entity. An Agentive Entity is an entity which has a capacity to carry out Actions. Any Collective Action *hasForAgent* a Collective (i.e. a group of humans unified by a joint intention to form a group capable of acting [19]). Each Business Process *hasForAgent* an Organization. An Organizational Activity which *hasForAgent*, either a Human or an Organization Unit.

■ *Behavioural Perspective*: *When and how process elements are performed?* (*Control flow*) This perspective basically describes the control flow and the logical sequence of elements to be executed in a process. It includes synchronization, iterations, complex

---

[2] With respect to our notation, the informal labels on BPM4KI concepts appear in the text in the Courrier new font with First Capital Letters for the concepts and a javaLikeNotation for relations. The same conventions apply for the COOP ontology [11] presented in the paper.

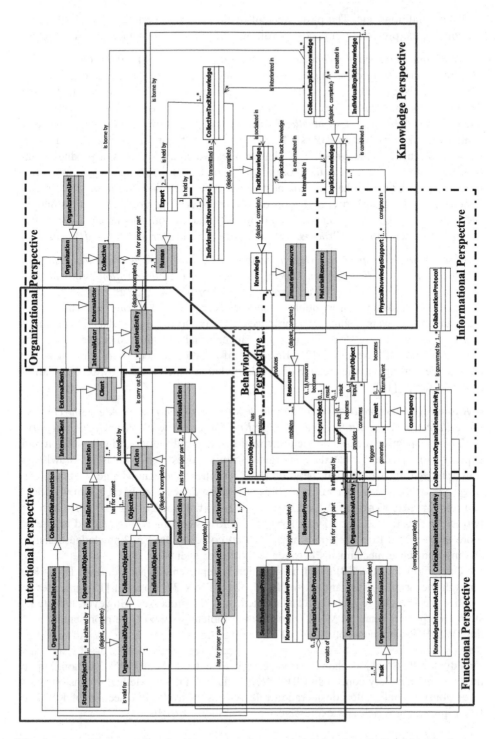

Fig. 1. A multi-perspective business process meta-model for knowledge identification – BPM4KI

decision conditions, etc. The basic element of this perspective is `Control Object` (as constraints, pre-conditions, post-conditions, triggers, performance indicators, etc.).

■ *Intentional Perspective*: *What are the BP characteristics? (Intention, Objective)*
The Intentional Perspective (also called BP context perspective [14]) describes major BP characteristics and captures BP context information, in order to ensure the BP flexibility. A `Business Process` meets an `Organizational Objective` intentionally defined and *ControlledBy* a `Distal Intention` (which plans and controls the action). Then a BP is a `Deliberate Action` [11]). Each `Business Process` must provide a result (i.e., deliverables which are services or products) that is useful for a `Client`. It is therefore a `Culminated Process` [11]. Moreover, a `Business Process` has a certain process type classified according to several dimensions (like granularity, affiliation of operant agents, perceptible value, strategic and repetition [11]). For example, according to the affiliation dimension, we can distinguish between *Internal Process*, *External Process* and *Partial ExternalProcess*.

■ *Informational Perspective*: *What informational entities are generated/exchanged? (Material Resource, Event, Input Object, Output Object, Collaboration Protocol)*
For its accomplishment, an `Organizational Activity` uses `Input Objects` (like data or information), mobilizes `Material Resources` (like informational and software resources) to produce `Output Objects` (as data, information, services or results). A `Contingency` is an external and unpredictable event that influences the BP execution [7]. Besides, information object form the basis for knowledge sharing and the creation of new knowledge which is done by externalization or combination [13].

■ *Knowledge Perspective:* *What and how knowledge and knowledge sources are shared/created (Tacit Knowledge, Explicit Knowledge, Physical Knowledge Support)*
This perspective focuses on the organizational and individual knowledge (tacit and explicit) mobilized and created by an organization's processes as well as the knowledge flow proceeding within and between BP activities and knowledge sources. An `Organizational Activity` mobilizes and produces `Explicit Knowledge` and `Tacit Knowledge`. A `Collective Tacit Knowledge` *isHeldBy* at least two `Experts` (a `Collective`). `Explicit Knowledge` is often stored in one or more `Physical Knowledge Support` that could be used as knowledge transfer, sharing and reuse mechanisms between activities or agents.

Once modeled, the SBPs can be graphically represented (according to BPM4KI) through the best-known standard for BPM, the BPMN 2.0 notation [17], in order to localize the knowledge mobilized and created by these processes. However, despite its strength representation, BPMN 2.0 does not completely support the representation of the BPM4KI core concepts (as `Collective Action`, `Critical Activity`, `Tacit Knowledge`, `Collective Explicit Knowledge`, `Expert`, etc.). To remedy for this lack, it should be necessary to extend the BPMN 2.0 notation to integrate the new key concepts of BPM4KI, including the knowledge dimension. This extension must take into consideration relevant issues at the intersection of KM and BPM, providing a rich and expressive representation of SBPs.

5 A Practical Example

In order to evaluate the potential of the BPM4KI meta-model in providing the knowledge identification as well as an adequate understanding and representation of a SBP, a medical process model in the context of the Association of Protection of the Motor-disabled of Sfax-Tunisia (ASHMS) was illustrated using BPMN 2.0. We are particularly interested in the early care of the disabled children with cerebral palsy (CP). An in depth analysis of this care has been made by [12]. In fact, the knowledge mobilized and created during this care is very important, heterogeneous and recorded on various scattered sources. The created knowledge stems from the interaction of a large number of healthcare professionals from several specialties and located on geographically remote sites. The raised problem concerns on the one hand, the insufficiency and the localization of medical knowledge necessary for decision-making, and on the other hand, the loss of knowledge held by these experts during their scattering or their departure at the end of the treatment. Thus, the ASHMS risks losing the acquired know-how and transferring this knowledge to new novices if ever no capitalization action is considered. Our main objective consists in improving the localization and identification of pragmatic medical knowledge necessary to the conduct of the medical care process for children with CP, which is a SBP.

In this study, we take into consideration the results of experimentation of the multi-criteria methodology SOPIM proposed by [5] for the early care of children with CP. We have opted for the SBP « Process related to the neuropediatric care of a child with CP » to evaluate the potential of the BPM4KI with regard to its applicability and capability of making all relevant knowledge embedded in a SBP explicit. Figure 2 depicts an excerpt from the BPMN model of the neuro-pediatric care process enriched with the knowledge dimension. The produced BPMN SBP model is the result of many individual meetings for review and validation with the Neuro-Pediatrician. In order to overcoming the shortcomings of BPMN 2.0 to represent explicitly the knowledge aspects, we have defined some specific graphical icons relating to each new additional concept to highlight the BPMN extension of (cf. Fig. 2).

During our experimentation, we have identified tacit and explicit medical knowledge required to perform activities as well as created as a result of BP activities. For instance, the knowledge A_2K_{p1} related to « Synthesis of neurological abnormalities related to motor, somatic and sensory development of the young child at risk and the different clinical signs » is produced by the critical activity A_2 « Clinical neurological examination ». This materialized knowledge is created as a result of the activity execution by the Neuro-pediatrician, during which he interacts with information related to the child at risk to generate and communicate his own knowledge.

A_2K_{p1} is stored in the following physical media: the neurological assessment, the sensitive assessment sheet and the neuro-motor assessment sheet. These assessments are recorded in the personal medical records and in the overall clinical picture of the child. These physical media of knowledge are located internally within the Neonatology service in the University Hospital Hedi Chaker, precisely in the various archives drawers or patients' directories. A_2K_{p1} is of a scientific, technical and measure nature which is related to patients. It represents a collective explicit knowledge, part of which

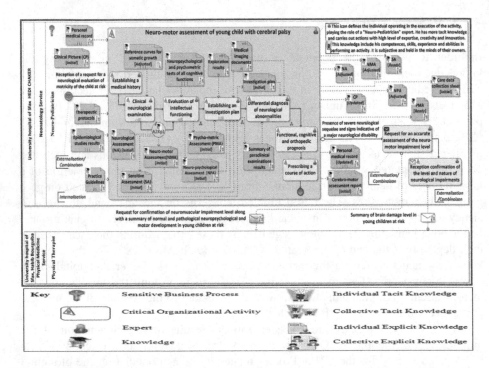

Fig. 2. An extract of the graphical representation model of the neuropediatric care process of a child with CP using ARIS Express 2.4 tool [20]

can be represented as an individual explicit knowledge recorded on the care data collection sheet of the Neuro-pediatrician. This knowledge is mobilized by the activity A_3 « Evaluation of intellectual functioning ».

Therefore, extending BPMN BP models with the knowledge dimension provides some benefits such as: (i) illustrating the knowledge and its sources that are mobilized, generated and/or modified by the critical activities, their localization (where knowledge can be obtained and clearly stated), their degree of formalization and their organizational coverage, (ii) illustrating experts who hold knowledge, (iii) illustrating the knowledge flow and the different knowledge conversion, and (iv) characterizing the identified knowledge to determine which one is more crucial.

6 Conclusion and Future Works

In this paper, we have proposed a meta-model of sensitive business processes modeling for knowledge management, called « BPM4KI », to localize and identify the crucial knowledge. BPM4KI comprises concepts from several perspectives (of BPM and KM) that are crucial for an adequate understanding and representation of a SBP, namely: functional, organizational, behavioral, intentional, informational and knowledge perspectives. The proposed meta-model is semantically rich and well founded on the core ontology COOP [11]. It is evaluated through a real SBP scenario from medical domain using the BPMN 2.0 standard.

This paper identifies several directions for future research. Firstly, we consider improving the definition of some BPM4KI concepts so as to facilitate their understanding, as well as adding new elements to cover missing issues and deepen the characterization of concepts related to SBP notion. Hence, it should be necessary to position itself in relation to other process types in this area, such as agile BPs and knowledge-intense processes (where flexibility, collaboration and knowledge are the central aspects). Secondly, we consider improving the extended « knowledge perspective » of BPM4KI, relying on core ontology, such as COOK, a core ontology of know-how and knowing-that proposed by [21] in order to deeply bridge the gap between KM and BP models. Thirdly, we consider justifying the choice of the most suitable approach for SBPs representation, taking the conceptualization defined by BPM4KI as a baseline for a deeper comparative analysis of the adequacy of BPM approaches. Finally, BPM4KI may be adopted as a basis to enhance existing BPM approaches. Thus, we consider proposing an extension of BPMN 2.0 for KM. In practice, this extension will be integrated into a more general framework supporting the SBPs modeling which advocates a model driven engineering approach (MDE).

References

1. Gronau, N., Korf, R., Müller, C.: KMDL-capturing, analyzing and improving knowledge-intensive business processes. J. Univ. Comput. Sci. **11**, 452–472 (2005)
2. Heisig, P.: The GPO-WM® method for the integration of knowledge management into business processes. In: International Conference on Knowledge Management, Graz, Austria, pp. 331–337 (2006)
3. Zhaoli, Z., Zongkai, Y.: Modeling knowledge flow using Petri net. In: IEEE International Symposium on Knowledge Acquisition and Modeling Workshop, China, pp. 142–146 (2008)
4. Papavassiliou, G., Mentzas, G.: Knowledge modelling in weakly-structured business processes. J. Knowl. Manag. **7**(2), 18–33 (2003)
5. Supulniece, I., Businska, L., Kirikova, M.: Towards extending BPMN with the knowledge dimension. In: Bider, I., Halpin, T., Krogstie, J., Nurcan, S., Proper, E., Schmidt, R., Ukor, R. (eds.) BPMDS 2010 and EMMSAD 2010. LNBIP, vol. 50, pp. 69–81. Springer, Heidelberg (2010)
6. Businska, L., Supulniece, I., Kirikova, M.: On data, information, and knowledge representation in business process models. In: The 20th International Conference on Information Systems Development (ISD 2011), pp. 24–26. Springer, Edinburgh, Scotland (2011)
7. Netto, J.M., Franca, J.B.S., Baião, F.A., Santoro, F.M.: A notation for knowledge-intensive processes. In: IEEE 17th International Conference on Computer Supported Cooperative Work in Design (CSCWD 2013), vol. 1, pp. 1–6 (2013)
8. Grundstein, M.: From capitalizing on company knowledge to knowledge management. In: Morey, D., Maybury, M. (eds.) Knowledge Management, Classic and Contemporary Works, Chapter 12, pp. 261–287. The MIT Press, Cambridge (2000)

9. Saad, I., Grundstein, M., SABROUX, C.: Une méthode d'aide à l'identification des connaissances cruciales pour l'entreprise. Revue SIM, vol 14, N° 3 (2009)
10. Turki, M., Saad, I., Gargouri, F., Kassel, G.: A business process evaluation methodology for knowledge management based on multi-criteria decision making approach. In: Saad, I., Sabroux, C.R., Gargouri, F. (eds.) Information systems for knowledge management. Wiley-ISTE, Chichester (2014a). ISBN 978-1-84821-664-8
11. Turki, M., Kassel, G., Saad, I., Gargouri, F.: COOP: a core ontology of organization's processes for group decision making. J. Decis. Syst. **23**(1), 55–68 (2014)
12. Turki, M., Saad, I., Gargouri, F., Kassel, G.: Towards identifying sensitive processes for knowledge localization. In: International Conference on Collaboration Technologies and Systems (CTS 2011), pp. 224–232 (2011)
13. Nonaka, I., Takeuchi, H.: Knowledge-Creating Company: How Japanese Companies Create the Dynamics of Innovation. Oxford University Press, New York (1995)
14. List, B., Korherr, B.: An evaluation of conceptual business process modelling languages. In: ACM Symposium on Applied Computing (SAC 2006). ACM Press, France (2006)
15. OMG: unified modeling language (UML), version 2.0 (2011a). http://www.uml.org/
16. Schlenoff, C., Gruninger, M., Tissot, F., Valois, J.: The process specification language (PSL) overview and version 1.0 specification (2000). http://www.mel.nist.gov/psl/
17. OMG: business process modeling and notation (BPMN), version 2.0 (2011b). http://www.bpmn.org/
18. Nurcan, S.: A survey on the flexibility requirements related to business processes and modeling artifact. In: Proceedings of the 41st Hawaii International Conference on System Sciences, IEE, Hawaii, USA, p. 378, 7–10 Jan 2008
19. Kassel, G., Turki, M., Saad, I., Gargouri, F.: From collective actions to actions of organizations: an ontological analysis. In: Symposium Understanding and Modelling Collective Phenomena (UMoCop), University of Birmingham, Birmingham, England (2012)
20. The IDS-Scheer website (2013). http://www.ids-scheer.com/
21. Ghrab, S., Saad, I., Kassel, G., Gargouri, F.: An ontological framework for improving the model of contribution degree of knowledge. In: International Conference on Knowledge Management, Information and Knowledge Systems (KMIKS 2015), Tunisia, pp. 45–58 (2015)

Mobility, Privacy and Security

Partial Order Preserving Encryption
Search Trees

Kyriakos Ispoglou[1], Christos Makris[1], Yannis C. Stamatiou[2,3(✉)],
Elias C. Stavropoulos[1,4], Athanasios K. Tsakalidis[1],
and Vasileios Iosifidis[1]

[1] Computer Engineering and Informatics Department, University of Patras,
265 04 Rio, Patras, Greece
[2] Department of Business Administration, University of Patras,
265 04 Rio, Patras, Greece
[3] Computer Technology Institute and Press ("Diophantus"),
University of Patras, 265 04 Rio, Patras, Greece
stamatiu@ceid.upatras.gr
[4] Business Administration Department, Technological Educational
Institute of Western Greece, 263 34 Koukouli, Patras, Greece

Abstract. As Internet services expand and proliferate, service users' data show
an increase in volume as well as geographical dispersion mainly due to the large
number of personalized services that users often access today on a daily basis.
This fact, however, presents a user privacy and user data security challenge for
service providers: how to protect theirs users' data from unauthorized access. In
this paper we present a new tree-based data structure for storing encrypted
information in order to support fast search, update, and delete operations on the
encrypted data. The data structure relies on exposing limited ordering information of the data in order to locate them fast. After showing that a totally order
preserving encryption scheme is not secure, we describe a new tree data
structure and assess its security and efficiency.

1 Introduction

There is a number of proposals in the literature with respect to our problem with
trade-offs between efficiency and security. The work in [5] is one of the earliest and
simplest efforts towards an order preserving encryption scheme (OPE). Unfortunately,
it work only for integers. However there are some security problems with this scheme
(as with any OPE scheme) as it does not exhibit perfect secrecy.

In [1] a scheme is proposed that allows direct comparison of data in their encrypted
forms. The scheme also supports update and insert operations without requiring
modifications in the encrypted data. However, an adversary is not allowed to insert
values on his own, i.e. perform a chosen plaintext attack.

In this paper we propose a simple data structure based on binary search trees that
supports privacy preserving data operations. We first show that it is not possible to
have a secure *fully* order preserving scheme. Thus, we relax the full data ordering
property and settle for partial ordering in order to support fast and secure search and
update operations on data in their encrypted forms. Our approach differs from the ones

Q. Chen et al. (Eds.): DEXA 2015, Part II, LNCS 9262, pp. 49–56, 2015.
DOI: 10.1007/978-3-319-22852-5_5

examined above, as well as similar approaches, in that it can be implemented on any common data structure used for storing searching ordered data, it can use any underlying data encryption algorithm, and can keep the data structure balanced most of the time in order to support fast operations.

2 Fully Order Preserving Encryption Schemes and Their Main Disadvantages

Let us assume that we have a fully order preserving encryption scheme. This means that we can use one of the normal data structures for totally ordered data sets, e.g. a binary search tree, in order to store the encrypted values. Since the encryption preserves the order, we can perform the search and update operations fast on the encrypted data: we first locate the sought data item and then decrypt it if some other operation is needed (e.g. update or delete). Although this solution appears to be suitable in most cases, there are some limitations (it cannot support range queries) and security issues (the problem of duplicates[1]).

Totally order preserving encryption schemes, i.e. schemes in which plaintext order is preserved in the ciphertexts, can support range queries, but they suffer from chosen plaintext attacks. The attack resembles a binary search process and starts by, first, locating the "middle" value of the value range for M bits (where the largest value in the tree is 2^M) and submits it for insertion. The database inserts this plaintext in encrypted form. The adversary sees where his ciphertext c_A has landed with respect to the target ciphertext c using the order preserving property of the encryption algorithm. If $c_A = c$ then the adversary has discovered the sought plaintext. If $c_A > c$ he repeats this process on the left half of the current value range and if $c_A < c$ he continues with the right subrange. Finally, after about $\log M$ (all logarithms are in base 2) steps at most, the adversary will manage to decrypt the target ciphertext c.

In what follows, in order to overcome the deficiencies of fully order preserving encryption schemes, we propose a *partially* order preserving scheme which handles the duplicates problem and supports limited range queries. Partially order preserving here means that two plaintexts can be potentially separated by other plaintexts in the ciphertext domain who are not in the correct order.

3 Relaxing the Full Order Preserving Property

In this section we will assume that the data encryption scheme is not fully preserving. If we dispense with this property, the security risks discussed in the previous section are not applicable and we can design a data structure to handle duplicates as well as range queries without revealing much information about the ciphertexts.

---

[1] *The Duplicates problem*: The same plaintext values are always encrypted to the same ciphertext values. Thus an adversary can get meaningful information when the total number of possible plaintext values is small (e.g. Votes in an election). If we use a randomized encryption function, then same data item will appear in many different encrypted forms, thus preventing us from performing fast data operations.

3.1   Dealing with Duplicates

The proposed data structure is very similar to a Digital Tree or Trie. However the organization of the data is done according to their *hashed* versions and not the encrypted values. That is, each data item is hashed before it is inserted in the tree and the placement is organized along bit positions of the hashed versions. When the place of insertion of the hashed values), then the encrypted version of the data is placed there. This technique, i.e. combination of hash function for locating the position of the data in the tree encryption function for actually storing the data, allows the change of the encryption function without changing the tree.

Figure 1 shows the proposed data structure with all of the different node types in the tree. We briefly describe them in what follows: (a) Root node: This is the root of the tree (simply a pointer to locate the tree). (b) Inner nodes: These are nodes that direct the search process much like it is done in binary search trees. Each such node has exactly two children. (c) Outer nodes: These nodes have as children only leaf nodes (see right below), which hold the encrypted data. (d) Leaf nodes: Each of these nodes holds an encrypted data item (ciphertext).

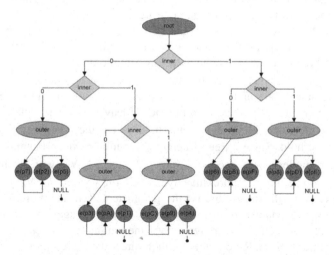

Fig. 1. The different types of nodes in the tree

We will define this new structure through the insertion process. The first step is to pass the value to be inserted in the tree through an *R-box*, which is a randomizing component. The purpose of an R-box is to accept an input of user data and compute a pseudorandom output on it. In other words, it randomizes the input. The output of an R-box will have a fixed length and its functionality can be performed by any simple hash function such as the SHA-1.

Each of the inner nodes checks a specific bit of the R-box output and directs the search either to the left or the right subtree. Let us assume that this binary search process led us to a specific outer node. The encrypted values in each such node are stored in encrypted form using the *CBC mode* of the chosen block. This has the advantage of eliminating the duplicates problem discussed earlier. Each outer node can hold at most B leaves. The value of this parameter, that will be discussed later, plays an important role, along with the height of the tree, in keeping the time complexity of the tree operations logarithmic in the number of data items in the tree.

If the search process leads the new data item to an outer node which has, already, B data items, then a split operation must be performed on this outer node, which is composed of the following steps: (a) Replace the outer node with an inner node and two new outer nodes. (b) Decrypt each leaf, get its original value, pass is through the R-box. Let D be the first bit position (starting from the Least Significant Bit of the R-box value length) that has not been used on the path from the root to the outer node. (c) Replace the outer node by an inner node labeled D, which directs the search left or right depending on the value of the D-th bit of the R-box value. (d) After re-encrypting them in CBC mode, place the data items of the deleted outer node in the two new outer nodes to which the new inner node points depending on the value of the D-th bit.

We will now focus on the determination of the parameter B. Let us assume that there are m outer nodes in the tree. Then these m outer nodes hold CBC chains of encrypted data, as described above. Our goal is to achieve search times of logarithmic order in the number of data items in the tree, which we denote by N. The search time in the tree is composed of two components: (a) searching along the path of inner nodes in order to locate the correct outer node, and (b) searching through the CBC chain of encrypted data by decrypting them and comparing them with the searched item. Our goal is to keep both of these components of logarithmic order.

We set $B = \log^k N$, where N is the current number of leaves (i.e. encrypted data items) in all the outer nodes and k is a real constant $k > 0$ (we will further constrain its values below). Since B is the maximum number of leaves an outer node can hold and we need to accommodate all N data items in the tree, the inequality $m * B \geq N \rightarrow m \log^k N \geq N$ must hold. Since a tree of height h (counting from 0) cannot have more than 2^h outer nodes, we need to have $2^h \log^k N \geq N$, which implies that $h \geq \log(N/\log^k N)$ or, approximately, $h \geq logN - kloglogN$. Obviously, $k \leq N/loglogN$. We will further discuss the parameters B, h and k in Sect. 3.3 with respect to the security analysis of the proposed data structure against a weak adversary (to be defined in that section). Moreover, since the height of the tree also defines the required bit positions of the R-box output which direct the search process, this implies that the R-box output should be at least $logN - kloglogN$.

Returning to the tree operations, if we want to delete a value we first locate the outer node in which it resides using the process outlined above. Then, we decrypt all the leaves residing in this outer node and if the value we seek is found, we delete it taking care to encrypt in CBC the leaves next to the node that was deleted. However, deletions need special care to ensure the preservation of the inequality $B \leq \log^k N$. If many deletions are performed, then N decreases. As N is decreased, it will reach a certain value N' such that $\log^k N' < \log^k N$. Then all outer nodes with $B = \log^k N$ leaves fail to satisfy the new inequality $B \leq \log^k N'$. This necessitates the splitting of this outer node into two new ones, much like we did when an outer node became full during an insertion operation.

Another issue is that the tree may become imbalanced during delete operations. In order to remedy this we introduce two types of rebalancing operations: (a) Rebalancing with two outer nodes. It applies to outer nodes which have less than $B/3$ leaves, and (b) Rebalancing with one outer and one inner node.

Let us, first, examine the case (a) of rebalancing with two outer nodes. All we have to do is to delete the last inner node, take all the leaves together, decrypt them, pass them through the R-box and reinsert them in a one outer node according to the last inner node bit, just above the inner node that was discarded. The resulting outer node has less than $2B/3$ nodes.

In the upper portion of Fig. 2 we consider an example of rebalancing with an outer and an inner node (case b). For concreteness, we label the inner nodes with the bit position they check. The rebalancing operation consists in putting all leaves of the outer node as well as the two outer nodes of the inner node, discarding the inner node leading to the outer and the inner node (i.e. in this case the node checking the 5th bit position), and redistributing the leaves according to the bit checked by the inner node, i.e. the 63th bit. In the two outer nodes on the left, labeled 00 and 01, the first bit 0 corresponds to the 5th bit of the R-box values of the stored data. In the lower portion of Fig. 2, we see the new tree, which has its height reduced by 1.

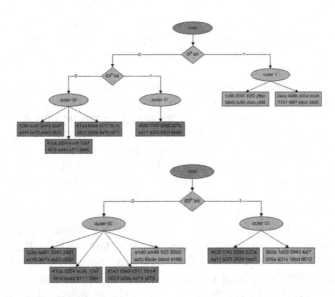

Fig. 2. Rebalancing with one outer and one inner node

3.2 Supporting Limited Range Queries on Encrypted Data

We start by relaxing the full order preserving property as follows: *(partial order preserving encryption)* An encryption scheme is called partially order preserving with neighborhood factor λ if for any two neighboring plaintext values p_i and p_j there can be at most λ other ciphertext values between their corresponding ciphertexts $e(p_i)$ and $e(p_j)$, respectively.

In a fully order preserving scheme, we have $\lambda = 0$, while in the scheme we will define in this section $\lambda = 2(logN - 1)$. We should remark that in our scheme the encrypted values are not fully order preserving and the ordering of the resulting ciphertext depends on their placement in a tree structure which guarantees that, in this (partial) ordering there will be at most $2(logN - 1)$ other values between them. The tree structure we propose is called *Partial-Zero-Knowledge-Range-Tree*, or PZKRT for short. The term "Zero-Knowledge" comes from an advanced feature of the tree to be introduced in Section 3.4, after we discuss its basic version, which is based on the cryptographic primitives of *Privacy-ABCs* (see, for instance, [2, 3] for descriptions of Privacy-ABCs). This technology will enhance our tree with the advanced feature of *privacy preserving*, Zero-Knowledge based, comparisons, i.e. comparisons which can be performed without revealing the actual values involved, which increases the security of the stored information as well as the search operations. The PZKRT is a leaf-oriented tree which stores encrypted data at its leaves and allows range queries on them with small information leakage. The search is directed by range information stored at the internal nodes of the tree.

Let us assume that we want to store data items each of which is W bits long. The data ranges defined (and searched for) by the tree are determined by consecutive bits starting from the Most Significant Bit. Each inner tree node defines a range covered by the sub-tree rooted at this node. Also, the tree partitions this range in two consecutive ranges, the left sub-range and the right sub-range. This is a trie-type tree with main difference that each internal node covers a range and not a single bit. All plaintexts are of the form $p_0, p_1, \ldots p_{w-1}$.

In Fig. 3 we see an instance of a PZKRT that can cover data with $W > 128$ since it is based on the 7 left-most bits of plaintexts. In the first internal note labeled $p_0 - p_2$, the range of values covered are defined by the first three bits (always starting from the MSB). This range, as defined by the bits $p_0 - p_2$, is divided into the two sub-ranges shown on the left and right of this internal node. These two sub-ranges are, in principle, defined arbitrarily with the only condition to form a partition of the range assigned to the internal node (as defined by the consecutive bits assigned to this node). However, for tree balancing reasons, we partition the internal node range so as to have about half data items stored in the left sub-tree and about half data items on the right sub-tree.

The left and right sub-ranges of the first node are defined by [000,101] and [110,111] respectively, covering the data ranges [000XXXX, 101XXXX] and [110XXXX, 111XXXX] respectively. Thus, data items having the three left-most bits in the range [000,101] are placed in the leaves of the left sub-tree while data items having the three left-most bits in the range [110,111] are placed in the leaves of the right sub-tree.

Let us, now, proceed to the left sub-tree of the node covering the bits $p_0 - p_2$, which covers the bits $p_0 - p_6$. This range is partitioned (arbitrarily, again, but with care about balancing if desired) into the two sub-ranges [000 0000, 001 0110] and [001 0111, 101 1111]. Since the bits covered by the tree (7 bits) are exhausted, the two corresponding sub-trees are just leaves holding encrypted data values whose 7 left-most bits belong to the corresponding ranges. This discussion also applies to its right sub-tree. The leaves, finally, store the encrypted data, at random order. All that is known about them is that they fall within the range defined by their ancestor internal nodes.

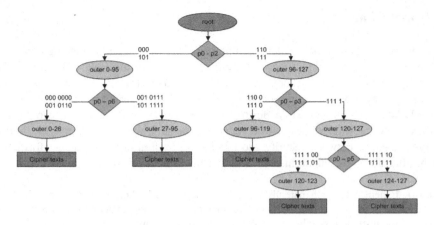

Fig. 3. An instance of a partial-zero-knowledge-range-tree

3.3 Securing the Scheme Against a Strong Adversary

A powerful -or "strong"- adversary can see both the leaves and the internal nodes of the tree, which direct the range search, and obtain information about the corresponding data items in the outer nodes. The range bits are kept encrypted, due to a cryptographic primitive, called Privacy-ABCs.

Privacy-ABCs is a new cryptographic construction based on Attribute Based Credentials (see, e.g., [2, 3]). Privacy-ABCs are issued just like normal electronic credentials from a PKI using a secret signature key owned by the credential issuer. However, the distinguishing feature of this technology is that the user can transform his/her credentials into a special *presentation token* that reveals only the personal information which is required by the service policy and nothing else. The verification of this token is performed with the issuer's public key. The idea, here, is to encode the range values of our tree as credentials and use the comparison functionality for directing the search process. Privacy-ABCs encode the range values so as to allow the execution of comparison operations with them without ever revealing them. Thus, the adversary only sees that an inner node directs the search either to the left or the right but he never sees what range values are really checked by the inner node.

More specifically, the data structure, i.e. the search tree that supports partial range queries, acts as the *issuer*. To each created node that contains a value range (internal node) it issues a credential certifying the range values (upper and lower values of the interval) assigned to the node. Note that these values are encrypted in the credential and cannot be accessed by any process (e.g. insertion and deletion) that attempts to access the tree. Before insertion, each data item is augmented so as to contain a credential containing its exact key value, which also is not accessible. During the insertion process the value to be inserted is compared, without revealing any information, with the ranges in the internal node credentials. In this way, the insertion process does not leak any information about the range in which the inserted data item belongs.

4   Conclusions

In this work we studied the problem of supporting fast search and update operations in encrypted data items. Taking as a departure point the fact the fully order preserving encryption schemes are not secure, we defined a new data structure which is simple and can support fast operations on encrypted data, including limited range queries. All inserted data are randomized to destroy any information about the distribution of plaintexts and the final values can be encrypted by any algorithm. Our work tried to achieve a balance between security and efficiency, by limiting the leaked information which, after all, is inevitable in order to have fast operations on encrypted data.

Acknowledgment. Research partially done while the third author was on Sabbatical leave (Spring-Summer 2015) at the Department of Business Informatics, Faculty of Economics and Business Administration, Goethe University, Frankfurt, Germany.

References

1. Agrawal, R., Kiernan, J., Srikant, R., Xu, Y.: Order preserving encryption for numeric data. In: Proceedings of the 2004 ACM SIGMOD International Conference on Management of Data (SIGMOD 2004), pp. 563–574, ACM (2014)
2. Bjones, R., Krontiris, I., Paillier, P., Rannenberg, K.: Integrating anonymous credentials with eIDs for privacy-respecting online authentication. In: Preneel, B., Ikonomou, D. (eds.) APF 2012. LNCS, vol. 8319, pp. 111–124. Springer, Heidelberg (2014)
3. Brands, S.: Rethinking Public Key Infrastructures and Digital Certificates: Building in Privacy, 1st edn. The MIT Press, Cambridge (2000)
4. Camenisch, J., Groß, T.: Efficient attributes for anonymous credentials. In: Proceedings ACM Conference on Computer and Communications Security, pp. 345–356, ACM (2008)
5. Ozsoyoglu, S.C., Singer, D.: Anti-tamper databases: querying encrypted databases. In: Proceedings of the 17th Annual IFIP WG 11.3 Working Conference on Database and Applications Security (2003)

mobiSurround: An Auditory User Interface for Geo-Service Delivery

Keith Gardiner[(✉)], Charlie Cullen, and James D. Carswell

Digital Media Centre, Dublin Institute of Technology,
Dublin, Ireland
{keith.gardiner,charlie.cullen,jcarswell}@dit.ie

Abstract. This paper describes original research carried out in the area of Location-Based Services (LBS) with an emphasis on Auditory User Interfaces (AUI) for content delivery. Previous work in this area has focused on accurately determining spatial interactions and informing the user mainly by means of the visual modality. *mobiSurround* is new research that builds upon these principles with a focus on multimodal content delivery and navigation and in particular the development of an AUI. This AUI enables the delivery of rich media content and natural directions using audio. This novel approach provides a hands free method for navigating a space while experiencing rich media content dynamically constructed using techniques such as phrase synthesis, algorithmic music and 3D soundscaping. This paper outlines the innovative ideas employed in the design and development of the AUI that provides an overall immersive user experience.

Keywords: Auditory user interface · Phrase synthesis · Content delivery · Geo-services · Navigation

1 Introduction

In this paper the delivery of high quality media content and navigational information via multiple modalities is investigated. The idea of presenting geo-referenced information both visually and aurally to the user in a collective way is the primary focus. The visual modality is presented using a graphical user interface (GUI) in a state-of-the-art mobile app that incorporates and builds on many of the ideas developed in previous works in the area of mobile spatial interaction [1–8]. The aural modality is presented using an auditory user interface (AUI), which is the main innovation in terms of delivering context sensitive information using focus independent (eyes free) means. The AUI is intended to be a non-visual interface that can be used in combination with a GUI, or not. The idea is that all information in the app, including content, directions, events, etc. can be delivered without the user ever needing to interact with the mobile device [9].

There are two main parts to this research. First, the use of an AUI for the delivery of context sensitive geo-tagged content is investigated and secondly, the contextual

© Springer International Publishing Switzerland 2015
Q. Chen et al. (Eds.): DEXA 2015, Part II, LNCS 9262, pp. 57–72, 2015.
DOI: 10.1007/978-3-319-22852-5_6

data modelling process used to structure the data for effective delivery is described. In previous work [10–13], several difficulties were encountered because of the large quantities of content required to achieve this task and as a direct result, the size of the content was an issue. In many cases, the level of redundancy in these systems is very high. In this research, because the content used by the system is primarily audio, the approach taken aims to prevent this level of redundancy by introducing the concept of phrase synthesis to location-based services (LBS). Phrase synthesis is the process by which a dictionary of recorded words is used to create phrases or sentences "on-the-fly" based on linguistic rules. Using this technique we attempt to deliver content and navigational instructions via the AUI resulting in significant reductions in size and redundancy, thus reducing the overheads of the app. This approach is then extended to produce soundscaping where multiple audio channels are sequenced to provide ambient background audio and sound effects that contribute to an overall immersive experience.

In Sect. 2, previous work in the areas of location-based services and mobile spatial interaction is outlined. Section 3 describes the *mobiSurround* AUI and the innovations that provide natural directions for real-time navigation, intelligent phrase synthesis, algorithmic music sequencing and adaptive virtualized 3D soundscapes. In Sect. 4, conclusions and proposed further work is described.

2   Previous Work

Research into location-based information retrieval and delivery has been on going for more that 4 decades now [14–16]. Since the first methods of creating digital maps and cityscapes, points-of-interest (POI) locations and geo-tags, the idea of providing information based on a users location and supported by an underlying geographic map have become increasingly common and sophisticated. It is this work that forms the foundation for research in this area and is the basis for our contributions in the areas of location-based information retrieval and mobile spatial interaction.

2.1   Location-Based Services

With advances in mobile computing technology (e.g. spatially enabled smartphones), implementation of effective location-based services has become a reality in many ways. Initial research in [13] describes work that simulated LBS in a 3D environment using a users virtual position and orientation for the discovery of cultural heritage artefacts. Further research focussed on demonstrating the concept of LBS in real-world environments [17, 18] identifying several fields that required further investigation. One of these areas was visibility and the exact nature of human information retrieval based on the senses, i.e. how we retrieve and process information presented to us; essentially looking at the different modalities for content consumption. The visual modality was studied and in particular the accurate representation of a user's line-of-sight was investigated. In [12, 19], the horizontal field-of-view was considered as the search space, representing what the user can see. This required significant accuracy in terms of positioning and orientation, which was not yet available on mobile devices. Subsequent research in [20], demonstrated the use of location and orientation for mobile spatial search. The point-to-discover service used a customised mobile device to perform spatial selection queries of cultural artefacts demonstrating mobile spatial interaction.

2.2 Mobile Spatial Interaction

Mobile spatial interaction in this context is the study of the interaction between a mobile user and a world of spatial information, i.e. both the physical built environment and all it's connected sensors and related attributes. In [5, 7, 8, 21, 22], the effect direction has on query results presented to users on mobile devices was studied and the idea of an egocentric view of the world was investigated in a mobile context. The applications developed as part of this research operated on contemporary COTS (commercial-off-the-shelf) mobile devices and thus were capable of being used in real-world situations. The results of this research proved successful in addressing the problem of information overload, where too much information gets returned to the mobile device causing display clutter and confusion.

Extending this research further, in [23, 24], the idea of creating a more accurate search space is investigated. The horizontal position and orientation of the user is used to determine the interaction between a ground map of buildings and a user's position and orientation; resulting in a more realistic representation of the user's viewport. In addition, an Isovist [25] was used to create a 360-degree visibility window. In [26] the idea of 3D directional queries was introduced and resulted in the development of the *3DQ* query processor which enabled a user to point a mobile device at a building, for example, to obtain information about the building or more specifically at a particular floor.

Following considerable work investigating novel applications for location-based services and mobile spatial interaction, the need for an improved approach to data modelling and organisation [16] for the delivery of high quality content was apparent [3, 6]. This led to the present approach described in Sect. 3 of an efficient media content delivery engine that can deliver geo-referenced media in an immersive way, using multiple modalities while reducing data redundancy [21, 27]. This new work produced a spatialised auditory user interface (AUI) for LBS with contributions in the areas of location/data modelling and adaptive content production and delivery [16].

3   mobiSurround

The *mobiSurround* engine is an auditory user interface (AUI) for location-based services (LBS) that models context and location to provide a number of exciting new innovations in the areas of content delivery, mobile navigation, and narrative context. More specifically the engine provides the following features:

- An intelligent content phrase synthesiser
- A natural directions model for real time navigation

- An algorithmic music sequencer
- Adaptive virtualised 3D soundscapes
- A full demonstration tour application based on Dublin Zoo

The central component of the *mobiSurround* engine is based on a 3-stage data modelling process that informs the various parts of the application that feed both the visual and aural presentation modalities. The geodata modelling (GM) task is a process by which the geographic data that defines POIs, navigational waypoints and regions are collected. This data informs the rest of system and is the basis of the content modelling strategy. The user modelling (UM) task is where user preferences are modelled, essentially defining the type of user and their context. This data has a scaling effect on the content modelling strategy, where varying amounts of content is presented based on a user's profile. The content modelling (CM) task is where the structure of the media content is defined for each node based on the geo-data (GM) and application preferences data (UM). It is this strictly controlled process of data modelling from multiple perspectives that ensures that data in the system is structured correctly, enabling the dynamic loading and delivery of media content in a coherent manner (Fig. 1).

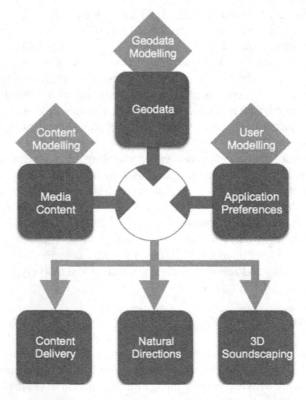

Fig. 1. 3-stage data modelling process includes user, context and content modelling for effective data management and content organization

The following sections outline our innovative approach to modelling for geo-services and describe how the 3-stage process integrates to deliver the functionality described. In particular, aspects relating to content modelling and delivery are outlined that describe; how a space is mapped, how content is modelled around the geo-data, and finally how the content is delivered based on spatial interactions.

3.1 Geodata Mapping and Delivery Matrix

To provide these services a model for the structure and narrative of the content to be delivered was developed. The *mobiSurround* concept focuses on the idea of adaptive narrative that may be used in (though not limited to) tours and exhibitions, where the user may experience a space in their own way, in any direction and in any order. This has been achieved in *mobiSurround* with the development of a node/waypoint/region model for the mapping of data in a given environment (Fig. 2).

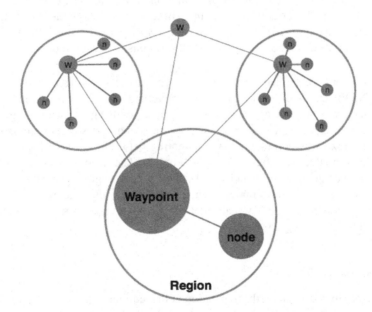

Fig. 2. Data mapping structure describing node, waypoint and region nodes for navigation

Nodes. A node is considered a basic unit of data in the *mobiSurround* engine and is used to represent a POI, service, or facility. Each node has a nodeID and contains information about its 4 neighbouring nodes, its closest waypoint and the region it belongs to. Using the nodeID, the engine determines what content to load on arrival, whether or not to change region and what the next node is based on the users path.

Each node has a number of associated media assets that includes images, text and audio. The phrase synthesizer dynamically loads segments of recorded speech, each containing separate facts about the POI. The phrase synthesizer is used in all node content delivery to allow a relatively small set of commentary phrases (delivered by the narrator) to be reused for a given node based on a set of linguistic rules. In this manner, the content for each node is built at runtime to give the user the impression of listening to a dynamic documentary type interview between narrator and expert. This is illustrated in Figs. 4 and 5.

Waypoints. Waypoints are the main navigational unit in the *mobiSurround* engine. They can be considered as reference points for all other nodes in a space, where each node essentially belongs to or is governed by a waypoint such that navigation around a space is achieved by specifying directions between waypoints. Once navigation between waypoints has been achieved, directions from the destination waypoint to all of its associated nodes are accessible. Specific directions are then provided to a destination node. Significantly, waypoints are defined as natural landmarks that exist in the area making them perfectly suitable for delivering natural directions. For example: *"walk down as far as the Big Tree and take a left; the shop is on your right hand side"*. In this instance the Big Tree is the destination waypoint and the shop is the destination node.

Regions. Regions are defined in this context as functional areas that divide a space, for example, Dublin Zoo is divided into four characteristic regions representing different "corners" of the world. Each region has a set of boundaries that give it a unique geospatial identity. Regions can encompass multiple waypoints and nodes essentially helping to define the context in which they exist. Navigating between regions triggers a region change, which in effect signifies a change in context. The result is a managed "outro" transition of all contextual audio and sound effects for that region followed by an introduction to the new region and a managed intro transition of all contextual audio for that region. This subtle effect is a key feature of the adaptive virtualised 3D soundscaping described in the following section.

3.2   Content Modelling

With a delivery matrix in place, the translation of the required functionality into a content model for authoring is undertaken for each of the 3 elements in the matrix (nodes, waypoints and regions). These 3 elements are then used to define and structure the content for the following areas:

- **Content Delivery**- the information relating to a specific node in the space
- **Navigation**- the information needed to traverse between points in the space using waypoints
- **Narrative Context**- changes between defined regions allow the narrative to adapt to the context of the user

Listening Model. The *mobiSurround* content delivery model is based on a background to foreground listening paradigm [28, 29]. Aligning these elements based on the listening paradigm prioritises speech over background audio and music. This ensures that content delivery and directions are always heard and coherent. This is illustrated in Fig. 3.

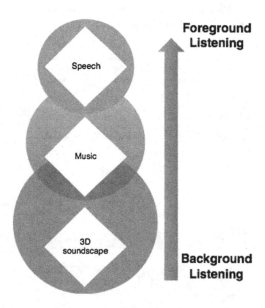

Fig. 3. mobiSurround background to foreground listening model that prioritises speech over background audio and music

The listening model illustrates how content is delivered to the user. A 3D soundscape is produced to recreate environmental aspects of the current region. The 3D soundscape is primarily composed of ambient synth pad sounds and virtualised transient sound effects (SFX). Together, they produce a background harmony and atmosphere as a result of very long attack and decay times with extended sustains [30]. This is layered with algorithmic music composed from melodies, rhythms, and harmonics that are randomly selected and organised at runtime and that the user can detect but not prioritize directly. Speech content (node information or directions) is then layered and will always be foreground content. This layered approach of using 3D Soundscaping, algorithmic music and prioritised speech helps convey broader conceptual variations within the narrative of the space.

The following sections describe the key components and software processes that enable the *mobiSurround* engine to deliver the innovative features described.

Intelligent Content Phrase Synthesizer. One of the major contributions of the *mobiSurround* engine is the intelligent phrase synthesizer implemented to deliver speech content. The synthesizer was developed to provide multimodal content delivery on constrained devices that preclude large amounts of audio content being included in a wireless network download. The phrase synthesizer is used for all node content delivery to allow a relatively small set of commentary phrases (delivered by the narrator) to be reused for a given node based on a set of linguistic rules. In this manner, node content is dynamically assembled at runtime to give the user the impression of listening to a documentary interview between narrator and expert. A multi-channel audio graph (*AUGraph*) is used to provides 16 individually controlled busses, much like a traditional mixing desk [31]. This is illustrated in Fig. 4.

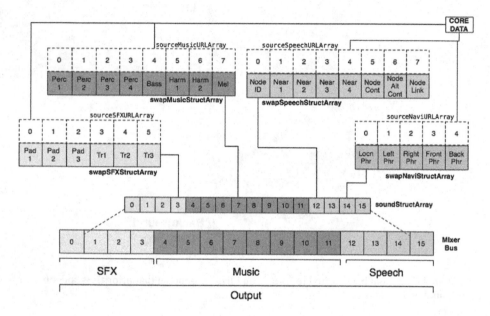

Fig. 4. Mixerbus multichannel audio graph controls all audio playback in the application based on a set of rules

The phrase synthesiser has the exclusive use of 4 busses in the *AUGraph* with each bus capable of managing 2 sets of files at once. Full control of each bus is maintained as data is swapped in and out in real-time. This is all managed dynamically by a set of rules determined during the data-modelling phase. The geodata and content for each node is combined with a narration template defining the overall structure of a node. When the context changes, the graph is dynamically controlled by a set of decision rules that are stored on Core Data. These rules are described in Table 1.

Figure 5 describes how the phrase synthesizer constructs node, region, help and direction content sentences in real time using busses 1–4. Each bus (Intro, Content, altContent, Outro) is dynamically loaded with audio using a handle and a content phrase. The handle is randomly selected from a predefined library and the content relates to the nodeID of the currently selected node. An example of a synthesized sentence is as follows:

[You are now arriving at] + [the Snow Leopard exhibit] + [Lets ask john to give us some information] + [Snow Leopards are large cats that live way up high in the mountains ranges of central Asia] + [Wow that's interesting] + [FACT2] + [Cool can you give us another fact] + [FACT3] + [Interesting, what else can you tell us about it] + [FACT4] + [Well now you have heard about it lets move on to] + [the red river hog exhibit] + [DIRECTIONS]

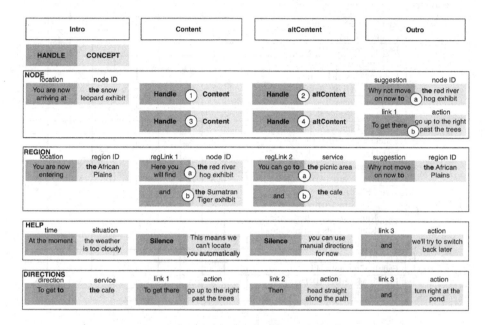

Fig. 5. Phrase synthesis structure for node, region and directions content

Each node has two narrators associated with it, the main narrator and a domain expert. Facts 1 and 2 are initially loaded into the *Content* and *altContent* busses, and then they are swapped out for Facts 3 and 4. The phrases are constructed in real-time and loaded into the *AUGraph* where they are organised sequentially, initialised with the required bus control data from the database and played in order. The result is a seamless set of sentences played back in real-time describing the current node (POI). When the node information is finished, all 4 busses are faded out and reset awaiting the discovery and subsequent delivery of new node information. All node, region information, help and directions content are constructed in this manner, significantly reducing the overhead of recording and playing large static files.

The conditional logic that controls the mixerbus is described in the decision table below. Each time conditions change; the graph control data is loaded from core data and applied.

Table 1. Decision table describing the rules, conditions and actions that control loading, management and playback of the mixerbus AUGraph

		Rules									
Conditions	New POI Detected	Y		Y			Y		Y		Y
	New Region Detected		Y	Y			Y		Y		Y
	New Event Occuring				Y						
	Directions Request					Y	Y				
	Random SFX Change							Y	Y		
	Random Music Change									Y	Y
Actions	Prepare speechURLArray	X		X	X		X		X		X
	Prepare musicURLArray		X	X					X	X	X
	Prepare SFXURLArray		X	X			X		X		X
	Prepare naviURLArray		X	X	X	X	X		X		X
	FadeOut Channel 12-15	X			X	X					
	FadeOut Channel 0-3										
	FadeOut Channel 4-11										
	FadeOut MixerBus (0-16)		X	X			X		X		X
	FadeOut Channel Index							X		X	
	swapSpeechStructArray	X		X	X				X		X
	swapMusicStructArray		X	X					X		X
	swapSFXStructArray		X	X					X		X
	swapNaviStructArray		X	X	X	X	X		X		X
	swapRandomSFXStructElement							X			
	swapRandomMusicStructElement									X	
	FadeIn Channel 12-15	X			X	X	X				
	FadeIn Channel 0-3										
	FadeIn Channel 4-11										
	FadeIn MixerBus (0-16)		X	X					X		X
	FadeIn Channel Index							X		X	

Natural Directions Model for Real-Time Navigation. Natural directions are a recent innovation in navigation technology. Instead of providing measurements and compass headings, landmarks are used to give directions. The advantages of natural directions are speed (humans do not need to translate data) and adaptability (there is no need to hold a compass in a certain way to obtain a heading) that allow them to be delivered hands free. The disadvantage of natural directions is the size of the content model required, where each direction is a unique narrative in itself.

To address this, the *mobiSurround* natural directions model uses a combination of node and waypoint information to control the phrase synthesizer, providing a set of natural directions (which can also be displayed on a map if required) to any given point in the space. The model uses the concept of waypoint linkage to create a lookup table of directions between waypoints that is then completed by directions from the last waypoint to the node required. In this manner, natural points of focus within the space can be defined as waypoints that are used to script and create the directions needed to link them together. When a specific node is queried, the only information required is its nearest waypoint, which in turn allows a lookup table to be used to configure the phrase synthesizer.

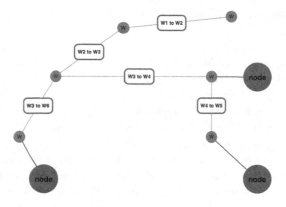

Fig. 6. Waypoint linkage example for node and waypoint navigation

Figure 6 demonstrates how waypoint linkages can be used to provide directions to any node in the space. For the 3 example nodes shown, a common set of waypoint linkage directions is used to get to waypoint 3, where different waypoint combinations then lead to nodes 3, 4 and 5. By building a set of waypoint linkages in both directions (getting from W2 to W1 is not the same as going from W1 to W2) each node is linked to its nearest waypoint, which is then linked by a lookup table to every other waypoint. The phrase synthesizer is then loaded with the entries from the lookup table and provides the directions to the user as required.

Location Manager with Adaptive Resolution Monitoring. The *mobiSurround* engine considers spatialised auditory user interfaces, but the concept of LBS requires suitably accurate positioning in order to deliver an immersive and hands free user experience. The engine is designed to be fully automated in its response to changes in location so a Location Manager was developed (Fig. 7).

Fig. 7. *mobiSurround* location manager (demonstration using Dublin Zoo tour).

The manager defines both current region and nearest node in real-time, with a location resolution or proximity monitor being used to determine when the engine can deliver content for a given node. In addition, the manager can determine different types of node (e.g. POI, facilities, events) at once, as demonstrated in the Dublin Zoo demo tour (Fig. 7).

Accurate positioning (particularly in urban areas) is still an issue with COTS devices, with changes in weather conditions and urban canyons due to the presence of buildings and other large objects preventing a GPS receiver from delivering an accurate reading in real-time. Although *mobiSurround* does not directly seek to innovate in positioning accuracy, the location manager employs an adaptive resolution monitor to specify reasonable bounds for positioning accuracy. This ensures that content delivered by the engine is always appropriate, and also that future advances in positioning (such as EGNOS or NFC) to provide sub-meter accuracy are compatible with the current engine.

In any condition where positioning accuracy is too low to deliver content effectively, an intelligent audio housekeeping manager updates the user in real-time by informing them of the condition and asking them to switch to manual mode, where all nodes are listed in the app for triggering by manual selection (Fig. 8). In this manner, the engine will ideally not deliver the wrong content to the user and so provide them with a far higher degree of usability and flexibility. As an additional development to this, the location manager has incorporated live weather updates into its accuracy assessment (GPS accuracy is often linked to cloud cover) that can be broadcast to the user by the housekeeping manager.

Fig. 8. mobiSurround GUI used in manual mode enables access to POI, facility and event information when accurate positioning data is unavailable

Adaptive Virtualized 3D Soundscapes. The *mobiSurround* engine is capable of delivering immersive soundscaping based on the region context of the user. Transitions between regions are dealt with automatically as fades between atmospheric 3D background sounds as the user moves between different regions in a space. Fades are timed to function at walking pace (though this can be varied programmatically) so the user perceives a natural spatial transition between one context and another.

Each region has a specifically authored soundscape comprising of background synth pad sounds and foreground transient SFX, where a transient (such as a bird singing) is delivered using real-time transient positioning. This gives a listener the sense of being in a live environment, giving the impression of hearing a lion roar just over your right shoulder, for example. This effect is achieved using a Digital Theatre Systems (DTS) virtualiser to give the impression of movement within the soundscape, allowing sounds relating to non-stationary objects to be delivered algorithmically as part of the narrative.

The engine automatically adapts to changes in location and context from a top down perspective, so transient sound effects (SFX) never clash with more important foreground sounds such as node content or directions.

The virtualiser provided by DTS Audio contains a low bitrate version of the code for potential implementation on constrained devices. A virtualization algorithm takes a head related impulse response (HRIR) for a given azimuth (measured using a binaural dummy head) and convolutes this with the incident signal to deliver an output signal which is perceptually similar to one as heard coming from that azimuth. In the case of the DTS virtualiser, azimuths are provided for a common 5.1 multichannel audio setup that defines rear left (rL) and rear right (rR) positions. This virtualiser was performance tested using the iOS *AUGraph* render model to determine whether a real-time Fast Fourier Transform (FFT) convolution could be performed using the iOS render callback model.

The results (Fig. 9) define a cardioid response pattern for rear content when delivered using the virtualizer. This leads to a practical constraint of locationing where idealised radius r (translating to distance) is replaced by r1 in the rear condition and r2 for frontal cues in the virtualised case. Taking the desired effect into account, these results are satisfactory and produce a realistic directional listener experience.

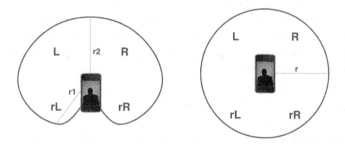

Fig. 9. Visual summary of virtualiser perception test results (Comparing ideal and actual)

Algorithmic Music Sequencer. The *mobiSurround* engine contains a full multipart algorithmic music sequencer, where regions are given a specific musical score containing several parts that vary automatically over time. The sequencer swaps content based on certain compositional rules provided to ensure that the user never becomes bored by a continuous loop of music in a similar manner to algorithmic music found in video games. In *mobiSurround* the sequencer also responds to contextual changes, swapping all music for each region to compliment the underlying soundscape using the same fading strategy (Table 1). In this manner, the user may move wherever they wish, with SFX and music seamlessly adapting to their current context. Music and background synth pads are juxtaposed by placing them at the front and rear of the virtualised soundscape respectively. This provides the maximum spatial distance possible between content elements, providing a wider auditory vista within the interface.

4 Conclusions and Future Work

The *mobiSurround* prototype demonstrates significant advances in many areas of spatialised multi-modal content delivery. In addition to a traditional graphical user interface (map), an auditory user interface (AUI) is used to deliver rich, interactive content using audio as the primary modality. The AUI comprises novel work in the areas of natural directions (providing directions using landmarks as opposed to cardinal directions) and content delivery using an intelligent content phrase synthesiser controlled by user location, orientation, and previous visits to the space. In addition, an immersive audio experience is delivered using region detection and an algorithmic music sequencer. The result is an audio tour that provides timely information about the current environment (i.e. Dublin Zoo), 3D soundscaping that delivers immersive background audio and transient sounds based on location and natural directions between nodes on the tour. All of the media content in the app is assembled on-the-fly, with music, sound effects and speech audio being constructed dynamically using the music sequencer and phrase synthesizer. This approach has proven to significantly reduce the amount of redundancy in the system enabling a far greater degree of efficiency overall.

Future work (as a result of testing) will also consider how best to further reduce the application footprint. This will include the manipulation of non-interleaved audio files (all virtualised content is stereo) to better function with the iPhone *AUGraph*. In addition a comprehensive user trial is planned to test and provide feedback in terms of functionality and usability of the AUI for delivering content. Development of an authoring tool for planning, structuring, and combining content is also underway. This tool will help streamline the process of audio production as outlined in the data-mapping matrix and listening model.

References

1. Fröhlich, P., Simon, R., Baillie, L., Anegg, H.: Comparing conceptual designs for mobile access to geo-spatial information. In: Proceedings of the 8th Conference on Human-Computer Interaction with Mobile Devices and Services – MobileHCI, p. 109 (2006)
2. Persson, P., Espinoza, F., Fagerberg, V., Sandin, A., Cöster, R.: GeoNotes: a location-based information system for public spaces. In: Höök, K., Benyon, D., Munro, A.J. (eds.) Designing Information Spaces: The Social Navigation Approach Computer Supported Cooperative Work, pp. 151–173. Springer, London (2003)
3. Robinson, S., Eslambolchilar, P., Jones, M.: Point-to-GeoBlog: gestures and sensors to support user generated content creation. In: Proceedings of the 10th International Conference on Human Computer Interaction with Mobile Devices and Services. ACM, Amsterdam, The Netherlands (2008)
4. Simon, R., Frohlich, P.: A mobile application framework for the geospatial web. In: Proceedings of the 16th International Conference on World Wide Web, pp. 381–390. ACM, Banff, Alberta, Canada (2007)
5. Simon, R., Fröhlich, P., Gerhard Obernberger, G., Wittowetz, E.: The point to discover GeoWand. In: Proceedings of the 9th International Conference on Ubiquitous Computing, Innsbruck, Austria (2007)

6. Simon, R., Frohlich, P., Grechenig, T.: GeoPointing: evaluating the performance of orientation-aware location-based interaction under real-world conditions. J. Locat. Based Serv. **2**(1), 24–40 (2008)
7. Frohlich, P., Simon, R., Baillie, L.: Mobile spatial interaction. Pers. Ubiquit. Comput. **13**(4), 251–253 (2009)
8. Strachan, S., Murray-Smith, R.: Bearing-based selection in mobile spatial interaction. Pers. Ubiquit. Comput. **13**(4), 265–280 (2009)
9. McGee, J., Cullen, C.: Vocate: auditory interfaces for location-based services. In: Proceedings of the 23rd Conference on Computer Human Interaction, Cambridge, United Kingdom (2009)
10. Carswell, J.D., Gardiner, K., Bertolotto, M., Rizzini, A., Mandrak, N.: A web-based and mobile environmental management system. J. Environ. Inform. **12**, 9–20 (2008)
11. Carswell, J.D., Gardiner, K., Neumann, M.: Wireless spatio-semantic transactions on multimedia datasets. In: Proceedings of the 2004 ACM Symposium on Applied Computing, pp. 1201–1205. ACM, Nicosia, Cyprus (2004)
12. Gardiner, K., Carswell, J.D.: Viewer-based directional querying for mobile applications. In: Third International Workshop on Web and Wireless Geographical Information Systems W2GIS. IEEE CS Press, Rome, Italy (2003)
13. Carswell, J.D., Eustace, A., Gardiner, K., Kilfeather, E., Neumann, M.: An environment for mobile context-based hypermedia retrieval. In: Proceedings 13th International Workshop on Database and Expert Systems Applications (2002)
14. Salton, G., McGill, M.J.: Introduction to Modern Information Retrieval. McGraw-Hill Computer Science Series, xv, p. 448. McGraw-Hill, New York (1983)
15. Schiller, J., Voisard, A.: Location Based Services. Morgan Kaufmann Publishers Inc., San Francisco (2004)
16. Jiang, B., Yao, X.: Location-based services and GIS in perspective. Comput. Environ. Urban Syst. **30**, 712–725 (2006)
17. Mohapatra, D., Suma, S.B.: Survey of location based wireless services. In: IEEE International Conference on Personal Wireless Communications (ICPWC) (2005)
18. Rao, B., Minakakis, L.: Evolution of mobile location-based services. Commun. ACM Mob. Comput. Oppor. Chall. **46**(12), 61–65 (2003)
19. Chan, L.-W., Hsu, Y.-Y., Hung, Y.-P., Hsu, J.Y.-J.: Orientation-aware handhelds for panorama-based museum guiding system. In: UbiComp 2005 Workshop: Smart Environments and their Applications to Cultural Heritage (2005)
20. Simon, R., Kunczier, H., Anegg, H.: Towards orientation-aware location based mobile services. Inf. Syst. J. p. 1–8 (2007)
21. Wilson, A., Shafer, S.: XWand: UI for intelligent spaces. In: Proceedings of the SIGCHI Conference on Human Factors in Computing Systems, pp. 545–552. ACM, Ft. Lauderdale, Florida, USA (2003)
22. Strachan, S., Williamson, J., Murray-Smith, R.: Show me the way to Monte Carlo: density-based trajectory navigation. In: Proceedings of the SIGCHI Conference on Human Factors in Computing Systems, pp. 1245–1248. ACM, San Jose, California, USA (2007)
23. Gardiner, K., Yin, J., Carswell, J.D.: EgoViz – a mobile based spatial interaction system. In: Carswell, J.D., Fotheringham, A., McArdle, G. (eds.) W2GIS 2009. LNCS, vol. 5886, pp. 135–152. Springer, Heidelberg (2009)
24. Carswell, J.D., Gardiner, K., Yin, J.: Mobile visibility querying for LBS. Trans. GIS **14**, 791–809 (2010)
25. Benedikt, M.L.: To take hold of space: isovists and isovist fields. Environ. Plan. **6**, 47–65 (1979)

26. Carswell, J.D., Gardiner, K., Yin, J.: 3DQ: threat dome visibility querying on mobile devices. GIM Int. **24**(8), 24 (2010)
27. Williamson, J., Murray-Smith, R., Hughes, S.: Shoogle: excitatory multimodal interaction on mobile devices. In: Conference on Human Factors in Computing Systems, p. 121 (2007)
28. Truax, B.: Soundscape, acoustic communication and environmental sound composition. In: Contemporary Music Review, pp. 49–65 (1996)
29. Misra, A., Cook, P.R., Wang, G.: A new paradigm for sound design. In: Proceedings of the International Conference on Digital Audio Effects, pp. 1–6 (2006)
30. Schoenberg, A.: Fundamentals of Music Composition. Faber and Faber, London (1967)
31. Apple. iOS developer library (Core audio), 2015 February. https://developer.apple.com/library/ios/documentation/MusicAudio/Conceptual/CoreAudioOverview/CoreAudioEssentials/CoreAudioEssentials.html. Accessed 2015

A Diversity-Seeking Mobile News App Based on Difference Analysis of News Articles

Keisuke Kiritoshi [(⊠)] and Qiang Ma [(⊠)]

Kyoto University, Kyoto 606-8501, Japan
kiritoshi@db.soc.i.kyoto-u.ac.jp, qiang@i.kyoto-u.ac.jp

Abstract. To support the efficient gathering of diverse information about a news event, we propose a diversity-seeking mobile news app on smart mobile devices. At first, by extending our previous work, based on the entity-oriented news analysis, we propose three measures for searching and ranking news articles from perspectives of difference in opinions, difference in details, and difference in factor coverages. Then, by utilizing these measures, we develop a news app on mobile devices to help users to acquire diverse reports for improving news understanding. One of the notable features of our system is a context-aware re-ranking method for enhancing the diversity of news reports presented to users by considering the access history. The experimental results demonstrate the efficiency of our methods.

Keywords: News app · Diversity · Difference analysis · Context aware re-ranking · Crowdsource experiment

1 Introduction

In some sense, news is never free from bias due to the intentions of editors and sponsors. To helping users to understand news events, considering diversity of news articles is important [1,2,5,8]. With the spreading of smart phones and tablets, news apps, the news reading applications on smart mobile devices, have been widely used. However, a news app usually provides only one article per each topic. In addition, mobile search is more difficult than general web one due to the limitations of environment and devices. As a result, a user may lose the chance to obtain information from multi-viewpoints to avoid biased impression.

In this paper, to improve the users' experiences of reading news on smart mobile devices, we propose a novel news app with the function of helping users to seek diverse information on a news event.

At first, we propose three measures to search and rank news articles by extending our previous work [1]. In [1], based on a user survey, we have proposed four ranking measures, (*relatedness, diversity, difference in opinion*, and *difference of detailedness*). The experimental results described in [1] clarify that although the difference between articles is important, they should be related to each other at first. Based on this observation, we refine these ranking measures. The refined measures are summarized as follows.

© Springer International Publishing Switzerland 2015
Q. Chen et al. (Eds.): DEXA 2015, Part II, LNCS 9262, pp. 73–81, 2015.
DOI: 10.1007/978-3-319-22852-5_7

- *DC (Difference in Factor Coverage)* is the extent of how many different things are described in two news articles reporting the same event.
- *DO (Difference in Opinion)* is the extent of difference of subjective descriptions in two news articles reporting the same event.
- *DD (Difference in Details)* is the extent of difference of details of two articles reporting the same event.

By utilizing these measures, we develop a news system effectively providing different reports on same events to support users' news understanding on smart mobile devices. The major contributions of this paper can be as follows:

- We propose three entity-oriented measures for ranking news articles by extending our previous work (Sect. 3) to help us to search and rank news articles by focusing on the difference in news reports.
- We propose a diversity-seeking news app (Sect. 4). As one of the notable features of our system, we propose a context-aware re-ranking method for enhancing the diversity of news reports provided to users. (Sect. 4.3).
- We conducted crowdsource experiment to validate the effectiveness of our ranking measures (Sect. 5.1). The re-ranking method is validated by a simulation experiment (Sect. 5.2).

2   Related Works

To help users' better understanding of news articles, news browsing systems which visualize and highlight the differences between news articles have been proposed. Ogawa et al. [3] study the analysis of differences between news articles by focusing on named entities and they propose a stakeholder mining mechanism. They extract stakeholders who are mentioned and present a graph constructed based on description polarity. The target news of Ogawa et al. is text news, while Xu et al. study on stakeholder mining on multimedia news [4].

NewsCube [2] presents various aspects of a news event and presents these using an aspect viewer to facilitate understanding of the news. TVBanc [5] compares news articles based on a notion of topic structure. TVBanc gathers related news from various media and extracts pairs of topics and viewpoints to reveal the diversity and bias of news reports on a certain news event.

In contrast, we are studying on searching and ranking related news articles by estimating the difference between articles. Our method focuses on how to provide users diverse reports different from the ones users have already read. In addition, our system is the first attempt trying to effectively provide diverse news reports on smart mobile devices.

3   Entity-Oriented Ranking Measures of News Articles

In our previous work [1], the experimental results reveal that although the difference between articles is important to obtain diverse information, a news article reporting totally different event is not useful to help users' news understanding. Based on this observation, in this paper, we refine the definitions and propose three entity-oriented ranking measures: *DC (Difference in Factor Coverage)*, *DO (Difference in Opinion)*, and *DD (Difference in Details)*.

In this section, at first, we brief the method of extracting entities and entity-related descriptions from news articles (refer to [1] and [3] for the details). Then, we introduce the entity-oriented ranking measures.

3.1 Extraction of Entities and Entity-Related Descriptions

We use a language tool StanfordCoreNLP[1] to extract named entities and apply the method proposed by Ogawa et al. [3] to generate a tree structure (Ogawa et al. call it Relationship Structure). We consider descriptions on named entities as sets of sub-trees of relationship structure, the root of each being a verb and its descendants containing the target named entities.

Another important notion in our entity-oriented news analysis is core entity [1]. Intuitively, core entities in an event are the named entities mentioned frequently in the articles reporting that event.

3.2 Ranking Measures

As mentioned before, our entity-oriented ranking measures are defined based on comparing named entities and their descriptions in news articles.

DC (Difference in Factor Coverage): DC is the degree of how many different things are described in two news articles reporting the same event. We estimate DC from two aspects, (1) how many different factors mentioned in these articles, and (2) whether these articles are related to the same event or not. For aspect (1), we can simply compare the entities mentioned in articles. The more different entities two articles describe, the higher difference in factor coverage the two articles are. For aspect (2), we compare the entity mentioned in each article with core entity set of the given event. Let E_{core} be the set of core entities of a certain news event. DC between articles a and o, $dc(a, o)$ is calculated as follows:

$$dc(a, o) = rel_{eve}(E_{core}, a) \times div_{dif}(a, o) \tag{1}$$

$$div_{dif}(a, o) = |E_a - E_o| \tag{2}$$

$$rel_{eve}(E_{core}, a) = \frac{|E_a \cap E_{core}|}{|E_{core}|} \tag{3}$$

where, E_a and E_o are sets of named entities in mentioned by a and o respectively.

---

[1] http://nlp.stanford.edu/software/dependenciesmanual.pdf

DO (Difference in Opinion): *DO* denotes the different extent of opinions between two articles. We compare the description polarities (positive, negative, and neutral) on named entities in articles. If two news articles report the same entities while their polarities are different, we regard these articles are different from opinion. *DO* of two news articles a and o, $do(a, o)$ is calculated as follows.

$$do(a, o) = rel_{mut}(a, o) \times sup(a, o) \tag{4}$$

$$sup(a, o) = w_{do} \times \sum_{e \in \{E_a \cup E_o\}} |sup_a(e) - sup_o(e)| \tag{5}$$

$$rel_{mut}(a, o) = \frac{|E_a \cap E_o|}{|E_a \cup E_o|} \tag{6}$$

where, $sup_a(e)$ and $sup_o(e)$ are polarities of named entity e in articles a and o respectively. w_{do} is the weight of core entities and is calculated as follows:

$$w_{do} = \begin{cases} w_{core,do} & (e \in E_{core}) \\ 1 - w_{core,do} & (others) \end{cases} \tag{7}$$

DD (Difference in Details): *DD* denotes the different degree of details provided by two news articles. We compare named entity related descriptions in articles to estimate *DD* following *difference of detailedness* in our previous work. After extraction of named entity related descriptions, we apply LDA (Latent Dirichlet Allocation) [6] to detected topics from the description of named entities. Then, we compare topic coverage and topic word on each named entity between two news articles. *DD* includes weight w_{dd} calculated by the same as formula (7).

4   Diversity-Seeking Mobile News App

By utilizing the proposed ranking measures, we develop a system to help users to acquire diverse reports of news events on a smart mobile device.

Users may use mobile devices to check news when users have a little time during moving or waiting. However, due to the time limitation and features of mobile devices, it is not easy to find diverse reports on a certain news event even the users may be interested in the event. In addition, the typical news apps present only one article per each news event. These are reason why we focus on news app on smart mobile device. The system consists of a news server and news client and the main functions are explained Sects. 4.1–4.3.

4.1   News Server: Gathering and Analyzing News Articles

Our diversity-seeking news system targets English news articles. We gather top news articles and their related article by using Google News Realtime Coverage[2] (one of the Google News' functions to present articles related to top news

---

[2] https://support.google.com/news/answer/2602970

articles). After gathering news articles, the news server analyzes and ranks news articles to top news articles per each event with the method introduced in Sect. 3. Differences between news articles are quantified by DC, DO, and DD, respectively. Each top news article and its related articles with top rank of each measure will be delivered to news clients and then be presented. The server will carry out context-aware re-ranking every time a user history increases as Sect. 4.3.

4.2 News Client: Presenting News Articles

The news client of our diversity-seeking news system has two view modes. One is the *top news view* to list today's top news. The other is *article details view*, which presents the content of a news article with links to its three top-ranked related articles. Figure 1 illustrates the running examples of these two views.

Top News View Mode   Article Details View Mode

Fig. 1. View nodes of news client

Top news view shows top news articles gathered from Google News per each category. A user can click one article to view its details in *article details view*.

Article details view presents the details of a news article. There are three link buttons named "Opposite", "Wide", and "Deep", corresponding to the top ranked article of DO, DC and DD, respectively. By clicking these buttons, a user can access diverse reports on the same event as the current one.

4.3 Context-Aware Re-ranking

It is possible that the presented top-ranked different articles are the ones a user have read already and are similar to each other because the rank is decided by comparison with the current viewing article. In order to present more diverse information from different articles, we propose a context-aware re-ranking method. That is, the compare target to ranking does not only include the current article, but also includes the articles have been accessed before.

Suppose that A is the set of related news articles about a certain news event and $H(H \subset A)$ are the articles the user has read before. We update the ranking score $d(b, H)$ of article $b \in \{A - H\}$ regarding H as follow.

$$d(b, H) = \frac{1}{|H|} \sum_{h_i \in H} d(b, h_i) \qquad (8)$$

where, $d()$ represent $dc()$, $do()$, and $dd()$.

5   Experiments

We carried out two experiments to evaluate our methods. One is for the ranking measures. The other is for the context-aware re-ranking method.

5.1   Crowdsource Experiment on Ranking

To evaluate our refined ranking measures by various people, we conducted a crowdsource experiment on the platform provided by CrowdFlower[3]. In the experiment, we gathered news articles with Google News US edition (Top stories and the Realtime Coverage) from June 17th 2014 to June 24th 2014. The number of news events is twenty and fourteen news articles in average were selected randomly per each topic. In each topic, we selected one article as the current article and the others as its related articles.

In our experiment, we asked workers to compare the pair of current article and one of its related articles. As a result, we have thirteen pair-comparisons per each news event in average. In each pair comparison, we asked workers score the related article according to the five grades evaluation system from three viewpoints, DC, DO, and DD respectively.

We compared ranking by the average scores assigned by crowdsource workers with that by our proposed ranking methods. nDCG (Normalized Discounted Cumulative Gain) [7] is the evaluation measure. We varied each parameter w_{do} (Formula 7), and w_{dd} (Sect. 3.2) and calculated the average of nDCG for the top k ranking results ($k = 6$) of the 20 topics (events). The nDCG results of each measure are shown in Table 1.

In Table 1, DCB and DOB denote the comparative method, which calculates DC and DO without considering relatedness between the target articles, respectively. These scores are calculated as $DCB = div_{dif}(a, o)$ and $DOB = sup(a, o)$.

In the nDCG results of DD, the highest evaluation value was 0.8964 and then $w_{dd} = 0.9$. These values are acceptably high. In contrary, it is hard to say which method is better for calculating DC and DO. One of the considerable reasons is that the articles used in the experiment are strongly related to each other. To confirm this assumption, we conduct a small additional experiment. We manually insert some noisy articles to a news event in our data set and test the ranks of these noisy articles. The ranking results with noisy articles are shown in Tables 2 and 3. These tables show the noisy articles are high-ranked by DCB and DOB while DC and DO decrease their ranks significantly. Therefore, we can say, although the proposed method cannot outperform previous method, it works well and is independent on the accuracy of gathering related news articles.

---

[3] http://www.crowdflower.com/

Table 1. nDCG of DD, DC, and DO ($k = 6$)

Parameter w_{dd}	0.1	0.2	0.3	0.4	0.5	0.6	0.7	0.8	0.9
DD	0.8934	0.8949	0.8953	0.8952	0.8945	0.8938	0.8938	0.8944	**0.8964**
Parameter w_f	0.1	0.2	0.3	0.4	0.5	0.6	0.7	0.8	0.9
DOB	0.9186	0.9204	0.9251	**0.9273**	0.9225	0.9229	0.9206	0.9222	0.9236
DO	0.9086	0.9076	0.9086	0.9088	0.9087	0.9130	0.9142	**0.9159**	0.9142
DCB	**0.9332**								
DC	**0.9242**	Note : DC and DCB have no parameter.							

Table 2. Ranking of DC with noisy articles

Article	DCB	DC
Noisy article1	3	5
Noisy article2	2	1
Noisy article3	1	8

Table 3. Ranking of DO with noisy articles

Article	DOB	DO
Noisy article1	2	18
Noisy article2	1	18
Noisy article3	6	18

5.2 Experiment on Context-Aware Re-Ranking Method

We conducted experiment to evaluate our context-aware re-ranking method by simulating the news reading sequences. We randomly selected fifteen news events that are used in our crowdsource experiment. The simulation based experiment is carried out by repeating the following steps per each news event.

1. Let a and a_b be the current news article. Let A be the related article set of a certain news event. Let accessed articles $H = a$.
2. Rank the articles in A respectively from three aspects: DC, DO and DD. Store these ranks in R_c, R_o, and R_d, respectively.
3. Perform the context-aware re-ranking method to create ranks of articles in A respectively from three aspects.
4. Randomly choose one aspect x with the limitation that each aspect should not be selected more than twice. Choose the top-ranked article t from the chosen aspect x. Also, choose the top-ranked article t_b from R_x.
5. One evaluator reads article t and t_b, and then evaluates how much difference information t and t_b have comparing H in five grades, respectively.
6. $a \leftarrow t$. $a_b \leftarrow t_b$. $A = A - t$, $H = H \cup t$. $R_c = R_c - t_b$, $R_o = R_o - t_b$, $R_d = R_d - t_b$.
7. If A is not empty, go to step 3; else stop.

Where, a and t are used for our context-aware re-ranking method while a_b and t_b are used for the baseline method.

The five grades evaluation has been conducted to estimate how much different information can be obtained from the new article comparing with the articles have been read before from three viewpoints in user survey [1]: relevance, opinion, and the amount of additional information. The evaluator scored articles

Table 4. Experimental results of re-ranking method

Event	Event 1	Event 2	Event 3	Event 4	Event 5	Event 6
Baseline	3.00	2.67	3.67	3.83	2.17	3.00
Re-ranking	3.50	3.33	3.67	3.83	2.17	3.17

comprehensively from these three viewpoints. There are six news events' results of baseline and our re-ranking methods are different. We calculated the average of user scores of these six events.[4] The results are shown in Table 4.

Among the six events, the user scores of re-ranking method are greater than or equal to those of baseline. This result reveals the availability of our re-ranking method to enhance users to read more diverse information.

6   Conclusion

In this paper, we propose three entity-oriented ranking measures to support users obtaining diverse reports on the same news event. As one application of these ranking measures, we develop a diversity-seeking news app on smart mobile devices. A context-aware re-ranking method is also proposed to provide more diverse information. We conducted crowdsource experiment to validate our entity-oriented ranking measures. The context-aware re-ranking method is validated by a simulation-based analysis.

As one of the important future work, we need conduct user study of our news app and re-ranking method. In addition, we plan to carry out experiments to investigate the change of user's media literacy by using our system.

Acknowledgement. This work is partly supported by KAKENHI(No. 25700033) and SCAT Research Funding.

References

1. Kiritoshi, K., Ma, Q.: Named entity oriented related news ranking. In: Decker, H., Lhotská, L., Link, S., Spies, M., Wagner, R.R. (eds.) DEXA 2014, Part II. LNCS, vol. 8645, pp. 82–96. Springer, Heidelberg (2014)
2. Park, S., Kang, S., Song, S.C.J.: Newscube: delivering multiple aspects of news to mitigate media bias. In: CHI 2009, pp. 443–452 (2009)
3. Ogawa, T., Ma, Q., Yoshikawa, M.: News bias analysis based on stakeholder mining. IEICE Trans. **94–D**, 578–586 (2011)
4. Xu, L., Ma, Q., Yoshikawa, M.: A cross-media method of stakeholder extraction for news contents analysis. In: Chen, L., Tang, C., Yang, J., Gao, Y. (eds.) WAIM 2010. LNCS, vol. 6184, pp. 232–237. Springer, Heidelberg (2010)

---

[4] Notice the other nine news events have same scores.

5. Ma, Q., Yoshikawa, M.: Topic and viewpoint extraction for diversity and bias analysis of news contents. In: Li, Q., Feng, L., Pei, J., Wang, S.X., Zhou, X., Zhu, Q.-M. (eds.) APWeb/WAIM 2009. LNCS, vol. 5446, pp. 150–161. Springer, Heidelberg (2009)
6. Blei, D.M., Ng, A.Y., Jordan, M.I.: Latent dirichlet allocation. J. Mach. Learn. Res. **3**, 993 (2003)
7. Jarvelin, K., Kekalainen, J.: Evaluation, cumulated gain-based, of IR techniques. ACM Trans. Inf. Syst. **20**, 422–446 (2002)
8. Ma, Q., Nadamoto, A., Tanaka, K.: Complementary information retrieval for cross-media news content. Inf. Syst. **31**(7), 659–678 (2006)

K^{UR}-Algorithm: From Position to Trajectory Privacy Protection in Location-Based Applications

Trong Nhan Phan[1(✉)], Josef Küng[1], and Tran Khanh Dang[2]

[1] FAW Institute, Johannes Kepler University Linz, Linz, Austria
{nphan,jkueng}@faw.jku.at
[2] HCMC University of Technology, Ho Chi Minh City, Vietnam
khanh@cse.hcmut.edu.vn

Abstract. Some obfuscation techniques may fail to protect user privacy because of the moving context as well as the background knowledge of adversaries. In this paper, we propose a novel scheme to distinctly protect user privacy not only from user position but also from user trajectory. Furthermore, we present k^{UR}-algorithm, which is context-aware and can be employed as either an independent method or a supportive technique, to give the high-level user privacy protection against privacy disclosure and privacy leak. Last but not least, we analyse other potential privacy problems which usually emerge as outliers and show how well our proposed solution overcomes these scenarios.

Keywords: User privacy · K-anonymity · Location-based services · Space-filling curve · Obfuscation · Dummies · Context-awareness

1 Introduction

With the rapid development of Geographic Information Systems (GIS), mobile devices, and Internet, Location-based Services (LBSs) have emerged to support human's life. There is, however, still a concern in that end-users are uncomfortable to use these services once they have no idea how their information is used. An adversary can exploit one's position to re-identify who he or she is, and then his or her private information will be disclosed. For instance, position information can reveal one's point of interests indicating his or her interests and trends. Thus, no one is happy to see what he or she tries to keep by himself or herself is revealed to other people. As a consequence, user privacy becomes a challenge along with the development of location-based services. In the meantime, protecting user position might lead to privacy leak by time. An adversary can, time by time, collect a set of user positions which will form a trajectory of a user. A trajectory can reveal more things about the user such as hobbies, trends, beliefs, interests, personal problems, and so on [3, 10, 16]. Consequently, preserving user privacy demands high-level protection that can keep both user position and user trajectory safe. While most of state-of-the-arts work primarily focus on location privacy [5, 11, 12, 18–21], few aim at trajectory privacy but still leave the privacy leak behind [4, 7, 14, 15, 22]. Besides, we

© Springer International Publishing Switzerland 2015
Q. Chen et al. (Eds.): DEXA 2015, Part II, LNCS 9262, pp. 82–89, 2015.
DOI: 10.1007/978-3-319-22852-5_8

argue with the paper in [1] whereas promote the paper in [9] that user privacy may be assured even if a user walks alone. This paper is our next step taking the privacy leak into account. Our main contributions are generally summarized as follows:

1. We propose a novel scheme to protect user sensitive information not only from user position but also from user trajectory.
2. We present k^{UR}-algorithm to deal with user privacy as well as privacy leak. Furthermore, two supportive strategies called hoppers and fan-outs promote the adaptive obfuscation when combined with the algorithm.
3. We show our privacy analysis to how our methods further resolve other potential privacy problems which popularly emerge because of privacy leak.

The rest of the paper is organized as follows. Section 2 introduces our related work. Next, we present some preliminaries that we employ in our approach in Sect. 3. After that, our privacy solution will be detailed in Sect. 4. Then our privacy analysis is given in Sect. 5 before our remarks in Sect. 6.

2   Related Work

There are different levels and approaches [16] where privacy-preserving techniques may be developed to protect user trajectory privacy. Gao et al. [6] propose a trajectory privacy-preserving framework for participatory sensing. The authors classify a trajectory into sensitive and insensitive trajectory segments in that only the former is protected by a mix-zone model and a pseudonym technique. Meanwhile, You et al. [22] exploit a dummy strategy to hide the real trajectory. The basic idea is to confuse adversaries with the other false trajectories. On the other hand, Divanis et al. [4] focus on trajectory privacy while users typically move in a network-confined environment. The authors provide two types of k-anonymity to generate anonymity regions, which depends on whether a user is in the frequent route or not. Besides, Ghinita [7] proposes a method to prevent users' trajectories from re-construction. In order to achieve that goal, individual location data are distorted. Having another approach, Nergiz et al. [14] also introduce a method to group trajectories according to k-anonymity and find a representative among them to communicate with LBS providers. Meanwhile, Phan et al. [15] introduce PPST-tree, the database-centric approach protects user privacy by obfuscating a user's road segments with the ones nearby sharing by *(k-1)* other users. The privacy leak can be easily seen in most of existing work mainly because a user movement becomes an outlier in his or her context. Our goal is, therefore, to find a way resolving this issue.

3   Preliminaries

It is worth noting that our proposed method in this paper is the next step of our developing privacy-preserving solution. We make our method work as a mutual coupling component with PPST-tree [15] to reinforce user privacy protection in an effort of avoiding possible privacy attacks. The base structure of PPST-tree is like that of B-tree but with added-values in each node. As illustrated in Fig. 1, each node has the form of $< P_1, G_1,$

$O_1, K_1, P_2, G_2, O_2, K_2, ..., P_{n-1}, G_{n-1}, O_{n-1}, K_{n-1}, P_{n-1}, G_{n-1}, O_{n-1}, K_{n-1} >$, where P_i is a tree pointer which points to the next lower-level node, G_i is the number of groups which have the key K_i, also known as a search key value, in common, and O_i is the number of moving objects which belong to the key K_i.

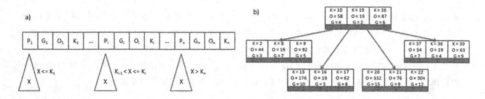

Fig. 1. PPST-tree [15]; (a) Index structure; (b) An example

When given a query, PPST-tree performs its search over the nodes and returns the anonymized road segments that satisfy k-anonymity. As defined by Samarati et al. [17], k-anonymity refers to k-anonymous usage of user information. In LBS, a mobile user is considered k-anonymous with respect to his or her location information if and only if the information sent from that user to a LBS provider is indistinguishable from the location information of at least *(k-1)* other users.

4   The Proposed Solution

4.1   The Whole Scheme

In this paper, we mainly classify LBSs into two classes as follows: (1) LBSs issued by either the past or current user positions; and (2) LBSs provided with anticipated user positions. There are lots of work focusing on the former [5, 11, 12, 18–21] whilst the latter is still a promising field [2, 13, 15]. Hence, we devote much our effort to protecting user privacy in location-anticipated applications. In general, the whole scheme comprehends two main phases including *Phase I – Movement Prediction* and *Phase II – Privacy Enhancing Obfuscation*. The former is to predict where a moving object should be in the future time according to its current context whereas the latter is to aim at how to efficiently reinforce privacy protection by overcoming disadvantages of obfuscation techniques and diminishing the attack probability of malicious adversaries.

Phase I – Movement Prediction. In order to conjecture the expected future positions of a moving object within a certain amount of time, we employ the prediction model as presented in [15]. The model treats a group of nearby moving objects instead of each individual due to the fact that those which are close to one another and have the same movement (i.e., speed, direction, and moving on the same lane) should share the same prediction result. This strategy helps reducing the costs arisen in moving object management. On the other hand, we constrain the result of the prediction model with context data as illustrated in Fig. 2a to make sure that the output is suitable for realistic movements and not in outlier cases easily eliminated by adversaries. The context data that affect to the movement of a moving object may vary from time to time, and we temporarily classify them into two categories known as static and dynamic context data. Static context data indicate what are less frequently changed such as road network infrastructure (e.g., bridges, tunnels, highway or mountain pass, intersections) and regulations for movement (e.g., unidirectional or bidirectional paths, the available maximum velocities) whilst dynamic context data reflect what are much frequently changed at the time of movement such as means of transportation, velocity, direction, weather, and traffic exceptions (e.g., traffic congestions, traffic accidents, infrastructure upgrade). Though the static context data are basically used to protect user privacy, but the dynamic context data will cause privacy leak and even worse deactivate the protection.

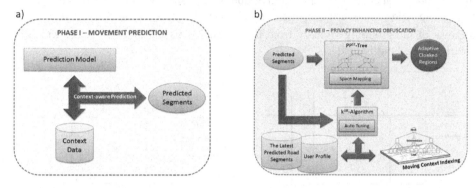

Fig. 2. The two main phases; (a) Phase 1 – Movement prediction; (b) Phase 2 – Privacy enhancing obfuscation

Phase II – Privacy Enhancing Obfuscation. As illustrated in Fig. 2b, the context-aware predicted segments, generated in Phase I, will be the input of Phase II to produce cloaked regions for location-based services. Here we use Hilbert curve theory [8] as space mapping indicator and PP^{ST}-tree from the research in [15] to not only support some kinds of location-based future-time services like *"How many vehicles will there be in this area in the next 10 min?"* or *"Where is this moving object going to be within the next 12 min?"* but also ensure user privacy in terms of both position and trajectory. Nevertheless, PP^{ST}-tree works well with ordinary cases when moving objects move in cities with the dense road network while leaves exceptional circumstances on the edge of privacy risk that can be exploited by adversaries to break the privacy protection and violate user private information. Consequently, we also propose k^{UR}-algorithm as a mutual coupling component with PP^{ST}-tree to reinforce user privacy protection in an effort of avoiding possible privacy attacks. In our scenario, an attacker is assumed to have base knowledge about at least four aspects as following: (1) the road network; (2) user profiles; (3) privacy-preserving algorithms; and (4) the cloaked regions collected each time his victim requests location-based services. Furthermore, the algorithm is further designed to be self-tuning in order to avoid the obfuscation booming, which indicates the unexpected scenario when a number of obfuscations are too many for a LBS provider. More details will be discussed turn-by-turn in the next sections.

4.2 K^{UR}-Algorithm

At first, k^{UR}-algorithm, known as k-User group sharing Road segments, employs k-anonymity concept [17] as its skeleton for both users and road segments. The idea behind the algorithm is to generate cloaked regions that are compatible with the current movement of a moving object and enhance user privacy protection by diminishing the probability attack of adversaries on both position and trajectory identification. In order to do that, the algorithm takes predicted segments as its original input and tries to make them well-obfuscated. Moreover, hoppers and fan-outs are the two supportive strategies for k^{UR}-algorithm, which will be presented later on in this section. It is worth noting that a moving object, apart from users' original queries, considered in the proposed scheme as a group representative object (i.e., on behalf of its group) instead of an individual because of the grouping strategy from the prediction model and PP^{ST}-tree, which is clearly pointed out in [15].

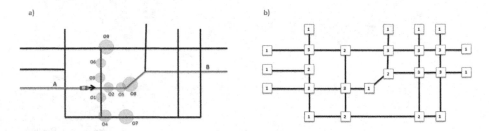

Fig. 3. (a) An example of hoppers; (b) An example of fan-outs

Hoppers. Assuming that a moving object is traveling from A to B, and it starts to request a location-based future-time service at the second intersection as illustrated in Fig. 3a. According to the privacy mechanism, three obfuscations $O1$, $O2$, and $O3$ are generated and then sent to location-based service providers. If the result satisfies the user, there is no further processing and the service is ended. However, in case the user keeps requesting the service or makes another request right afterwards, only the obfuscation $O5$ is generated and returned to the LBS providers as usual due to the fact that there is no other predicted road segments nearby. An adversary, therefore, knows right away his target object is going to move on that road. As a consequence, there is an essential need to resolve this issue. Once the request falls into this exceptional situation, the latest predicted road segments of the moving object from the previous-using time, together with the current predicted segments, are used to generate next obfuscations. In other words, there are two additional obfuscations $O4$ and $O6$, along with $O5$, generated. Again, keeping protecting user privacy during service-requesting time, three obfuscations $O7$, $O8$, and $O9$ are generated, and we have obfuscation hoppers. Finally, finishing the service ends the hoppers. By doing this, attackers have no idea about where the moving object is going to move on. On the other hand, how to generate hoppers decides whether user privacy is leaked or the protection is broken. For instance, using the same velocity parameter for prediction or taking the same information to generate obfuscation makes no sense to the real world due to the fact that each movement on each road segment is different from others and affected by its own conditions. Thanks to the context-aware prediction model, hoppers should be aware of the current context as much as possible.

Fan-outs. We represent the road network as a graph with V vertices (junctions) and E edges (road segments). The fan-out or branch factor of each road segment is defined as the number of new road segments a moving object may move in from its current road. In other words, it is the number of edges connected to the current edge on which the object is moving. The fan-outs are identified at each junction respectively from the road network in advance. In case there is no or only one new road segment, the fan-out is considered as 1, for these is no choice for confusing adversaries from the next itinerary of the moving object. Figure 3b illustrates the sample road network fulfilled with the fan-out information.

5   Privacy Analysis

The contributions of the prediction model and PPST-tree to user privacy concerns known as stopping re-identifying a user's current position and protecting position and trajectory privacy have been showed in [15]. Hence, we further analyze and discuss other potential privacy problems which popularly emerge not only in the proposed solution but also in most of the existing work because of privacy leak. The well-known privacy leak is the exceptional case originating from the sparse road network. Once a moving object is on a road segment without its neighbors, the predicted segment resulted in the prediction phase is always one. Figure 4 gives an example of this case in that *O10*, *O11*, and *O12* are discrete obfuscations for the service request at time *(t-1)* while *O13* and *O14* are obfuscations for the service request at time *t*. Thus, returning an obfuscation either containing only the predicted segment as *O13* or widening its area as *O14* let the attacker notice about the exception.

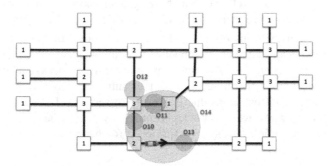

Fig. 4. Privacy leak from sparse road network

Thanks to k^{UR}-algorithm, this privacy break is not a problem anymore in our method. On the one hand, let P_1^{pos} the attack probability an adversary can identify one's position and P_1^{tra} the attack probability an adversary can identify one's trajectory when k^{UR}-algorithm is not applied. On the other hand, let P_2^{pos} and P_2^{tra} be the ones when k^{UR}-

algorithm is activated. When an end-user requesting LBSs moves in a single road, we have the attack probabilities as follows:

$$P_1^{pos} = \frac{1}{kProj_rS} \frac{1}{\sum_{i=1}^{m} G_i \sum_{j=1}^{n} K_j} \tag{1}$$

$$P_1^{tra} = 1 \tag{2}$$

In the Eqs. (1) and (2), k is the number of moving objects for each group, $Proj_r$ is the projection on the road network and S is the area of the cloaked region, $\sum_{i=1}^{m} G_i$ is the total number of groups joining the shared road segment, and $\sum_{j=1}^{n} K_j$ is the total added-key values when the obfuscation is expanded by the nearest neighbors in PP^{ST}-tree. As a consequence, trajectory privacy does not work in this case, for an adversary exactly knows which road the user is moving. The K^{UR}-algorithm, therefore, contributes its important role to decrease the attack probability from both user position and user trajectory. Let us say $\sum_{p=1}^{q} h_p$ the number of hoppers and $\sum_{o=1}^{l} f_o$ the total numbers of fan-outs of the road segments where the hoppers are in active, the attack probabilities are now diminished as follows:

$$P_2^{pos} = \frac{1}{kProj_rS} \frac{1}{\sum_{i=1}^{m} G_i \sum_{j=1}^{n} K_j} \frac{1}{\sum_{p=1}^{q} h_p} \frac{1}{\sum_{o=1}^{l} f_o} \tag{3}$$

$$P_2^{tra} = \frac{1}{k} \frac{1}{\sum_{p=1}^{q} h_p} \frac{1}{\sum_{o=1}^{l} f_o} \tag{4}$$

6   Summary

In this paper, we propose a novel scheme to thoroughly protect user sensitive information not only from position privacy but also from trajectory privacy. In addition, we present k^{UR}-algorithm, which contributes to a high-level user privacy protection. Moreover, our method is context-aware of the environment of moving objects so that adaptive obfuscations give not so many chances to adversaries, even with their background knowledge, to violate user privacy. Furthermore, we set up a tuning parameter in order to control the booming effect of obfuscations due to quality of services. Last but not least, our privacy analysis shows how well our methods entirely overcome the privacy leak though privacy-preserving techniques have been applied beforehand.

References

1. Abul, O., Bonchi, F., Nanni, M.: Never walk alone: uncertainty for anonymity in moving objects databases. In: Proceedings of the IEEE 24th International Conference on Data Engineering, pp. 376–385 (2008)
2. Chen, J., Meng, X.: Indexing future trajectories of moving objects in a constrained network. J. Comput. Sci. Technol. **22**(2), 245–251 (2007)

3. Chow, C.Y., Mokbel, M.F.: Trajectory privacy in location-based services and data publication. SIGKDD Explor. Newsl. **13**(1), 19–29 (2011)
4. Divanis, A.G., Verykios, V.S., Mokbel, M.F.: Identifying unsafe routes for network-based trajectory privacy. In: Proceedings of the SIAM International Conference on Data Mining, pp. 942–953 (2009)
5. Fung, B.C.M., Chen, K.W.R., Yu, P.S.: Privacy-preserving data publishing: a survey of recent developments. ACM Comput. Surv. **42**(4), 1–53 (2010)
6. Gao, S., Ma, J., Shi, W., Zhan, G., Sun, C.: TrPF: a trajectory privacy-preserving framework for participatory sensing. IEEE Trans. Inf. Forensics Secur. **8**(6), 874–887 (2013)
7. Ghinita, G.: Private queries and trajectory anonymization: a dual perspective on location privacy. Trans. Data Priv. **2**(1), 3–19 (2009)
8. Hilbert, D.: Über die stetige Abbildung einer Linie auf ein Flächenstück. Math. Ann. **38**, 459–460 (1891)
9. Huo, Z., Meng, X., Hu, H., Huang, Y.: You can walk alone: trajectory privacy-preserving through significant stays protection. In: Proceedings of the 17th International Conference on Database Systems for Advanced Applications, pp. 351–366 (2012)
10. Kaplan, E., Pedersen, T.B., Savas, E., Saygın, Y.: Discovering private trajectories using background information. J. Data Knowl. Eng. **69**(7), 723–736 (2010)
11. Lin, D., Bertino, E., Cheng, R., Prabhakar, S.: Location privacy in moving-object environments. Trans. Data Priv. **2**(1), 21–46 (2009)
12. Liu, L.: From data privacy to location privacy: models and algorithms. In: Proceedings of the 33rd International Conference on Very Large Databases, pp. 1429–1430 (2007)
13. Liu, X., Karimi, H.A.: Location awareness through trajectory prediction. In: Proceedings of Computers, Environment and Urban Systems, pp. 741–756 (2006)
14. Nergiz, M.E., Atzori, M., Saygin, Y.: Towards trajectory anonymization: a generalization-based approach. In: Proceedings of the SIGSPATIAL ACM GIS 2008 International Workshop on Security and Privacy in GIS and LBS (2008)
15. Phan, T.N., Dang, T.K.: A novel trajectory privacy-preserving future time index structure in moving object databases. In: Proceedings of the 4th International Conference on Computational Collective Intelligence Technologies and Applications, Part I, pp. 124–134 (2012)
16. Phan, T.N., Dang, T.K., Küng, J.: User privacy protection from trajectory perspective in location-based applications. In: Proceedings of the 19th Interdisciplinary Information Management Talks, Jindřichův Hradec, Czech Republic, pp. 281–288 (2011)
17. Samarati, P., Sweeney, L.: Protecting privacy when disclosing information: k-anonymity and its enforcement through generalization and suppression. Technical report, SRI-CSL-98–04, SRI Computer Science Laboratory (1998)
18. Shokri, R., Theodorakopoulos, G., Le Boudec, J.-Y., Hubaux J.-P.: Quantifying location privacy. In: Proceedings of the 2011 IEEE Symposium on Security and Privacy, pp. 247–262 (2011)
19. To, Q.C., Dang, T.K., Küng, J.: A Hilbert-based framework for preserving privacy in location-based services. Int. J. Intell. Inf. Database Syst. **7**(2), 113–134 (2013)
20. Xu, T., Cai, Y.: Exploring historical location data for anonymity preservation in location-based services. In: Proceedings of the 27th IEEE International Conference on Computer Communications (2008)
21. Yigitoglu, E., Damiani, M.L., Abul, O., Silvestri, C.: Privacy-preserving sharing of sensitive semantic locations under road-network constraints. In: Proceedings of the IEEE 13th International Conference on Mobile Data Management, pp. 186–195 (2012)
22. You, T.H., Peng, W.C., Lee, W.C.: Protecting moving trajectories with dummies. In: Proceedings of the 2007 International Conference on Mobile Data Management, pp. 278–282 (2007)

Data Streams, Web Services

Candidate Pruning Technique for Skyline Computation Over Frequent Update Streams

Kamalas Udomlamlert$^{(\boxtimes)}$, Takahiro Hara, and Shojiro Nishio

Graduate School of Information Science and Technology,
Osaka University, Suita, Japan
{kamalas.u,hara,nishio}@ist.osaka-u.ac.jp

Abstract. Skyline query processing reveals a set of preferable results based on the competitiveness of many criteria among all data objects. This is a very useful query for multi-attribute decision making. Moreover, monitoring and tracing skyline over time-series data are also important not only for real-time applications (e.g., environmental monitoring) but also historical time-series analysis (e.g., sports archives, historical stock data). In these applications, considering consecutive snapshots, a large fraction of the fixed number of observing objects (e.g., weather stations) can change their values resulting to the possibility of complete change in the previous skyline. Without any technique, computing skyline from a scratch is unavoidable and can be outperformed some traditional skyline update methods. In this paper, we propose an efficient method to compute skyline sets over data update streams. Our proposed method uses bounding boxes to summarize consecutive data updates of each data object. This technique enables the pruning capability to identify a smaller set of candidates in skyline computation resulting in faster total computation time. We conduct some experiments through both synthetic and real-life datasets. The results explicitly show that our proposed method significantly runs faster than the baseline in various parameter studies.

Keywords: Skyline queries · Continuous queries · Update streams

1 Introduction

Recently, many query methods have been developed and gained a lot of attentions in database researches in order to deliver most satisfactory results to various classes of end-users. Considering the dominance relations among objects (the competitiveness of each object), skyline computation [1], which represents a result set which each result item is not worse than others, is also one of popular query methods so far. An example usage of this query is commonly referred to to multi-criteria decision making of hotel selection [1].

In this paper, we focus on skyline monitoring queries which deliver the up-to-date skyline answers over frequent update streams. We stress this problem on the frequent update stream where a large portion of observing data objects change their values in each timestamp (interchangeably called a snapshot).

© Springer International Publishing Switzerland 2015
Q. Chen et al. (Eds.): DEXA 2015, Part II, LNCS 9262, pp. 93–108, 2015.
DOI: 10.1007/978-3-319-22852-5_9

The data streams like this can be often seen in many real-life situations, for example, environmental monitoring and stock market analysis.

In this paper, we propose an efficient method based on the properties of a bounding box (a minimum bounding rectangle in the case of 2 dimensions). We utilize bounding boxes to capture and prune unnecessary data candidates as well as neglect no-effect data updates. Therefore, we can identify a smaller candidate set in skyline computation in consecutive data snapshots resulting in saving overall execution time.

In summary, the contributions of this paper are as follows:

- We formulate the problem definition of skyline computation on frequent data update streams as well as illustrate example applications of this problem.
- We propose an efficient algorithm and index structures to identify a smaller set of data candidates before skyline calculation, and the cost of maintenance is paid in according with degree of data changes (pay as you go).
- We conduct some experiments in various settings by using both synthetic and real datasets to show that our proposed method can run faster than the baseline and the comparison methods.

2   Related Work

Skyline computation in database research was firstly introduced in [1]. The authors proposed two skyline algorithms including BNL skyline and D & C skyline algorithms. After that, numerous research papers tried to enhance the performance by using more complicated index structures such as Branch-and-Bound skyline algorithm [12].

Apart from the traditional skyline processing in databases, skyline processing for distributed systems has been studied described in the survey [5]. Moreover, many interesting variants of skyline processing methods have also been studied for example, reverse skyline query [3], fragmented skyline [13], subspace skyline [16], uncertain skyline [4] and interval skyline [8].

The papers in [2,7,9] studied skyline monitoring over moving objects. They assume a kinetic model of moving data objects aiming to find the skyline objects (static attributes) when some dynamic attributes (locations or query points) are movable. These are quite different from ours, and their techniques are not suitable to solve our context's problems because a kinetic model is not assumed.

Some existing works [6,10,11,15,18] proposed efficient methods to continuously calculate skyline results over data streams. These aim to efficiently monitor the latest skyline set in the sliding window where window-range, data arrival time and data expiration time are given. Some algorithms together with indexing techniques are used to deal with data insertion and deletion into the assumed sliding windows. For the case of data modification assumed in this paper, a single data modification (a data update) can be taken as two consecutive operations - insertion then deletion. This can incur a very large cost because there can be multiple data updates at each timestamp.

In [17] assumed a very close problem to ours according to its data model. This work tries to monitor the latest modification of the skyline set when each data object is updated by the information from update streams. Its main contribution relies on allocating data into grid cells and consider the dominance relations between those grids to prune unnecessary candidates from skyline calculation. Therefore, our experiment also adopts this technique as well as the method in [11] as the representative of its group to compare to our proposed method.

3 Preliminaries

3.1 Data Model and Its Update Model

We assume that the system analyzes a skyline set over a fixed number of data objects. Each data object is comprised of m-numerical values as attributes and data id id as an object's identifier. Data attributes of data object i at the initialization (the first snapshot in the historical archive, snapshot 0) are represented as a tuple $p_i^0 = (p_i^0[1], p_i^0[2], \ldots, p_i^0[m])$. In this paper, we assume that lower values are more preferable. Let N be the number of all data objects in the system, a set of all data objects at snapshot t is denoted by $P^t = \{p_1^t, p_2^t, \ldots, p_N^t\}$, where $t = \{0, 1, \ldots, T\}$ and T is the total number of snapshots in the archive.

Definition 1. *(Point dominance) A data point p_i^t dominates p_j^t ($p_i^t \prec p_j^t$) if and only if $\forall k \in \{1, 2, \ldots, m\} : p_i^t[k] \leq p_j^t[k]$, and $\exists l \in \{1, 2, \ldots, m\} : p_i^t[l] < p_j^t[l]$.*

In this research, we assume that at each timestamp (snapshot) t, only a partial set of data objects changes their attributes' values from the previous timestamp $t-1$. An update model like this can be often found in pull-based data delivery model that the server pulls new updates from data sources periodically.

How data change their values is described by an update tuple which can be defined in many ways based on the applications, for example, a new value update defined by a 3-tuple $u = (id, t, (p[1], p[2], \ldots, p[m]))$ that means $p_{id}^t = (p[1], p[2], \ldots, p[m])$ and a modification update defined by a 3-tuple $u = (id, t, (\Delta p[1], \Delta p[2], \ldots, \Delta p[m]))$ that means $p_{id}^t = (p_{id}^{t-1}[1] \otimes \Delta p[1], p_{id}^{t-1}[2] \otimes \Delta p[2], \ldots, p_{id}^{t-1}[m] \otimes \Delta p[m])$ where $\otimes$ is an operator, for example, addition, multiplication and average. This changes the corresponding data object p_{id}^{t-1} to p_{id}^t.

A list of updates of snapshot t (update streams) is a list of update tuples $U^t = \{u_1^t, u_2^t, \ldots\}$ where $|U| \leq N$. Therefore, the data objects which are not modified by any update tuples remain the same values that is $p_{id}^t = p_{id}^{t-1}$.

In this research, we aim to continuously calculate a set of skyline (SK^t) efficiently from P^t at each consecutive snapshot t.

Definition 2. *(Skyline set) Given a set of data points at snapshot t (P^t), $p_i^t \in P^t$ is included in the skyline set SK^t if and only if $\forall p_j^t \in (P^t \backslash \{p_i^t\}), p_j^t$ does not dominate p_i^t ($p_j^t \nprec p_i^t$).*

3.2 Summarizing Consecutive Data Snapshots with Minimum Bounding Rectangles (MBRs)

A minimum bounding rectangle (MBR) is the smallest oriented rectangle enclosing a set of points which is a 2-dimensional case of a minimum bounding box in a coordinate system. Our proposed solution can deal with any number of dimensions by using the same idea. According to all examples in this paper illustrated in a 2-dimensional space, for simplicity, we use the term *MBRs* to refer to this expression in general.

While each data object possibly changes its attributes' values at every timestamp t, those tracing data points can be seen as a set of points which can be summarized (represented) as an MBR. Therefore, we use an MBR to summarize the space that a data object changes its values between consecutive snapshot a to b, i.e., $\{p_i^a, p_i^{a+1}, \ldots, p_i^b\}$ where $a \leq b$.

An MBR of data object id from consecutive snapshot a to b is represented by a 3-tuple $M_{id}(a, b) = (id, ub, lb)$ where $ub[l] = \max_{j \in [a,b]} p_{id}^j[l]$ and $lb[l] = \min_{j \in [a,b]} p_{id}^j[l]$ when $l \in \{1, 2, \ldots, m\}$. We represent a list of MBRs which ends at snapshot t as $M^t = \{M_1(a_1, t), M_2(a_2, t), \ldots, M_N(a_N, t)\}$ where a_{id} is a number of the start snapshot of MBR id. Figure 1a exemplifies 7 consecutive data snapshots of a data object ($id = 1$) denoted by $\{p_1^0, p_1^1, \ldots, p_1^6\}$ while the arrows express their trajectories between two consecutive data snapshots. Their MBR is a box $M_1(0, 6)$ shown in the figure. For short, M_{id}^t refers to the latest MBR of object id at timestamp t regardless of the beginning timestamp ($M_{id}(*, t)$).

3.3 Dominance Region and Anti-dominance Region

A dominance region of an MBR ($M_i^t.DR$) is a region where point $x \in M_i^t.DR$ such that $x[l] \geq M_i^t.ub[l]$ for all $l \in \{1, 2, \ldots, m\}$. Any data points or MBRs that fully fall within this region will be dominated by M_i^t. In the contrary, an anti-dominance region of MBR ($M_i^t.ADR$) is a region where $x \in M_i^t.ADR$ such that $x[l] \leq M_i^t.lb[l]$ for all $l \in \{1, 2, \ldots, m\}$. Any data points or MBRs that fully fall within this region dominate M_i^t.

Definition 3. *(MBR Dominance) An MBR M_i^t dominates M_j^t ($M_i^t \prec M_j^t$) if and only if $\forall l \in \{1, 2, \ldots, m\}$: $M_i^t.ub[l] \leq M_j^t.lb[l]$ and $\exists l \in \{1, 2, \ldots, m\}$: $M_i^t.ub[l] < M_j^t.lb[l]$, i.e., $M_i^t.ub \prec M_j^t.lb$ (M_j^t fully falls in $M_i^t.DR$).*

Due to $M_{id}^t.lb[l] = \min_{j \in [a,b]} p_{id}^j[l]$ when $l \in \{1, 2, \ldots, m\}$, we conclude that a point $M_{id}^t.lb$ dominates every point p_{id}^k where $k = \{a, a+1, \ldots, b\}$ except point p where $p_{id}^j[l] = M_{id}^t.lb[l]$ for all $l \in \{1, 2, \ldots, m\}$. We further define a definition of a set of skyline MBRs at snapshot t (SKM^t).

Definition 4. *(Skyline MBR) Given a set of MBRs at snapshot t (M^t), $M_i^t \in M^t$ is included in the set of skyline MBRs, SKM^t, if and only if $\forall M_j^t \in (M^t \backslash \{M_i^t\}), M_j^t$ does not dominate M_i^t.*

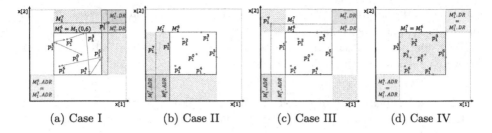

| (a) Case I | (b) Case II | (c) Case III | (d) Case IV |

Fig. 1. MBR updates before including p_1^7 and after including p_1^7

In addition, we denote a set of MBRs that do not belong to SKM^t as a set of non-skyline MBRs ($NSKM^t = M^t \backslash SKM^t$).

Lemma 1. $\forall M_i^t \in NSKM^t$: *there must be at least one MBR which is inside* $M_i^t.ADR$.

Proof. (Proof by contradiction) Assume that $M_i^t \in NSKM^t$, but there is no MBR inside $M_i^t.ADR$. Due to $M_i^t \in NSKM^t$ and Definition 4, there exists at least one MBR M_j^t dominating M_i^t. As a result, $M_j^t.ub[l] \le M_i^t.lb[l]$ for $\forall l \in \{1, 2, \ldots, m\}$ and $M_j^t.ub[l] < M_i^t.lb[l]$ for $\exists l \in \{1, 2, \ldots, m\}$. From the explanation in Sect. 3.3, this leads to the contradiction because M_j^t must be contained in $M_i^t.ADR$.

3.4 Pruning Candidates for Skyline Calculation Using MBRs

At each snapshot, instead of finding a skyline set from all data points P^t, we find the skyline by using only candidates in the skyline MBRs ($\{p_i^t / M_i^t \in SKM^t\}$). The cardinality of SKM^t is likely to be much smaller than that of P^t, so skyline calculation can be computed faster.

Lemma 2. *Skyline calculation from* $\{p_i^t / M_i^t \in SKM^t\}$ *produces the correct skyline set* (SK^t) *as same as using all data points* P^t.

Proof. (Proof by contradiction) In order to produce incorrect SK^t, there must be at least a point p_i^t where $M_i^t \in NSKM^t \wedge p_i^t \in SK^t$. By Lemma 1, there exists $M_j^t : M_j^t \prec M_i^t$. This leads to contradiction that $p_i^k \notin SK^t$ because $\forall k : M_j^t.ub \preceq p_i^k$.

3.5 Changes of MBRs When Considering a New Data Snapshot

We start considering how a list of new updates at the next snapshot (U^{t+1}) affects the current MBRs M_i^t. Certainly, including a new snapshot of data can make M_i^t changed in size as well as their properties, i.e., lb, ub, DR and ADR. Consider an MBR of data object id at snapshot t and its update tuple u_j^{t+1}

where $u_j^{t+1}.id = id$ and $l \in \{1, 2, \ldots, m\}$, $M_i^{t+1}.ub[l] = \max(M_i^t.ub[l], p_i^{t+1}[l])$ and $M_i^{t+1}.lb[l] = \min(M_i^t.lb[l], p_i^{t+1}[l])$.

The effects of the data updates to M_{id}^{t+1} can be classified into 4 cases as follows:

(Case I) $M_{id}^{t+1}.lb = M_{id}^t.lb$ and $M_{id}^{t+1}.ub \succ M_{id}^t.ub$
This case happens when p_{id}^{t+1} falls in gray-shaded area at the right-top corner illustrated in Fig. 1a. $M_{id}^{t+1}.ADR$ remains the same as M_{id}^t, but $M_{id}^{t+1}.DR$ becomes smaller.
(Case II) $M_{id}^{t+1}.lb \prec M_{id}^t.lb$ and $M_{id}^{t+1}.ub = M_{id}^t.ub$
This case happens when p_{id}^{t+1} falls in the gray-shaded area at the left-bottom corner illustrated in Fig. 1b. In addition, $M_{id}^{t+1}.ADR$ becomes smaller than that of M_{id}^t while $M_{id}^{t+1}.DR$ remains the same as M_{id}^t.
(Case III) $M_{id}^{t+1}.lb \prec M_{id}^t.lb$ and $M_{id}^{t+1}.ub \succ M_{id}^t.ub$
If p_{id}^{t+1} falls in the gray-shaded areas at the right-top and left-bottom corners illustrated in Fig. 1c., this changes both lb and ub. In the same way, both $M_{id}^{t+1}.DR$ and $M_{id}^{t+1}.ADR$ are degraded compared to that of M_{id}^t.
(Case IV) $M_{id}^{t+1}.lb = M_{id}^t.lb$ and $M_{id}^{t+1}.ub = M_{id}^t.ub$
M_{id}^{t+1} and M_{id}^t are identical if p_{id}^{t+1} falls inside M_{id}^t (gray-shaded area) illustrated in Fig. 1d.

According to Definition 4, the membership of MBRs in SKM^t and $NSKM^t$ possibly no longer holds for snapshot $t+1$ due to the changes of its DR and ADR. Therefore, our proposed method introduces an efficient method to maintain the consistency in order to identify SKM^{t+1} as well as $NSKM^{t+1}$. We describe details in the next section.

4 Proposed Algorithms

4.1 Overview

From the preliminaries in Sect. 3, we proved that only the data objects whose current MBR belongs to SKM^t are a sufficient candidate set for skyline calculation of each snapshot t. This can significantly reduce the cardinality of data candidates to be calculated in skyline computation. Therefore, we propose an efficient method to maintain those MBRs by keeping two separated lists including SKM^t and $NSKM^t$ where $M^t = SKM^t \cup NSKM^t$ and $SKM^t \cap NSKM^t = \phi$.

Our proposed method can be roughly divided into 3 steps.

1. **Pre-computation Maintenance (PRE)**
   According to new data updates from the stream at t, the MBRs at $t - 1$ can be possibly changed in terms of physical MBRs and their pruning capability. This step tries to identify the correct SKM^t by paying low maintenance cost as much as necessary before skyline calculation.
2. **Skyline Calculation (SKY)**
   This process is straight-forward. We calculate a final skyline result (SK^t) by one of many state-of-the-art skyline computation methods but using a smaller set of candidates.

Algorithm 1. Brief algorithm at snapshot $t > 0$

Data: The list of update tuples at snapshot t (U^t)
Result: The skyline set at snapshot t (SK^t)

1 $VL \leftarrow \emptyset$
2 **foreach** *Update tuple $u_i^t \in U^t$* **do**
3 $\quad$ Update M_i^t
4 $\quad$ Add some MBRs needed to be further verified to VL

5 **foreach** M_i^t *in VL* **do**
6 $\quad$ **if** *there exists $M_j^t \in SKM^t : M_j^t \prec M_i^t$* **then**
7 $\quad\quad$ $NSKM^t \leftarrow NSKM^t \cup \{M_i^t\}$
8 $\quad$ **else**
9 $\quad\quad$ $SKM^t \leftarrow SKM^t \cup \{M_i^t\}$

10 Calculate a skyline set SK^t from $\{p_i^t / M_i^t \in SKM^t\}$
11 Check the false positives from SKM^t and SK^t
12 Reconstruct MBRs in SKM^t that incur too many consecutive false positives

---

3. Post-computation Maintenance (POST)

Regarding to the final skyline result, we are able to detect some data objects whose MBR belongs to SKM^t but does not appear in SK^t. In other words, these MBRs produce unpleasant false positives degrading overall the pruning capability. In this process, we propose a heuristic rule to solve this problem.

A brief pseudo-code of our proposed algorithm is shown in Algorithm 1. We explain the details of pre-computation maintenance (lines 1–9) in Subsect. 4.3 and post-computation maintenance (lines 11–12) in Subsect. 4.4.

4.2 Intialization $(t = 0)$

At the initialization $(t = 0)$, we have to construct the initial MBRs (M^0) of all data objects (P^0). That means, in each MBR M_i^0, $M_i^0.lb = M_i^0.ub = p_i^0$ for all $i \in \{1, 2, \ldots, N\}$. Hence, we calculate the first skyline set of P^0 and classify MBRs into 2 lists as $SKM^0 = \{M_i^0 / p_i^0 \in SK^0\}$ and $NSKM^0 = M^0 \backslash SKM^0$.

Moreover, we additionally introduce two important elements to help easily identifying the relations between MBRs in SKM^t and $NSKM^t$ including an *id* list of MBRs in a dominance region of each MBR $(M_i^t.DRM)$ and a single *id* of MBR which is in an anti-dominance region $(M_i^t.adrm)$. These relations can be established while executing skyline calculation by using the following rules:

1. If M_i^t is dominated by M_j^t on skyline computation, then $M_i^t \in NSKM^t$, $M_i^t.admr = j$ and $i \in M_j^t.DRM$.
2. If M_i^t is not dominated by any other MBRs, then $M_i^t \in SKM^t$ and $M_i^t.admr = nil$ (not applicable).

According to Definition 4 and Lemma 1, we conclude that $\forall M_i^t \in NSKM^t$: $M_i^t \neq nil$ while $\forall M_i^t \in SKM^t : M_i^t = nil$, and as long as M_j^t where $j = M_i^t.admr$ dominates M_i^t, M_i^t must belong to $NSKM^t$.

4.3   Data Updates at Snapshot $t > 0$

When receiving or considering a list of data updates $U^t = \{u_1^t, u_2^t, \ldots\}$, each update tuple describes how a new data tuple p^t can be generated while the rest (not indicated in U^t) $p_i^t = p_i^{t-1}$. For those unchanged data tuples, their MBRs are also unchanged, so the system does nothing in such a case.

However, we need to verify some affected MBRs both in SKM^{t-1} and $NSKM^{t-1}$ whether they are still in the correct lists at snapshot t due to new data updates. Therefore, we create an additional list called a verification list (VL) containing MBRs whose elements are needed to be further investigated. This process included in the pre-calculation maintenance checks the following conditions.

1. $M_i^{t-1} \in NSKM^{t-1}$ and M_i^t affected by p_i^t fall in either Case II or Case III (referred to Sect. 3.5).
   Because $M_i^t.ADR$ is deteriorating, $M_{M_i^t.admr}^{t-1}$ may no longer dominate M_i^t. Therefore, we check if it still dominates, and if not, M_i^t is pushed to VL.
2. $M_i^{t-1} \in SKM^{t-1}$ and M_i^t affected by p_i^t fall in either Case I or Case III (referred to Sect. 3.5).
   Because the dominance capability of M_i^t ($M_i^t.DR$) has been reduced, some MBRs in $M_i^{t-1}.DRM$ may no longer be dominated by M_i^t. We move $j \in M_i^{t-1}.DRM$ which is not dominated by M_i^t to VL.
3. Otherwise: No change of MBRs' status (Neglected).

Verification of MBRs in VL. We can see that all listed MBRs in VL used to be in $NSKM^{t-1}$, but possibly they are no longer able to be in $NSKM^t$. This will increase the number of MBRs in SKM^t resulting in increasing in the number of candidates in skyline calculation. At this process, we try to check these MBRs again whether there exists at least an MBR in SKM^t dominating them before including them into SKM^t. Therefore, for each $M_i^t \in VL$, we search for any first $M_j^t \in SKM^t$ that dominates M_i^t. If found, M_i^t is pushed back to $NSKM^t$ and set $M_i^t.admr = j$. Otherwise, it is swapped to SKM^t.

It is noted that there may be more than one M_j^t dominating M_i^t. In best practice, M_j^t to be chosen should be an MBR in SKM^t which gives the longest distance between $M_i^t.lb$ and $M_j^t.ub$ ($M_i^t.admr = \arg\max_{\forall j: M_j^t \prec M_i^t} d(M_i^t.lb, M_j^t.ub)$). However, to do this approach consumes more time because we cannot avoid scanning the entire list of SKM^t. Heuristically, we may choose M_j^t having more pruning power than the others, i.e., $M_j^t.ub$ near the origin. Using the Manhattan distance, we choose $M_i^t.admr = \arg\min_{\forall j: M_j^t \prec M_i^t} \sum_{l=1}^m M_j^t.ub[l]$. Maintaining a list of SKM^t sorted by $\sum_{l=1}^m M_j^t.ub[l]$, we can simply choose the first found dominating MBR in the list without scanning the entire list.

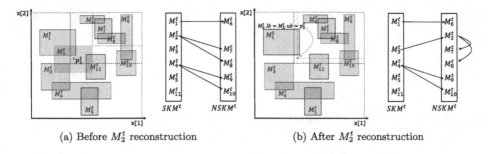

(a) Before M_2^t reconstruction   (b) After M_2^t reconstruction

Fig. 2. Running example of an MBR reconstruction

4.4 Post-computation Maintenance

After SK^t has been calculated, it is possible that some of candidates from SKM^t do not finally belong to SK^t (false positives). If $M_i(*, t) = M_i^t$ usually incurs a false positive for a long period (too many consecutive snapshots), it is worth considering paying maintenance cost to reconstruct and newly start an MBR from the current snapshot t ($M_i^t = M_i(t, t)$) because of the possible higher gains of pruning capability in the next iteration.

Lemma 3. *The dominance and anti-dominance regions of a newly-reconstructed MBR $M_i' = M_i(t, t)$ are not smaller than the old $M_i = M_i^t(a, t)$ where $a < t$.*

Proof. The dominance region of M_i ($M_i.DR$) is $x[l] \geq M_i.ub[l]$ for $l \in \{1, 2, \dots, m\}$. However, $M_i'.ub = p_i^t$ and $p_i^t[l] \leq \max_{k \in [a,t]} p_i^k[l] = M'.ub$. Hence, $M_i'.DR$ must not be smaller than that of M_i. In the same way, the anti-dominance region of M_i ($M_i.ADR$) is $x[l] \leq M_i.lb[l]$ for $l \in \{1, 2, \dots, m\}$. However, $M_i'.lb = p_i^t$ and $p_i^t[l] \geq \min_{k \in [a,t]} p_i^k[l] = M'.lb$. Hence, $M_i'.ADR$ must not be smaller than that of M_i.

MBR Reconstruction Strategy. MBR M_i^t that belongs to SKM^t but p_i^t is not included in SK^t for a long period should be lowered the rank to $NSKM^t$ to decrease the cardinality of skyline calculation in each snapshot. In this section, we discuss about a heuristic rule to decide which MBR should be reconstructed followed by a running example.

Firstly, a record of the number of consecutive false positives of each MBR should be tracked by adding a new MBR attribute called $M_i^t.cfp$. At each iteration we calculate this parameter for all MBRs in SKM^t as follows:

$$M_i^t.cfp = \begin{cases} M_i^{t-1}.cfp + 1 & ; M_i^t \in SKM^t \wedge p_i^t \notin SK^t \\ 0 & ; Otherwise \end{cases}$$

We decide to reconstruct an MBR M_i^t when

$$M_i^t.cfp \geq cfp_{ths} \tag{1}$$

where cfp_{ths} is a false positive tolerance threshold, i.e., $M_i^t : \forall k \in [t - cfp_{ths}, t] : M_i^k \in SKM^k \wedge p_i^k \notin SK^k$.

Lemma 4. *If M_i^t is reconstructed, a list of MBRs M_i^t dominates ($M_i^t.DRM$) remains the same.*

Proof. Due to Lemma 3, the dominance region of M_i^t does not become smaller. Therefore, all MBRs that M_i^t dominated before reconstruction are still dominated by M_i^t after the reconstruction.

Lemma 5. *After $M_i^t \in SKM^t$ is reconstructed, M_i^t may change its membership to $NSKM^t$*

Proof. According to Lemma 3, the anti-dominance region of M_i^t may become larger. Therefore, it is possible that some $M_j^t \in SKM^t$ can dominate M_i^t.

Running example Fig. 2 illustrates an example of an MBR reconstruction. In Fig. 2a, there are 11 different MBRs in the space which can be classified to SKM^t and $NSKM^t$. An arrow from M_i^t to M_j^t shows the relation that M_i^t dominates M_j^t, i.e., $M_j^t \in M_i^t.DRM$ and $M_j^t.admr = M_i^t$. Assume that M_2^t is decided to be reconstructed at snapshot t and the recent data point p_2^t is as shown in Fig. 2a. After M_2^t's reconstruction (Fig. 2b), both dominance and anti-dominance region of M_2^t have changed, and M_2^t is no longer in SKM^t because it is dominated by M_3^t while $M_2^t.DMR$ ($\{7,8\}$) remains the same (no additional cost of finding). Note that $M_i^t \in NSKM^t$ can be dominated by some $M_j^t \in NSKM^t$ (not only limited to $M_j^t \in SKM^t$).

In summary, while the process in pre-calculation maintenance swaps some MBRs from $NSKM^t$ to SKM^t, the process in post-calculation maintenance dynamically swaps back some MBRs from SKM^t to $NSKM^t$. This responses to behaviors of data movement in an adaptive way.

5   Performance Evaluation

In this section, we conducted some experiments by implementing algorithms using C# on a single commodity PC. We tested our proposed algorithm as well as other competitive methods on the same environment. The system was to process a large file dataset which contains a series of $(D^0, U^1, U^2, \ldots, U^T)$ and find the desired output which is $(SK^0, SK^1, \ldots SK^T)$. We evaluated the performance by measuring the wall clock time of total execution time in each method. In the results, only synthetic datasets were generated 3 times in each case, and the results report the average case of them.

5.1   Datasets

In this experiment, we use both synthetic and real datasets to simulate and show our proposed method's performance.

1. **Synthetic Dataset (SYN):** Firstly, each data record p_i^0 is uniformly random on each dimension as a point on the m-dimensional data space $[0, 100]^m$. We model a data value on each dimension as a Gaussian random walk pattern

following $p_i^t[l] = p_i^{t-1}[l] + u_i^t[l]$ where $u_i^t[l] = \lambda_i^t \cdot e_t[l]$, $e_t[l] \sim N(0, 0.5)$ (normal distribution), $1 \le l \le m$ and

$$\lambda_i^t = \begin{cases} 1, & \text{with probability } p. \\ 0, & \text{with probability } 1 - p. \end{cases} \tag{2}$$

The synthetic datasets are generated by varying each parameter from default setting. We assign the default setting for parameters as follows, $N = 5000$, $m = 3$, $T = 10000$, and p (in Eq. 2) $= 0.05$.

2. **Stock Dataset (STK):** This stock dataset aggregated from Yahoo! Finance[1] consists of the daily information of all stocks in NYSE between 2004 and 2013. For the scenario that we want to form a defensive investment portfolio, stocks that have a lower beta (not fluctuate with the market) and a trend of increasing in price than other stocks for a long period of time in the market are preferable. Therefore, we extract only 2 attributes including 200-day beta (β) and a 200-day slope of a regression line of close price (2 decimal precision). The system analyzes daily which stock acts or holds this characteristic for a long period of time and no better other choices in the market (skyline).

3. **NBA Dataset (NBA):** This NBA dataset aggregated by the authors in [14] consists of all historical NBA information on both game plays and player statistics between 1991 and 2004. We aim to find the skyline of players being active over that time period. By taking the end of each game as one snapshot, we need to compute the skyline after every game play. In summary, there are 1225 players to be monitored, 16423 matches played (snapshots) and 312086 update tuples in total. We selected 5 useful attributes from a record of each player in each match including play time in minutes (MIN), points made (PTS), total rebounds (TOT), field goal made (FGM) and field goal attempts (FGA). However, we extracted 3 attributes to evaluate players including $PTSA = \sum PTS / \sum MIN$, $TOTA = \sum TOT / \sum MIN$ and $FGR = \sum FGM / \sum FGA$.

5.2 Comparison Methods

We implemented the following methods for comparing with our proposed method.

1. **Naive Method (Baseline):** Compute the skyline by the default skyline algorithm using entire data objects in every snapshot ($D^0, D^1, \ldots, D^T$).
2. **Grid-based Method (Grid-n):** We implemented the technique in [17] to prune by dividing the m-dimensional data space into n^m cells. Each cell is large $(c_w)^m$ units. In our experiments, we define $c_w = \max p_i^t[k]/n; \forall t \in \{0, 1, \ldots, T\}, \forall i \in \{1, 2, \ldots, N\}$ and $\forall k \in \{1, 2, \ldots, m\}$. We varied n from 10, 30 and 50 respectively in each experiment. Then the candidates only in the grid cells that are not dominated by other cells are computed for skyline

---

[1] Yahoo! Finance: http://finance.yahoo.com/.

calculation while the rest can be safely pruned. Also, this method requires the knowledge of data distribution, types of data attributes, data space to define proper values of grid cells' granularity and the space size. As in our experiments on real datasets, we had to find the maximum possible value in each attribute in advance to normalize the attribute values in order to fit in the specified data space. However, doing like this is not applicable in some real-life and real time applications, because this can be difficult and impossible for open-bounded value attributes. On the contrary, our proposed method has no restriction about the data space.

3. **LookOut** [11]: This method uses R*-tree index to efficiently compute skylines on every data insertion and data deletion occurring in the sliding-window. We implemented this method by maintaining all current data objects and taking an update tuple as 2 consequent operations – old data deletion and new data insertion.

5.3   Results of the Synthetic Datasets

Parameter Tuning. The false positive tolerance threshold (cfp_{ths}) is only one system parameter in the proposed method. Here, as a preliminary experiment, we studied an effect of this parameter to decide a suitable value. At low cfp_{ths}, it incurred frequent MBR reconstructions while the size of candidates is reduced because of less false positives in SKM^t, and vice versa. It shows that setting low cfp_{ths} around 10–100 gives more preferable outcome than higher cfp_{ths} (1000–10000). In other words, paying some maintenance cost to renew non-potential MBRs in SKM^t is worthy and able to reduce the overall computation time. Therefore, we assign cfp_{ths} equal to 50 as the default parameter in all experiments.

Impact of N. Increasing the number of objects, N, to be monitored directly affects the skyline computation time because of more candidates to be processed in each snapshot. As shown in Fig. 3a, our proposed method runs faster than the others. As expected, the LookOut method performed worst than the others by a number of magnitudes, so we left out this method for the rest of the results. Even though Grid-10 can prune some candidates, their maintenance time is high due to the large number of blocks to be maintained. Its result turned poorer than the naive method. Regardless of the setting of the grid-based method, our proposed method processes less candidates (results omitted) compared with other methods and still outperforms the other methods in the total computation time. This result ensures that our proposed method is scalable on a large number of data objects.

Impact of m. Normally, the cardinality of a skyline set is increasing exponentially with the number of dimensions, m, resulting in slower skyline computation. In Fig. 3b, our proposed method outperforms the grid-based method because increasing the dimensionality also increases exponentially the number of maintained cells in the grid-based method, i.e., n^m. We also tried experimenting on higher dimensionality than 5, but the grid-based method faces the errors

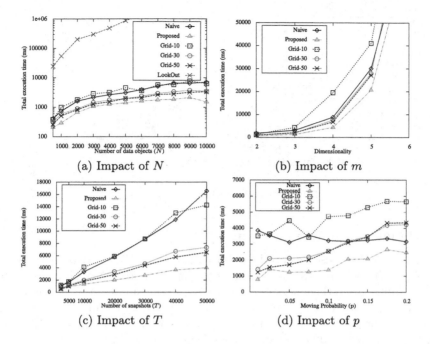

Fig. 3. Results of the synthetic dataset

due to the overflow of grid index number. At $m = 7$ (not shown in the figure), our proposed method still saves total computation time compared to the naive method by 20.5 %.

Impact of T. In this setting, we simulated the result of all methods when using for long periods of time by increasing the number of snapshots, T. Normally, the total computation time grows linearly with this factor. However, in practice, it also depends on the cardinality of output affected by data distribution and data updates. The results show that our proposed method is more efficient than the others in long-term usage denoted by more gradual slope in Fig. 3c.

Impact of p. In this setting, we study the effect when many data objects change their attribute values in each snapshot by increasing the probability of data object moving in a consecutive snapshot, p. According to the random walk model, increasing this probability scatters and deviates more data further from the initial points and produces different skyline output in each setup. However, this puts more work on the grid-based method to verify the correctness of membership in each cell as well as their cell dominance relationships and on our proposed method to verify the changes of MBRs in shape and dominance relationships. This maintenance cost worsens the grid-based method, and they become worse than the naive method as p increasing.

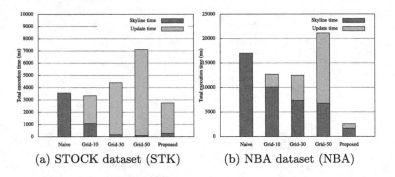

(a) STOCK dataset (STK)   (b) NBA dataset (NBA)

Fig. 4. Results of the real datasets

5.4 Results of the Real Datasets

We examine our proposed method's performance on 2 different real-life datasets, NBA and STK, which have totally different characteristics. NBA is quite larger than STK in terms of the number of snapshots and the number of dimensions, but the moving probability (update ratio in each snapshot) is quite low averagely at only 2 % for each snapshot while STK is at 47.6 %. In this subsection, the results are reported in more details about the skyline computation time and the maintenance time.

In STK dataset, the data objects are frequently changed (many update tuples at a day tick) by its nature, and the moving probability is quite high averagely at 47 %. Therefore, very high cost of maintenance for the grid-based method and our proposed method should be anticipated. As expected, the result of this dataset in Fig. 4a reports the huge cost of maintenance time (Update time) in our proposed method and especially in the grid-based method. However, it is obvious that the more grid partitions, the higher cost of maintenance. Grid-30 and Grid-50 incur high cost of maintenance (Update time shown in Fig. 4a), so the total execution time for these 2 methods eventually cannot beat the naive method. Nevertheless, our proposed method still saves the computation cost by 20 % compared with the naive method.

Unlike STK, the NBA dataset consists of 3 attributes and more snapshots with lower average update rate. In Fig. 4b, our proposed method achieves the best result among other comparison methods in both skyline calculation time (Skyline time) and maintenance time (Update time). Using the grid-based method can help pruning a lot in the STK dataset, but the pruning capability in this dataset is quite poor due to higher dimensionality and data distribution. In spite of lower number of candidates to be computed, the cost of maintenance of Grid-50 makes it underperformed even the naive method.

6 Conclusion

In this paper, we proposed an efficient method for skyline calculation when there are many data updates at each snapshot (timestamp). This is useful for analyzing historical (time-series) data archive as well as skyline computation on data update streams. In the assumed historical data series, the changes of data between consecutive timestamps are expressed and kept as update tuples. In practice, data insertion, deletion or any modification of a single data object between timestamp can totally change the final skyline set. Therefore, the naive method for this problem is to re-compute the new skyline set every timestamp (snapshot). This can be very expensive and time-consuming.

Our proposed method makes use of bounding boxes, i.e., MBRs to summarize and represent a series of data snapshots of each data object. Due to the properties of MBRs and our technique to manage them, we can identify a smaller set of candidates for skyline computation by pruning non-potential data objects while the accuracy can be guaranteed. Moreover, we also discuss about the maintenance of our technique which is adaptive to the data changes in terms of temporal updates and data distribution.

We compared the performance of our proposed method through the experiments by using both synthetic dataset and 2 real datasets. The results obviously showed the benefits of our proposed method over the other methods by measuring the total execution time.

This research is partially supported by the Grant-in-Aid for Scientific Research (A)(26240013) of MEXT, Japan.

References

1. Börzsönyi, S., Kossmann, D., Stocker, K.: The skyline operator. In: ICDE, pp. 421–430 (2001)
2. Cheema, M.A., Lin, X., Zhang, W., Zhang, Y.: A safe zone based approach for monitoring moving skyline queries. In: EDBT, pp. 275–286, ACM (2013)
3. Dellis, E., Seeger, B.: Efficient computation of reverse skyline queries. In: VLDB, pp. 291–302, VLDB Endowment (2007)
4. Ding, X., Lian, X., Chen, L., Jin, H.: Continuous monitoring of skylines over uncertain data streams. Inf. Sci. **184**(1), 196–214 (2012)
5. Hose, K., Vlachou, A.: A survey of skyline processing in highly distributed environments. VLDB J. **21**(3), 359–384 (2012)
6. Hsueh, Y.-L., Zimmermann, R., Ku, W.-S.: Efficient updates for continuous skyline computations. In: Bhowmick, S.S., Küng, J., Wagner, R. (eds.) DEXA 2008. LNCS, vol. 5181, pp. 419–433. Springer, Heidelberg (2008)
7. Huang, Z., Lu, H., Ooi, B.C., Tung, A.: Continuous skyline queries for moving objects. IEEE TKDE **18**(12), 1645–1658 (2006)
8. Jiang, B., Pei, J.: Online interval skyline queries on time series. In: ICDE, pp. 1036–1047, IEEE Computer Society, Washington, DC, USA (2009)
9. Lee, M.-W., Hwang, S.-W.: Continuous skylining on volatile moving data. In: ICDE, pp. 1568–1575 (2009)
10. Lee, Y.W., Lee, K.Y., Kim, M.H.: Efficient processing of multiple continuous skyline queries over a data stream. Inf. Sci. **221**, 316–337 (2013)
11. Morse, M., Patel, J.M., Grosky, W.I.: Efficient continuous skyline computation. Inf. Sci. **177**(17), 3411–3437 (2007)
12. Papadias, D., Tao, Y., Fu, G., Seeger, B.: Progressive skyline computation in database systems. TODS **30**(1), 41–82 (2005)

13. Papapetrou, O., Garofalakis, M.: Continuous fragmented skylines over distributed streams. In: ICDE, pp. 124–135 (2014)
14. Sultana, A., Hassan, N., Li, C., Yang, J., Yu, C.: Incremental discovery of prominent situational facts. In: ICDE, pp. 112–123, IEEE (2014)
15. Sun, S., Huang, Z., Zhong, H., Dai, D., Liu, H., Li, J.: Efficient monitoring of skyline queries over distributed data streams. Knowl. Inf. Syst. **25**(3), 575–606 (2010)
16. Tao, Y., Xiao, X., Pei, J.: Subsky: efficient computation of skylines in subspaces. In: ICDE, pp. 65–65, IEEE (2006)
17. Tian, L., Wang, L., Li, A.-P., Zou, P., Jia, Y.: Continuous skyline tracking on update data streams. In: Chang, K.C.-C., Wang, W., Chen, L., Ellis, C.A., Hsu, C.-H., Tsoi, A.C., Wang, H. (eds.) APWeb/WAIM 2007. LNCS, vol. 4537, pp. 192–197. Springer, Heidelberg (2007)
18. Xin, J., Wang, G., Chen, L., Zhang, X., Wang, Z.: Continuously maintaining sliding window skylines in a sensor network. In: Kotagiri, R., Radha Krishna, P., Mohania, M., Nantajeewarawat, E. (eds.) DASFAA 2007. LNCS, vol. 4443, pp. 509–521. Springer, Heidelberg (2007)

Mining Frequent Closed Flows Based on Approximate Support with a Sliding Window over Packet Streams

Imen Brahmi[1]([✉]), Hanen Brahmi[1], and Sadok Ben Yahia[2]

[1] Faculty of Sciences of Tunis, Computer Science Department,
Campus University, 1060 Tunis, Tunisia
imen.brahmi@gmail.com
[2] Institut Mines-TELECOM, TELECOM SudParis,
UMR CNRS Samovar, 91011 Evry Cedex, France
sadok.benyahia@fst.rnu.tn

Abstract. Due to the varying and dynamic characteristics of network traffic, the analysis of traffic flows is of paramount importance for network security. In this context, the main challenge consists in mining the traffic flows with high accuracy and limited memory consumption. In this respect, we introduce a novel algorithm, which mines the approximate closed frequent patterns over a stream of packets within a sliding window model. The latter is based on *a relaxation rate* parameter as well as an *approximate support* concept. Our experiment results show the robustness and efficiency of our new algorithm against those in the literature.

Keywords: Stream of packets · Sliding windows · Relaxation rate

1 Introduction

Within many applications data arises in the form of a continuous stream. A particular problem of interest motivated by network traffic analysis concerns the analysis of packet streams with a focus on newly arrived data and frequently appearing packet types [5,9].

Unfortunately, there always exist a huge amount of network flows, which results in overwhelming and unpredictable memory requirements for packet streams mining [5]. Consequently, this fact hampers the pattern mining over packet streams. In addition, network traffic patterns are believed to obey the power law, implying that most of the bandwidth is consumed by a small set of heavy users [3,7]. Therefore, to make such analysis meaningful, bandwidth usage statistics should be kept for only a limited amount of time before being replaced with new measurements. Failure to remove stale data leads to statistics aggregated over the entire lifetime of the stream, which are unsuitable for identifying recent usage trends [3].

© Springer International Publishing Switzerland 2015
Q. Chen et al. (Eds.): DEXA 2015, Part II, LNCS 9262, pp. 109–116, 2015.
DOI: 10.1007/978-3-319-22852-5_10

A solution for removing stale data is to periodically reset all statistics. This gives rise to the *sliding window model*, which expires old items as new items arrive. Indeed, it has been shown that the sliding windows are a typical method of flow control, especially for network packet pattern mining [1,3,5,6,9].

Alongside, the concepts of closed frequent patterns usually can help in accelerating the mining process and compressing the memory usage [1,2,6,8]. However, it is unfeasible to find the exact frequency of network traffic patterns over packet streams [1,3,7]. The main difficulty lies in the high complexity of maintenance, taking into account the dynamic changing of packet streams [8]. Consequently, it is sufficient to mining only approximated frequent closed patterns instead of those patterns in full precision [1,4,8].

In this paper, we introduce a single-pass approximate algorithm, called ACL-SWIN (*A*pproximate *CL*osed frequent pattern mining in *S*liding *Win*dow) that provides a condensed representation of network packet streams. The algorithm uses a sliding window model for data processing to ensure that the novel mining results are integrated. Through extensive carried out experiments on a real network traffic traces, we show the significance of our approach in comparison with those fitting in the same trend.

The remainder of the paper is organized as follows. We scrutinize, in Sect. 2, the related work. In Sect. 3, we introduce the ACL-SWIN algorithm. We also report the encouraging results of the carried out experiments in Sect. 4. Finally, we conclude by resuming the strengths of our contribution and sketching future research issues.

2   Related Work

Due to its high practical relevance, the researches that focus on mining a set of closed frequent patterns over the data stream has grasped a lot of attention in recent years [1,2,8].

One of the studies worth of mention was the algorithm MOMENT proposed by Chi et *al.* [2]. MOMENT was regarded as the first to find frequent closed patterns from data streams within a sliding window. The main moan that can be addressed to such approach stands in the fact that the memory usage of MOMENT will be inefficient, if there are a large number of closed frequent patterns [1,8].

In order to improve the MOMENT algorithm, various algorithm are proposed. Unfortunately, this trend of approaches still follows the traditional definition of closed patterns. The latter permit to remove sub-patterns which have the same support as some of their super-patterns. However, the compression by the closed pattern approach may not be so effective since there often exist slightly different counts between super- and sub- patterns. The problem worsen due to the dynamic changing of packet streams and any little concept drift happenings can lead to the changing of closed patterns [8].

In this respect, Song et *al.* [8] introduced the CLAIM algorithm, which allows the mining of closed approximate patterns. The CLAIM adopts bipartite graph model to store frequent closed patterns and builds a compact hash based tree structure to maintain the relaxed patterns and support mining process. Likewise, Cheng et al. [1] proposed the INCMINE algorithm to approximately mine closed frequent patterns in sliding window over data streams. INCMINE is developed by taking advantages of relaxed closed patterns to control the accuracy of mining. Within the context of packet streams, various approaches, such as [3,7], follows the trend of counter-based algorithms. These approaches consist of mining frequently occurring patterns whose estimated frequencies exceed $(ms - \epsilon)$ over each window, where ϵ is an error parameter ϵ and ms is a minimum support threshold.

Due to the varying and dynamic characteristics of network traffic including fast transfer, huge volume, incestimable and infinite, flows mining with respect to a reduced memory space still present a thriving and a compelling issue. In this respect, under the consideration of the distribution and fluidity of network flows, we propose a novel algorithm, called ACL-SWIN, to mine approximate closed frequent patterns using a sliding window over packet streams. In fact, it has been shown that the extraction of the closed patterns requires less memory within the stream mining context.

3   ACL-SWin: Closed Flows Mining Over a Sliding Window

Clearly, it has been shown that, it is unfeasible to find the exact frequency of network patterns using memory resources sub-linear to their number [3,7]. In this respect, the ACL-SWIN algorithm follows an estimation mechanism allowing the mining of the approximate frequent closed patterns.

3.1   The Estimation Mechanism

Due to the dynamically characteristic of traffic stream, a pattern may be infrequent at some point in a stream but becomes frequent later. Since there are exponentially many infrequent patterns at any point in a stream, it is infeasible to keep all infrequent patterns. In this respect, to estimate the supports of the latters, the counter-based algorithms [3,7] use the error parameter, ϵ. However, even the use of small ϵ allows more accurate approximation, it results in a large number of patterns to be processed and maintained. Consequently, this fact drastically increases the memory consumption and severely degrades the processing efficiency.

To palliate this drawback, we propose the use of ϵ as a *relaxed minimum support threshold* ms_{relax}, such as $0 \leq ms_{relax} \leq ms \leq 1$. In addition, we propose to progressively increase the value of ϵ for a pattern as it is retained longer in the window. Thus, the relaxed minimum support threshold allows us to effectively identify and drop the unpromising patterns, thereby drastically reducing the number of patterns that need to be kept and processed.

Definition 1 *(Relaxation Rate).* *The relaxed minimum support threshold* ms_{relax} *is equal to* $r \times ms$, *where* $r(0 \leq r \leq 1)$ *is the relaxation rate and* ms *is a minimum support threshold.*

All patterns whose support is lower than ms_{relax} are discarded. Thus, if a pattern X is frequent within the next window, the obtained support is estimated. The latter is defined as *an approximate support* which allows to better ensure the accuracy of mining.

Definition 2 *(Approximate Support).* *The approximate support of a pattern* X *over a time unit* t *is defined as*

$$\widetilde{SUP(X,t)} = \begin{cases} 0, & \text{if } sup(X,t) < r \times \ ms; \\ sup(X,t), & \text{otherwise.} \end{cases} \tag{1}$$

3.2   The Data Structures of ACL-SWin

With the consideration of time and space limitation, the proposed algorithm uses two in-memory data structures which are called CITABLE (*Closed Incremental Table*) and CILIST (*Closed Identifier List*) respectively. In addition, it employs a *hash table*, called $Temp_{New}$, to put the patterns that have to be updated whenever a new packet arrives. In fact, the rationales behind such in-memory data structures are: (*i*) saving storage space; and (*ii*) reducing the cost of the incremental maintenance of patterns.

The CITABLE is used to keep track of the evolution of closed patterns. Each record of the CITABLE represents the information of a closed pattern. It consists of three fields: CID, CLOS and COUNT. Each closed pattern was assigned a unique closed identifier, called CID. The CID field is used to identify closed patterns. Given a CID, the ACL-SWIN algorithm gets corresponding closed patterns in the CLOS field. The support counts are stored in the COUNT field.

Example 1. According to the database shown by Table 1, the CITABLE is sketched by Table 2.

The CILIST is used to maintain the items and their cidsets. It consists of two fields: the Item field and the cidset field. The cidset of an item X, denoted as cidset(X), is a set which contains all cids of X's super closed patterns.

Table 1. A snapshot of network traffic data.

ID	Packets
p_1	src_IP1,protocol
p_2	src_port,dst_port
p_3	src_port,dst_port,src_IP1
p_4	src_port,dst_port,src_IP1
p_5	src_port,src_IP1,protocol

Table 2. Example of CITABLE.

CID	CLOS	COUNT
0	{0}	0
1	{src_IP1 protocol}	2
2	{src_port dst_port}	3
3	{src_port dst_port src_IP1}	2
4	{src_IP1}	4
5	{src_port src_IP1 protocol}	1
6	{src_port}	4
7	{src_port src_IP1}	3

Table 3. Example of the CILIST.

Item	cidset
src_port	{2, 3, 5, 6, 7}
dst_port	{2, 3}
src_IP1	{1, 3, 4, 5, 7}
protocol	{1, 5}

Example 2. According to Table 1 and the CITABLE shown by Table 2, {src_port dst_port} is closed and its CID is equal to 2. Thus, 2 will be added into cidset(src_port) and cidset(dst_port) respectively. Table 3 illustrates a CILIST. It maintains the items and their corresponding superset cids shown by Table 2.

3.3   The ACL-SWin Algorithm

The ACL-SWIN algorithm attempts to mine a concise representation of patterns that delivers approximate closed frequent flows. The pseudo-code of the ACL-SWIN algorithm is shown by Algorithm 1.

Algorithm 1. The ACL-SWIN algorithm

Input: $\mathcal{T}$, r, ms
Output: Updated CITABLE
1 **begin**
2 | Scan SW;
3 | **foreach** $p_{New} \in SW$ **do**
4 | | //Phase 1
5 | | $Temp_{New} := (p_{New}, 0)$;
6 | | $SET(\{p_{New}\}) = cidset(i_1) \cup ... \cup cidset(i_k)$;
7 | | **foreach** $Cid(i) \in SET(\{p_{New}\})$ **do**
8 | | | $\mathcal{IR} := $ Null;
9 | | | $\mathcal{IR} := p_{New} \cap Clos[i]$;
10 | | | **if** $\mathcal{IR} \in Temp_{New}$ **then**
11 | | | | **if** $SUP(\widetilde{Clos[i]}) > SUP(\widetilde{Clos[z]})$ **then**
12 | | | | | replace $(\mathcal{IR}, i)$ with $(\mathcal{IR}, z)$ in $Temp_{New}$
13 | | | **else**
14 | | | | $Temp_{New} := Temp_{New} \cup (\mathcal{IR}, i)$
15 | | //Phase 2
16 | | **foreach** $(X, c) \in Temp_{New}$ **do**
17 | | | **if** $X == Clos[c]$ **then**
18 | | | | $SUP(\widetilde{Clos[c]}) := SUP(\widetilde{Clos[c]}) + 1$;
19 | | | **else**
20 | | | | $j := j+1$;
21 | | | | CITABLE $:= $ CITABLE $\cup$ $(j, X, SUP(\widetilde{Clos[c]}) + 1)$;
22 | | | | **foreach** $i \in p_{New}$ **do**
23 | | | | | $cidset(i) := cidset(i) \cup j$

Indeed, the algorithm takes on input a network traffic trace $\mathcal{T}$, a minimum support threshold ms and a relaxation rate r. It starts by reading a sliding window of packets, SW (line 2). Although, the sliding window is a powerful solution to handle the network flows, in order to mine the recent patterns. It suffers from the high requirement on memory and storage space caused by the frequent and the continuous updates [3,5,9]. Consequently, the ACL-SWIN algorithm adopts a basic windows strategy, which divides the sliding windows on equally-sized partitions. The latter facilitates the continuous monitoring of changes in the stream. Moreover, it can be used to palliate the drawback of unbounded memory over the packet streams [3,5]. In addition, whenever a new packet p_{New} arrives, the

ACL-SWIN algorithm consists of two phases. During the first one, the algorithm finds all patterns that need to be updated with their closures, and puts them into $Temp_{New}$ (lines 4−13). Within the second phase, the ACL-SWIN algorithm updates their supports, CITABLE and CILIST (lines 15−22). Consequently, the updated closed patterns can be obtained without multiple scans of whole search spaces, *i.e.*, by scanning the CITABLE once.

4   Experimental Results

To evaluate the effectiveness and efficiency of our proposed algorithm ACL-SWIN, we carried out extensive experiments. Therefore, we used a real network traffic trace collected from the gateway of a campus network with 1500 users in China[1]. Table 4 sketches dataset characteristics used during our experiments.

Figure 1 (A, B, C, D, E, and F) plot the comparison of the ACL-SWIN, CLAIM and INCMINE algorithms in terms of precision and memory consumption. We take into account three different perspectives, including: (1) different support thresholds; (2) different relaxation rates; and (3) different sliding windows sizes.

Table 4. The considered real traffic traces at a glance.

Traffic traces	
Source	Campus network
Date	2009-08-24
	16:20-16:35
Packets	49,999,860
Unique flows	4,136,226

During the first type of experiment, we assess the effect of minimum support threshold variations, while fixing r as 0.5. Figure 1(A) verifies that ACL-SWIN achieves 100 % precision in all cases of ms. Clearly, ACL-SWIN generates a high accuracy mining results than the other algorithms, over the network flows. This result confirms the effectiveness of the ACL-SWIN estimation of the approximate support. Figure 1(D) shows that ACL-SWIN can achieve better memory usage than both CLAIM and INCMINE algorithms. For instance, when ms is equal to 0.2 %, the memory usage achieves 45MB for ACL-SWIN, against 66MB for CLAIM and 160MB for INCMINE, which shows that our approximate support can condense frequent patterns efficiently. Finally, we conclude that ACL-SWIN is able to correctly mine the network patterns and consumes much less memory than do CLAIM and INCMINE.

---

[1] We thank Mr. Q. Rong [7] for providing us with the real network traffic trace.

The second simulation is to evaluate the effect of relaxation rate variations, while setting $ms = 0.4\%$ and $SW = 5 \times 10^3$. In this respect, according to Fig. 1(B), we remark that the ACL-SWin algorithm achieves 100 % precision representing the optimal case. Whereas, the precision of both CLAIM and INCMINE drops linearly to be less than 98 % as far as r increases. Generally, ACL-SWin always improves both CLAIM and INCMINE algorithms by discovering a lower number of false positive patterns. In this respect, we conclude that the estimation mechanism of both INCMINE and CLAIM algorithms relies on the relaxation rate parameter r to control the accuracy. Compared to these two algorithms, ACL-SWin is much less sensitive to r and is able to significantly achieves high accurate approximation results by increasing r. Moreover, Fig. 1(E) shows that ACL-SWin achieves a roughly constant memory consumption of no more than 150MB. Considering the three algorithms, the memory consumption of ACL-SWin is lower than that of both INCMINE and CLAIM. This results are due to our data structure mechanisms.

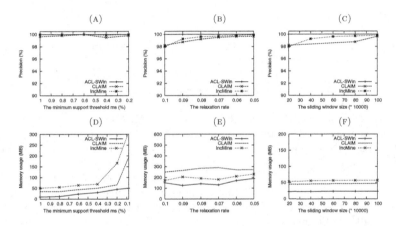

Fig. 1. The precision and the memory consumption of ACCL-SWin *vs.* CLAIM and INCEMINE.

Finally, within the third type of experiments we test the impact of the sliding window size variations. To this end, we present Fig. 1 (C) and (F) where the ms is fixed as 0.4 % and r as 0.1 %. Figure 1 (C) indicates that the ACL-SWin algorithm allows high accuracy mining results. Moreover, Fig. 1(F) shows that ACL-SWin achieves a roughly constant memory consumption of no more than 22MB, while CLAIM and INCMINE consume, respectively 45MB and 53MB memory. The results of these figures can be explained by the basic window strategy adopted by the sliding window model within the ACL-SWin algorithm. Clearly, it has been shown that the basic window strategy allows the reduction of space storage and memory usage [3,5]. Thus, the mechanism adopted by the ACL-SWin is more effective than that adopted by both the INCMINE and CLAIM algorithms, related to the network packet streams context. In this respect, both latters algorithms are hampered by the ineffectiveness of the sliding windows, which may be too large to fit in main memory within the network traffic mining.

5 Conclusion

In this paper, we introduced a novel stream mining algorithm called ACL-SWIN to approximately mine the frequent closed patterns in sliding window over packet streams. The carried out experimental results highlighted the effectiveness of the introduced algorithm ACL-SWIN *vs.* the pioneering algorithm in literature.

Future issues for the present work mainly concern: (*i*) The consideration of the intrusion detection over online packet streams and the mining of closed frequent patterns from flows for network monitoring; (*ii*) Study of the extraction of "generic streaming" association rules based on the ACL-SWIN algorithm.

References

1. Cheng, J., Ke, Y., Ng, W.: Maintaining frequent closed itemsets over a sliding window. J. Intell. Inf. Syst. **31**(3), 191–215 (2008)
2. Chi, Y., Wang, H., Yu, P.S., Muntz, R.R.: Catch the moment: maintaining closed frequent itemsets over a data stream sliding window. Knowl. Inf. Syst. (KAIS) **10**(3), 265–294 (2006)
3. Golab, L., DeHaan, D., Demaine, E., López-Ortiz, A., Munro, J.I.: Identifying frequent items in sliding windows over online packet streams. In: Proceedings of the 3rd ACM SIGCOMM Conference on Internet Measurement, IMC 2003, Miami Beach, FL, USA, pp. 173–178 (2003)
4. Li, H., Lu, Z., Chen, H.: Mining approximate closed frequent itemsets over stream. In: Proceedings of the 9th ACIS International Conference on Software Engineering, Artificial Intelligence, Networking, and Parallel/Distributed Computing SNPD 2008, pp. 405–410 (2008)
5. Li, X., Deng, Z.H.: Mining frequent patterns from network flows for monitoring network. Expert Syst. Appl. **37**(12), 8850–8860 (2010)
6. Nori, F., Deypir, M., Sadreddini, M.: A sliding window based algorithm for frequent closed itemset mining over data streams. J. Syst. Softw. **86**(3), 615–623 (2013)
7. Rong, Q., Zhang, G., Xie, G., Salamatian, K.: Mnemonic lossy counting: an efficient and accurate heavy-hitters identification algorithm. In: Proceedings of the 29th IEEE International Performance Computing and Communications Conference, Albuquerque, United States (2010)
8. Song, G., Yang, D., Cui, B., Zheng, B., Liu, Y., Xie, K.-Q.: CLAIM: an efficient method for relaxed frequent closed itemsets mining over stream data. In: Kotagiri, R., Radha Krishna, P., Mohania, M., Nantajeewarawat, E. (eds.) DASFAA 2007. LNCS, vol. 4443, pp. 664–675. Springer, Heidelberg (2007)
9. Zhang, Z., Wang, B., Chen, S., Zhu, K.: Mining frequent flows based on adaptive threshold with a sliding window over online packet stream. In: Proceedings of the International Conference on Communication Software and Networks, ICCSN 2009, Macau, United States, pp. 210–214 (2009)

D-FOPA: A Dynamic Final Object Pruning Algorithm to Efficiently Produce Skyline Points Over Data Streams

Stephanie Alibrandi$^{(\boxtimes)}$, Sofia Bravo, Marlene Goncalves,
and Maria-Esther Vidal

Universidad Simón Bolívar, Caracas, Venezuela
{salibrandi,sbravo,mgoncalves,mvidal}@ldc.usb.ve

Abstract. Emerging technologies that support sensor networks are making available large volumes of *sensed data*. Commonly, sensed data entries are characterized by several *sensed attributes* and can be related to *static attributes*; both sensed and static attributes can be represented using Vertically Partitioned Tables (VPTs). *Skyline-based* ranking techniques provide the basis to distinguish the entries that best meet a user condition, and allow for the pruning of the space of potential answers. We tackle the problem of efficiently computing the skyline over sensed and static data represented as Vertically Partitioned Tables (VPTs). We propose an algorithm named D-FOPA (Dynamic Final Object Pruning Algorithm), a rank-based approach able to dynamically adjust the skyline by processing changes on values of sensed attributes. We conducted an empirical study on datasets of synthetic sensed data, and the results suggest that D-FOPA is not only able to scale up to large datasets, but reduces average execution time and number of comparisons of state-of-the-art approaches by up to one order of magnitude.

1 Introduction

Due to the increasing amount of data available on the Web, users can retrieve information satisfying their requests. For example, given *sensed data* that state the status of bike parking slots and *static data* that represent their location, a user may be interested in retrieving the information of bike terminals close to a given Metro station in Valencia that have free parking slots to park a bike. For many applications, it is important to efficiently identify the best response that satisfies a given query representing similar requests. These applications must be designed to provide relevant answers quickly. The elements that best meet the requirements of a user, modeled as multi-objective functions, are identified by executing *skyline queries* against sensed and static data. We address the problem of calculating the *skyline* in a dynamic environment over sensed and static data, and based on state-of-the-art techniques, we devise a solution where tuples that best meet a given user request correspond to the skyline tuples.

© Springer International Publishing Switzerland 2015
Q. Chen et al. (Eds.): DEXA 2015, Part II, LNCS 9262, pp. 117–133, 2015.
DOI: 10.1007/978-3-319-22852-5_11

The problem of computing the skyline has been extensively treated in the literature, where existing approaches focus on minimizing the number of comparisons during a *skyline query* execution against *static* or *sensed* data [2,5,6,8–14]. For example, the Final Object Pruning Algorithm (FOPA) proposed by Alvarado *et al.* [2], represents *static data* in a distributed way, using Vertically Partitioned Tables called VPTs [1]; this representation allows for reducing the data *storage space* and for an *efficient computation* of muti-objective functions. Further, the approach proposed by Tao and Papadias [15] addresses the problem of computing the skyline set in a *sensed data* environment, i.e., that maintains the skyline set through the entire query execution in a *lazy* fashion. Although experimental studies reported in the literature [2,15] suggest that FOPA, and Tao and Papadias algorithms are efficient, both suffer of the following drawbacks: (i) FOPA is not able to work with sensed data, and (ii) Tao and Papadias's strategies may perform a large number of comparisons depending on the size of the dataset.

We address the problem of efficiently executing multi-objective queries over sensed and static data. To overcome the limitations of FOPA and the *lazy* approach in [15], we define a novel algorithm named D-FOPA (for Dynamic Final Object Pruning Algorithm) that empowers FOPA with the capability of managing both sensed and static data. Additionally, we extend the *lazy* strategy proposed by Tao and Papadias to manage different criteria of sensed data expiration; we named this algorithm Lazy+. To empirically evaluate the performance of our approach, we conducted a comparative study between D-FOPA and Lazy+. Sensed and static data are synthetically generated using a variation of the data generator implemented by Börzsönyi *et al.* [5]. The experimental study indicates that D-FOPA, is scalable with respect to Lazy+, and reduces average execution time and number of comparisons up to one order of magnitude.

To summarize, the main contributions of this paper are the following:

- The Dynamic Final Object Pruning Algorithm (D-FOPA), which efficiently identifies entries that best meet a skyline query over sensed and static data. Answers produced by D-FOPA are complete while the number of comparisons is reduced.
- An extensive empirical study where D-FOPA and Lazy+ are compared in terms of average execution time, number of comparisons, throughput, and answer completeness; results suggest that D-FOPA overcomes Lazy+ in *large* datasets.

This paper is composed of six additional sections. Section 2 briefly describes the preliminaries. Section 3 illustrates a motivating example. Section 4 formalizes D-FOPA and defines the main properties of Lazy+, the modified *lazy* strategy from [15]. Section 5 details an experimental study where we report the performance of D-FOPA in comparison with Lazy+. Section 6 summarizes existing state-of-the-art approaches. Finally, we conclude in Sect. 7 with an outlook to future work.

First Segment	arr_1	arr_2	arr_3	arr_4	arr_5	arr_6	arr_7	arr_8	arr_9

Second Segment	arr_1	arr_2	arr_3	arr_4	arr_5	arr_6	arr_7	arr_8	arr_9

Third Segment	arr_1	arr_2	arr_3	arr_4	arr_5	arr_6	arr_7	arr_8	arr_9

Fig. 1. Example of a sliding window of size three, over a data stream.

2 Preliminaries

We present the basic terminology required to understand the problem addressed
in this paper. We start with the definition of *dominance* derived from the Pareto
Optimality [3].

Definition 1. *Let* $a = (a_1, a_2, ..., a_n)$ *and* $b = (b_1, b_2, ..., b_n)$ *be multi-
dimensional entries. We say that* a *dominates* b *if:*

$$\forall i \in [1, n](a_i \geq b_i) \land \exists j \in [1, n](a_j > b_j) \tag{1}$$

Secondly, suppose that a sensor network exists for monitoring some dynamic
attributes updates, named *sensed attributes*. Each entry, also called *tuple*, is
stored in a *stream of data*, which is defined as an ordered pair (s, Δ) where s is a
sequence of elements, and Δ is a sequence of positive real numbers representing
time intervals. The operator named *sliding window*, is used to produce an answer
of a query through a *window* or segment of the data stream. This segment slides
to cover all the arrivals arr_i in the data stream as shown in Fig. 1. According to
Lin and Wei [10], the sliding window implements a natural technique to produce
a solution to a stream query. Thus, during the computation of the skyline set over
a data stream, the N most recent elements are considered, where N represents
the window size. According to Chan and Yuan [6], there are two models regarding
the sliding window size: the *count-based* model and the *time-based* model. The
time-based and *count-based* models are defined as follows:

Definition 2. *Let* S *be a data stream, and* s *be a possible subset of* S. *Let* t_0
and t_n *be two arrival times associated with* s. *It is said that an element* d *belongs
to* s, *if* d *arrival time,* t_d, *holds that* $t_0 \leq t_d \leq t_n$. *In this case,* s *is a sliding
window time-based.*

Definition 3. *Let* S *be a data stream, and* s *a possible subset of* S. *Let* n *be a
fixed value that represents the cardinality of* s. *It is said that an element* d *belongs
to* s *if there is no more than* $n - 1$ *elements unprocessed in* S, *whose arrival time
is less than* d *arrival time. In this case,* s *is a sliding window count-based.*

3 Motivating Example

Suppose a tourist is visiting a city for the first time. The tourist is interested
in visiting the restaurants that meet the following conditions: *(i)* serve the best

Table 1. Running example. Restaurant attribute values denoted by ids from r_1 to r_9. Static attributes are **Quality** and **Popularity**, and sensed attributes are **Parking spaces** and **Tables**.

ID	Parking spaces	Tables	Quality	Popularity
r_2	18	0	3	5
r_4	2	10	1	2
r_5	3	2	1	2
r_3	25	5	4	1
r_1	20	8	3	5
r_6	5	8	5	2
r_9	18	0	5	5
r_7	5	8	5	3
r_8	12	4	3	2

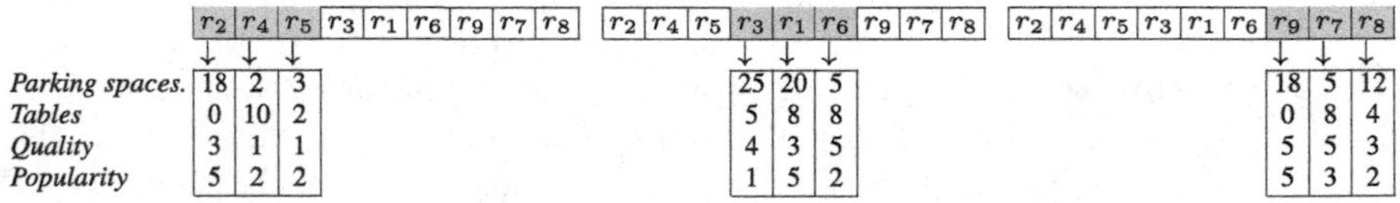

Fig. 2. Sliding window of size three that slides to cover all the arrivals in the stream corresponding to: first r_2, r_4, r_5, second, r_3, r_1, r_6, and finally, r_9, r_7, r_8.

food, *(ii)* are the most popular, *(iii)* have available tables, and *(iv)* have parking spaces available. All of these specifications are equally important for the tourist. Following our tourist's criteria, a restaurant will be taken into account if there is no other restaurant with better quality, popularity, tables available, and parking spaces available. Quality and popularity are (quasi-)static attributes, while the amount of tables and parking spaces available are attributes continuously updated i.e., sensed attributes; sensed data updates are represented in a data stream (as shown in Fig. 1). For the purpose of this example, the sliding window is of size three. For each one of the window slides, a response will be presented to the tourist containing the best restaurants under his criteria. Formally, the set of restaurants preferred by the user is composed of those that are not dominated by any other restaurant. Given A and B restaurants, A dominates B, if A has the same values in quality, popularity, tables, and parking spaces available as B, and it has a better value in at least one of them. For example, $A = (5, 10, 5, 3)$ dominates $B = (5, 9, 4, 3)$.

Suppose that at time t, nine restaurant arrivals are produced, where r_i represents the restaurant with $id = i$, as shown in Table 1. The sensed attributes *Parking spaces* and *Tables* are expected to constantly change, unlikely the static attributes, *Quality* and *Popularity*. A restaurant arrival occurs when, at any time k, a new value is received for any of its sensed attributes, while values of the static attributes remain the same.

As the window size is three, initially, only the restaurants r_2, r_4, and r_5 are processed. The skyline set is calculated, obtaining as a result that the non-dominated restaurants are r_2, r_4, and r_5. Then, just like in Fig. 2, the window slides in order to process new restaurants, particularly those with ids r_1, r_3, and r_6. At this time, the skyline set is recalculated for the up-to-date values, i.e., the non-expired items of the restaurants r_1, r_2, r_3, r_4, r_5, and r_6. The skyline set is: r_1, r_3, r_4, and r_6. Finally, to cover all restaurants arrived at time t, the sliding window must be rolled again, as shown in Fig. 2; thus, the restaurants included in the calculation are r_7, r_8, and r_9. With restaurants r_1, r_2, r_3, r_4, r_5, r_6, r_7, r_8, and r_9, the skyline set is recalculated. The result is that the non-dominated restaurants are r_1, r_3, r_4, r_7, and r_9. Similarly, the skyline set calculation is performed for updates whose arrivals occur in time t', where $t' > t$.

The Final Object Pruning Algorithm by Alvarado *et al.*, tackles the problem of computing the skyline set for static data represented in a distributed way using Vertically Partitioned Tables (VPTs) [1]. Within the existing approaches that calculate the skyline set over data streams, we can mention the algorithm proposed by Tao and Papadias [15]. In [15], they developed two general frameworks: an *eager* strategy that minimizes memory consumption, and a *lazy* strategy that delays most computational work until certain events occur. The goal of these strategies is to continuously return the changes that occur in the skyline set. Although experimental studies reported in the literature [2,15] suggest that both FOPA and *lazy* are efficient, these algorithms suffer from the different drawbacks. First, FOPA does not consider the arrival times, i.e., only last values will be taken into account for the sensed attributes; this would be equivalent to process the restaurant values $r_1 - r_9$ after all the arrivals in the sliding window have been produced. Contrary, *lazy* considers the arrival times but, it does not differentiate between sensed and static data; thus, it may maintain redundant information about data that arrival times of data that will not change.

4   Our Approach

We describe the D-FOPA and Lazy+ approaches, and illustrate the main features of these approaches using our running example.

4.1   D-FOPA: The Dynamic Final Object Pruning Algorithm

D-FOPA assumes that data is stored following a Vertically Partitioned Table representation, where each aR VPT is formed by $aValue$, indexed by *tupleId*, and ordered in aR based on the values of a. D-FOPA identifies the skyline set in one phase for each processed *slided window*. In one phase, D-FOPA reads the tuples holding the same dimension value in a given VPT, discarding the dominated ones within this group (lines 6 through 11 of Algorithm 1) this is the first prune performed. Secondly, in order to reduce the number of comparisons, D-FOPA maintains a *dimensional skyline* (a skyline set for each dimension), where the non-dominated tuples, seen in the VPT being traversed, will be added (lines 22 and 23 of Algorithm 1). Finally, once the algorithm execution is finished, the dominated entries within the union of the dimensional skylines are expunged, and the skyline set is returned.

D-FOPA keeps the last scored values of each dimension in the historical entry H, to guide the search for the *most promising final object mpfo*. D-FOPA computes c_{val} and f_{val}; f_{val} corresponds to the distance between the current position, represented by the historical entry, and the *mpfo*. Similarly, c_{val} represents the distance between the current position and the current tuple being processed.

$$f_{val} = \sum |H_i - mpfo_i| \qquad (2)$$

$$c_{val} = \sum |H_i - c_i| \qquad (3)$$

D-FOPA initializes the *most promising final object* (*mpfo*) with one of the first non dominated tuples among the group from which it was read. Whenever a new tuple is neither discarded nor dominated, c_{val} is computed, and f_{val} recalculated. Then, *mpfo* will be updated whenever $c_{val} < f_{val}$. D-FOPA finalizes to process a window slide when all the values of the most promising final object *mpfo* are seen, or when any of the VPTs is completely traversed. The execution stops when the data stream is fully covered by the sliding window. D-FOPA pseudo-code is presented in Algorithm 1. To handle dynamism presented in sensed data, D-FOPA uses two sets of VPTs: *oldVPT* and *newVPT*. The *oldVPT* tables are used to store the non-expired entries from previous slided windows, while the *newVPT* tables are used to store the new tuples to be incorporated into the skyline calculation, obtained from each window slide. These sets of VPTs will be traversed alternately according to the multi-objective function (lines 5–10 of Algorithm 1). As a result of the traversal, a unique set of VPTs is produced by merging both *newVPT* and *oldVPT*. This unique set of VPTs will be used in future window slides as the *oldVPT* tables. Additionally, D-FOPA does not maintain the skyline set through the whole query execution (line 2 of Algorithm 1); instead, the skyline set is recalculated for each window slide. Additionally, D-FOPA effectively eliminates an expired tuple by removing it from the structures; so, it will not be considered for future slided windows, reducing thus the number of comparisons.

At first, when the entries belonging to the first slided window are received (see Fig. 2), the behavior of D-FOPA is analogous to FOPA [2], i.e., the first slided window is treated as static data, it does not differentiate between a static dataset and a sensed one. At the moment that D-FOPA receives the tuples belonging to another slided window, it will be able to distinguish expired data entries from new arrivals. This occurs because entries in the first slided window are stored in the *newVPT* but they are not part of any *oldVPT* table (as shown in Table 2). Note that static attributes will moved to *oldVPTs* and remain there until the Algorithm finalizes. Following our running example, at the end, when the calculation of the skyline terminates for the first slided window, the skyline set is formed by tuples r_2, r_4, and r_5. By sliding the window (see Table 3) the

Algorithm 1. D-FOPA

```
1: for window in stream of data do
2:      Sk, P, mpfo ← ∅
3:      H ← best values of oldVPT and newVPT
4:      i, c_val, f_val ← 0
5:      while mpfo not found in all VPT and VPT_i not traversed do
6:          if oldVPT_i ≠ ∅ and newVPT_i ≠ ∅ then
7:              G ← elements of oldVPT_i and newVPT_i with the same value
8:          else if oldVPT_i = ∅ then
9:              G ← elements of newVPT_i with the same value
10:         else if newVPT_i = ∅ then
11:             G ← elements of oldVPT_i with the same value
12:         Ignore those elements that e ∈ P and e ∉ Sk_i
13:         for e in G do
14:             if e notDominated(not_dominated) then
15:                 not_dominated ← not_dominated + {e}
16:                 P ← P + {e}
17:         for e in not_dominated do
18:             if e ∉ P then
19:                 P ← P + {e}
20:             if first_iteration then
21:                 mpfo ← e
22:             if e notDominated(Sk_i) then
23:                 Sk_i ← Sk_i + {e}
24:             if e ∈ Sk_i then
25:                 H_i ← e_i
26:                 f_val ← Σ|H_i − mpfo_i|
27:                 c_val ← Σ|H_i − e_i|
28:                 if c_val < f_val then
29:                     mpfo ← e
30:         if mpfo found in all VPT then
31:             break
32:         else
33:             i ← dimension where mpfo has not been found
34:     return not dominated elements in Sk_i
```

Table 2. Running example. $newVPT$ containing data from new arrivals.

id	Parking spcs.	id	Tables	id	Quality	id	Popularity
r_2	18	r_4	10	r_2	3	r_2	5
r_4	3	r_5	2	r_4	1	r_4	2
r_5	2	r_2	0	r_5	1	r_5	2

skyline set S_k is flushed to be recalculated. D-FOPA initializes the historical entry H with the best values of each dimension between both VPT sets (i.e., $oldVPT$ and $newVPT$ in Table 3), then $H = < 25, 10, 5, 5 >$. Next, the first tuple from *Parking spaces* is read $r_3 = (25, 5, 4, 1)$; because there is no other entry in the current VPT set with the same value, r_3 is added to the seen entries P (line 13–15 of Algorithm 1) and since is the first iteration, it becomes the *most promising final object mpfo*. The entry r_3 has value 25 in *Parking spaces*, so the historical entry is not updated. Since no other entries have been read so far, r_3 is not pruned by D-FOPA and is added to the skyline set $S_{Parking\ spaces}$. Next, D-FOPA sums up the differences between values of r_3 and H, and computes f_{val}; $f_{val} = 10$, which in this first iteration will be the same as c_{val} (line 26 of Algorithm 1).

Subsequently, D-FOPA considers one of the tables where r_3 has not been found. Suppose the tables $oldVPT$ and $newVPT$ of the sensed attribute *Tables* are explored. The element $r_4 = (3, \mathbf{10}, 1, 2)$ is read because it has the best value among both $oldVPT$ and $newVPT$ of the sensed attribute $Tables^1$, r_4 is added to P, and since S_{Tables} is empty, r_4 is added to S_{Tables} as well. The historical entry H is not updated. Next c_{val} and f_{val} are computed, being $c_{val} = 30$ and $f_{val} = 10$. Since $c_{val} > f_{val}$, $mpfo$ remains the same. D-FOPA continues exploring the VPTs of the attribute $Tables$, and it reads entries $r1 = (20, \mathbf{8}, 3, 5)$ and $r6 = (5, \mathbf{8}, 5, 2)$; H is updated to $H = < 25, \mathbf{8}, 5, 5 >$, and r_1, r_6 are added to P. Since r_1 and r_6 are incomparable, they are evaluated with respect to the entries in S_{Tables} (specifically, with r_4); not being dominated, both entries r_1 and r_6 become part of the dimensional skyline S_{Tables}. Finally, c_{val} and f_{val} are calculated for r_1, resulting $c_{val} = 7$ and $f_{val} = 8$. Because $c_{val} < f_{val}$, $mpfo$ is now r_1. The same procedure is performed for r_6 where $c_{val} = 23$, since $c_{val} > f_{val}$, $mpfo$ remains the same.

Table 3. Running example. $oldVPT$ containing data from previous window slides and $newVPT$ containing data from new arrivals.

$oldVPT$	id	Parking spcs.	id	Tables	id	Quality	id	Popularity
	r_2	18	r_4	10	r_2	3	r_2	5
	r_4	3	r_5	2	r_4	1	r_4	2
	r_5	2	r_2	0	r_5	1	r_5	2

$newVPT$	id	Parking spcs.	id	Tables	id	Quality	id	Popularity
	r_3	25	r_1	8	r_6	5	r_1	5
	r_1	20	r_6	8	r_3	4	r_6	2
	r_6	5	r_3	5	r_1	3	r_3	1

Table 4. Running example. $oldVPT$ containing data from previous window slides and $newVPT$ containing data from new arrivals.

$oldVPT$	id	Parking spcs.	id	Tables	id	Quality	id	Popularity
	r_2	18	r_4	10	r_2	3	r_2	5
	r_4	3	r_5	2	r_4	1	r_4	2
	r_5	2	r_2	0	r_5	1	r_5	2
	r_3	25	r_1	8	r_6	5	r_1	5
	r_1	20	r_6	8	r_3	4	r_6	2
	r_6	5	r_3	5	r_1	3	r_3	1

$newVPT$	id	Parking spcs.	id	Tables	id	Quality	id	Popularity
	r_9	18	r_9	8	r_9	5	r_9	5
	r_8	12	r_8	8	r_7	5	r_7	3
	r_7	5	r_7	0	r_8	3	r_8	2

Given that the new *most promising final object* $mpfo$ is r_1, D-FOPA considers the tables where r_1 has not been found, like *Quality*. It proceeds to read entries r_6 and r_3, because these entries have already been seen, they will not be processed again. Now, D-FOPA reads entries $r_2 = (18, 0, \mathbf{3}, 5)$ and $r_1 = (20, 8, \mathbf{3}, 5)$; r_2 is added to the seen entries P. Given that r_1 dominates r_2, the former will not be compared against $S_{Quality}$. As the $mpfo$, i.e., r_1, is found in *Quality*,

---

[1] The best value corresponds to the greatest amount of tables available.

D-FOPA now examines another VPTs where r_1 has not been seen so far, e.g., *Popularity*. Then, entries r_1 and r_2 are read once again. Because $r_1 \in P$ and $r_2 \in P$, they will not be processed again. Since r_1 was found in *Quality*, D-FOPA now explores the only VPTs where it has not been seen, i.e., the *oldVPT* and *newVPT* of the sensed attribute *Parking spaces*, where the first entry read is r_1. The execution of the algorithm ends since *mpfo* has been found in all VPTs[2]. At the end of the execution, one last dominance check is made to obtain the actual skyline set. At this moment, the dimensional skylines are $S_{Parking\ spaces} = \{r_3\}$, $S_{Tables} = \{r_4, r_6, r_1\}$, $S_{Quality} = \{\}$ and $S_{Popularity} = \{\}$, now entries r_1, r_3, r_4, and r_6 are compared with each other, obtaining as a result that the skyline set comprises all four entries. Once the execution of this iteration terminates, the window slides one last time, here entries r_9, r_7, and r_8 are incorporated. The *oldVPT* and *newVPT* tables for this case are shown in Table 4. Similarly, D-FOPA will perform the same statements executed for sensed data in Table 3. The skyline set for this window slide is composed of: r_1, r_3, r_4, r_7, and r_9.

The time complexity of D-FOPA is as follows. Let n be the number of entries in a slided window. Let d be the number of attributes (i.e., dimensions) of these entries. It is known that for each dimension exists a VPT. Given this, the best case of D-FOPA occurs when the *final object* is found in the first position of all the VPTs. In this case the cost of finding the skyline set is to explore all dimensions d, i.e., $O(|d|)$. The worst case scenario for D-FOPA occurs when the *final object* is in the last position of all VPTs. Consequently, all VPTs must be traversed completely. In this last case, the cost is $O(n \cdot |d|)$. Now, let S be a data stream and v the sliding window size, it is said that the number of times the window must slide is:

$$number\ of\ slides = \left\lceil \frac{|S|}{v} \right\rceil \qquad (4)$$

As v decreases, the number of slided windows increases in order to cover S. If these iterations are incorporated to the previous calculations, the new time complexity order of D-FOPA in its best case is $O\left(|d| \cdot \left\lceil \frac{|S|}{v} \right\rceil\right)$; while $O\left(n \cdot |d| \cdot \left\lceil \frac{|S|}{v} \right\rceil\right)$ is D-FOPA worst case. In conclusion, the behavior of D-FOPA improves as the sliding window size increases. Therefore, the goal of our experimental study is to study the impact of the sliding window size in different scenarios.

4.2 Lazy+: The Modified Lazy Strategy

We extend the *lazy* algorithm from [15] with the following features:

1. In the original *lazy* algorithm, a sliding window time-based model was used. This was changed by the count-based model, in order to control the amount of data processed in the experimental study.

---

[2] The *mpfo* was found in *newVPT* and *oldVPT* in all dimensions, i.e., *Parking spaces*, *Tables*, *Quality* and *Popularity*.

2. In [15], an entry expiration time is known when an entry update occurs. Contrary, Lazy+ does not assume an expiration time for the entries; entries are expired when either an *expiration event*[3] is triggered, or a new version of the entry is received.
3. In Lazy+, all intermediate skyline sets are computed by FOPA [2].

Just like the original strategy, Lazy+ maintains two sets: DB_{sky} and DB_{rest}. DB_{sky} stores the skyline set with respect to the elements within the sliding window. DB_{rest} contains the elements that are not part of the skyline, but can become part of it in the future when an item in DB_{sky} expires. We illustrate how Lazy+ operates in our running example. At first, both $DB_{sky} = \{\}$ and $DB_{rest} = \{\}$, when the first restaurant $r_2 = (18, 0, 3, 5)$ arrives, it becomes part of DB_{sky} because it is empty. Then, $r_4 = (3, 10, 1, 2)$ is processed and compared to every restaurant in DB_{sky}. Since r_4 is not dominated by r_2, now $DB_{sky} = \{r_2, r_4\}$. Similarly to r_4, $r_5 = (3, 2, 1, 2)$ becomes part of the skyline and $DB_{sky} = \{r_2, r_4, r_5\}$. The window slides over the data stream to process the restaurants r_3, r_1, and r_6. As a result of comparing these restaurants with the ones currently in DB_{sky}, we obtain that $r_3 = (25, 5, 4, 1)$ belongs to the skyline, while r_5 (dominated by r_3) becomes part of DB_{rest}. The behavior is analogous with $r_1 = (20, 8, 3, 5)$ and r_2. Finally, $r_6 = (5, 8, 5, 2)$ is processed, and since it is not dominated, it is added to DB_{sky}. Now, $DB_{sky} = \{r_4, r_3, r_1, r_6\}$ and $DB_{rest} = \{r_2, r_5\}$. The window slides one last time over the data stream to process the remaining elements r_9, r_7, and r_8. Thus, $r_9 = (18, 0, 5, 5)$ is not dominated by any other restaurant in $DB_{sky} = \{r_4, r_3, r_1, r_6\}$, and r_9 is added to DB_{sky}. The restaurant $r_7 = (5, 8, 5, 3)$ dominates r_6, but it is not dominated by any other restaurant, i.e., r_4, r_3, or r_1; therefore, r_6 is expunged from DB_{sky} and r_7 is added to DB_{sky}. The restaurant $r_8 = (12, 4, 3, 2)$ is dominated by r_1, and r_8 is added to DB_{rest}. Finally, restaurants r_4, r_3, r_1, r_7, and r_9 belong to DB_{sky}, and restaurants r_2, r_5, r_6, and r_8 belong to DB_{rest}.

5   Experimental Study

The goal of the experimental study is to analyze the impact of the size of sensed datasets on our approach. We empirically compare D-FOPA and Lazy+, and analyze throughput and performance in terms of execution time and number of comparisons.

5.1   Experiment Configuration

Datasets and queries: We extended the data generator from Börzönyi *et al.* [5] to generate the synthetic sensed and static data used in our experiments. The extended generator is able to produce the arrival and expiration time of the sensed data entries following an Exponential distribution; consequently, the overall entry arrival follows a Poisson distribution. Different correlations between

---

[3] Expiration events are sent from sensor network.

data arrivals can be selected, e.g., anti-correlated, correlated, or uniform. We generate four datasets; each dataset simulates sensed data produced by a sensor network in a 10-minute interval. During this interval, arrivals and expiration occurred following the conditions: *(i)* an arrival may occur when an entry appears for the first time in the stream; *(ii)* an arrival may occur when an entry updates some of its attributes; *(iii)* an expiration happens when an *expiration event* is received; and *(iv)* an expiration also happens when an entry is updated. The static data is generated with the setup originally provided by the tool.

Evaluation Metrics: We measure performance and effectiveness. Performance is reported as the *average execution time*, the *average number of comparisons*, and *throughput*. The *average execution time* is computed as the average of the elapsed time, since the sliding window is filled until the response for that query is received. The *average number of comparisons* corresponds to the average of the number of comparisons performed during the execution of a skyline query. Finally, the *throughput* is calculated as the average of:

$$throughput = \frac{|skyline|}{execution\ time} \qquad (5)$$

for each slided window. Effectiveness is measured as the completeness and correctness of the response. An output is correct if all the output entries belong to the skyline of the data comprised in the slided windows. An answer is complete, if all the entries in the skyline of the data that comprised the slided windows are part of the output. The gold standard of our evaluation is computed using FOPA [2], i.e., FOPA computed the skyline for each slided window as if data were static. In all the experiments, D-FOPA and Lazy+ produce 100 % correct and complete responses.

Implementation: D-FOPA and Lazy+ were implemented in Python version 2.7.3. For D-FOPA, data was stored in Vertically Partitioned Tables, for Lazy+ data was stored into hash tables. All data was stored in main memory. Experiments were executed on Debian 3.2.60, architecture amd64, equipped with Intel(R) Core(TM) i7-3770 processors running at 3.40 GHz, and 8 GB of RAM.

Independent Studied Parameters: For each experiment, two simulations were executed using the same dataset. Thus, the behavior of D-FOPA and Lazy+ was evaluated under the same conditions. In order to prove that the algorithm is capable of handling sensed and static data, half of the dataset dimensions were generated as sensed attributes and the other half as static attributes.

Goals and Hypotheses of the Experiments: *Experiment 1.* The goal of this experiment is to study the impact of the sliding window size on the execution time. Our hypothesis HP_1 is that as the sliding window size increases, the algorithm runs fewer iterations; therefore, it decreases the overall execution time. Seven window sizes were defined, each representing a percentage of the dataset size, which is 500,000. The percentages used were as follows: 2 %, 4 %, 6 %, 8 %, 10 %, 12 % to 14 %.

Experiment 2. The goal of this experiment is to study the impact of the sensed data updates on the execution time. The hypothesis HP_2 is: given that the number of updates increases, the number of expiration will increase as well; thus, as a result the average execution time will be negatively affected. Three number of updates were defined: 3, 2, and 1 in a 10-minute time interval.

Experiment 3. The goal of this experiment is to study the impact of the number of dimensions on the execution time. As demonstrated in [2], the complexity of the query affects more the execution time than the size of the dataset (Hypothesis HP_3). Thus, we validate hypothesis HP_3. Three parameters for the number of dimensions were defined: 6, 8, and 10, where half of the dimensions are static, and the other half are sensed.

Experiment 4. The goal of this experiment is to study the impact of distribution of updates on execution time. Three parameters for the distribution of updates were used: uniform, correlated, and anti-correlated. Our hypothesis HP_4 is that if updates are distributed in an anti-correlated way, the result will be a skyline set with high variability, increasing in this way, the average execution time, the average number of comparisons and decreasing the throughput of the algorithm. Otherwise, if updates are distributed in a correlated way the skyline set will be less volatile. Thus, diminishing the average execution time, the average number of comparisons, and incrementing the throughput of the algorithm.

When evaluating a specific parameter, the other are defined with an average value as follows: size of the dataset 250, 000, number of updates 2, number of dimensions 8 (i.e., 4 sensed attributes and 4 static attributes), and a uniform distribution of updates.

5.2   Experiment Results

Experiment 1 (Sliding window size): The plots of Fig. 3 (where the X axis represents the sliding window size) allows us to conclude, that D-FOPA overcomes Lazy+ up to one order of magnitude in the following metrics: average execution time and average number of comparisons. The registered throughput for both algorithms shows that D-FOPA overcomes Lazy+ in at least 30 %. As a result it can be concluded that D-FOPA outperforms Lazy+ in all metrics as

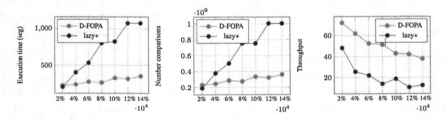

Fig. 3. Execution time, number comparisons and throughput in terms of the window size. The X and Y axis represents the sliding window size, and the evaluated metrics: average execution time, average number of comparisons, and throughput, respectively

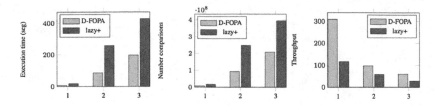

Fig. 4. Execution time, number comparisons and throughput in terms of the number of updates. The X and Y axis represents the number of updates, and the evaluated metrics: average execution time, average number of comparisons, and throughput, respectively

the size of the sliding window increases. This seems to validate our hypothesis HP_1, and can be explained by the fact that D-FOPA traverses the whole dataset in the worst case scenario, when the most promising final object is found at last. In contrast, Lazy+ needs to traverse the skyline set for each element arriving to the sliding window. Although the dataset is greater in size than the skyline set, D-FOPA needs to traverse it just once. Lazy+ instead, needs to traverse the skyline set n times for n entries in the sliding window. It tends to represent greater cost than the required for D-FOPA, as the sliding window increases.

Experiment 2 (Number of updates): As evidenced in plots of Fig. 4 (where the X axis represents the number of updates), both algorithms experienced a performance decrease with an average of three updates. As the number of updates increases, the number of expiration events increases as well. D-FOPA handles updates substituting the old value of the entry with the new one. Lazy+ instead, handles both events in the same manner: it expires the old entry value to insert it once again with the new value. The results obtained regarding the average execution time and number of comparisons between both algorithms, support the conclusion that D-FOPA overcomes Lazy+ in at least 33 % and 37 %, respectively. With respect to the throughput, D-FOPA achieves better results than Lazy+ regardless the number of updates. The techniques implemented by D-FOPA, support this behavior, where the number of comparisons is reduced. Additionally, D-FOPA expunges the expired values before calculating the skyline, while Lazy+ processes an expiration event when it is triggered, incurring in higher cost than D-FOPA. Thus, the hypothesis HP_2 holds: in each case the average execution time is increased; however, D-FOPA exhibits better performance than Lazy+.

Experiment 3 (Number of dimensions): The results presented in plots of Fig. 5 (where the X axis represents the number of dimensions) evidences that D-FOPA is capable of producing a response regardless the number of dimensions. The most expensive scenario for D-FOPA occurs with 10 dimensions, but even in this case, it is still possible to produce a response. If the number of dimensions is large, the complexity of the query increases (as demonstrated in [2]) augmenting

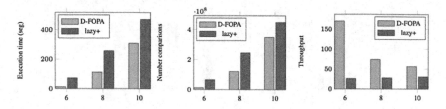

Fig. 5. Execution time, number comparisons and throughput in terms of the number of dimensions. The X and Y axis represents the number of dimensions, and the evaluated metrics: average execution time, average number of comparisons, and throughput, respectively

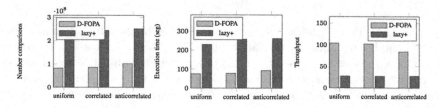

Fig. 6. Execution time, number comparisons and throughput in terms of the data distribution. The X and Y axis represents the update distribution, and the evaluated metrics: average execution time, average number of comparisons, and throughput, respectively

the number of comparisons between tuples as well. Since the most expensive task for both algorithms, D-FOPA and Lazy+, is the performance of comparisons, both algorithms experienced a decrease in general performance metrics, when the number of dimensions is 10. As evidenced D-FOPA is most affected by the number of dimensions than by the size of the dataset; this can be observed in the increases in terms of performance metrics suffered by the algorithm in Fig. 5. However, regarding to the average execution time and average number of comparisons between both algorithms, support the conclusion that D-FOPA overcomes Lazy+ in at least 17 % in average execution time, 19 % in average number of comparisons, and 86 % in throughput. Thus, the hypothesis HP_3 holds: the complexity of the query affects more the metrics being evaluated than the size of the dataset. However, D-FOPA outperforms Lazy+.

Experiment 4 (Distribution of updates): In the results presented in plots of Fig. 6 (where the X axis represents the used distribution of the entry updates) similar results are observed. With respect to the comparison of average execution time and number of comparisons between both algorithms, the results support the conclusion that D-FOPA overcomes Lazy+ by up to one order of magnitude. The throughput registered allows for the conclusion that D-FOPA overcomes Lazy+ in 300 % of the throughput. The behavior of D-FOPA in this case produces similar results for all evaluated metrics, regardless of the entry updates distribution. Given that D-FOPA does not maintain the skyline set, it is not affected by the variability suffered by the skyline set. Lazy+ produces similar results for all evaluated metrics. Nonetheless, the strategy of maintaining the skyline set, causes Lazy+ to incur in higher costs in calculations, regardless the distribution followed by the entry updates. Thus, these experimental results *falsifies* hypothesis HP_4: D-FOPA and Lazy+ are not affected by the entry updates distribution. This can be explained by the operations performed by both algorithms when an expiration event is triggered. These operations must be performed regardless the entry updates distribution.

6   Related Work

The problem of computing the skyline set over static data has been widely studied [2,5,8,9,11,13,14]. Initially Börzsönyi *et al.* [5] introduced two solutions. The first solution is based on the *divide and conquer* principle, where data is partitioned, so it can be processed in main memory, and then merged. Later, a final calculation is conducted to produce the whole skyline set. The second solution proposed by Börzsönyi *et al.* [5], is based on the *Block Nested Loop* (BNL) algorithm; where each one of the tuples is compared with the rest and the element is returned only if is not dominated by any other. Additional algorithms for computing the skyline set over static data are: *Sort-First-Skyline* [8], the bitmap based algorithm proposed by Tan *et al.* [14], and the approach proposed by Kossmann *et al.* [9] and Papadias *et al.* [11] that relies on the *Nearest-Neighbor Search* to obtain the skyline set.

Furthermore, approaches proposed by Balke *et al.* [4], Chen *et al.* [7], and Alvarado *et al.* [2] assume that static data is distributed on tables. Moreover, the experimental results reported by Alvarado *et al.* [2], suggest that FOPA outperforms existing approaches, i.e., in contrast to the algorithms IDSA and RSJFH [2], FOPA identifies the skyline set in only one phase or iteration. FOPA maintains information about the values seen so far, and uses this information to prune elements within a group read in a given iteration. This allows FOPA to assure that the skyline set has been found as soon as it finds the final object or a dimension is completely scanned. FOPA ensures completeness, while reducing the number of comparisons with respect to RSJFH and IDSA. However, FOPA is not able to manage dynamic data.

Recent investigations focus on processing queries maintaining the skyline set dynamically over data streams [6,10,12,13]. The main goal is to monitor skyline changes through a sliding window containing the most recent tuples. The research proposed by Lin and Wei [10] is performed under the precondition that the attributes have constraints. An algorithm called *Constrained Skyline Computing* is defined, with the objective of efficiently computing the skyline over sensed data with constrained attributes. Pu *et al.* [12] introduce a skyline query based on a user location, the skyline is separated by the nature of the attributes: i.e., static or sensed. Thus, this algorithm takes advantage of the fact that the static skyline should not be recalculated. However, only one sensed attribute is considered, i.e., the user location; therefore, multi-objective queries cannot be executed. In contrast, Chan and Yuan [6] present another approach through partially ordered domains and introduce several modifications of an existing algorithm named STARS with the objective to improve its performance. In [15], Tao and Papadias describe a solution that allows for the computation of the skyline set over a data stream, considering only the elements arriving into a time-based sliding window. The objective is to maintain the skyline continuously returning the changes that occur in the window. Particularly, two strategies are developed: a *lazy* strategy that delays all the computational work until the expiration of a tuple, and an *eager* strategy that takes advantage of pre-processing to minimize memory usage. Both frameworks maintain two sets, DB_{sky} and DB_{rest} where the tuples that belong or does not belong to the skyline are stored, respectively. When a tuple expires, any point in DB_{rest} can be promoted to become part of the skyline set. Although these approaches efficiently address the issue of computing the skyline, they are not able to deal with both static and sensed data.

7   Conclusions and Future Work

We have proposed a skyline based algorithm able to compute the skyline set over sensed and static data. We studied and analyzed the performance of D-FOPA with respect to Lazy+ on synthetic data; the experimental obtained results suggest that D-FOPA overcome Lazy+, under certain conditions. In the future, we plan to consider an approach able to independently compute the skyline for sensed and static data, and merge the resulting skylines once the sliding window has been traversed completely. We hypothesize that this improvement may result on speeding up the skyline computation.

References

1. Abadi, D., Marcus, A., Madden, S., Hollenbach, K.: Scalable semantic web data management using vertical partitioning. In: 32th International Conference of Very Large Data Bases (VLDB) (2007)
2. Alvarado, A., Baldizán, O., Vidal, M., Goncalves, M.: Fopa: A Final Object Pruning Algorithm to Efficiently Produce Skyline Points. Springer, Heidelberg (2013)
3. Balke, W., Guntzer, U.: Multi-objective query processing for database systems (2004)
4. Balke, W.-T., Güntzer, U., Zheng, J.X.: Efficient distributed skylining for web information systems. In: Bertino, E., Christodoulakis, S., Plexousakis, D., Christophides, V., Koubarakis, M., Böhm, K. (eds.) EDBT 2004. LNCS, vol. 2992, pp. 256–273. Springer, Heidelberg (2004)
5. Börzsönyi, S., Kossmann, D., Stocker, K.: The skyline operator. In: ICDE, pp. 334–348 (2001)
6. Fang, Y., Chan, C.-Y.: Efficient skyline maintenance for streaming data with partially-ordered domains. In: Kitagawa, H., Ishikawa, Y., Li, Q., Watanabe, C. (eds.) DASFAA 2010. LNCS, vol. 5981, pp. 322–336. Springer, Heidelberg (2010)

7. Chen, L., Gao, S., Anyanwu, K.: Efficiently evaluating skyline queries on RDF databases. In: Antoniou, G., Grobelnik, M., Simperl, E., Parsia, B., Plexousakis, D., De Leenheer, P., Pan, J. (eds.) ESWC 2011, Part II. LNCS, vol. 6644, pp. 123–138. Springer, Heidelberg (2011)
8. Chomicki, J., Godfrey, P., Gryz, J., Liang, D.: Skyline with presorting. In: International Conference on Data, Engineering, pp. 717–719 (2003)
9. Kossmann, D., Ramsak, F., Rost, S.: Shooting stars in the sky: an online algorithm for skyline queries. In: International Conference on Very Large Databases (VLDB), pp. 275–286 (2003)
10. Lin, J., Wei, J.: Constrained skyline computing over data streams. In: IEEE International Conference on e-Bussiness Engineering (2008)
11. Papadias, D., Tao, Y., Fu, G., Seeger, B.: An optimal and progressive algorithm for skyline queries, pp. 467–478. ACM SIGMOD (2003)
12. Pu, Q., Lbath, A., He, D.: Location based recommendation for mobile users using language model and skyline query. Inf. Technol. Comput. Sci. (IJITCS) **4**, 19–28 (2012)
13. Sarkas, N., Das, G., Koudas, N., Tung, A.K.: Categorical skylines for streaming data. In: SIGMOD, pp. 239–250 (2008)
14. Tan, K.N., Eng, P.K., Ooi, B.C.: Efficient proggressive skyline computation. In: International Conference on Very Large Databases (VLDB), pp. 301–310 (2001)
15. Tao, Y., Papadias, D.: Maintaining sliding window skylines on data streams. IEEE Trans. Knowl. Data Eng. **18**(3), 377–391 (2006)

GraphEvol: A Graph Evolution Technique for Web Service Composition

Alexandre Sawczuk da Silva[✉], Hui Ma, and Mengjie Zhang

School of Engineering and Computer Science,
Victoria University of Wellington, Wellington, New Zealand
{Alexandre.Sawczuk.Da.Silva,Hui.Ma,Mengjie.Zhang}@ecs.vuw.ac.nz

Abstract. Web service composition can be thought of as the combination of reusable functionality modules available over the network to create applications that accomplish more complex tasks, and Evolutionary Computation (EC) techniques have been applied with success to this problem. Genetic Programming (GP) is a traditionally employed EC technique in this domain, and it encodes solutions as trees instead of their natural Directed Acyclic Graph (DAG) form. This complicates the enforcement of dependencies between service nodes, which is much easier to accomplish in a DAG. To overcome this we propose GraphEvol, an evolutionary technique that uses DAGs directly to represent and evolve Web service composition solutions. GraphEvol is analogous to GP, but it implements the mutation and crossover operators differently. Experiments were carried out comparing GraphEvol with GP for a series of composition tasks, with results showing that GraphEvol solutions either match or surpass the quality of those obtained using GP, at the same time relying on a more intuitive representation.

1 Introduction

A Web service can be defined as a software module that accomplishes a specific task and that is made available for requests over the Internet [6]. The fundamental benefit of such modules is that they can be interwoven with new applications, preventing developers from rewriting functionality that has already been implemented. Service-Oriented Architecture (SOA) is a paradigm that expands on this idea, advocating that the main atomic components of a software system should be Web services, since this maximizes code reuse and information sharing [4,9]. As services are typically made available through standard interfaces, the possibility arises to create approaches capable of combining them automatically according to the final desired system, in a process known as *Web service composition* [8]. The objective of these approaches is to produce a workflow, i.e. a directed acyclic graph (DAG), stipulating the sequence in which each atomic service should be executed, as well as the output-input connections between services. Many approaches to Web service composition have been proposed in the literature, from variations on AI planning techniques [5,13] to the employment of integer linear programming solvers [1,15]. In particular, promising results

© Springer International Publishing Switzerland 2015
Q. Chen et al. (Eds.): DEXA 2015, Part II, LNCS 9262, pp. 134–142, 2015.
DOI: 10.1007/978-3-319-22852-5_12

have been achieved with the use of Evolutionary Computation (EC) techniques [12,14], though these approaches require the composition workflow to be encoded in linear or tree forms. The objective of this paper is to present and analyse *GraphEvol*, an evolutionary computation technique for Web service composition where each candidate is represented as a DAG and modified while remaining in that form. The main advantage of GraphEvol is that it represents each solution in an intuitive and direct way, allowing for a more powerful way of ensuring that solutions meet correctness constraints.

2   Related Work

Two of the pioneering GP composition approaches [2,10] use workflow constructs as the non-terminal tree nodes and Web service candidates as the terminal nodes, where workflow constructs represent the output-input connections between two services. An example of a tree generated by these techniques is shown in Fig. 1a. In addition to GP, an approach using PSO has also been shown to be a suitable method for fully automated Web service composition [11]. In this technique, candidates encoded as a vectors of weights (particles), each ranging from 0 to 1. The central idea of this technique is to utilise a greedy algorithm to extract non-redundant functional solutions from a graph showing all possible connections between all services, using the weights as guides that prioritise the choice of certain edges and nodes of the structure.

The biggest disadvantage concerning these two GP approaches is that their initial populations do not have a high degree of output-input matches, meaning that the solutions in their initial populations correspond to compositions that

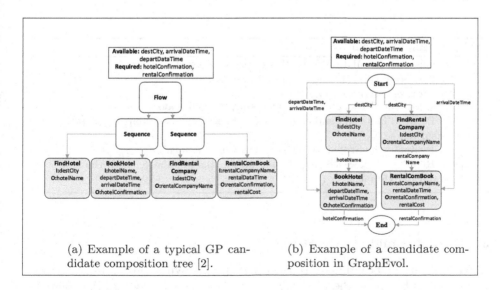

(a) Example of a typical GP candidate composition tree [2].

(b) Example of a candidate composition in GraphEvol.

Fig. 1. Examples of tree and graph-based composition representations.

are not fully executable due to a lack of correct service inputs. In the case of [2], even after the evolutionary process some of the solutions produced could not be fully executed. The use of a grammar by [10] does help with this issue, but at the cost of increased execution complexity and repeated tree adjustments. As for the PSO approach, its biggest drawback is that it requires the decoding a solution from a graph every time its fitness need to be calculated, causing the performance of the approach to decrease significantly as this graph grows [11]. These issues could be successfully addressed by representing solutions directly as graphs, since they do not require any form of decoding and can be easily built using algorithms that always generate fully executable compositions. This is the main motivation for the creation of the approach presented in this work.

3   GraphEvol

GraphEvol, an evolutionary computation approach proposed in this work, bears many similarities with the GP approaches discussed above. However, as opposed to representing candidate compositions as trees that correspond to underlying graph structures, GraphEvol represents them directly as graphs with Web service nodes. Figure 1b shows a graph representation example that is equivalent to the candidate tree shown in Fig. 1a. As a consequence of this direct graph representation, the mutation and crossover operators must be implemented differently, and so must the fitness function. Another important aspect of GraphEvol is that it uses a graph-building algorithm for creating new solutions, and for performing mutation and crossover. The general procedure for GraphEvol is shown in Algorithm 1, however each fundamental aspect of the proposed technique is explored in more detail in the following subsections.

Algorithm 1. GraphEvol algorithm.

---
1. Initialise the population using the graph building algorithm.
2. Evaluate the fitness of the initialised population.
while *max. generations not met* **do**
> **3.** Select the fittest graph candidates for reproduction.
> **4.** Perform mutation and crossover on the selected candidates, generating offspring.
> **5.** Evaluate the fitness of the new graph individuals.
> **6.** Replace the lowest-fitness individuals in the population with the new graph individuals.

3.1   Graph Building Algorithm

The initialisation of candidates is performed by employing a graph-building algorithm that is based on the planning graph approach discussed in the composition literature [5,13]. Algorithm 2 begins by adding the *start* node to the graph,

---

Algorithm 2. Generating a new candidate graph.

    Input   : I, O, $relevant$
    Output: candidate graph G
1: $start.outputs \leftarrow \{I\}$;
2: $end.inputs \leftarrow \{O\}$;
3: $G.edges \leftarrow \{\}$;
4: $G.nodes \leftarrow \{start\}$;
5: $seenNodes \leftarrow \{start\}$;
6: $currEndInputs \leftarrow \{start.outputs\}$;
7: $candidateList \leftarrow \texttt{findCands}(seenNodes, relevant)$;
8: **while** $end.inputs \not\sqsubseteq currentEndInputs$ **do**
9:     $cand \leftarrow candidateList.next()$;
10:    **if** $cand \notin seenNodes \wedge cand.inputs \sqsubseteq currEndInputs$ **then**
11:       $\texttt{connectNode}(cand, G)$;
12:       $currEndInputs \leftarrow currEndInputs \cup \{cand.outputs\}$;
13:       $seenNodes \leftarrow seenNodes \cup \{cand\}$;
14:       $candidateList \leftarrow candidateList \cup \texttt{findCands}(seenNodes, relevant)$;
15: $\texttt{connectNode}(end, G)$;
16: $\texttt{removeDangling}(G)$;
17: **return** G;

18: **Procedure** $\texttt{connectNode}(n, G)$
19:     $G.nodes \leftarrow G.nodes \cup \{n\}$;
20:     $inputsToFulfil \leftarrow \{cand.inputs\}$;
21:     **while** $|inputsToFulfil| > 0$ **do**
22:       $graphN \leftarrow G.nodes.next()$;
23:       **if** $|n.inputs \sqcap graphN.outputs| > 0$ **then**
24:         $inputsToFulfil \leftarrow$
            $inputsToFulfil - (n.inputs \sqcap graphN.outputs)$;
25:         $G.edges \leftarrow G.edges \cup \{graphN \rightarrow n\}$;

---

marking it as one of the *seenNodes*, and adding some initial candidates to the *candidateList* to be considered for connection. These candidates are identified using the *findCands* function, which discovers elements from the *relevant* set having at least some of their input satisfied by the nodes already in the graph (i.e. the *seenNodes*). Then the building process begins, continuing as long as the composition outputs represented in *end.inputs* have not been fulfilled by the currently available graph outputs in *currentEndInputs*. In this process, a candidate *cand* is selected at random from the *candidateList*. If it has not already been used in the graph and all of its inputs can be fulfilled by the currently available graph outputs, then it is connected to the graph using the *connectNode* function. This function identifies a random, minimal set of edges connecting the new node (n) to already existing nodes in the graph so that the inputs of this new node are fully satisfied.

After connecting *cand* to the graph, it is added to the set of *seenNodes* and the *candidateList* is updated to include services that may be fulfilled by the outputs of *cand*. Finally, once the composition's required output has been reached, the *end* node is connected. This particular graph building algorithm often results in *dangling nodes*, which are chains of nodes that are connected to the graph but whose output is not used to fulfil any other nodes. Because of this, a routine to remove such chains (*removeDangling*) is executed on the graph before the completed structure is returned.

3.2   Mutation

Intuitively, the mutation operator for a graph candidate implements the same idea of the corresponding tree operator [2]: a subpart of the candidate should be removed and replaced with a new randomly generated fragment, while maintaining the functional properties of the original subpart (i.e. correct output-input matches where the subpart connects with the main part of the candidate). The difference is that the multiple dependency points in a graph require more careful consideration. To perform a graph mutation, we begin by randomly selecting a node in the graph (excluding the end node) to act as the 'root' of the subpart to be replaced. Subsequently, all nodes that are directly or indirectly dependent on the outputs of this root are also identified, all the way to the end node, as shown in Fig. 2a. These nodes are removed from the graph, and all of its connections (edges) to the main part of the graph are severed. Finally, the incomplete graph is fed into Algorithm 2, but beginning execution from line 8. This completes the graph and results in an offspring with the same main part as its parent, but with a distinct subpart.

3.3   Crossover

The idea of *merging* and *extracting* graphs has been employed in the implementation of this operator to ensure that the correctness of connections between services is maintained. The basic intuition is to select two candidate graphs, merge them into a single structure, and then extract a new candidate out of this merged structure. The merging process is depicted in Fig. 2b, and consists of combining any two nodes that represent the same service into a single node, maintaining all original dependencies from both graphs but resulting in the presence of redundant nodes and edges. Once the merge has taken place, an offspring can be extracted from the structure to obtain a new non-redundant solution. A modified version of Algorithm 2 is used for this task, where instead of adding candidates to the *candidateList* from the entire set of *relevant* nodes, only nodes from the merged structure can be considered. In particular, whenever a node is added to the extracted solution graph, only the service nodes it connects to (through an outgoing edge) in the merged graph structure are added to the *candidateList*. Figure 2b highlights one of the possible solutions that could be extracted from the merged structure.

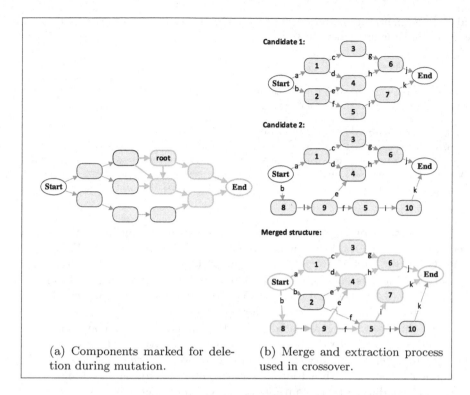

(a) Components marked for dele-   (b) Merge and extraction process
tion during mutation.   used in crossover.

Fig. 2. Examples of node and edge manipulation for genetic operators

3.4   Fitness Function

The fitness function employed in the evolutionary process seeks to produce solutions with the smallest possible number of service nodes and with the shortest possible paths from the start node to the end node. The rationale behind this decision is that it encourages features indicative of the quality of the overall composition [10]. The fitness function is formally described as follows:

$$fitness_i = \omega_1 \cdot \frac{1}{runPath_i} + \omega_2 \cdot \frac{1}{\#atomicService_i} \qquad (1)$$

where $\omega_1 + \omega_2 = 1$, $runPath_i$ is the longest path from the start node to the end node of a solution i (measured using a longest path algorithm), and $\#atomicService_i$ is the total number of service nodes included in a solution i.

4   Experiments

Experiments were carried out to compare the performance of GraphEvol against the traditional GP approach presented in [10]. The datasets employed in this experiments were OWL-S TC V2.2 [7] and WSC 2008 [3], both of which present

Table 1. Mean execution times, longest path lengths, and number of service nodes for each task in GraphEvol and GP [10].

Task	GraphEvol			GP Approach		
	Time (ms)	runPath	#atomService	Time (ms)	runPath	#atomService
OWL-S-1	469.60 ± 108.10	1.00 ± 0.0	1.00 ± 0.0	749.00 ± 364.10	1.00 ± 0.0	1.00 ± 0.0
OWL-S-2	326.73 ± 29.07	1.00 ± 0.0 ↓	1.00 ± 0.0 ↓	484.50 ± 139.20	2.00 ± 0.0	2.00 ± 0.0
OWL-S-3	517.17 ± 99.98	2.00 ± 0.0	2.00 ± 0.0	473.60 ± 76.19	2.00 ± 0.0	2.00 ± 0.0
OWL-S-4	674.23 ± 107.50	2.00 ± 0.0 ↓	4.00 ± 0.0 ↓	3010.20 ± 422.91	2.20 ± 0.40	5.70 ± 1.19
OWL-S-5	472.23 ± 46.33	1.00 ± 0.0	3.00 ± 0.0 ↓	1098.30 ± 240.72	1.00 ± 0.0	3.30 ± 0.46
WSC-1	699.43 ± 93.79	3.00 ± 0.0 ↓	10.00 ± 0.0 ↓	6919.70 ± 1612.99	6.00 ± 1.26	15.8 ± 5.71
WSC-2	734.63 ± 102.84	3.00 ± 0.0 ↓	5.00 ± 0.0 ↓	11137.20 ± 3106.75	3.50 ± 0.67	6.00 ± 0.89
WSC-5	918.40 ± 120.13	8.00 ± 0.0 ↓	20.00 ± 0.0 ↓	95390.20 ± 43521.30	9.20 ± 2.96	49.90 ± 16.84

service collections of varying sizes. Tasks 1–5, which are outlined in [10], were used when to test GraphEvol with the OWL-S TC dataset; tasks WSC 2008-1, WSC 2008-2, and WSC 2008-5 were used for testing with the WSC 2008 dataset. Testing was conducted using a personal computer with an Intel Core i7-4770 CPU (3.4 GHz), and 8 GB RAM. To match the GP approach, a population of 200 candidates was evolved during 20 generations for each composition task, and this process was repeated over 30 independent runs. The fitness function weights ω_1 and ω_2 were both set to 0.5, the mutation probability to 0.05, and the crossover probability to 0.5. Individuals were chosen for breeding using tournament selection with a tournament size of 2, mirroring the design of the experiment conducted by the authors of the GP approach [10].

Experiment results for GraphEvol are presented in columns 1–3 of Table 1, along with the previously published GP results. Column 1 displays the mean execution time for GraphEvol; column 2 presents the mean length of the longest path in a run's best solution; column 3 shows the mean number of service nodes included in a run's best solution. Means are accompanied by their respective standard deviations. Since both approaches were tested using the same datasets and tasks, as well as employing equivalent fitness functions during the evolutionary process, it is possible to perform a direct comparison on the longest path lengths and the overall number of nodes of the solutions produced. Unpaired t-tests at 0.05 significance level were conducted to verify whether there are statistically significant differences between the results produced by each technique. Such differences are denoted using ↓ for significantly lower results. The tests revealed that GraphEvol yielded solutions with equivalent or significantly smaller longest paths and numbers of nodes, that is, the quality of the solutions produced by GraphEvol always matched or surpassed that of the solutions produced by GP. These results establish the GraphEvol approach as a powerful alternative when performing fully automated Web service composition.

5   Conclusions

This work presented GraphEvol, an evolutionary computation technique aimed at performing fully automated Web service composition using graph representations for solutions, as opposed to encoding them into tree or vector representations. A graph building algorithm was proposed for generating the initial population, and variations of it were employed during the evolutionary process. The traditional mutation and crossover operations were modified to work with graph candidates, involving graph merging and traversal procedures. Finally, experiments were conducted comparing GraphEvol to an analogous GP approach. Results showed that the quality of the results produced by GraphEvol always matched or surpassed those produced by GP. A future work possibility is to extend this approach to optimise solutions based on Quality of Service attributes.

References

1. Ardagna, D., Pernici, B.: Adaptive service composition in flexible processes. IEEE Trans. Softw. Eng. **33**(6), 369–384 (2007)
2. Aversano, L., Di Penta, M., Taneja, K.: A genetic programming approach to support the design of service compositions. Int. J. Comput. Syst. Sci. Eng. **21**(4), 247–254 (2006)
3. Bansal, A., Blake, M.B., Kona, S., Bleul, S., Weise, T., Jaeger, M.C.: Wsc-08: continuing the web services challenge. In: 2008 10th IEEE Conference on E-Commerce Technology and the Fifth IEEE Conference on Enterprise Computing, E-Commerce and E-Services, pp. 351–354, IEEE (2008)
4. Channabasavaiah, K., Holley, K., Tuggle, E.: Migrating to a service-oriented architecture. In: IBM DeveloperWorks, 16 Dec 2003
5. Chen, M., Yan, Y.: Qos-aware service composition over graphplan through graph reachability. In: 2014 IEEE International Conference on Services Computing (SCC), pp. 544–551, IEEE (2014)
6. Gottschalk, K., Graham, S., Kreger, H., Snell, J.: Introduction to web services architecture. IBM Syst. J. **41**(2), 170–177 (2002)
7. Kuster, U., Konig-Ries, B., Krug, A.: Opossum-an online portal to collect and share sws descriptions. In: 2008 IEEE International Conference on Semantic Computing, pp. 480–481, IEEE (2008)
8. Milanovic, N., Malek, M.: Current solutions for web service composition. IEEE Int. Comput. **8**(6), 51–59 (2004)
9. Perrey, R., Lycett, M.: Service-oriented architecture. In: 2003 Symposium on Applications and the Internet Workshops, Proceedings, pp. 116–119, IEEE (2003)
10. Rodriguez-Mier, P., Mucientes, M., Lama, M., Couto, M.I.: Composition of web services through genetic programming. Evol. Intell. **3**(3–4), 171–186 (2010)
11. da Silva, A., Ma, H., Zhang, M.: A graph-based particle swarm optimisation approach to qos-aware web service composition and selection. In: 2014 IEEE Congress on Evolutionary Computation (CEC), pp. 3127–3134, July 2014
12. Su, K., Liangli, M., Xiaoming, G., Yufei, S.: An efficient parameter-adaptive genetic algorithm for service selection with end-to-end qos constraints. J. Comput. Inf. Syst. **10**(2), 581–588 (2014)
13. Wang, A., Ma, H., Zhang, M.: Genetic programming with greedy search for web service composition. In: Decker, H., Lhotská, L., Link, S., Basl, J., Tjoa, A.M. (eds.) DEXA 2013, Part II. LNCS, vol. 8056, pp. 9–17. Springer, Heidelberg (2013)

14. Wang, L., Shen, J., Yong, J.: A survey on bio-inspired algorithms for web service composition. In: 2012 IEEE 16th International Conference on Computer Supported Cooperative Work in Design (CSCWD), pp. 569–574, IEEE (2012)
15. Yoo, J.J.W., Kumara, S., Lee, D., Oh, S.C.: A web service composition framework using integer programming with non-functional objectives and constraints. Algorithms 1, 7 (2008)

Distributed, Parallel
and Cloud Databases

Can Data Integration Quality Be Enhanced on Multi-cloud Using SLA?

Daniel A.S. Carvalho[1], Plácido A. Souza Neto[3](✉), Genoveva Vargas-Solar[4],
Nadia Bennani[2], and Chirine Ghedira[1]

[1] MAGELLAN, IAE, Université Jean Moulin Lyon 3, Lyon, France
{daniel.carvalho,chirine.ghedira-guegan}@univ-lyon3.fr
[2] CNRS INSA-Lyon, LIRIS, UMR5205, Villeurbanne, France
nadia.bennani@insa-lyon.fr
[3] Instituto Federal do Rio Grande do Norte, Natal, Brazil
placido.neto@ifrn.edu.br
[4] CNRS, LIG-LAFMIA, Saint Martin D'Hères, France
genoveva.vargas@imag.fr

Abstract. This paper identifies trends and open issues regarding the use of SLA in data integration solutions on multi-cloud environments. Therefore it presents results of a Systematic Mapping [3] that analyzes the way SLA, data integration and multi-cloud environments are correlated in existing works. The main result is a classification scheme consisting of facets and dimensions namely (i) data integration environment (cloud; data warehouse; federated database; multi-cloud); (ii) data integration description (knowledge; metadata; schema); and (iii) data quality (confidentiality; privacy; security; SLA; data protection; data provenance). The proposed classification scheme is used to organize a collection of representative papers and discuss the numerical analysis about research trends in the domain.

Keywords: Systematic mapping · Service level agreement · Data integration · Multi-cloud environment

1 Introduction

The emergence of new architectures like the cloud opens new opportunities for data integration. The possibility of having unlimited access to cloud resources and the "pay as U go" model make it possible to change the hypothesis for processing big data collections. Instead of designing processes and algorithms taking into consideration limitations on resources availability, the cloud sets the focus on the economic cost implied when using resources and producing results.

Integrating and processing heterogeneous huge data collections (i.e., Big Data) calls for efficient methods for correlating, associating, and filtering them according to their "structural" characteristics (due to data variety) and their quality (veracity), e.g., trust, freshness, provenance, partial or total consistency. Existing data integration techniques must be revisited considering weakly

© Springer International Publishing Switzerland 2015
Q. Chen et al. (Eds.): DEXA 2015, Part II, LNCS 9262, pp. 145–152, 2015.
DOI: 10.1007/978-3-319-22852-5_13

curated and modeled data sets provided by different services under different quality conditions. Data integration can be done according to (i) quality of service (QoS) requirements expressed by their consumers and (ii) Service Level Agreements (SLA) exported by the cloud providers that host huge data collections and deliver resources for executing the associated management processes. Yet, it is not an easy task to completely enforce SLAs particularly because consumers use several cloud providers to store, integrate and process the data they require under the specific conditions they expect. For example, a major concern when integrating data from different sources (services) is privacy that can be associated to the conditions in which integrated data collections are built and shared [5]. Naturally, a collaboration between cloud providers becomes necessary [1] but this should be done in a user-friendly way, with some degree of transparency.

In this context, the main contribution of our work is a classification scheme of existing works fully or partially addressing the problem of integrating data in multi-cloud environments taking into consideration an extended form of SLA. The classification scheme results from applying the methodology defined in [3] called *systematic mapping*. It consists of dimensions clustered into facets in which publications (i.e., papers) are aggregated according to frequencies (i.e., number of published papers). According to the methodology, the study consists in five interdependent steps including (i) the definition of a research scope by defining research questions; (ii) retrieving candidate papers by querying different scientific databases (e.g. IEEE, Citeseer, DBLP); (iii) selecting relevant papers that can be used for answering the research questions by defining inclusion and exclusion criteria; (iv) defining a classification scheme by analyzing the abstracts of the selected papers to identify the terms to be used as dimensions for classifying the papers; (v) producing a systematic mapping by sorting papers according to the classification scheme.

The remainder of this paper is organized as follows. Section 2 describes our study of data integration perspectives and the evolution of the research works that address some aspects of the problem. Section 3 gives a quantitative analysis of our study and identifies open issues in the field. Section 4 concludes the paper and discusses future work with reference to the stated problem.

2   Data Integration Challenges: Classification Scheme

The aim of our bibliographic study using the systematic mapping methodology [3] is to (i) categorize and quantify the key contributions and the evolution of the research done on *SLA-guided data integration in a multi-cloud environment* and (ii) discover open issues and limitations of existing works. Our study is guided by three research questions:

RQ1: Which are the SLA measures that have been mostly applied in the cloud? This question identifies the type of properties used for characterizing and evaluating the services provided by different clouds.

RQ2: *How have published papers on data integration evolved towards cloud topics?* This question is devoted to identify the way data integration problems addressed in the literature started to include issues introduced by the cloud.

RQ3: *In which way and in which context has data integration been linked to Quality of Service (QoS) measures in the literature?* The objective of this question is to understand which QoS measures have been used for evaluating data integration and to determine the conditions in which specific measures are particularly used.

2.1 Searching and Screening Papers

According to our research questions and our expertise in data integration we chose a set of keywords to define a complex query to be used for retrieving papers from four target publication databases: IEEE[1], ACM[2], Science Direct[3] and CiteSeerX[4]. We used the following conjunctive and disjunctive general query which was completed with associated terms from a thesaurus and rewritten according to the expression rules of advanced queries in each database:

("Service level agreement" AND ("Data integration" OR "Database integration") AND ("Cloud" OR "Multi-cloud"))

We retrieved a total of 1832 publications. As a result of the filtering process proposed by the systematic mapping methodology [3] we excluded 1718 publications. The number of papers included for building the final collection were 114 publications[5].

2.2 Defining Classification Facets

We analyzed the titles and abstracts of the papers derived in the previous phase using information retrieval techniques to identify frequent terms. We used these terms for proposing a classification scheme consisting of three facets that group dimensions. The following lines define the facets and dimensions of the classification scheme we propose.

Data Integration Environment: This facet groups the dimensions that characterize the architectures used for delivering data integration services (*data warehouse* and *federated database*) and architectures used for deploying these services (*cloud* and *multi-cloud*).

---

[1] http://ieeexplore.ieee.org/.
[2] http://dl.acm.org/.
[3] http://www.sciencedirect.com/.
[4] http://citeseerx.ist.psu.edu/.
[5] List of references available in: https://github.com/danielboni/DEXA-2015-Can-Data-Integration-Quality-be-Enhanced-on-Multi-cloud-using-SLA.git.

Data Integration Description: This facet groups the dimensions describing the approaches used for describing the databases content in order to integrate them. Data integration can be done by using *meta-data, schema,* and *knowledge.*

Data Quality: This facet groups the dimensions representing data quality measures. Measures can be related directly to data for instance *confidentiality, privacy, security, protection and provenance* and to the conditions in which data is integrated and delivered (i.e., dimension *SLA*).

The original vision of our classification scheme is that of adding the notion of *quality* to data integration represented by the facets *data quality* and *SLA*. With these facets our classification scheme shows the aspects that must be considered when addressing data integration in the cloud taking into account (i) the quality of data, (ii) the systems that integrate data and (iii) the quality warranties that a data consumer can expect expressed in SLAs.

3   Quantitative Analysis

This section discusses the quantitative analysis presented in bubble charts that combine different facets. In order to observe the evolution of the publication trends we defined a time screen between the years 1998 and 2014 (see Fig. 1). SLA has emerged when Cloud issues started to be addressed around 2009. The number of publications has increased as cloud infrastructures have become more popular and accessible. It seems that data integration is an open issue when it is combined with SLA and cloud trends. Less recent papers seem to be devoted to the way data is described under schemata or knowledge representation strategies. This could be due to the fact that these strategies are consolidated today and to the emergence of NoSQL approaches with their schema-less philosophy [4].

We combined facets for answering the research questions proposed for guiding our study. The following lines discuss the answers.

RQ1: Which are the SLA measures that have been mostly applied in the cloud? The facets SLA expression, data integration description and contribution give elements for determining which SLA measures have been applied to the cloud (Fig. 2). The resulting bubble chart shows that most contributions propose SLA models and that *privacy* and *security* (11 papers - 9.65 %) are the most popular measures considered by SLA models for the cloud. These measures concern the network, information, data protection and confidentiality in the cloud. Most contributions propose SLA models (53 papers - 46.49 %) but some languages (8 papers - 7.02 %) have also emerged. *Data provenance* is also a measure that emerges but only in papers dealing with multi-cloud environments. Data integration is merely addressed by using schemata (12 papers - 10.53 %) and meta-data (4 papers - 3.51 %) particularly through models (34 papers - 29.82 %) and tools (25 papers - 21.93 %). Still, some works propose surveys (8 papers - 7.02 %).

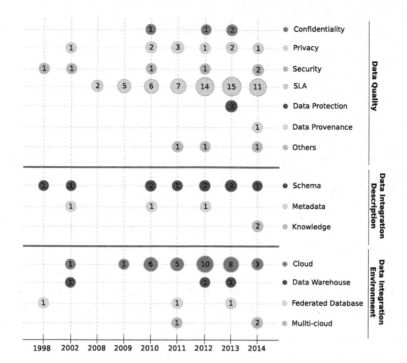

Fig. 1. Publications per year

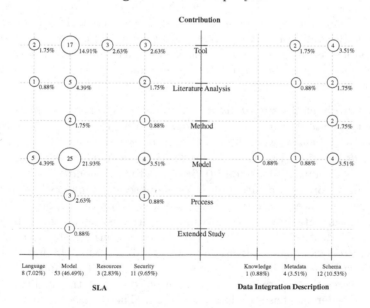

Fig. 2. Facets contribution, SLA and data integration description

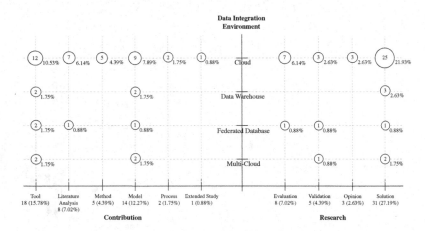

Fig. 3. Facets data integration environment, contribution and research

RQ2: How have published papers on data integration evolved towards cloud topics? Combining the facets data integration environment, contribution and research it is possible to observe the evolution of publications on data integration towards the cloud (Fig. 3). *Data warehouse* environments are the most common architecture. This can be explained by the increase of scientific and industrial applications needing to build integrated data sets for performing analysis and decision making tasks. The proposals are delivered as *models* (14 papers - 12.27 %) and *tools* (18 papers - 15.78 %) used for facilitating data integration, mostly done in the *cloud*. The most popular deployment environment of recent papers is the *cloud*. Given the importance and crucial need of data integration most papers present concrete solutions as algorithms, methods and systems (31 papers - 27.19 %).

RQ3: In which way and in which context has data integration been linked to QoS measures in the literature? We answered RQ3 by combining the facet *data quality* with the facets *data integration environment* and *data integration description* (Fig. 4). Data integration and QoS measures are associated within environments like cloud (9.68 %) and multi-cloud (4.39 %).

According to our quantitative analysis we observe that QoS has started to be considered for integrating data. The cloud is becoming a popular environment to perform data integration in which security issues are most frequently addressed. We identify a promising research area concerning the need of studying SLA which is currently addressed for the cloud as a whole [2] but that needs to be specialized for data integration aspects. Therefore, it is important to identify the measures that characterize the quality of data and the quality measures associated to different phases of data integration. These phases include selecting data services, retrieving data, integrating and correlating them and building a query result that can be eventually stored and that must be delivered. The data integration phases are implemented by greedy algorithms and generate interme-

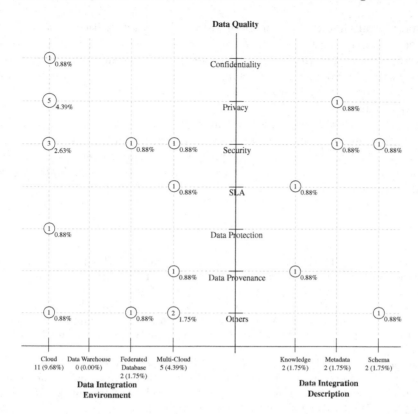

Fig. 4. Facets data quality, data integration environment and data integration description

diate data that can be stored for further use. Therefore they consume storage, computing, processing and communication resources that have an associated economic cost. These resources must ensure some QoS guarantees to data consumers. This problem seems to be open in the domain, and we believe that it must be part of a new vision of data integration. We believe that it is possible to add and enhance the quality of data integration by including SLAs.

4 Conclusion and Final Remarks

This paper introduces the challenge of integrating data from distributed data services deployed on different cloud providers guided by SLAs and user preferences statement. The data integration problem is stated as a continuous data provision problem that has associated SLAs and that uses techniques for ensuring different qualities of delivered data (freshness, precision, completeness). The problem statement was derived from a classification scheme that resulted from a study of existing publications identified by applying the systematic mapping method. Our contribution is the definition of a classification scheme that shows the aspects that characterize a modern vision of data integration done in multi-cloud environments and that can be enhanced by including SLAs in its process.

Current big data settings impose to consider SLA and different data delivery models. Given the volume and the complexity of query evaluation that includes steps that imply greedy computations, it is important to combine and revisit well-known solutions.

From the results of our systematic analysis, we identified trends and open issues in our research topic and proposed the general lines of an original data integration solution. We are also developing the strategies to better define a SLA extension and data consumers preferences description for guiding data integration in multi-cloud environments.

References

1. Hamze, M., Mbarek, N., Togni, O.: Self-establishing a service level agreement within autonomic cloud networking environment. In: 2014 IEEE Network Operations and Management Symposium (NOMS), pp. 1–4. IEEE (2014)
2. Pedrinaci, C., Cardoso, J., Leidig, T.: Linked USDL: a vocabulary for web-scale service trading. In: Presutti, V., d'Amato, C., Gandon, F., d'Aquin, M., Staab, S., Tordai, A. (eds.) ESWC 2014. LNCS, vol. 8465, pp. 68–82. Springer, Heidelberg (2014)
3. Petersen, K., Feldt, R., Mujtaba, S., Mattsson, M.: Systematic mapping studies in software engineering. In: Proceedings of the 12th International Conference on Evaluation and Assessment in Software Engineering, EASE'2008, pp. 68–77, British Computer Society, Swinton (2008)
4. Sadalage, P.J., Fowler, M.: NoSQL Distilled: A Brief Guide to the Emerging World of Polyglot Persistence. Pearson Education, Essex (2012)
5. Yau, S.S., Yin, Y.: A privacy preserving repository for data integration across data sharing services. IEEE Trans. Serv. Comput. 1(3), 130–140 (2008)

An Efficient Gear-Shifting Power-Proportional Distributed File System

Hieu Hanh Le$^{1(\boxtimes)}$, Satoshi Hikida^2, and Haruo Yokota^2

^{1} Center for Technology Innovation, R&D Group, Hitachi Ltd., Kanagawa, Japan
hanhlh@de.cs.titech.ac.jp
^{2} Department of Computer Science, Tokyo Institute of Technology,
Meguro, Tokyo, Japan
hikida@de.cs.titech.ac.jp, yokota@cs.titech.ac.jp

Abstract. Recently, power-aware distributed file systems for efficient big data processing have increasingly moved toward power proportional designs. However, inefficient gear-shifting in such systems is an important issue that can seriously degrade their performance. To address this issue, we propose and evaluate an efficient gear-shifting power proportional distributed file system. The proposed system utilizes flexible data placement that reduces the amount of reflected data and has an architecture that improves the metadata management to achieve high-efficiency gear-shifting. Extensive empirical experiments using actual machines based on the HDFS demonstrated that the proposed system gains up to 22 % better throughput-per-watt performance. Moreover, a suitable metadata management setting corresponding to the amount of data updated while in low gear is found from the experimental results.

1 Introduction

Commercial off-the-shelf-based distributed file systems (DFS) have been widely used for cloud applications for their fast deployment and easy scaling. Among these systems, power-aware DFS have increasingly moved toward power proportional designs [1]. To realize such systems, current data placement methods commonly divides the nodes into a set of small and separated groups [2–4]. These groups are then configured to operate in multiple "gears" where each gear contains a different number of groups, and offers a different level of parallelism and aggregate I/O throughput [2].

However, the current methods do not fully consider the effects of the reflection of updated data during gear-shifting on the performance. For example, in the morning, the system may have to update the datasets modified in a low gear while a subset of the nodes was powered off overnight. When the system moves to a higher gear to gain a better performance by reactivating inactive nodes, it must replicate the updated data to the reactivated nodes to share the load among all the active nodes for better performance. Inefficient reflection of updated data

---

This work was done when the author was at Tokyo Institute of Technology.

© Springer International Publishing Switzerland 2015
Q. Chen et al. (Eds.): DEXA 2015, Part II, LNCS 9262, pp. 153–161, 2015.
DOI: 10.1007/978-3-319-22852-5_14

with large amounts of retransferred data is believed to degrade the performance of such power proportional systems greatly during gear-shifting.

Moreover, metadata management in the DFS is believed to play an important role during gear-shifting because the metadata management will be more complex. In the low gear, the system generally creates log records specifying the locations of updated data. When changing to a higher gear, it must identify the replicated data from the log records, access their metadata, transfer the data to the appropriate nodes, and update the corresponding metadata for later references. Carrying out this process effectively with efficient distributed metadata management is vital in realizing power proportionality DFS.

To provide efficient gear-shifting for power proportional DFS, an integration of distributed metadata management and data placement is further important because they are so closely related to each other. By leveraging both actions, the carefully designed integration will greatly increase the efficiency of gear-shifting with less throughput performance degradation.

In this paper, we propose a novel DFS that efficiently combines both of our previous works, Accordion [5,7] and NDCouplingHDFS [6], to provide high throughput performance during gear-shifting. Although the amount of retransferred data is reduced in Accordion, efficient metadata management is required for better power proportional throughput performance. In the proposed system, this is achieved with support from NDCouplingHDFS, which distributes the metadata management cost efficiently to multiple nodes with small overhead.

The contributions of this paper are as follows.

- We propose a DFS for efficient gear-shifting to maintain high power proportional results during gear-shifting.
- We evaluate the effectiveness of the proposed system through empirical experiments. The experiments show that the proposed system gained up to 22 % better power proportional performance than the base system configured with Accordion and the default HDFS.
- It is observed that the proposed system gains better performance for large amounts of updated data under a heavy metadata load; and for small amounts of updated data under a light metadata load.

The remainder of this paper is organized as follows. The related work is reviewed in Sect. 2. The proposed system is described and evaluated in Sects. 3 and 4. The conclusions of this paper are discussed in Sect. 5.

2   Related Work

Rabbit [4] was the first method to provide power proportionality to an Hadoop Distributed File System (HDFS) by focusing on the read performance by utilizing an equal-work data layout policy based on data replication on organized nodes. Sierra [3] also organizes the replicas of the dataset such that each replica is stored in a group of nodes. However, Sierra differs from Rabbit in that each replica of the dataset is evenly distributed to all the nodes in each group.

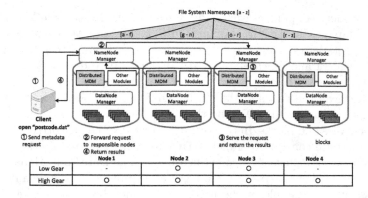

Fig. 1. The NDCoupling HDFS architecture and data flow.

We also proposed Accordion [5], a flexible data placement method based on data replication to reduce the amount of retransferred data during gear-shifting by differentially considering the locations of primary data. As the primary data are located at all nodes, when the modified dataset is updated (or appended) in low gear, part of the primary data in the updated dataset is already stored on the active nodes. Hence, only the remainder of the updated dataset, which should have been written to the deactivated nodes, must be retransferred when the system shifts to a higher gear. Although Accordion improves the power proportional performance by 30 % compared with Rabbit and Sierra [7], the metadata management in this study is still constrained by the single metadata node in the system.

We also previously presented an architecture known as NDCouplingHDFS [6] to facilitate the efficient reflection of updated data in a power proportional HDFS. NDCouplingHDFS focuses on coupled metadata management and data management on each HDFS node, which reduces the cost of managing the metadata generated during changes in the system configuration. However, the effect of NDCouplingHDFS on the throughput performance during gear-shifting was not considered in the earlier study.

3   System Description

We confidently expect that utilizing distributed metadata management, NDCouplingHDFS can improve further the effectiveness of Accordion, because of the very close relationship between metadata management and data placement. In this section, we describes the NDCouplingHDFS architecture, the Accordion data placement then presents the updated data reflection process.

3.1   NDCouplingHDFS Architecture

In this paper, because we focus on the locality of metadata management for improving the efficiency of reflecting the updated data in gear-shifting, we have applied equivalent coupling as each node contains both NameNode Manager and DataNode Manager. In the NDCouplingHDFS, NameNode Manager includes the distributed metadata management (Distributed MDM) and other modules such as Data Placement and Block Mapping, as at the NameNode in a default HDFS. The difference from a default HDFS is that the namespace of the file system is divided among all the nodes in the cluster while taking locality into consideration. The local Distributed MDM and the Block Mapping only manage the metadata for local files and blocks. In this system, we utilize the Fat-Btree [8] method, which is an update-conscious parallel B-Tree structure to maintain the metadata of the system whose efficiency was verified in [6]. The DataNode Manager module at each node is the DataNode Manager at DataNode in the default HDFS. Figure 1 shows an example of the architecture and the data flow of NDCouplingHDFS in the four-node system. This system operates in two gears; the Low Gear requires two active nodes Node 2 and Node 3 and the High Gear requires all four active nodes.

3.2 Accordion Data Placement

Accordion [7] is designed to provide power proportionality in distributed file systems that use commodity computer servers such as the HDFS or the Google File System. In Accordion, the files are divided into a large number of blocks and a number of replicas of each data block are distributed among the nodes of the cluster. Like other approaches, Accordion aims to control the power consumption of the system by dividing the nodes into several separate groups. An Accordion-based system can then operate in a multiple-gear mode where higher gears have more groups of nodes. In Accordion, the nodes are arranged geometrically in a horizontal array because the nodes that belong to lower groups are bounded by the nodes of higher groups.

At first, the primary data in the dataset are distributed to all the nodes in the system. This means that each node stores the same amount of primary data. Then, starting with the highest group, the data stored in this group are replicated to the next lower group. To guarantee the data reliability in the lowest gear, the chained declustering policy is applied to the smallest group. Each node replicates its data to its neighbor node, which guarantees that all of the data in the dataset are replicated in the two neighbor nodes. In the example in Fig. 1, all the data from Node 1 and Node 4 are replicated to Node 2 and Node 3 accordingly. Then, the data of Node 2 are replicated to Node 3 and vice versa.

3.3 Gear Controller

For easy implementation, there is one master Gear controller at a node, which is assumed to be always active and is responsible for any request related to controlling the gear of the system from the administrator such as down gear or up gear. Here, the master Gear controller will communicate with other Gear controllers to fulfill requests. Other approaches such as allowing any Gear controller among the nodes of the lowest group to be the master Gear controller are possible.

3.4   Updated Data Reflection Process

In this section, we refer to Fig. 2 and describe the behavior of the proposed system in serving data update requests in low gear and reflecting the updated data when the system changes to a higher gear by reactivating a subset of nodes. In the default HDFS, basically all the operations are similar; however, because there is only a single NameNode that is in charge of metadata management, all the metadata operations are processed at the NameNode.

Step 1: Issue a gear change command. When receiving the gear-shifting command from the administrator, the Gear controller at the master node (Node 2) sends the command to the Gear controllers at all other active nodes.

Step 2: Issue update metadata commands. After receiving the commands, active nodes will respond according to their roles. Nonoffload nodes that are not affected by the gear-shifting simply delay I/O requests from the clients. Offload nodes (node Node 2 and Node 3) that store the updated data issue the update metadata command to the Metadata Management.

Step 3: Transfer updated metadata. The Metadata Management modules that receive the command check the log files and transfer only the changed metadata to the intended nodes specified in the log files. When the updated metadata transfers have finished, both the offload nodes and the intended nodes are ready to process the I/O requests from the clients, including any requests queued during the data reflection process, and send a "finished" indication to the Gear controller at the master node. After gathering all the finished indications from offload nodes, the Gear controllers forward this indication to all the nonoffload nodes. When the nonoffload nodes receive the indication, they are ready to process I/O requests from clients. Concurrently, in the background, the updated data reflection process continues with Step 4.

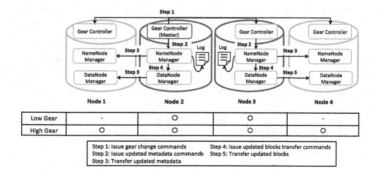

Fig. 2. Flow of the updated data reflection process.

Step 4: Issue block transfer commands. Next, the Metadata Management searches the log records for updated file blocks and issues block transfer commands with pairs of blocks and intended node identifiers to the local DataNode Manager. After each **heartbeat_interval**, the DataNode Manager receives a command and transfers the blocks to the intended nodes.

Step 5: Transfer updated blocks. When the DataNode Manager receives the command issued by the Metadata Management, for better efficiency it sends the blocks to the intended nodes in a batch manner, which is called **batch transfer method**. When the DataNode Manager knows all the blocks it must transfer, the cost of opening a new network connection can be reduced by sending all the relevant blocks through a single network connection. The current implementation of the HDFS requires opening a new connection for each block.

After receiving the updated data, the DataNode Managers at reactivated nodes (Node 1 and Node 4) notify the newly arrived data information to the responsible Metadata Management as in the default HDFS.

4   Experimental Evaluation

We conducted an empirical experiment using actual machines to verify the efficiency of the proposed system described in Sect. 3 during gear-shifting. For the evaluation, we chose a system that deploys Accordion with the default HDFS architecture as the base system.

4.1   Experimental Method

The workloads generated were close to the actual operation of multiple-gear DFS like the HDFS. We assumed that initially the file system was operated in a High gear and stored an initial dataset. Then, the system shifted to a Low gear for a specified power proportional service agreement. During this period, this dataset was updated as new files were appended from the clients. Here, the dataset that contains all these new files is called the update dataset. Next, the system was shifted to the High gear to satisfy the higher throughput performance on reading the whole dataset from the clients. At this time, the system must serve read requests from the client while performing updated data reflection in the background. As we focused on the applications on the DFS like the HDFS, we chose the method of updating the dataset as appending new files to the dataset and the method for reading the dataset as scanning all the files in the dataset. The sizes of the reading dataset, which includes both the initial and the update dataset, is fixed to 26880 [MB]. The sizes of the initial and the updated dataset, which are used in the evaluation, are varied as in Table 2.

4.2   Framework of the Experiments

Our test-bed for the experiments comprised dozens of commodity nodes based on the HDFS. We were focused on energy-aware commodity systems so we used low

Table 1. Node specification

CPU	TM8600 1.0 GHz
Memory	DRAM 4 GB
NIC	1000 Mbps
OS	Linux 3.0 64 bit
Java	JDK-1.7.0

Table 2. Sizes of datasets [MB]

Configuration	Without update	Small	Medium	Large
Updated dataset	0	4480	8960	13440
Initial dataset	26880	22400	17920	13440

power consumption ASUS Eeebox EB1007 machines, the specifications for which are provided in Table 1. In the base system, there is one NameNode besides the cluster of DataNodes in each gear (2, 8 and 20 nodes). However, in the proposed system, the numbers of nodes in each gear are limited to 2, 8 and 20.

4.3 Experimental Results

In this section, experimental results are reported for four cases relating to the load of the metadata, in which the size of the files is set to 64 MB, 16 MB, 4 MB, and 1 MB. In this experiment, because of the same data placements in both the base system and the proposed system, the sizes of the reflected data are the same. However, because of the file size, the number of blocks varies between the systems, hence the cost for metadata management changes. Table 3 describes the file size and the number of blocks for the four cases Light, Medium–Light, Medium–Heavy, and Heavy.

Figures 3(a) shows the experimental results of the average throughput-per-watt when the system changes from Gear 2 to Gear 3 with scanning the dataset in four configurations: Without Update, Small, Medium, and Large configurations. Note that in the Without Update configuration, the performance of the scanning dataset workload was not affected by the update data reflection process. The effectiveness of NDCouplingHDFS is confirmed as the throughput-per-watt performance of the proposed system was better than the base system in the Medium and Large configurations, by approximately 10 % and 22 %, respectively. This is explained by the advantages of the coupling architecture in NDCouplingHDFS employed in the proposed system compared with the normal HDFS in the base system. However, we can also see from Fig. 3(a) that NDCouplingHDFS was

Table 3. Settings

Case	File size [MB]	#blocks
Light	64	420
Medium–Light	16	1680
Medium–Heavy	4	6720
Heavy	1	26880

Table 4. Numbers of updated blocks

Case	Without update	Small	Medium	Large
Light	0	42	84	126
Medium–Light	0	168	336	504
Medium–Heavy	0	672	1344	2016
Heavy	0	2688	5376	8064

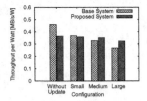 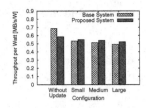 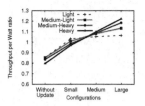

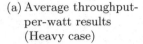

(a) Average throughput-per-watt results (Heavy case)

(b) Average throughput-per-watt results (low case)

(c) Throughput-per-watt ratio of the proposed system and the base system

Fig. 3. Experiment results

not effective in Without Update and Small configuration. We suggest that the default HDFS showed better results in such situations because the cost of reflecting updated data is small. Table 4 shows the number of updated blocks of four cases in all configurations.

Figure 3(b) show the results for the average throughput-per-watt results in Light case. In contrast with the Heavy case, the effectiveness of the coupling architecture NDCouplingHDFS in the proposed system was difficult to observe as the throughput-per-watt performance of the proposed system was at most 6 % better than the performance of the base system (Large configuration). The main reason was the light load in metadata management as the numbers of updated blocks in the Light cases were extremely small showing in Table 4.

The effect of the metadata load on the effectiveness of the proposed system is evaluated by comparing the throughput-per-watt results of the proposed system with the base system for our four cases (Light, Medium–Light, Medium–Heavy, Heavy). Figure 3(c) presents the throughput-per-watt results of the proposed system divided by those for the base system. We observe that the effect of the metadata load (number of blocks) depends on the amount of updated data. In the Small configuration, the smaller number of blocks is better for the proposed system as the Light case gave the best result. However, in the Medium and Large configurations in Heavy case where the amount of updated data is greater, the heavier metadata load cases delivered the better result.

5   Conclusion and Future Work

We have demonstrated that the distributed metadata management in NDCouplingHDFS is effective for smooth gear-shifting in systems applying the Accordion data placement. Our experiments showed that the proposed system integrating Accordion and NDCouplingHDFS could achieve up to 22% better power proportionality than the base system configured with Accordion and the default HDFS. The efficiency of the proposed system is expected to be increasing when the metadata load is higher and the amount of updated data is larger. We would like to further evaluate the proposed system with other data placement methods.

References

1. André, B.L., Urs, H.: The case for energy-proportional computing. Computer **40**, 33–37 (2007)
2. Charles, W., Mathew, O., Jin, Q., Andy, W.A.I., Peter, R., Geoff, K.: PARAID: a gear-shifting power-aware RAID. Trans. Storage **3**(3), 13:1–13:33 (2007)
3. Thereska, E., Donnelly, A., Narayanan, D.: Sierra: practical power-proportionality for data center storage. In: Proceedings of 6th European Conference on Computer Systems, EuroSys 2011, pp. 169–182. ACM (2011)
4. Amur, H., Cipar, J., Gupta, V., Ganger, G.R., Kozuch, M.A., Schwan, K.: Robust and flexible power-proportional storage. In: Proceeding of the 1st ACM Symposium on Cloud Computing, SoCC 2010, pp. 217–228 (2010)
5. Le, H.H., Hikida, S., Yokota, H.: Efficient gear-shifting for a power-proportional distributed data-placement method. In: Proceedings 2013 IEEE International Conference on Big Data, pp. 76–84. IEEE (2013)
6. Le, H.H., Hikida, S., Yokota, H.: NDCouplingHDFS: a coupling architecture for a power-proportional hadoop distributed file system. IEICE Trans. Inf. Syst. **E97–D**(2), 213–222 (2014)
7. Le, H.H., Hikida, S., Yokota, H.: Accordion: an efficient gear-shifting for a power-proportional distributed data-placement method. IEICE Trans. Inf. Syst. **E98–D**(5), 1013–1026 (2015)
8. Yokota, H., Kanemasa, Y., Miyazaki, J.: Fat-Btree: an update conscious parallel directory structure. In: Proceedings of the 15th International Conference on Data Engineering, ICDE 1999, pp. 448–457. IEEE Computer Society (1999)

Highly Efficient Parallel Framework:
A Divide-and-Conquer Approach

Takaya Kawakatsu[1]([⊠]), Akira Kinoshita[1], Atsuhiro Takasu[2], and Jun Adachi[2]

[1] The University of Tokyo, 2-1-2 Hitotsubashi, Chiyoda, Tokyo, Japan
{kat,kinoshita}@nii.ac.jp
[2] National Institute of Informatics, 2-1-2 Hitotsubashi, Chiyoda, Tokyo, Japan
{takasu,adachi}@nii.ac.jp

Abstract. Coupling a database and a parallel-programming framework reduces the I/O overhead between them. However, there will be serious issues such as memory bandwidth limitations, load imbalances, and race conditions. Existing frameworks such as MapReduce do not resolve these problems because they adopt *flat* parallelization, i.e., partitioning a task without regard to its structure. In this paper, we propose a recursive divide-and-conquer-based method for spatial databases which supports high-throughput machine learning. Our approach uses a tree-based task structure, which improves the reference locality, and load balancing is realized by setting the grain size of tasks dynamically. Race conditions are also avoided. We applied our method to the task of learning a hierarchical Poisson mixture model. The results show that our approach achieves strong scalability and robustness against load-imbalanced datasets.

Keywords: Divide-and-conquer · Load balancing · EM algorithm · Parallelization

1 Introduction

Today, there is growing interest in the data mining of huge datasets against a backdrop of inexpensive but high-performance parallel-computing environments, such as shared memory architectures and distributed memory clusters. Among the many approaches to data mining, statistical approaches that involve mixture models such as the Gaussian mixture model (GMM) and the Poisson mixture model (PMM) are the most popular techniques. The approaches learn statistical parameters such as the mean and covariance of the probabilistic model that the observation data follow. In these models, the density function of the probabilistic model is described by a weighted linear sum of various component distributions. The observation data are represented by multidimensional numeric tuples called feature vectors, and items can be classified as coming from the component that is most likely to have generated the record. This is a simple example of multi-class classification. Mixture models are particularly versatile, with their applications including object recognition [1], speech recognition [2,3], and unreliable sensor

© Springer International Publishing Switzerland 2015
Q. Chen et al. (Eds.): DEXA 2015, Part II, LNCS 9262, pp. 162–176, 2015.
DOI: 10.1007/978-3-319-22852-5_15

networks [4]. Recent research on traffic incident detection has attracted much attention, and mixture models have been utilized in this work. For example, in our research group, Kinoshita et al. proposed a new detection technique using a hierarchical PMM [5]. When applying mixture models, it is necessary to know their parameter values such as weight and mean. The expectation–maximization (EM) algorithm is a popular method for the parameter estimation of mixture models. The EM algorithm comprises an *E-step* and an *M-step*. In the E-step, the algorithm calculates a posterior probability for each pair of an observation item and a mixture component that indicates how likely it is that the item was generated by the component. In the M-step, the posteriors are summed to estimate revised parameter values. The E-step is then repeated using the revised parameter values. This iteration continues until the likelihood function, which indicates how well the model regenerates the dataset, converges to a maximum.

The quantity of observation data can be extremely large and it is usual to involve a database management system, with the separate estimation program publishing the *select* queries and retrieving the required records. However, such a process can be quite inefficient, because an I/O bottleneck between the database and machine-learning program is inevitable, even though the retrieved data is eventually discarded. Coupling the database and parallel-processing frameworks will relieve the I/O overhead between them. However, there will be serious issues to consider, such as memory bandwidth limitations, load imbalances among the computing nodes, and possible race conditions. Fortunately, modern computers can have huge memories, with more than tens of gigabytes per core. Therefore, the memory size limitation may not continue to be such a severe problem by itself. Nevertheless, a memory bandwidth problem is inevitable, for which the reference locality becomes very important. In some probabilistic models, the computation cost per observation item is not uniform and load imbalances may occur, causing serious performance degradation. For example, the hierarchical PMM involves two-level parameter estimation, namely means common to all road segments and weights particular to each road segment. The segments may have variable amounts of probe-car records. To avoid load imbalance, dynamic scheduling may be a solution. However, the traditional first-in first-out (FIFO) schedulers supported by existing frameworks such as OpenMP and MapReduce adopt *flat* partitioning tactics, which divide and allocate the data to each core without careful consideration of the hierarchical structure of the data. If the dataset has a tree structure, such as road segments stored in an R-tree-based spatial database, a flat scheduler cannot optimize the grain size for each subtask or a flat data structure such as a matrix. Another characteristic problem in parallel processing is the existence of race conditions. If several cores attempt to read from or write to the same memory address at the same time, the integrity of the calculation can be compromised. A memory lock such as a semaphore or mutex can be used to avoid such cases, but that is computation intensive.

In this paper, we propose a novel solution for these issues by bringing together a work-stealing scheduler and a divide-and-conquer-based parallel framework for machine learning. The divide-and-conquer method gives good reference locality, and is beneficial for dynamic load balancing, because of its dynamic optimization of the grain size of each subtask allocated to cores. In addition, it dramatically reduces the number of critical sections, making it more efficient than standard shared-memory-based frameworks. We demonstrate its superior scalability and robustness against load-imbalanced datasets by experiment and then consider its future development.

2 The EM Algorithm

2.1 Hierarchical PMM

Kinoshita et al. used a PMM to detect traffic incidents [5]. They assumed that probe-car records follow a hierarchical PMM and that each road segment has its own local parameters. In their model, the probability of a single record x in a segment s is described as follows:

$$p(x|s) = \sum_{k=1}^{K} w_{sk} \mathcal{P}(x; \mu_k), \tag{1}$$

where w_{sk} is the k-th Poisson distribution's weight in segment s, and μ_k is the k-th Poisson distribution's mean. w_{sk} is particular to the segment, whereas μ_k is common to all segments. The log-likelihood $\mathcal{L}(\theta)$ is defined as follows:

$$\mathcal{L}(\theta) = \sum_{s=1}^{S} \sum_{n=1}^{N_s} \log \sum_{k=1}^{K} w_{sk} \mathcal{P}(x_{sn}; \mu_k), \tag{2}$$

where N_s is the number of records in segment s. As for GMMs, we must calculate the posterior probability q_{snk} that the n-th record x_{sn} in segment s is generated by the k-th Poisson distribution for all pairs of (s, n, k) in each E-step:

$$q_{snk} = \frac{w_{sk} \mathcal{P}(x_{sn}; \mu_k)}{\sum_{k=1}^{K} w_{sk} \mathcal{P}(x_{sn}; \mu_k)}. \tag{3}$$

In the M-step, the weight w_{sk} and mean μ_k are recalculated:

$$\hat{w}'_{sk} = \frac{1}{N_s} \sum_{n=1}^{N_s} q_{snk}, \tag{4}$$

$$\hat{\mu}'_k = \frac{\sum_{s=1}^{S} \sum_{n=1}^{N_s} q_{snk} x_{sn}}{\sum_{s=1}^{S} \sum_{n=1}^{N_s} q_{snk}}. \tag{5}$$

Each road segment has a massive number of records, but the actual number varies greatly from segment to segment, implying that we must take measures against load imbalance.

2.2   Parallelization Approach

To parallelize the EM algorithm, we must divide each E-step and M-step into a number of subtasks and assign them to processing cores. For GMMs, the axis of the observation dataset n and the axis of mixtures k are available for this division. In general, the number of observation items is much greater than the number of mixtures. As shown in Fig. 1, we can divide each step into arbitrary rectangles. However, for maximum reference locality, we should divide them into squares or into rectangles that are as nearly square as possible. This is because the total numbers of observation items and mixture parameters to load for calculating the subtable will be smallest for a square, thereby giving an optimal cache hit ratio. Therefore, we must divide the n axis into more segments than the k axis.

The EM algorithm includes a summation process for estimating parameter values. When we parallelize the M-step, protection against race hazards between cores working with each other is a critical problem. In general, such a condition is resolved by using mutual exclusion via semaphores or mutexes. However, those solutions are not recommended because they involve much blocking time. The best solution is for each core to use its own local memory for the summation of an intermediate result, thereby computing that sum independently. A single core can then retrieve these intermediate results and compute the total.

3   Related Work

3.1   Shared Memory

Nonuniform memory access (NUMA) is a shared-memory architecture in which a processor core and local memory form a pair called a *node*, with nodes being assembled into a mother board. The overall memory address space is sequential, enabling each node to access another node's local memory as for its own local memory. In a NUMA architecture, all cores share the threads of a process and a thread may be executed in any core. Because threads share a common memory address space, we can easily write NUMA programs without explicit message passing among cores. However, noting that NUMA nodes will be interconnected via a bus, a processor will access nonlocal memory more slowly than it can access

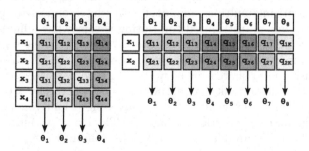

Fig. 1. Cubic division of the posterior table.

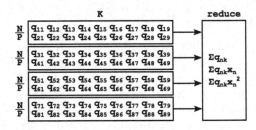

Fig. 2. Kwedlo's parallel EM implementation with P processors.

its own local memory. Therefore, programmers should contrive to maximize the reference locality so that a core accesses nonlocal memory as little as possible.

Kwedlo proposed a parallel implementation of the EM algorithm for GMMs in a NUMA architecture [6]. He employed OpenMP[1] as a framework for shared memory parallelization and proposed two techniques to improve the reference locality, namely summation buffering and thread fixing. When we parallelize an EM algorithm, we normally partition the observation dataset into a number of subsets and assign them to worker threads. In the E-step, there is no dependence between observation items, which means that race conditions will never arise. However, parameter recalculation in the M-step requires summations over the observation data, for which we must avoid race conditions. When several cores "simultaneously" read from and write to a common memory address, the result of the parameter update may differ from the correct result. To avoid this condition, Kwedlo arranged a separate array for each core, and each core calculated a partial sum for its own array. The partial sums are then integrated by a single thread at the end of the M-step. OpenMP provides safe parallel reduction operations via its `reduction` clause. Using the EM algorithm on a mixture model requires simultaneous multiple reductions. Reduction into an array is therefore needed. However, the `reduction` keyword does not support arrays. Accordingly, Kwedlo implemented his own array buffering as shown in Fig. 2.

3.2 Distributed Shared Memory

Distributed shared memory (or simply distributed memory) is a parallel machine architecture comprising multiple computer *nodes* interconnected via a network. The meaning of *node* in distributed computing is somewhat different from that used in the description of the NUMA architecture. For distributed memory, each node is a shared-memory motherboard usually adopting the NUMA architecture. That is, the distributed shared memory architecture has at least two levels of memory hierarchy, namely internode and intranode hierarchy. Communication among nodes is realized by explicit message-passing primitives, with internode latency being greater than intranode latency. Therefore, we must consider the reference locality more carefully than in NUMA computing.

---

[1] http://www.openmp.org.

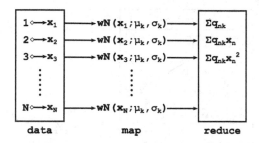

Fig. 3. The MapReduce programming model.

The message-passing-based parallel-programming model is applicable to both the distributed and shared-memory computers. It rarely depends on the detailed machine architecture, with its application being wider than that of the shared memory programming model. For this reason, the message-passing model is more popular with programmers using high-level languages. In high-level languages, a programmer can use highly abstract concurrent-execution statements such as `fork` and `join`, with the background scheduler then assigning those tasks to the appropriate cores automatically. Hadoop[2] is one of the most popular frameworks for distributed programming. Its MapReduce programming model [7], shown in Fig. 3, is a well-known parallel-computing model. MapReduce offers simple but powerful abstraction. However, it is too abstract to allow control of the reference locality. In MapReduce, the dependency on `map` is removed completely, and its performance tuning is quite difficult. Therefore, MapReduce might be convenient but can have poor throughput. Its handling of hard-disk I/O latency is another principal reason for its poor performance [8].

Currently, there are several parallel-processing frameworks based on message passing, such as Spark [8], Piccolo [9], and GraphLab [10]. Spark is a framework that aligns data in memory to reduce hard-disk I/O overheads, and provides its own distributed, immutable collection framework called the resilient distributed dataset (RDD)[11], even while making use of Hadoop functions. Because RDD data are in memory, Spark runs faster than Hadoop MapReduce, which must read observation data from a hard disk each time they are required. Piccolo is a distributed in-memory hash-table framework that runs parallel applications with high efficiency, similarly to RDD. GraphLab is a distributed machine-learning framework in which a programmer describes calculations and data flows by using directed graphs. Each node behaves as if it were a local MapReduce, and the many local MapReduce instances run concurrently as a whole.

In the world of low-level programming, the mainstream approach is hybrid parallelization that uses a thread implementation such as OpenMP inside the node and a message-passing framework such as MPI[3] between the nodes. Yang et al. [12] proposed a parallel implementation of the EM algorithm using hybrid

---

[2] http://hadoop.apache.org.
[3] http://www.mpi-forum.org.

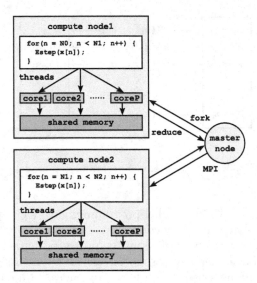

Fig. 4. Hybrid parallel implementation.

parallelization that "divides and conquers" the observation data and integrates subsummations in parallel. The implementation has the hierarchical structure shown in Fig. 4. First, the master node assigns observation data subsets to each calculation node by passing messages. Next, each calculation node partitions its subset into smaller subsets and assigns them to worker threads running on NUMA cores. At the end of an M-step, each calculation node calculates its local subsummation and the master node collects these to calculate the total sum. Their approach achieves good locality because it does not exchange any data until the intranode calculation is completely finished. Moreover, because the summation follows the divide-and-conquer approach, the overhead of the serial summation part is substantially concealed. The authors insist that they achieved a reduction in context-switching overheads because of the benefits of static scheduling. However, this claim is not essential for probabilistic models other than GMMs, as we discuss below.

4   Load Balancing

4.1   FIFO Scheduler

For the EM algorithm applied to the hierarchical PMM described in Sect. 2.1, each segment may have a different quantity of probe-car data. Therefore, load balancing across the cores will be required if idle cores are to be avoided. However, predicting the processing time for each core is difficult and we must therefore introduce the concept of dynamic load balancing. OpenMP has support for dynamic load balancing via the `dynamic` directive, which equalizes the load as shown in Fig. 5. The program is partitioned into fine-grained subtasks that are

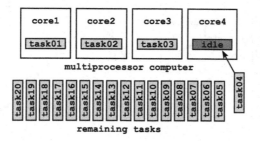

Fig. 5. Dynamic load balancing.

appended to a FIFO queue in the shared memory. When a core becomes idle, it polls the next task from the FIFO queue and starts to process it. While this may appear to be a good solution, it can cause fine-grained task switching and this overhead will reduce the benefits of using a multicore computer. To avoid this, each subtask must be defined carefully, because if any tasks are too large, dynamic load balancing will be less effective.

4.2   Work-Stealing Scheduler

Mohr et al. [13] introduced a dynamic load-balancing technique called *lazy task creation*. They focused on the property of the divide-and-conquer algorithm that the size of each node's task can be halved. As shown in Fig. 6, the lazy task creation scheduler first divides a task into a minimum number of subtasks. When a core later becomes idle, the scheduler divides another core's task into halves and reassigns one half to the idle core. In this way, tasks are always divided into the smallest number of subtasks required, thereby achieving a minimum number of context switches. Many existing systems, including state-of-the-art languages and frameworks such as Cilk [14] and Intel Threading Building Blocks[4], support a *work-stealing* scheduler similar to lazy task creation. A typical work-stealing scheduler is constructed using existing multithreading libraries such as pthreads. With respect to the order of task execution, there are two styles, namely breadth first and depth first. Cilk employs the breadth-first execution approach, whereas MassiveThreads [15] adopts the depth-first approach. In a depth-aware scheduler, each *worker thread* is fixed to a core and has its own local *deque* or double-ended queue to hold its remaining tasks that are ready to execute. Tasks are popped by the owner core and executed one by one in first-in last-out (FILO) fashion. When there are no idling cores, each worker thread behaves as a serial processing program independent of the others. When a core becomes idle, i.e., its deque is empty, the core checks other workers' deques until it finds a remaining task and steals it in FIFO fashion, as shown in Fig. 6. With the divide-and-conquer algorithm, a remaining task may create new tasks by executing `fork` statements. Such tasks are pushed into the deque in FILO

---

[4] https://www.threadingbuildingblocks.org.

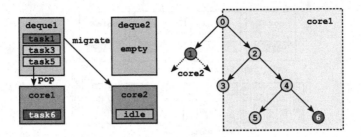

Fig. 6. Work-stealing scheduler.

fashion, with each deque playing the role of a call stack. Therefore, the remaining tasks in each deque are stored in ascending age order. That is, an idle core steals the oldest remaining task, which will be the one nearest to the root of the divide-and-conquer tree. This is the reason for the work-stealing strategy to be able to achieve a minimum number of context switches. Its overhead will be much smaller than that of pure round-robin scheduling.

4.3   Divide and Conquer EM

To utilize the work-stealing scheduler, we need to rewrite the EM algorithm in divide-and-conquer form. This is not difficult because we only have to divide the observation data recursively and compute the posterior probabilities in parallel in the E-step. In the M-step, we partition the dataset again to recalculate the parameter values and sum recursively as shown in Algorithm 1. This version of the EM algorithm is effective at dynamic load balancing, while simultaneously achieving good reference locality. The divided observation-data subset and its counterpart are aligned closely in the memory address space. In particular, with distributed memory, the number of internode I/O transactions will be reduced dramatically. In addition, because the summation is performed in parallel, the overhead of the sequential parts of the program will be reduced. Note that the observation dataset should not be divided into single data points, because the calculation cost per observation item is not great, and too many conditional branch instructions may disturb the instruction pipeline. Therefore, we introduce a *grain size* as the minimum size for a subset of observation data. Once the recursively divided subsets reach the grain size, no further division is allowed. We do not discuss selection methods for the grain size here. However, it should be a multiple of the length of the SIMD vector.

The divide-and-conquer solution has another important merit, namely the avoidance of race hazards. Because there are no critical sections, there is no need for mutual exclusion, which makes much faster execution possible. On the other hand, the divide-and-conquer solution may have the disadvantage of requiring excessive memory. There is also the possibility that the second half of a divided subset is processed on the same core as the first half. In this case, the no-race condition would occur even if the arrays sum_1 and sum_2 were not allocated individually. Therefore, we investigated two versions of our EM implementation, namely *byneed* and *strict*, as explained in the next section.

Algorithm 1. Proposed EM algorithm for a hierarchical PMM.

Input: x_{sn}
Output: w_s, μ_{sk}

repeat
    for each segment s do
        Estep($x_1, .., x_N$)
    end for
    for each segment s do
        Mstep($x_1, .., x_N$)
    end for
until likelihood converges
procedure ESTEP(s, chunk of x_{snk})
    if chunk size is large then
        Estep(1st half of chunk)
        Estep(2nd half of chunk)
    else
        for each item x_{sn} do
            calculate q_{snk}
        end for
    end if
end procedure

procedure MSTEP(s, chunk of x_{snk})
    if chunk size is large then
        sum_1 = Mstep(1st half of chunk)
        sum_2 = Mstep(2nd half of chunk)
return $sum_1 + sum_2$
    else
        $sum^0 = [0, 0, .., 0]$ for all k
        $sum^1 = [0, 0, .., 0]$ for all k
        for each item x_n do
            for each class k do
                $sum_k^0 = sum_k^0 + q_{snk}$
                $sum_k^1 = sum_k^1 + q_{snk}x_{sn}$
            end for
        end forreturn sum^0, sum^1
    end if
end procedure

4.4 Joint Spatial Database

Among the many approaches to spatial databases, R-tree [16] is one of the most popular algorithms for the construction of a tree-based index. When a range query is accepted, the database searches recursively and finds child nodes whose rectangle intersects with the range query. When the search algorithm reaches a leaf node, it retrieves the several items of data whose rectangle intersects with the query and are contained in the leaf node. The found data are gathered into an array and, when the dataset becomes too large, serialized into a suitable file format. However, we know that the serialized file will be used only once and discarded after the machine-learning process finishes, which is inefficient. An I/O bottleneck between the database and the learning framework will be inevitable. Even though the spatial query processing is parallelized, its benefit will not be realized. Fortunately, a typical spatial database adopts a tree-based structure, and we conceived a joint tree-based scheduler that undertakes range query processing at the internal nodes and machine learning at the leaf nodes, as shown in Algorithm 2.

5 Experimental Results

To demonstrate the superior scalability and robustness against load imbalance of our divide-and-conquer-based EM algorithm, we rewrote the EM algorithms

Algorithm 2. Range query handler at the leaf node.

    procedure HANDLER(segment position r, segment s)
        if Estep **then**
            ESTEP(s, records in s)
        else
            MSTEP(s, records in s)
        end if
    end procedure

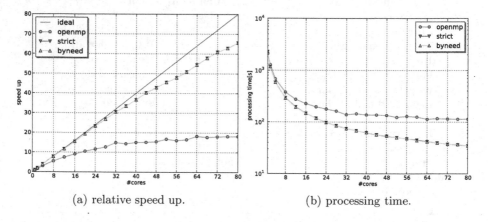

(a) relative speed up.   (b) processing time.

Fig. 7. Strong scalability on a hierarchical PMM.

for the hierarchical PMM in divide-and-conquer fashion using the programming language Chapel[5]. We adopted MassiveThreads [15] as Chapel's tasking layer that implements work stealing, and used the `cobegin` statement to parallelize the divide-and-conquer algorithm dynamically. To make the maximum use of dynamic load balancing, we tested two different versions. The version named *byneed* does not allocate memory for the second half of divided subtasks until it has migrated to another core, whereas the *strict* version always allocates in advance, regardless of whether the second half will actually migrate to another core. We also implemented an OpenMP version to retest the FIFO scheduling strategy. Unfortunately, MassiveThreads does not support a distributed memory environment. Therefore, these experiments involved a NUMA computer, which had 80 cores with eight NUMA nodes, as described in Table 1.

We conducted three kinds of measurement, namely the strong and weak scalability of the parallel divide-and-conquer EM algorithm as a standalone unit and the strong scalability of the algorithm with a range query on an R-tree-based spatial database implemented in Chapel. The database stores each road segment with its spatial location represented by a simple two-dimensional rectangle. For the experiments, we used randomly generated datasets for the hierarchical PMM. The dataset for measuring the strong scalability had 4,096 road segments and

---

[5] http://chapel.cray.com.

Table 1. Experimental environment.

Processor	Model	Intel Xeon E7-4870
	Freq.	2.395 GHz
	Cores	80 (40 + 40)
Cache	L1d	32 kB
	L2	256 kB
	L3	30,720 kB

16,777,216 probe-car records, with each record involving two dimensions and generated by eight Poisson mixtures. The number of records in each road segment followed a uniform distribution. The dataset for measuring the weak scalability had two different versions, one with a fixed number of segments regardless of the number of cores and the second having a number of segments proportional to the number of cores. The first version had total 4,096 segments and the second version had 64 segments per core. For both versions, the number of records in each segment followed a Gaussian distribution, with each core processing 31,768 or 131,072 two-dimensional records in total. The dataset for measuring with the spatial database had 8,192 segments and 20,971,520 two-dimensional probe-car records, retrieved by a parallel range query. In our experiments, each dimension of a feature vector was formatted as an eight-byte unsigned integer and the grain size was set to 1,024 for the first and second measurement and to 256 for the third measurement.

Figure 7 shows the strong scalability and processing times for 20 iterations on the first measurement. It achieved a near-linear speedup from 1 to 40 cores, and scaled linearly to 80 cores, although with a slightly lower growth rate. The reason why there was no significant difference between the *byneed* and *strict* versions is that the tasks often migrated to another core. In contrast, the OpenMP version

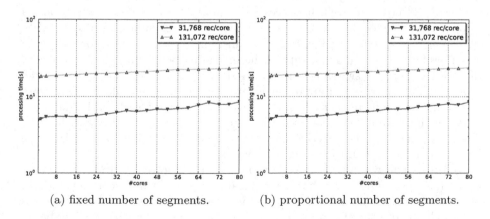

(a) fixed number of segments.   (b) proportional number of segments.

Fig. 8. Weak scalability on a hierarchical PMM.

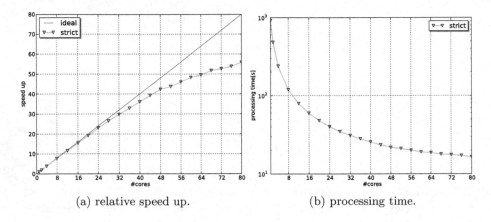

(a) relative speed up.   (b) processing time.

Fig. 9. Strong scalability on a dataset retrieved by a range query.

did not speed up to more than 18 times. Figure 8 shows the processing times for 20 iterations on the weak-scaling datasets. As shown in both graphs, the computational efficiency fell slightly as the number of cores increased. Note that the graphs almost exactly match each other, with the number of segments per core having little influence. Figure 9 shows the strong scalability and processing times for 20 iterations on the third measurement. It achieved near-linear speedup.

6   Discussion

Our divide-and-conquer-based EM algorithm achieved good performance for the cases of both strong and weak scalability. The number of probe-car records in each road segment was determined by a random function. That is testament to the high efficiency and robustness of the divide-and-conquer approach and the work-stealing strategy. In contrast, the flat-scheduling implementation using OpenMP did not scale well. It processed at most 18 times faster, which points to poor reference locality. In the OpenMP version, each subtask processed only one observation item by default and the memory access pattern for each core was therefore noncontinuous, with cache-hit ratios being quite low. Moreover, the overhead of task switching will have reduced the throughput of the parallel processing. In contrast, our approach ensured that each core could access at least 16 kB of continuous memory. The coupling of an R-tree-based spatial database with our divide-and-conquer-based EM algorithm also achieved fine scalability. That is evidence of the effective integration of the schedulers for the database and the parallel machine-learning framework.

In this paper, we tested our approach only for a hierarchical PMM. However, we would expect our approach to be applicable to any machine-learning method. In future work, we intend to evaluate more algorithms to prove the correctness of this claim. In other future work, we will develop a new work-stealing-based scheduler that supports a distributed memory cluster. However, when running a program in a distributed environment, the latency of communication between nodes may be a serious bottleneck. We must make the communication sufficiently coarse grained to reduce the overhead but the subtasks need to be finely "divided and conquered." To resolve this apparent contradiction, we must make use of the property of the divide-and-conquer algorithm that automatically optimizes the grain sizes for the task and the data. We are currently investigating a new memory model that is optimized automatically in conjunction with the scheduler. Existing implementations of a distributed memory model do not work together with the scheduler, which is inefficient. In many cases, the memory access pattern is closely related to the computation, particularly for the divide-and-conquer approach. Therefore, we should connect the memory management system to the scheduler to optimize the communication behavior. By utilizing the properties of the divide-and-conquer approach, we aim to construct a suitable distributed framework.

7 Conclusion

We have investigated a divide-and-conquer-based parallelization strategy for machine-learning algorithms and tree-based databases. Our approach not only reduces context switching dramatically, but also enables efficient dynamic load balancing. In addition, the divide-and-conquer algorithm greatly improves the reference locality, thereby improving cache-hit ratios and reducing internode I/O bottlenecks. Furthermore, because the summation operation is also divided, there is no possibility of race hazards, enabling our approach to achieve superior scalability without requiring mutual-exclusion techniques such as semaphores or mutexes. We tested the scalability of our approach on an 80-core NUMA machine and showed that the divide-and-conquer solution was superior to flat-scheduled parallelization.

Acknowledgment. This work was supported by the CPS-IIP Project (http://www.cps.nii.ac.jp) in the research promotion program for national challenges "Research and development for the realization of next-generation IT platforms" of the Ministry of Education, Culture, Sports, Science and Technology, Japan. The environment on which we conducted our experiment was provided by Assistant Prof. Hajime Imura at the Meme Media Laboratory, Hokkaido University.

References

1. Stauffer, C., Grimson, W.E.L.: Adaptive background mixture models for real-time tracking. In: IEEE Computer Society Conference on Computer Vision and Pattern Recognition, Jun 1999
2. Miura, K., Noguchi, H., Kawaguchi, H., Yoshimoto, M.: A low memory bandwidth gaussian mixture model (GMM) processor for 20,000-word real-time speech recognition FPGA system. In: International Conference on ICECE Technology, Dec 2008

3. Gupta, K., Owens, J.D.: Three-layer optimizations for fast GMM computations on GPU-like parallel processors. In: IEEE Workshop on Automatic Speech Recognition & Understanding, Dec 2009
4. Pereira, S.S., Lopez-Valcarce, R., Pages-Zamora, A.: A diffusion-based EM algorithm for distributed estimation in unreliable sensor networks. IEEE Signal Process. Lett. **20**(6), 595–598 (2013)
5. Kinoshita, A., Takasu, A., Adachi, J.: Traffic incident detection using probabilistic topic model. In: Proceedings of the Workshops of the EDBT/ICDT 2014 Joint Conference, Mar 2014
6. Kwedlo, W.: A parallel EM algorithm for Gaussian mixture models implemented on a NUMA system using OpenMP. In: 2014 22nd Euromicro International Conference on Parallel, Distributed and Network-Based Processing (PDP), Feb 2014
7. Dean, J., Ghemawat, S.: MapReduce: simplified data processing on large clusters. In: Proceedings of the 6th Conference on Symposium on Opearting Systems Design & Implementation, vol. 6, Dec 2004
8. Zaharia, M., Chowdhury, M., Franklin, M.J., Shenker, S., Stoica, I.: Spark: cluster computing with working sets. In: Proceedings of the 2nd USENIX Conference on Hot Topics in Cloud Computing, Jun 2010
9. Power, R., Li, J.: Piccolo: building fast, distributed programs with partitioned tables. In: Proceedings of the 9th USENIX Conference on Operating Systems Design and Implementation, Oct 2010
10. Low, Y., Bickson, D., Gonzalez, J., Guestrin, C., Kyrola, A., Hellerstein, J.M.: Distributed GraphLab: a framework for machine learning and data mining in the cloud. In: Proceedings of the VLDB Endowment, Apr 2012
11. Zaharia, M., Chowdhury, M., Das, T., Dave, A., Ma, J., MacCauley, M., Franklin, M.J., Shenker, S., Stoica, I.: Resilient distributed datasets: a fault-tolerant abstraction for in-memory cluster computing. In: Proceedings of the 9th USENIX conference on Networked Systems Design and Implementation, Apr 2012
12. Yang, R., Xiong, T., Chen, T., Huang, Z., Feng, S.: DISTRIM: parallel GMM learning on multicore cluster. In: 2012 IEEE International Conference on Computer Science and Automation Engineering (CSAE), May 2012
13. Mohr, E., Kranz, D.A., Halstead, Jr., R.H.: Lazy task creation: a technique for increasing the granularity of parallel programs. In: Proceedings of the 1990 ACM Conference on LISP and Functional Programming, May 1990
14. Blumofe, R.D., Joerg, C.F., Kuszmaul, B.C., Leiserson, C.E., Randall, K.H., Zhou, Y.: Cilk: an efficient multithreaded runtime system. In: Proceedings of the Fifth ACM SIGPLAN Symposium on Principles and Practice of Parallel Programming, Aug 1995
15. Nakashima, J., Nakatani, S., Taura, K.: Design and implementation of a customizable work stealing scheduler. In: 3rd International Workshop on Runtime and Operating Systems for Supercomputers, Jun 2013
16. Guttman, A.: R-trees: a dynamic index structure for spatial searching. In: Proceedings of the 1984 ACM SIGMOD International Conference on Management of Data, Jun 1984

Ontology-Driven Data Partitioning and Recovery for Flexible Query Answering

Lena Wiese [✉]

Institute of Computer Science, University of Göttingen, Goldschmidtstraße 7,
37077 Göttingen, Germany
lena.wiese@uni-goettingen.de

Abstract. Flexible Query Answering helps users find relevant information to their queries even if no exactly matching answers can be found in a database system. However, relaxing query conditions at runtime is inherently slow and does not scale as the data set grows. In this paper we propose a method to partition the data by using an ontology that semantically guides the query relaxation. Moreover, if several different partitioning strategies are applied in parallel, a lookup table is maintained in order to recover the ontology-driven partitioning in case of data loss or server failure. We tested performance of the partitioning and recovery strategy with a distributed SAP HANA database.

Keywords: Query relaxation · Anti-Instantiation · Recovery · Distributed database

1 Introduction

When storing large-scale data sets in distributed database systems, these data sets are usually *partitioned* into smaller subsets and these subsets are distributed over several database servers. When answering queries in such a distributed database system, it might be necessary to contact several servers to collect matching data records. It is hence worthwhile to improve *data locality* of the partitioning approach such that the amount of servers involved in answering a single query is reduced. In this paper we improve data locality for a partitioning that can be used in an *intelligent query answering* system. These intelligent query answering mechanisms are increasingly important to find relevant answers to user queries. Flexible (or cooperative) query answering systems help users of a database system find answers related to his original query in case the original query cannot be answered exactly. *Semantic* techniques rely on taxonomies (or ontologies) to replace some values in a query by others that are closely related according the taxonomy. This can be achieved by techniques of *query relaxation* – and in particular *query generalization*: the user query is rewritten into a weaker, more general version to allow for related answers.

In this paper we make the following contributions:

© Springer International Publishing Switzerland 2015
Q. Chen et al. (Eds.): DEXA 2015, Part II, LNCS 9262, pp. 177–191, 2015.
DOI: 10.1007/978-3-319-22852-5_16

- we introduce ontology-driven data partitioning that is based on a semantic clustering of values in a column,
- we apply different partitioning strategies (ontology-driven vs. round robin) to several columns of a database table with the aim to show improved data locality for flexible query answering in a distributed database,
- we describe a recovery procedure based on a replicated lookup table,
- we present performance tests of the query answering procedure as well as the recovery procedure based on the Medical Subject Headings (MeSH) in a distributed SAP HANA database that shows that ontology-driven partitioning leads to lower execution times for flexible query answering with less servers involved while still allowing for a fast recovery.

1.1   Organization of the Article

Section 2 introduces the main notions used in this article and gives an illustrative example. Section 3 describes ontology-driven query answering, Sect. 4 shows query answering with derived partitions, Sect. 5 analyzes the update behavior, Sect. 6 discusses deletions and Sect. 7 presents a recovery procedure. Related work is presented in Sects. 8 and 9 concludes this article with suggestions for future work.

2   Background and Example

2.1   Query Generalization

Query generalization has long been studied in flexible query answering [13]. Query generalization at runtime has been implemented in the CoopQA system [8] by applying three generalization operators to a conjunctive query. *Anti-Instantiation* (AI) is one query generalization operator that replaces a constant (or a variable occurring at least twice) in a query with a new variable y. In this paper we focus on replacements of constants because this allows for finding answers that are semantically close to the replaced constant.

As the query language we focus on conjunctive queries expressed as logical formulas. We assume a logical language $\mathscr{L}$ consisting of a finite set of predicate symbols (denoting the table names; for example, *Ill*, *Treat* or P), a possibly infinite set *dom* of constant symbols (denoting the values in table cells; for example, *Mary* or a), and an infinite set of variables (x or y). A term is either a constant or a variable. The capital letter X denotes a vector of variables; if the order of variables in X does not matter, we identify X with the set of its variables and apply set operators – for example we write $y \in X$. We use the standard logical connectors conjunction $\wedge$, disjunction $\vee$, negation $\neg$ and material implication $\rightarrow$ and universal $\forall$ as well as existential $\exists$ quantifiers. An atom is a formula consisting of a single predicate symbol only; a literal is an atom (a "positive literal") or a negation of an atom (a "negative literal"); a clause is a disjunction of atoms; a ground formula is one that contains no variables. The existential (universal) closure of a formula ϕ is written as $\exists\phi$ ($\forall\phi$) and denotes the closed formula obtained by binding all free variables of ϕ with the quantifier.

A query formula Q is a conjunction of literals with some variables X occurring freely (that is, not bound by variables); that is, $Q(X) = L_{i_1} \wedge \ldots \wedge L_{i_n}$. The Anti-Instantiation (AI) operator chooses a constant a in a query $Q(X)$, replaces one occurrence of a by a new variable y and returns the query $Q^{AI}(X, y)$ as the relaxed query. The relaxed query Q^{AI} is a deductive generalization of Q.

As a running example, we consider a hospital information system that stores illnesses and treatments of patients as well as their personal information (like address and age) in the following three database tables:

Ill	PatientID	Diagnosis
	8457	Cough
	2784	Flu
	2784	Asthma
	2784	brokenLeg
	8765	Asthma
	1055	brokenArm

Treat	PatientID	Prescription
	8457	Inhalation
	2784	Inhalation
	8765	Inhalation
	2784	Plaster bandage
	1055	Plaster bandage

Info	PatientID	Name	Address
	8457	Pete	Main Str 5, Newtown
	2784	Mary	New Str 3, Newtown
	8765	Lisa	Main Str 20, Oldtown
	1055	Anne	High Str 2, Oldtown

The query $Q(x_1, x_2, x_3) = Ill(x_1, Flu) \wedge Ill(x_1, Cough) \wedge Info(x_1, x_2, x_3)$ asks for all the patient IDs x_1 as well as names x_2 and addresses x_3 of patients that suffer from both flu and cough. This query fails with the given database tables as there is no patient with both flu and cough. However, the querying user might instead be interested in the patient called Mary who is ill with both flu and asthma. We can find this informative answer by relaxing the query condition *Cough* and instead allowing other related values (like *Asthma*) in the answers. An example generalization with AI is $Q^{AI}(x_1, x_2, x_3, y) = Ill(x_1, Flu) \wedge Ill(x_1, y) \wedge Info(x_1, x_2, x_3)$ by introducing the new variable y. It results in an non-empty (and hence informative) answer: $Ill(2748, Flu) \wedge Ill(2748, Asthma) \wedge Info(2748, Mary, \text{'}New\ Str\ 3, Newtown\text{'})$. Another answer obtained is the fact that Mary suffers from a broken leg as: $Ill(2748, Flu) \wedge Ill(2748, brokenLeg) \wedge Info(2748, Mary, \text{'}New\ Str\ 3, Newtown\text{'})$.

As can be seen from the example query Q^{AI}, query relaxation by anti-instantiation can go too far and lead to *overgeneralization*: while the first example answer (with the value asthma) is a valuable informative answer, the second one (containing broken leg) might be too far away from the user's query interest. Here we need semantic guidance to identify the set of relevant answers that are close enough to the original query.

2.2 Ontology-Driven Partitioning

In previous work [21], a clustering procedure was applied to partition the original tables into partitions based on single *relaxation attribute* chosen for anti-instantiation; whereas concomitant research [22] handles intelligent replication for *multiple* relaxation attributes. Partitioning is achieved by grouping (that is, *clustering*) the values of the respective table column (corresponding to the relaxation attribute) and then splitting the table into partitions according to the clusters found. The clustering relies on a similarity metrics which is derived from paths ("proximity") between any two terms in an ontology or taxonomy.

We assume that each of the clusterings (and hence the corresponding partitioning) is *complete*: every value in the column is assigned to one cluster and hence every tuple is assigned to one partition. We also assume that each clustering and each partitioning are also *non-redundant*: every value is assigned to exactly one cluster and every tuple belongs to exactly one partition (for one of the relaxation attributes); in other words, the partitions inside one partitioning do not overlap.

More formally, we apply the clustering approach described in [21] (or any other method to semantically split the attribute domain into subsets) on the relaxation attribute, so that each cluster inside one clustering is represented by a *head* term (also called prototype) and each term in a cluster has a similarity *sim* to the cluster head above a certain threshold α. We then obtain a clustering-based partitioning for the original table F into partitions as specified in the following definition.

Definition 1 (Clustering-Based Partitioning). *Let A be a relaxation attribute; let F be a table instance (a set of tuples); let $C = \{c_1, \ldots c_n\}$ be a complete clustering of the active domain $\pi_A(F)$ of A in F; let $head_i \in c_i$; then, a set of partitions $\{F_1 \ldots, F_n\}$ (defined over the same attributes as F) is a* clustering-based *partitioning if*

- *Horizontal partitioning: for every partition F_i, $F_i \subseteq F$*
- *Clustering: for every F_i there is a cluster $c_i \in C$ such that $c_i = \pi_A(F_i)$ (that is, the active domain of F_i on A is equal to a cluster in C)*
- *Threshold: for every $a \in c_i$ (with $a \neq head_i$) it holds that $sim(a, head_i) \geq \alpha$*
- *Completeness: For every tuple t in F there is an F_i in which t is contained*
- *Reconstructability: $F = F_1 \cup \ldots \cup F_n$*
- *Non-redundancy: for any $i \neq j$, $F_i \cap F_j = \emptyset$ (or in other words $c_i \cap c_j = \emptyset$)*

For example, clusters on the `Diagnosis` column can be made by differentiating between fractures on the one hand and respiratory diseases on the other hand. These clusters then lead to two partitions of the table `Ill` that can be assigned to two different servers:

Server 1:		
Respiratory	*PatientID*	*Diagnosis*
	8457	Cough
	2784	Flu
	2784	Asthma
	8765	Asthma

Server 2:		
Fracture	*PatientID*	*Diagnosis*
	2784	brokenLeg
	1055	brokenArm

Server 1 can then be used to answer queries related to respiratory diseases while Server 2 can process queries related to fractures. The example query $Q(x_1, x_2, x_3) = Ill(x_1, Flu) \wedge Ill(x_1, Cough) \wedge Info(x_1, x_2, x_3)$ will then be rewritten into $Q^{Resp}(x_1, x_2, x_3, y) = Respiratory(x_1, y) \wedge Respiratory(x_1, Cough) \wedge Info(x_1, x_2, x_3)$ and redirected to Server 1 where only the partition Respiratory is used to answer the query. In this way only the informative answer containing asthma is returned – while the one containing broken leg will not be generated. The most important advantage of the ontology-driven partitioning is that for answering the relaxed query only a single database server is contacted; with any other partitioning strategy that distributes data among the servers (that is, based on ranges or hash values or in round robin fashion), the relevant data might be distributed across several servers. If several tables are partitioned, *derived partitioning* can be used to store the matching tuples of the other tables (joinable by patient ID like the Info table) and hence improve data locality during processing of join queries.

2.3   Derived Partitioning

When having several tables that can be joined in a query, *data locality* is important for performance: Data that are often accessed together should be stored on the same server in order to avoid excessive network traffic and delays. If one table is chosen as the primary clustering table (like *Ill* in our example), partitioning of related tables (like **Treat** and **Info** in our example) can be derived from the primary partitioning. They are obtained by computing a semijoin between the primary table and the secondary table. Each derived partition should then be assigned to the same database server on which the primary partition with the matching join attribute values resides. Note that while the primary partitioning is usually non-redundant (each tuple of the original table is contained in exactly one partition), that might not be the case for derived partitioning: one tuple of a joinable tables might be contained in several derived partitions.

In the example, the entire partitioning for clusters on the **Diagnosis** column assigned to two servers then looks as follows:

Server 1:

Respiratory	PatientID	Diagnosis
	8457	Cough
	2784	Flu
	2784	Asthma
	8765	Asthma

Treat_resp	PatientID	Prescription
	8457	Inhalation
	2784	Inhalation
	8765	Inhalation
	2784	Plaster bandage

Info_resp	PatientID	Name	Address
	8457	Pete	Main Str 5, Newtown
	2784	Mary	New Str 3, Newtown
	8765	Lisa	Main Str 20, Oldtown

Server 2:

Fracture	PatientID	Diagnosis
	2784	brokenLeg
	1055	brokenArm

Treat_frac	PatientID	Prescription
	2784	Inhalation
	2784	Plaster bandage
	1055	Plaster bandage

Info_frac	PatientID	Name	Address
	2784	Mary	New Str 3, Newtown
	1055	Anne	High Str 2, Oldtown

3   Ontology-Driven Query Answering

To enable ontology-driven query answering, when a user sends a query to the database, the term (that is, constant) that can be anti-instantiated has to be extracted, the matching cluster has to be identified and then the user query has to be rewritten to return answers covering the entire cluster.

3.1   Metadata and Test Dataset

In order to manage the partitioning, several metadata tables are maintained:

- A **root** table stores an ID for each cluster (column *clusterid*) as well as the cluster head (column *head*) and the name of the server that hosts the cluster (column *serverid*).
- A **lookup** table stores for each cluster ID (column *clusterid*) the tuple IDs (column *tupleid*) of those tuples that constitute the clustered partition.
- A **similarities** table stores for each head term (column *head*) and each other term (column *term*) that occurs in the active domain of the corresponding relaxation attribute a similarity value between 0 and 1 (column *sim*).

Our prototype implementation – the OntQA-Replica system – runs on a distributed SAP HANA installation which is an in-memory database system and hence shows a fast execution without disk accesses. All runtime measurements shown below are taken as the median of several (at least 5) runs per experiment.

The example data set consists of a table (called *ill*) that resembles a medical health record and is based on the set of Medical Subject Headings (MeSH [20]). The table contains as columns an artificial, sequential *tupleid*, a random *patientid*, and a *disease* chosen from the MeSH data set as well as the *concept* identifier of the MeSH entry. We varied the table sizes during our test runs. The smallest table consists of 56,341 rows (one row for each MeSH term), a medium-sized table of 1,802,912 rows and the largest of 14,423,296 rows (obtained by duplicating the original data set 5 times and 8 times, respectively). A clustering is executed on the MeSH data based on the concept identifier (which orders the MeSH terms in a tree); in other words, entries from the same subconcept belong to the same cluster. One partitioning (the clustered partitioning) was obtained from this clustering and consists of 117 partitions which are each stored in a smaller table called *ill-i* where i is the cluster ID. To allow for a comparison and a test of the recovery strategy, another partitioning of the table was done using round robin resulting in a table called *ill-rr*; this distributes the data among the database servers in chunks of equal size without considering their semantic relationship; these partitions have an extra column called *clusterid*.

3.2   Identifying Matching Clusters

Flexible Query Answering intends to return those terms belonging to the same cluster as the query term as informative answers. Before being able to return the related terms, we hence have to identify the matching cluster: that is, the ID of the cluster the head of which has the *highest* similarity to the query term. We do this by consulting the similarities table and the root table. The similarities are derived by using the Unified Medical Language System as an ontology (or more precisely is-a hierarchy) and the similarity interface on top of it [12]. The relaxation term t is extracted from the query and then the top-1 entry of the similarities table is obtained when ordering the similarities in descending order:

```
SELECT TOP 1 root.clusterid FROM root, similarities
  WHERE similarities.term='t' AND similarities.head = root.head
  ORDER BY similarities.sim DESC
```

The query was tested on similarities tables of sizes 56341 entries, 14423296 entries and 72116480 entries. The runtime measurements in Fig. 1 show a decent performance of at most 125 ms impact even for the largest table size.

3.3   Query Rewriting Strategies

After having obtained the ID of the matching cluster, the original query has to be rewritten in order to consider all the related terms as valid answers. We tested and compared three query rewriting procedures:

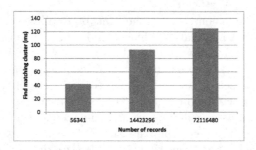

Fig. 1. Identify matching cluster

- lookup table: the first rewriting approach uses the lookup table to retrieve the tuple IDs of the corresponding rows and executes a JOIN on table *ill*.
- extra clusterid column: the next approach relies on the round robin table and retrieves all relevant tuples based on a selection predicate on the clusterid column.
- clustered partitioning: the last rewriting approach replaces the occurrences of the *ill* table by the corresponding *ill-i* table for clusterid i.

Assume the user sends a query
```
SELECT mesh,concept,patientid,tupleid
  FROM ill WHERE mesh ='cough'.
```

and 101 is the ID of the cluster containing cough. In the first strategy (lookup table) the rewritten query is

```
SELECT mesh,concept,patientid,tupleid
  FROM ill JOIN lookup
  ON (lookup.tupleid = ill.tupleid AND lookup.clusterid=101).
```
In the second strategy (extra clusterid column) the rewritten query is

```
SELECT mesh,concept,patientid,tupleid
  FROM ill-rr WHERE clusterid=101
```
In the third strategy (clustered partitioning), the rewritten query is

```
SELECT mesh,concept,patientid,tupleid FROM ill-101
```

In the small *ill* table with 56341 entries, 90 related answers are obtained, in the medium-sized *ill* table with 1802912 entries, 2880 related answers are obtained and in the large *ill* table with 14423296 entries, 23040 related answers are obtained. The runtime measurements in Fig. 2 in particular show that the lookup table approach does not scale with increasing data set size.

4   Query Answering with Derived Partitions

While the evaluation of a selection query on a single table shows a similar performance for all rewriting strategies, the evaluations of queries on two tables

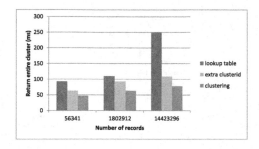

Fig. 2. Return entire cluster as related answers

using a distributed JOIN show a performance impact for the first two strategies when the secondary table is large. We tested a JOIN on the patient ID with a secondary table called *info* having a column *address*. The original query is

```
SELECT a.mesh,a.concept,a.patientid,a.tupleid,b.address
FROM ill AS a,info AS b
WHERE mesh='cough' AND b.patientid= a.patientid
```

We devised two test runs: test run one uses a small secondary table (each patient ID occurs only once) and test run two uses a large secondary table (each patient ID occurs 256 times). For the first rewriting strategy (lookup table) the secondary table is a non-partitioned table. For the second strategy, the secondary table is distributed in round robin fashion, too. For the last rewriting strategy, the secondary table is partitioned into a derived partitioning: whenever a patient ID occurs in some partition in the *ill-i* table, then the corresponding tuples in the secondary table are stored in a partition *info-i* on the same server as the primary partition.

In the first strategy (lookup table) the rewritten query is

```
SELECT a.mesh,a.concept,a.patientid,a.tupleid,b.address
FROM ill AS a,info AS b,lookup WHERE lookup.tupleid=a.tupleid
AND lookup.clusterid=101 AND b.patientid= a.patientid.
```

In the second strategy (extra clusterid column) the rewritten query is

```
SELECT a.mesh,a.concept,a.patientid,a.tupleid,b.address
FROM ill-rr AS a,info-rr AS b
WHERE a.clusterid=101 AND b.patientid=a.patientid.
```

In the third strategy (clustered partitioning), the rewritten query is

```
SELECT a.mesh,a.concept,a.patientid,a.tupleid,b.address
FROM ill-101 AS a JOIN info-101 AS b
ON (a.patientid=b.patientid).
```

As Fig. 3 shows, a small secondary table does not make much of a difference when executing the join operation (one matching tuple in the secondary table for each tuple in the primary table). However, for the larger secondary table (256 matching tuples in the secondary table for each tuple in the primary table), the impact of the lookup table access is huge in the case of the largest *ill* table.

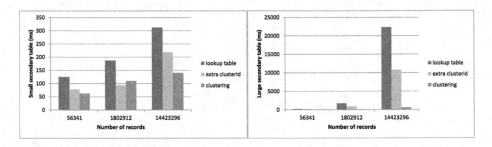

Fig. 3. Join on small and large secondary table

5   Insertions

We tested the update behavior for all three rewriting strategies by inserting 117 new rows (one for each cluster). Any insertion requires identifying the matching cluster i again (see Sect. 3.2). Each insertion query looks like this for mesh term m, concept c, patientid 1 and tupleid 1:

    INSERT INTO ill VALUES ('m','c',1,1).

In the first rewriting strategy, the lookup table has to be updated, too, so that two insertion queries are executed:

    INSERT INTO ill VALUES ('m','c',1,1).
    INSERT INTO lookup VALUES (i,1).

For the second rewriting strategy, the extra clusterid column contains the identified cluster i:

    INSERT INTO ill-rr VALUES ('m','c',1,1,i).

For the third rewriting strategy, the matching clustered partition is updated:

    INSERT INTO ill-i VALUES ('m','c',1,1).

As shown in Fig. 4, we only tested insertions for the largest table. Here the lookup table approach has a huge runtime impact due to the maintenance of the lookup table entries.

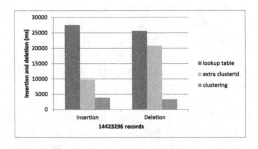

Fig. 4. Inserting one tuple into each cluster

6   Deletions

After the insertions we made a similar test by deleting the newly added tuples by issuing the query
DELETE FROM ill WHERE mesh='m'.
In the first rewriting strategy, the corresponding row in the lookup table has to be deleted, too, so that now first the corresponding tuple id of the to-be-deleted row has to be obtained and then two deletion queries are executed:
DELETE FROM lookup WHERE lookup.tupleid
  IN (SELECT ill.tupleid FROM ill WHERE mesh='m').
DELETE FROM ill WHERE mesh='m'
For the second rewriting strategy, no modification is necessary apart from replacing the table name and no clusterid is needed:
DELETE FROM ill-rr WHERE mesh='m'
For the third rewriting strategy, the matching clustered partition i is accessed which has to be identified first (as in Sect. 3.2):
DELETE FROM ill-i WHERE mesh='m'
As shown in Fig. 4, we only tested deletions for the largest table, where the time needed to identify the matching cluster is negligible. Even the round robin approach with extra clusterid does not perform well on this large data set.

7   Recovery

Lastly, we tested how long it takes to recover the clustered partitioning by either using the lookup table or the extra column ID. The recovery procedure was executed first on the original table and the lookup table by running for each cluster i:
INSERT INTO c_i SELECT * FROM ill JOIN lookup
  ON (lookup.tupleid=ill.tupleid) WHERE lookup.clusterid=i
for each cluster i. Then, the recovery procedure was executed on the round robin partitioned table with the extra clusterid column for each cluster i:
INSERT INTO c_i SELECT * FROM ill-rr WHERE clusterid=i
While for the smallest and the largest table the two approaches perform nearly identically, for the medium-sized table the extra cluster id approach offers some benefit (Fig. 5).

8   Related Work

We divide the related work survey into approaches for flexible query answering and approaches for data partitioning and replication.

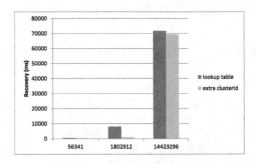

Fig. 5. Recovering the clustered partitioning

8.1 Flexible Query Answering

The area of flexible query answering (sometimes also called cooperative query answering) has been studied extensively for single server systems. Some approaches have used taxonomies or ontologies for flexible query answering but did not consider their application for distributed storage of data: CoBase [1] used a type abstraction hierarchy to generalize values; Shin et al. [18] use some specific notion of metric distance in a knowledge abstraction hierarchy to identify semantically related answers; Halder and Cortesi [5] define a partial order between cooperative answers based on their abstract interpretation framework; Muslea [14] discusses the relaxation of queries in disjunctive normal form. Ontology-based query relaxation has also been studied for non-relational data (like XML data in [7] or RDF data in [3]). Another form of query relaxation requires user input in an interactive query relaxation process [9,10] or analyse query relaxation based on a taxonomy with Bayesian Decision Theory [15].

All these approaches address query relaxation at runtime while answering the query. This is usually prohibitively expensive. In contrast, our approach precomputes the clustering and partitioning so that query answering does not incur a performance penalty. However user interaction might be needed during the clustering process in case some assigments of terms to clusters are ambiguous.

8.2 Data Partitioning and Replication

There are some approaches for fine-grained partitioning and replication on object/tuple level; however none of these approaches support the flexible query answering application aimed at in this paper. In contrast they are mostly workload-driven and try to optimize the locality of data that are covered in the same query. They only support exact query answering. In contrast to this, we do not consider workloads but a generic clustering approach that can work with arbitrary workloads providing the feature of flexible query answering by finding semantically related answers. Our results in this paper also show that the fine-grained lookup table approach – when applied to flexible query answering – is inherently slow due to the large amounts of JOIN operations and so it does not scale well even if lookup tables are replicated on all servers.

Curino et al. [2] represent database tuples as nodes in a graph. They assume a given transaction workload and add hyperedges to the graph between those nodes that are accessed by the same transactions. By using a standard graph partitioning algorithm, they find a database partitioning that minimizes the number of cut hyperedges. In a second phase, they use a machine learning classifier to derive a range-based partitioning. Then they make an experimental comparison between the graph-based, the range-based, a hash-based partitioning on tuple keys and full replication. Lastly, they also compare three different kinds of lookup tables to map tuple identifier to the corresponding partition: indexes, bit arrays and Bloom filters. Similar to them, we apply lookup tables to locate the replicated data; however we apply this to larger partitions and not to individual tuples.

Quamar et al. [16] also model the partitioning problem as minimizing cuts of hyperedges in a graph; for efficiency reasons, their algorithm works on a compressed representation of the hypergraph which results in groups of tuples. In particular, the authors criticize the fine-grained (tuple-wise) approach in [2] to be impractical for large number of tuples which is similar to our approach. The authors propose mechanisms to handle changes in the workload and compare their approach to random and tuple-level partitioning.

Tatarowicz et al. [19] assume three existing partitionings: hash-based, range-based and lookup tables for individual keys and compare those in terms of communication cost and throughput. For an efficient management of lookup tables, they experimented with different compression techniques. In particular they argue that for hash-based partitioning, the query decomposition step is a bottleneck. While we apply the notion of lookup tables, too, the authors do not discuss how the partitions are obtained, whereas we propose a ontology-driven partitioning approach here.

There is also related work on specifying resource management problems as optimization problems. An adaptive solution for data replication using a genetic algorithm is presented in [11]; they focus on fine-grained geo-replication for individual objects. They include an assumed number of reads and writes for each site as well as communication costs between sites. They reduce their problem to the Knapsack problem; they also consider transfer cost of replicas between servers. Load shedding in complex event processing systems is treated in [6]. Virtual machine placement is a very recent topic in cloud computing [4,17]. However, these specifications do not address the problem of overlapping resources as we need for the flexible query answering approach in this article.

9   Conclusion and Future Work

We presented an ontology-driven partitioning approach for the application of flexible query answering that finds related answers to a user query. We evaluated its performance on a distributed in-memory store. Due to the small size of the partitioned tables, the runtime performance is best for the clustered partitioning approach and the overhead of metadata management is negligible. It outperforms the lookup table approach that stores for each cluster the corresponding tuple IDs does not scale well as the data set size grows. In addition, the ontology-driven partitioning enables fine-grained load balancing and data locality: less servers have to be accessed when answering queries or updating tables. The idea of data locality can even be carried further by considering cluster affinity: if two clusters are accessed together frequently, their corresponding partitions can be placed on the same server. So far we did not address the dynamic adaptation of the clustering: whenever values are inserted or deleted, the clustering procedure on the entire data set might lead to different clusters. A particular problem that must be handled is the deletion of the head of a cluster: a new cluster head must be chosen before the current head can be deleted; in the simplest case, the term that is most similar to the previous head is chosen as the new head. Similarly, deletions and insertions lead to shrinking or growing partitions. Hence in some situation it might be useful to merge two smaller partitions that are semantically close to each other: we can merge two partitions when their heads are sufficiently similar to each other; or to repartition a larger partition into subpartitions based on a clustering of values of the relaxation attribute in the partition.

Acknowledgments. The author gratefully acknowledges that the infrastructure and SAP HANA installation for the test runs was provided by the Future SOC Lab of Hasso Plattner Institute (HPI), Potsdam.

References

1. Chu, W.W., Yang, H., Chiang, K., Minock, M., Chow, G., Larson, C.: CoBase: a scalable and extensible cooperative information system. JIIS **6**(2/3), 223–259 (1996)
2. Curino, C., Zhang, Y., Jones, E.P.C., Madden, S.: Schism: a workload-driven approach to database replication and partitioning. Proc. VLDB Endowment **3**(1), 48–57 (2010)
3. Fokou, G., Jean, S., Hadjali, A., Baron, M.: Cooperative techniques for SPARQL query relaxation in RDF databases. In: Gandon, F., Sabou, M., Sack, H., d'Amato, C., Cudré-Mauroux, P., Zimmermann, A. (eds.) ESWC 2015. LNCS, vol. 9088, pp. 237–252. Springer, Heidelberg (2015)
4. Goudarzi, H., Pedram, M.: Energy-efficient virtual machine replication and placement in a cloud computing system. In: IEEE 5th International Conference on Cloud Computing (CLOUD), pp. 750–757. IEEE (2012)
5. Halder, R., Cortesi, A.: Cooperative query answering by abstract interpretation. In: Černá, I., Gyimóthy, T., Hromkovič, J., Jefferey, K., Králović, R., Vukolić, M., Wolf, S. (eds.) SOFSEM 2011. LNCS, vol. 6543, pp. 284–296. Springer, Heidelberg (2011)
6. He, Y., Barman, S., Naughton, J.F.: On load shedding in complex event processing. In: 17th International Conference on Database Theory (ICDT), pp. 213–224 (2014)
7. Hill, J., Torson, J., Guo, B., Chen, Z.: Toward ontology-guided knowledge-driven xml query relaxation. In: Computational Intelligence, Modelling and Simulation (CIMSiM), pp. 448–453 (2010)

8. Inoue, K., Wiese, L.: Generalizing conjunctive queries for informative answers. In: Christiansen, H., De Tré, G., Yazici, A., Zadrozny, S., Andreasen, T., Larsen, H.L. (eds.) FQAS 2011. LNCS, vol. 7022, pp. 1–12. Springer, Heidelberg (2011)
9. Jannach, D.: Fast computation of query relaxations for knowledge-based recommenders. AI Commun. **22**(4), 235–248 (2009)
10. Kumaran, G., Allan, J.: Selective user interaction. In: Proceedings of the Sixteenth ACM Conference on Conference on Information and Knowledge Management, pp. 923–926. ACM (2007)
11. Loukopoulos, T., Ahmad, I.: Static and adaptive distributed data replication using genetic algorithms. J. Parallel Distrib. Comput. **64**(11), 1270–1285 (2004)
12. McInnes, B.T., Pedersen, T., Pakhomov, S.V.S., Liu, Y., Melton-Meaux, G.: Umls::similarity: Measuring the relatedness and similarity of biomedical concepts. In: Vanderwende, L., III, H.D., Kirchhoff, K. (eds.) Human Language Technologies: Conference of the North American Chapter of the Association of Computational Linguistics, pp. 28–31. The Association for Computational Linguistics, Stroudsburg, PA, USA (2013)
13. Michalski, R.S.: A theory and methodology of inductive learning. Artif. Intell. **20**(2), 111–161 (1983)
14. Muslea, I.: Machine learning for online query relaxation. In: Knowledge Discovery and Data Mining (KDD), pp. 246–255. ACM, New York (2004)
15. Pfuhl, M., Alpar, P.: Improving database retrieval on the web through query relaxation. In: Abramowicz, W., Flejter, D. (eds.) BIS 2009. LNBIP, vol. 37, pp. 17–27. Springer, Heidelberg (2009)
16. Quamar, A., Kumar, K.A., Deshpande, A.: Sword: scalable workload-aware data placement for transactional workloads. In: Guerrini, G., Paton, N.W. (eds.) Joint 2013 EDBT/ICDT Conferences, pp. 430–441. ACM, New York (2013)
17. Shi, W., Hong, B.: Towards profitable virtual machine placement in the data center. In: Fourth IEEE International Conference on Utility and Cloud Computing (UCC), pp. 138–145. IEEE (2011)
18. Shin, M.K., Huh, S.Y., Lee, W.: Providing ranked cooperative query answers using the metricized knowledge abstraction hierarchy. Expert Syst. Appl. **32**(2), 469–484 (2007)
19. Tatarowicz, A., Curino, C., Jones, E.P.C., Madden, S.: Lookup tables: fine-grained partitioning for distributed databases. In: Kementsietsidis, A., Salles, M.A.V. (eds.) IEEE 28th International Conference on Data Engineering (ICDE 2012), pp. 102–113. IEEE Computer Society, Washington, DC (2012)
20. U.S. National Library of Medicine: Medical subject headings. http://www.nlm.nih.gov/mesh/
21. Wiese, L.: Clustering-based fragmentation and data replication for flexible query answering in distributed databases. J. Cloud Comput. **3**(1), 1–15 (2014)
22. Wiese, L.: Horizontal fragmentation and replication for multiple relaxation attributes. In: Maneth, S. (ed.) BICOD 2015. LNCS, vol. 9147, pp. 157–169. Springer, Heidelberg (2015)

Information Retrieval

Detecting Near-Duplicate Documents Using Sentence Level Features

Jinbo Feng[1] and Shengli Wu[1,2(✉)]

[1] School of Computer Science, Jiangsu University, Zhenjiang 212013, China
[2] School of Computing and Mathematics, Ulster University,
Newtownabbey, BT37 0QB, UK
s.wu1@ulster.ac.uk

Abstract. In Web search engines, digital libraries and other types of online information services, duplicates and near-duplicates may cause severe problems if unaddressed. Typical problems include more space needed than necessary, longer indexing time and redundant results presented to users. In this paper, we propose a method of detecting near-duplicate documents. Two sentence level features, number of terms and terms at particular positions, are used in the method. Suffix tree is used to match sentence blocks very efficiently. Experiments are carried out to compare our method with two other representative methods and show that our method is effective and efficient. It has potential to be used in practice.

Keywords: Text documents · Information search · Near-duplicate detection · Suffix tree · Sentence-level features

1 Introduction

As the World Wide Web is increasingly popular, numerous copies of the same documents can be generated and spread very easily. A large amount of redundant web documents may cause severe problems for search engines: more space is needed to store the web documents; more time is needed for indexing; and the search results are less useful to users due to the large amount of redundant information. Sometimes it also causes problems that are related to copyright or intellectual property right. Thus, researches on detecting near-duplicate documents have gained attention in IR, WWW, digital libraries, and other related areas [1–4].

The definition of "near-duplicate" is documents that differ only slightly in content [4]: such as one document is a modification of the other via insertion, deletion or replacement of some terms. A near-duplicates detection algorithm compares two documents mainly based on syntactic similarity. For all the web documents, some prior treatment is necessary since they are usually very noisy. We need to extract the main content of a web document by removing html tags, navigation links, advertisements, and so on. The similarity of two documents can be calculated by comparing

© Springer International Publishing Switzerland 2015
Q. Chen et al. (Eds.): DEXA 2015, Part II, LNCS 9262, pp. 195–204, 2015.
DOI: 10.1007/978-3-319-22852-5_17

how many words or sentences are the same in one way or another. If the similarity between the two documents is greater than a given threshold, then we may regard them as near-duplicates.

Although quite a few methods for detecting near-duplicate documents have been proposed, it is still a challenge to have one that is both effective and efficient. One very efficient method was proposed by Wang and Chang in [4]. They take the number of terms in a sentence as the surrogate of that sentence. Thus each sentence is represented as a number. Two sentences are assumed to be identical if they have the same number of terms. In order to calculate the similarity of two documents, they use a fix-sized sliding window to decide how many sentences in the documents are to be compared. Obviously one problem of this method is its effectiveness since the number of terms in a sentence is not accurate enough to distinguish a sentence from others. Another problem is that using a fixed-size window is not able to find all possible matches unless the window is large enough to map the whole document. Effectiveness of the method is therefore further affected. However, using the number of terms in a sentence can be a good starting point, and errors in detection can be reduced by a few measures implemented at a relatively low cost. In this piece of work, we also apply the same feature (number of terms in the sentence) for the matching of a sequence of sentences. Specifically, we apply three measures to make the detection very effective and efficient:

- Instead of using fix-sized sliding windows, we use a suffix tree to compare two documents. By using a suffix tree we are able to find all possible pairs of identical sentences. This is good for improving effectiveness of the detection process. The detection process is also very efficient;
- To further mitigate the problem of relatively low accuracy caused by using the number of terms in a sentence as the only factor, we add a simple validation process after sentence sequence match, e.g., by comparing a few terms at particular positions in all the sentences involved.

Our experiment shows that these measures work and the proposed method is efficient and effective. Thus we believe the method has good potential to be used in practice.

The remainder of this paper is structured as follows: in Sect. 2, we describe some related work. Section 3 details the near-duplicate document detecting method including all the components. Section 4 presents the experimental results. Conclusions are provided in Sect. 5.

2   Related Work

Near-duplicate document detection can be used in different situations such as in Web search engines for duplicates removal [10, 11] or in digital libraries for document versioning and plagiarism detection [9, 12].

Two pieces of earliest work are from Garcia-Molina and his colleagues [5, 12]. Another early piece of work is done by Broder [1]. Broder used shingles to represent documents. A shingle is basically an n-gram of terms and a document is composed of a series of n-grams. For two documents, if they share more shingles, then they bear more similarity. However, it is a tedious task to compare all possible pairs of shingles between two documents, especially long ones. In Broder's work, not all shingles, but only a subset of them are considered and the hash function is used for speeding up the comparison process.

SCAM [5] takes the bag-of-words approach. It uses word as the unit of chunking to represent documents and stores chunks in the inverted index structure [14]. In [5], the Relative Frequency Model (RFM) is proposed for detecting overlap between two documents. RFM is different from the traditional Vector Space Model (VSM) [13].

Chowdhury et al. proposed a method, I-Match, by using statistics of the whole collection [2]. It ignores very infrequent terms or very common terms according to the IDF value of the collection.

SpotSigs [19] is one form of n-gram. But instead of using a fixed number of n terms, SpotSigs divides a sentence into short chains of adjacent content terms by using stop-words as separators. According to [19], Performance of SpotSigs is affected significantly by the stopwords used.

Lin et al. [3] proposed a supervised learning algorithm to detect near-duplicates using sentence-level features. A support vector machine (SVM) is adopted to learn a discriminate function from a training pattern set to calculate the similarity degree between any pair of documents.

Zhang et al. [6] presented an efficient algorithm for detecting partial-duplicates. This approach was explored with MapReduce [7] which was a framework for large-scale distributed computing and sequence matching algorithm.

Wang and Chang [4, 8] proposed a method that takes the number of terms in a sentence as the surrogate of that sentence, and then using a sliding window to compare two documents by a certain number of sentences inside the window. The size of the sliding window and the number of sentences that each time the sliding window moves forward affect the performance. Therefore, they empirically investigate many different combinations.

Similar to Wang and Chang's work, we use the same feature as they do. Apart from that, we made significant contributions in several different ways as mentioned at the end of Sect. 1. These measures ensure our method work efficiently and effectively at the same time.

3   Proposed Approach

Depending on the format of documents, some pre-processing is usually necessary before we can use a near-duplicate document detection method. In the following, we assume that all the documents have been treated properly and are ready for use as pure textual documents. The pseudo-code of our near-duplicate document detection method SL+ST (Sentence Length + Suffix Tree [15, 16]) working with two documents is shown in Fig. 1. If we need to do it with a collection of documents, then the extension is straightforward by comparing all different pairs.

In Fig. 1, there are 4 major steps (lines 1–9, 10–12, 13–19, 20–23). In the following let us discuss them one by one.

Algorithm 1. The near-duplicate document detection algorithm (SL+ST)
Input: a pair of documents (d_i,d_j), given threshold τ;
Output: A boolean variable Y, indicates if d_i and d_j are near-duplicates;
1: for each document d in (d_i,d_j) do
2:     SL(d) ← divide d into a list of sentences;
3:     for every sentence s_i in SL(d) do
4:         remove stopwords & stemming;
5:         store the first two terms as a feature;
6:         count the number of terms;
7:     end for
8:     generate a list L(d) indicating the number of terms of all the sentences;
9: end for
10: T ← build_suffix tree (L(d_i),L(d_j));
11: featureList ← traverse_suffix_tree( );
12: newFeatureList ← null;
13: for all feature f_i in featureList do
14:     if validate (f_i) = true then
15:         newFeatureList.add (f_i);
16:     else if split(f_i).length ≥ 2 then
17:         newFeatureList.add(split(f_i))
18:     end if
19: end for
20: similarity = calculate(newFeatureList)   //see Eq. 1 in Section 3.4
21: if (similarity ≥ τ) then {Y=true; return;}
22: else {Y=false; return;}
23: end if

Fig. 1. Pseudo-code of SL+ST

3.1 Generating Surrogates of Two Documents

In Algorithm 1, the first step is to represent each document as a string, in which each character indicates the number of terms in a given sentence. This task can be divided into three sub-tasks:

• First divide the whole documents into a list of sentences;
• Second remove stopwords and stem every term for every sentence in the document;
• Finally count the number of terms in each sentence and map each number to a character and put them together to form a string.

All these three sub-tasks can be done in one scan of the document. For the first sub-task, we need to set a group of predefined delimiters. Some punctuation marks such as period, question mark, exclamation mark, and so on, are good candidates for this. In the second sub-task, stopwords are removed and stemming is done to all remaining terms, thus every sentence is shorter than its original form. Thus it is more efficient to process them at later stages. In the third sub-task, we count how many terms are in each sentence, and then transform such information into a string in which one character is corresponding to a unique number (or regard each number as a character). After this, every document is represented as a string. The length of the string is the number of characters or the number of sentences in the whole document.

3.2   Finding Common Sentence Blocks in Two Documents

The second step is: for the two strings that are corresponding to the two documents, we compare them to see how many of the sentences are the same. This can be done by using a suffix tree.

A suffix tree is a trie structure built over all the suffixes of a string [15, 16]. For our purpose, we concatenate two strings together to form a long single string. In order to remember the initial two strings in the merged string, we add an extra character '#' at the end of the first string and '*' at the end of the second string. For example, suppose that we have two strings: "839767i" and "7839678i". Note that digits, letters, and so on comprise a string. Then these two strings are concatenated to form one long string "839767i#7839678i*". Its suffix tree is shown in Fig. 2. Every leaf node indicates the position of a given suffix string.

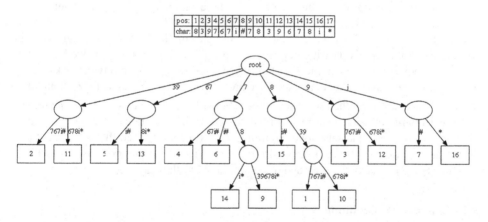

Fig. 2. The suffix tree of string "839767i#7839678i*"

When the suffix tree is built, we may traverse the suffix tree and obtain all the common substrings. Note that the number of terms in a sentence is a very obscure feature and not enough for identifying a sentence. Therefore, if we match all common substrings, then we may obtain a lot of false matches. Some sentences are not the same but are matched simply because they include the same number of terms. In order to reduce such false matches, we may increase the number of sentences needed for a match. Therefore, we define a series of sentences in a document as a sentence block and use it as a basic unit of match. It needs some consideration to decide how many sentences we should have in a sentence block. Nor a very big value neither a very small value can be good options. According to some observations and experiments, 2 appears to be a balanced option for this, although some other options (3 or more) are possible. In the above example, we obtain a set of 3 substrings: "39", "67" and "839".

Another problem is: in all the common substrings identified, some of them are substrings of others. In the above example, "39" is a substring of "839". In the above example, obviously it is not good to take "39" and "839" into consideration at the same time. Thus we only consider all longest common substrings whose length is above the given threshold.

In implementation, we represent each common substring identified as a triple $<n_1, n_2, n_3>$. Here n_1 is the end position of its first occurrence and n_2 the end position of its second occurrence and n_3 the length of the substring. Thus "39", "67" and "839" are represented as $<3, 4, 2>$, $<6, 6, 2>$ and $<3, 4, 3>$, respectively. We can find that $<3, 4, 2>$ is a substring of $<3, 4, 3>$ quite easily by comparing these two triples. When the suffix tree is built, finding all longest common sentence blocks can be done by traversing the suffix tree once.

3.3 Validation

This step is to validate our findings (all longest common substrings, or all longest common sentence blocks) in the previous step. For all those substrings obtained in step 2, we check if those corresponding sentences are the same. For every pair of sentences involved, we compare their first two terms. If one or two terms do not match in at least one sentence, then we discard the whole block or take smaller block(s) that can pass the validation step, depends on which option is appropriate. For example, if we have two sentence blocks of size 5 and the 3rd is different, then we split it into two blocks, each with 2 sentences. As another example, if we have two sentence blocks of size 4, and the 2nd and 3rd are different, then we discard the whole block. After this step, we obtain all the longest common sentence blocks in which each pair of sentences include the same number of terms and the same two beginning terms. It is regarded that such sentence blocks are identical. This step is helpful for us to reduce the number of errors, in which two blocks of sentences are regarded to be identical but they are not.

3.4 Similarity Calculation

Now we can calculate the degree of similarity between a pair of documents [17]. For documents d1 and d2, their similarity is measured by

$$\text{similarity}\left(d_1, d_2\right) = \frac{|d_1 \cap d_2|}{|d_1 \cup d_2|} \tag{1}$$

Here |d| denotes the number of sentences in d. We need to set a threshold τ. If similarity(d_1, d_2) is no less than τ, we claim that d_1 and d_2 are near duplicates; otherwise, we claim they are not [18].

3.5 Complexity Analysis

Suppose that document d_1 has $|d_1|$ sentences and t_1 terms and d_2 has $|d_2|$ sentences and t_2 terms. Step 1 can be done by one scan of d_1 and d_2. Thus the time needed is $O(t_1 + t_2)$. In Step 2, we first need to construct the suffix tree for d_1 and d_2. The suffix tree has a maximum of $|d_1| + |d_2| + 2$ leaf nodes, and the time needed for this is $O(|d_1| + |d_2|)$. We need to traverse the suffix tree once for common substring match, the time for this is also $O(|d_1| + |d_2|)$. Both Step 3 and Step 4 can be done in $O(|d_1| + |d_2|)$ time. Therefore, the time complexity of all the steps is $O(|d_1| + |d_2| + t_1 + t_2)$ and it shows that the method is very efficient.

4 Experiments

In this section, we evaluate our method empirically. All the experiments are carried out on a desktop computer, which has an Intel Core i7 quad-core CPU (3.4 GHz) and 32 GB of RAM. The data sets we use are two English document collections, AP90-S and Twitter. AP90-S is a subset of AP90 and Twitter is a subset of ClubWeb12-B. AP90 and ClubWeb12-B are two data sets used in TREC before. The AP90 document set includes copyrighted stories from the AP Newswire in 1990. In 1990's, NIST disseminated five discs and AP90 is on the third disk. Those five discs were used in the ad hoc tasks from TREC 1 to TREC 8. AP90-S is generated by the following method: first we manually select 6 documents {AP900629-0221, AP901120-0196, AP900326-0205, AP900524-0185, AP900706-0266, AP900806-0154} from AP90. Then we treat these 6 documents as queries to retrieve documents from AP90 using the Terrier retrieval system. Top 150 documents from each resultant list are put together with duplicates being removed (those having the same ID number). Thus we obtain AP90-S, which contains 891 documents and 56,432 sentences. ClueWeb12-B was disseminated in 2013 and was used for the web task. The Twitter documents are chosen from the 6 sub-folds of the ClueWeb12-B dataset. The information on the two data sets is summarized in Table 1.

Table 1. Summary of data sets used in our experiments

Collection	Source	Number of documents	Size
AP90-S	Disk 3-AP	891	5.2 MB
Twitter	ClubWeb12-B	2,716,306	116 GB

4.1 Effectiveness

Apart from our method SL+ST, we also test two other representative methods, 3-shingles [1] and SpotSigs [19], for comparison. The harmonic mean, or F1 [15], is used to evaluate all the methods involved. F1 is defined as

$$F_1 = 2 \frac{precision * recall}{precision + recall} \qquad (2)$$

Recall in Eq. 1, τ is the threshold that we use to determine if two documents are near-duplicates or not. After running those methods to obtain documents that are regarded as near-duplicates, human judgment is involved to decide if those documents are near-duplicates or not. It is also required to find out those missing near-duplicates in the collection that has not been identified by the detection program. Figure 3 shows the experimental results with varying τ values.

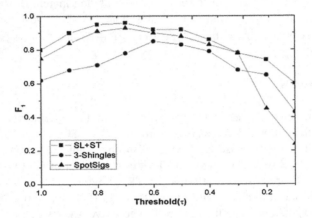

Fig. 3. Performance comparison of the three methods on AP90-S with different thresholds

From Fig. 3, we can see that the performance of all three methods vary with different τ values. Both SL+ST and SpotSigs achieve their best when $\tau = 0.70$, and 3-Shingles achieves its best when $\tau = 0.60$. All three methods do better when τ is close to neither 0 nor 1. This phenomenon is understandable, because F_1 is a measure that combines precision and recall. When τ is very close to 1, then all the methods would be good on precision but bad on recall; when τ is very close to 0, then all the methods would be good on recall but bad on precision. Referring to Eq. 2, F_1 obtains its highest value when precision and recall are equally good at the same time. In overall SL+ST performs better than the two others. SpotSigs is close to SL+ST when τ is no less than 0.3, but worse off when τ is smaller than 0.3.

Let us have a closer look at the points at which each method obtain its best F1 value. Table 2 lists both precision and recall values of all the three methods at that particular point.

Table 2. The best possible performance of each method at certain point

Method	τ	Precision	Recall	F_1
SL+ST	0.70	0.97	0.94	0.96
3-Shingles	0.60	0.82	0.88	0.85
SpotSigs	0.70	0.94	0.93	0.93

4.2 Efficiency

In this experiment, we compare the time needed for each of the three methods to do the task. Based on [19], pruning is conducted before applying each of the methods. The Twitter document collection is used in the experiment.

Figure 4 shows the time needed for each of the three methods: SL+ST, 3-Shingles and SpotSigs with different collection size. Those collections with different size are generated by choosing documents at certain intervals from the Twitter document collection.

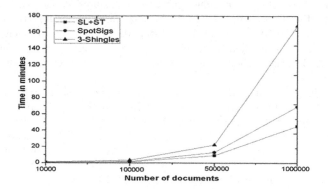

Fig. 4. Running time of SL+ST, SpotSigs and 3-Shingles with different collection sizes

In Fig. 4, the horizontal axis shows the size of the document collection used, while the vertical axis indicates the time (in minutes) needed for detecting all near-duplicates in the given document collection. When a collection of 100,000 documents are considered, it takes SL+ST 1.2 min to do the work, while the time needed for 3-Shingles is 3.3 min and for SpotSigs is 1.6 min. When the document collection adds up to 1,000,000 documents, the time needed for SL+ST, 3-Shingles and SpotSigs are 45 min, 168 min and 69 min, respectively. L+ST is always faster than the other two methods. The difference between them is even slightly larger when more documents are involved.

5 Conclusions

In this paper we have presented our method SL+ST of detecting near-duplicates. Experiments with two groups of documents show that our method is effective and efficient. In our experiments, it performs better than the two representative methods SpotSigs and 3-Shingles. Therefore, the method proposed has good potential to be used in practice.

One advantage of our method is that we are able to display the location of identical contents between the two documents compared, although this part has not been presented in this paper due to space limitation. This may be useful in certain applications.

References

1. Andrei, Z.B., Steven, C.G., Mark, S., Manasse, G.Z.: Syntactic clustering of the web. Comput. Netw. **29**(8–13), 1157–1166 (1997)
2. Chowdhury, A., Frieder, O., Grossman, D., McCabe, M.C.: Collection statistics for fast duplicate document detection. ACM Trans. Inf. Syst. **20**(2), 171–191 (2002)
3. Lin, Y.S., Liao, T.Y., Lee, S.J.: Detecting near-duplicate documents using sentence-level features and supervised learning. Expert Syst. Appl. **40**, 1467–1476 (2013)
4. Wang, J.-H., Chang, H.-C.: Exploiting sentence-level features for near-duplicate document detection. In: Lee, G.G., Song, D., Lin, C.-Y., Aizawa, A., Kuriyama, K., Yoshioka, M., Sakai, T. (eds.) AIRS 2009. LNCS, vol. 5839, pp. 205–217. Springer, Heidelberg (2009)
5. Shivakumar, N., Garcia-Molina, H.: SCAM: a copy detection mechanism for digital documents. In: Proceedings of the International Conference on Theory and Practice of Digital Libraries (1995)
6. Zhang, Q., Zhang, Y., Yu, H.M., Huang, X.J.: Efficient partial-duplicate detection based on sequence matching. In: Proceedings of ACM SIGIR, pp. 675–682 (2010)
7. Dean, J., Ghemawat, S.: MapReduce: simplified data processing on large clusters. In: Proceedings of 6th Symposium on Operating System Design and Implementation (2004)
8. Chang, H.C., Wang, J.H., Chiu, C.Y.: Finding event-relevant content from the web using a near-duplicate detection approach. In: Proceedings of the 2007 IEEE/WIC/ACM International Conference on Web Intelligence, pp. 291–294 (2007)
9. Hoad, T., Zobel, J.: Methods for identifying versioned and plagiarized documents. J. Am. Soc. Inf. Sci. Technol. **203–215**, 54 (2003)
10. Manku, G.S., Jain, A., Sarma, A.D.: Detecting near-duplicates for web crawling. In: Proceedings of the 16th International Conference on World Wide Web, pp. 141–150 (2007)
11. Schleimer, S., Wilkerson, D., Aiken, A.: Winnowing: local algorithms for document fingerprinting. In: Proceedings of the 2003 ACM SIGMOD, pp. 76–85 (2003)
12. Brin, S., Davis, J., Garcia-Molina, H.: Copy detection mechanisms for digital documents. In: Proceedings of ACM SIGMOD, pp. 388–409 (1995)
13. Salton, G.: The state of retrieval system evaluation. Inf. Process. Manage. **28**(4), 441–448 (1992)
14. Frakes, W.B., Baeza-Yates, R.A.: Information Retrieval: Data Structures and Algorithms. Prentice Hall, Englewood Cliffs (1992)
15. Baeza-Yates, R., Ribeiro-Neto, B.: Modern Information Retrieval: The Concepts and technology Behind Search. Pearson Education Limited, Harlow (2011)
16. Ukkonen, E.: On-line construction of suffix tree. Algorithmica **14**(3), 249–260 (1995)
17. Huang, L., Wang, L., Li, X.: Achieving both high precision and high recall in near-duplicate detection. In: Proceedings of ACM CIKM, pp. 63–72 (2008)
18. Yerra, R., Ng, Y.-K.: A sentence-based copy detection approach for web documents. In: Wang, L., Jin, Y. (eds.) FSKD 2005. LNCS (LNAI), vol. 3613, pp. 557–570. Springer, Heidelberg (2005)
19. Theobald, M., Siddharth, J., Paepcke, A.: SpotSigs: robust and efficient near duplicate detection in large web collections. In: Proceedings of ACM SIGIR, pp. 563–570 (2008)

A Dwell Time-Based Technique for Personalised Ranking Model

Safiya Al-Sharji$^{(\boxtimes)}$, Martin Beer, and Elizabeth Uruchurtu

Communication and Computing Research Institute, Sheffield Hallam University,
153 Arundel Street, Sheffield , S1 2NU, UK
Safiya.M.AlSharji@student.shu.ac.uk,
{M.Beer,E.Uruchurtu}@shu.ac.uk

Abstract. The aim of a Personalised Ranking Model (PRM) is to filter the top-k set of documents from a number of relevant documents matching the search query. Dwell times of previously clicked results have been shown to be valuable for estimating documents' relevance. The indexing structure of the dwell time is an important parameter. We propose a dwell time-based scoring scheme called *Dwell-tf-idf* to index text and non-text data, based on which search results are ranked. The effectiveness of incorporating into the ranking process the proposed *Dwell-tf-idf* scheme is validated by a controlled experiment which shows a significant improvement in the search results within the top-k rank.

Keywords: Dwell time · Search logs · Web search · Scoring scheme · Ranking

1 Introduction

While previous research [1–4] has incorporated the dwell time (referred to as dwell for simplicity) as an implicit feedback feature, a thorough exploration of its indexing structure remains unfortunately stagnant. The approach proposed here takes advantage of the keyword-based approach used in the classic Vector Space Model (VSM) and extends it by incorporating a new scoring scheme called *Dwell-tf-idf* to combine textual and non-textual retrieval (i.e. dwell) under a single indexing umbrella. Following on from [5], we present a framework of a *Dwell-tf-idf*-based PRM. The aim is to investigate the effects of the dwell and its relation in providing constant search results for individual users in the educational context.

Users' log files (see Table 1, detailed in [5]) are valuable implicit feedback based on which *tf-idf* standard measures are used to identify users' individual interests and preferences. This information and the dwells of the corresponding clicked documents are employed to infer relevant documents for the individual user. The proposed multi-strategy approach not only calculates the document's ranking score based on textual relevance, but also follows the tactic of [6] for retrieval and ranking mechanisms; uses the ideas of [7] for indexing data related to the dwell; adopts the frequency-sorted indexes, typically the retrieval of text-based information provided in [8] to access documents containing the query term.

© Springer International Publishing Switzerland 2015
Q. Chen et al. (Eds.): DEXA 2015, Part II, LNCS 9262, pp. 205–214, 2015.
DOI: 10.1007/978-3-319-22852-5_18

Table 1. Small sample of a user's log file

Keyword	URL clicked	Time stamp	URL position
Jupiter facts	http://space-facts.com/jupiter/	10:02:08	1
Insomnia	http://en.wikipedia.org/wiki/Insomnia	10:08:10	2
Insomnia	http://sleepfoundation.org/insomnia/home	10:12:56	8

In summary, the contributions of this paper include: (1) a general information retrieval (IR) approach to re-rank the search results returned by a non-personalised Search Engine (SE) based on a real life collection of search logs to provide a reliable personalised ranking model. (2) Of particular interest, a new scoring scheme called *Dwell-tf-idf*; makes a move beyond simple keyword matching to improve future retrieval applications for web browsers. The scoring scheme uses an efficient hybrid indexing technique to combine two different types of data.

2   Related Studies

Previous studies can be grouped into two main categories: (1) general PRM with search log-based analysis; (2) a dwell-based PRM to extend the *tf-idf* scheme of VSM aiming to provide static search results based on individual users' interests.

Jiang et al. [9] claimed that longer dwells indicate the importance of the page in the session; unfortunately, their claim is not empirically demonstrated in their report. Kelly et al. [1] used dwell time, named '*display time*', to present the results of a naturalistic study into how behaviour could be used as implicit feedback for the relevance of a document. Collins-Thompson et al. [3] incorporated the dwell parameter into computing document relevancy to investigate the reading proficiency of users. They noted that dwells provide a valuable new relevance signal for personalised Web search. Via search logs from a commercial Web SE, Hassan et al. [4] used dwells and result clicks to estimate users' satisfaction. They demonstrated that longer dwells are highly correlated with users' satisfaction as opposed to the findings of Guo et al. [10] who reported that post-click behaviour is more significant than dwell in inferring search result relevance.

While much of this research led to an improvement in PRM, their algorithms mine browsing history to analyse users' search activities in terms of user dwells rather than re-ranking the search results. The exception is Agichtein et al. [2] who measured a wide range of user behaviours to demonstrate that the ordering of top results in real web search setting can be significantly improved by using dwell as a user's implicit feedback. Xu et al. [6] also identified that the accuracy of existing algorithms is affected by the lack of a finer granularity in the representation of dwells. Their works are the closest to the current work and we all address these issues by employing user dwells captured at document level to derive users' dwells related to individual user's interests. However, while their works focused on the access representation side of the retrieval process, we focused on the indexing mechanism for a combination of textual and non-textual data types for the retrieval process.

3   Personalised Ranking Model

This section defines the problem of *Dwell-tf-idf* scoring scheme to re-rank the first three pages [11] of the search results returned by the Google Search Engine (GSE).

3.1   Pereliminaries

This section describes definition terms particular only to this project, the reader is referred to [12] for terms generally understood or/and details about the VSM.

Dwell Time.   Time stamp is defined as the time when the click occurred. Dwell time[1] is the document visiting time. This has previously been defined as both the interval between the page being loaded and the searcher leaving the page [13], and as the basic indicator of document relevance [10]. It is calculated in this paper by taking the time difference between two successive clicks as there is no direct means of measuring either.

Dwell-Keyword and Dwell Relevancy.   Assuming that $D = \{d_1, d_2, ...d_N\}$ represents a collection of N documents where each document d contains a set of keywords K_d and a dwell S_d corresponding to time the user will visit the document. A Dwell-Keyword is defined as $W = \{K_w, S_w\}$, where K_w is a set of keywords entered and S_w is the actual dwell found as mentioned above.c

   Dwell relevancy of a document d with respect to keyword w, denoted as $sRelw(d)$, is defined as the type of dwell relationship that exists between S_w and S_d, where S_w is the dwell for keyword w and S_d is the dwell for document d. A document d is said to be dwell relevant to keyword w if there exists a non-empty intersection between at least one of the keywords with one of the documents in that range; that is $S_w \cap S_d \neq \varnothing$. Literally, the larger the intersection, the more dwell relevant d and w are.

Textual Relevancy and Dwell-Keyword Relevancy.   A textual relevancy of a document d with respect to keyword w denoted as $kRelw(d)$ is the type of relationship that exists between document d and keyword w. A document d and keyword w are textually relevant if both d and w have at least one keyword in common; that is $K_w \cap K_d \neq \phi$. In other words, the more keywords w and d have in common, the more textually relevant they are.

   A Dwell-Keyword relevancy of a document d with respect to keyword w denoted as $skRelw(d)$ can be monotonically derived from a scoring function of both textual and

---

[1] A commonly-used threshold is a dwell of at least 30 s [10, 13], a manual check of our data set indicated the longest dwell to be less than 15 min - Time range used is thus 30"- 15'.

dwell relevancy. A document d is said to be Dwell-Keyword relevant to keyword w if it has both dwell relevance and textual relevance. The normalised function for the weighted sum of dwell relevancy and textual relevancy denoted as $skRelw(d)$ and previously developed in [5], can be represented as:

$$skRel_w(d) = \begin{cases} \alpha.sRel_w(d) + (1 - \alpha).kRel_w(d) & if \; sRel_w(d) > 0 \\ & andkRel_w(d) > 0 \\ 0 & otherwise \end{cases} \quad (1)$$

where α is a parameter that assigns relative weights to dwell and textual relevance.

Dwell-Keyword search. A Dwell-Keyword search identifies all the documents $D = \{d_1, d_2, ...d_N\}$ having *Dwell-Keyword* relevancy with the keyword w. The result list shows the top-k most Dwell-Keyword documents ranked based on the Dwell-Keyword Relevance scores.

3.2 Problem Definition

Our problem statement is set analogously to the Dwell-Keyword search above. Each document is given a weight score based on the statistical distributions of each user's Dwell-Keyword searches to reflect the user's interests based on which the order of the top-k documents is re-ranked. We formulate our problem statement similar to the Kaggle and Yandex 2014 competition[2] which was framed thus:

> *"participants need to personalise search using the long-term (user history-based) and short-term (session-based) user context. The evaluation relies on a variant of a dwell-time based model of personal relevance and is data-driven, as it is presently accepted in the state-of the-art research on personalised search".*

Through the proposed approach, we seek to answer two questions. Firstly, whether the dwells related to the previously visited documents can be used as predictor functions to provide a reliable PRM. Secondly, whether the dwell-based function scheme will guarantee the delivery of a lower level of irrelevant information (i.e. ensuring higher precision) while displaying most of the relevant information (i.e. ensuring higher recall) than is possible when using the standard *tf-idf* function.

To rank and identify relevant textual documents, all SEs use a weighting scheme [12] based on a similarity measure (i.e. variant and intuitive). This paper thus raises two interrelated issues for the representation of documents: weighting of terms and measuring of feature similarities. With respect to the weighting related to text, the *tf-idf* weighting scheme is based on the *tf*, the *idf* and the *ntf* empirical observations [14].

Dwell-Keyword Based Ranking. To identify and rank textual documents, the *tf*, *idf* and *ntf* [14] are used and the final weight is then based on the similarity measure between a document d and the query q using the inner product function [14]. To index and search

---

[2] http://www.kaggle.com/c/yandex-personalised-web-search-challenge

both textual and non-textual (i.e. dwell-based variable) data we define a new scoring mechanism called *Dwell-tf-idf* (see next section) to rank and identify documents which are Dwell-Keyword based. It uses the data of non-textual types from the input query [7] to identify the dwell relevance mapped onto the textual relevance, thus predicting those documents that best match the keywords [5] related to Dwell-Keyword.

Dwell-tf-idf. The dwell range can be partitioned into grid cells each of which is assigned a primary key to uniquely identify it. Each dwell of the keyword is thus associated with a set of cells. The intersection relationship being our concern, only the cells which intersect with the keyword are considered. In relation to the *tf* of a keyword in document *d*, given the existence of the intersection relationship and textual relevance described above, we can analogously define the frequency of a cell *p* in document *d*. It is the intersection of the keyword dwell S_d and the cell *p* in that document which can be used to provide the relevancy of that cell with respect to the document. A function $f_{d,p}$ for the frequency of the cell *p* in document *d* can thus be found using the intersection between the dwell S_d and the cell *p* divided by the cell range as shown in Eq. 2. This intersection between the cell and the document can provide the relevancy measure of that cell to the related document.

$$f_{d,p} = \frac{S_d \cap p}{p} \tag{2}$$

where S_d is the dwell and *p* is a cell of the dwell. Thus, the larger the intersection, the better the cell is related to the document dwell. By relating the term *tf*, *idf* and *ntf* to the dwell *tf*, *idf* and *ntf*, we formulate our *Dwell-tf-idf*. Recalling the above observations [14], cells having a larger intersection with the document are given more weight.

4 Experiment Results and Discussions

The findings of these experiments are related to the *Dwell-tf-idf* and its capability of combining textual and non-textual data types. Further results are reported in [5].

4.1 Data Set

Our data set originates from a set of log files via search logs collected from non-commercial individuals, who all work at the Educational Technology Centre (ETC) at the Nizwa College of Technology (NCT) in Oman. They are all professional researchers with web search experience; hence we were able to obtain rich search logs from their searching histories. This data sample was collected over a period of three weeks [4, 15] and it included 1004 clicked URLs with the keyword, URL position and time stamp of each. The dwell related to the last clicked URL (i.e. it often indicates the searcher's satisfaction) was calculated by averaging the dwells of a session.

Although we did not separate the respondents into two groups, the experiment conducted in this project can be considered a controlled experiment since we increased the reliability of our results with the PRM as our independent variable to test the rank of its returned results. The PRM was installed on each participant's machine and the collection of the search logs related to the GSE started two days later. The PRM is designed with two links. One link opens the GSE home page, and the other opens the proposed PRM. Participants were instructed to select the PRM on the first 2 days, then to select the GSE for the next 5 days, and again to use the PRM for the last two weeks. The idea of collecting evaluation feedback in this way was inspired by [16] although they designed their work in the form of a game.

The findings involving a sample of 7 subjects [1] could not be generalised to larger populations, but the range and quality of the data collected in the survey provided abundant insight into personalisation research. A Web-based personal information management survey presented in [17], involved 36 participants for 3 weeks. With 48 respondents involved in this study, this sample size is even more valid, especially in terms of search log data as opposed to a study related to a large-scale evaluation involving human labellers who are not query owners. The total consisted of three weeks of data collection, with 979 clicked URLs (i.e. useful data) out of which 322 were obtained from GSE and 657 from PRM resulting from 157 different keywords issued by these 48 participants including at least 2 predefined keywords per participant.

4.2 Data Analysis and Relevance Estimation Evaluation

The findings of these experiments assess the relationship between the proposed re-rank of our model and the individual perceived relevancy of our proposed re-rank when the dwell is incorporated. Since the PRM needs to use the individual users' search histories to learn their individual interests in order to re-rank the search results according to their personal needs, the PRM is not expected to produce a better performance than the GSE if used first - which forms our null hypothesis.

To measure the consistency of the judgment of the participants, they were required to score four statements [5] for also at least two keywords that they were asked to select from a set of predefined keywords. All of these keywords were of general interest (e.g. *Old Oman*, *USA wars*, *Jupiter facts* and *insomnia*). The average agreement derived from participants' feedback on these data was 66.78 %, which indicates that the participants' evaluations were reliable. Using the evaluation feedback, we were able to determine how the returned results matched the participants' search interests based on the overall average of the Mean Average Precision (MAP), precision, recall, and F-Measure[3] [12] of the corresponding keywords for both GSE and the proposed PRM. The obtained results are provided in Table 2 and are here referred to as *Second Experiment*. The results of PRM pre-dwell (identified by an *) are included in Table 2 solely for comparison purposes, but are derived from our previous experiment.

---

[3] The terms 'F-Measure' and 'F-Score' are used interchangeably for convenience throughout this paper as in some literature reviews.

Table 2. MAP, precision, recall and f-score values – second experiment

	MAP	Precision					Recall	F-Score
		@1	@3	@5	@10	@20		
*PRM Pre-dwell	0.85	0.85	0.89	0.88	0.84	0.77	0.64	0.72
Post-dwell (day 1-2)	0.74	0.63	0.71	0.73	0.82	0.80	0.54	0.63
GSE (Day 3-9)	0.74	0.63	0.71	0.73	0.81	0.80	0.55	0.63
PRM Post-dwell[a]	0.88	0.98	0.96	0.95	0.87	0.80	0.69	0.78

[a]Days 10–21, these numbers are omitted from Table 2 due to limited space.

A number of interesting observations were drawn. Firstly, a MAP falling between 0.74 (74 %) and 0.88 (88 %) in the retrospective experiments indicates a difference in SE performance - our PRM achieved better retrieval effectiveness (14 % improvement).

Secondly, we can see from the same table that all F-Scores of PRM are higher than those of the GSE, which indicates the PRM's improvement in accuracy. In the lower ranks, the precision and recall of the PRM are higher than those of the GSE, which indicates that the PRM returns not only more relevant results than GSE but also most of the relevant results. It can be seen that the PRM significantly improves precision @1, but it improves less significantly for precision @3–@20 indicating that PRM can be more effective in lower ranks than higher ranks. Precision @1–@5 values are constants to an extent for the PRM, which indicates static results for individual users. However, the GSE somehow fails to provide constant values, reflecting changes in results and thereby hindering the users' ability to find what they are looking for [18].

Finally, these differences are also noticeable when the PRM pre-dwell is compared with the PRM post-dwell which is consistent with the findings in [6, 13]. Considering the PRM post-dwell results from days 10–21 and using our null hypothesis that the PRM does not produce a better performance than the GSE, we were able to reject our null hypothesis. This is also clear from the post-dwell results on days 1–2 of our experiment (see Table 2), which indicate that the null hypothesis holds, since we obtained the same results from the GSE and PRM (apart from a difference of 0.01–1 % - in both precision @10 and average recall which might be due to feedback error when users abandon the clicks), an interesting property that shows the reliability of the ranking technique employed in our PRM.

To further assess the reliability of the evaluations of our PRM, we also conducted a 10-fold cross-validation [19] - the results are plotted in Fig. 1 - to ensure that it does not suffer over-fitting or under-fitting despite the fact that it might be computationally intensive and this is not often necessary in a small-scale evaluation. Irrespective of the subsets k-fold used, the results gave an average F-measure of 74.3 % and precision of 95.5 %, which indicates that the PRM is both reliable and achieved good results.

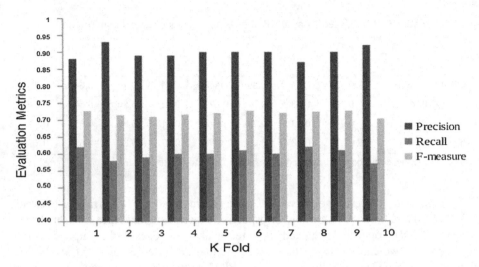

Fig. 1. K-Cross fold results

5 Conclusions

This paper has presented a novel personalised ranking model using dwell features to (1) identify individual users' interests and extract search information from web pages; (2) employ the Dwell-Keyword search as the representation of individuals' search information using a new scoring scheme called Dwell-tf-idf; (3) use the Dwell-Keyword as a feature to develop a reliable predictor function to determine the relevance ranking of PRM. The findings (see Table 2) were demonstrated to be effective enough to provide a personalised ranking model with a low level of non-relevant information (high precision) while displaying most of the relevant information (high recall) thus allowing users to directly retrieve what they need based on their queries.

We collected real life search logs returned by the GSE from consenting participants. These enabled us to identify users' individual preferences using their dwells inferred from dwells acquired through previously clicked documents in order to re-rank the top-k list of search results. The effectiveness of our PRM was validated through a controlled experiment which revealed a 14 % improvement in the overall average MAP in the top-k rank, interestingly consistent (14 % improvement in average F-Measure) with the results previously reported [5] for 3 queries.

These conclusions are based on our experimental data sets consisting of 979 documents, an IR scale-up limitation which must be taken into consideration for retrieval systems based on large-scale operational databases.

Acknowledgement. The authors extend their sincere thanks to the Dean, the Head of ETC and staff at the NCT in Oman for their cooperation and support during the data collection.

References

1. Kelly, D., Belkin, N.J.: Display time as implicit feedback: understanding task effects. In: Proceedings of the 27th Annual International SIGIR Conference on Research and Development in Information Retrieval, pp. 377–384, ACM (2004)
2. Agichtein, E., Brill, E., Dumais, S.: Improving web search ranking by incorporating user behaviour information. In: Proceedings of the 29th Annual International SIGIR Conference on Research and Development in Information Retrieval, pp. 19–26, ACM (2006)
3. Collins-Thompson, K., Bennett, P.N., White, R.W., De La Chica, S., Sontag, D.: Personalising web search results by reading level. In: Proceedings of the 20th International Conference on Information and Knowledge Management, pp. 403–412, ACM (2011)
4. Hassan, A., White, R.W.: Personalised models of search satisfaction. In: Proceedings of the 22nd International Conference on Information and Knowledge Management, pp. 2009–2018, ACM (2013)
5. Al Sharji, S., Beer, M., Uruchurtu, E.: Enhancing the degree of personalisation through vector space model and profile ontology. In: IEEE RIVF International Conference Computing and Communication Technologies, Research, Innovation, and Vision for the Future (RIVF), pp. 248–252, IEEE (2013)
6. Xu, S., Jiang, H., Lau, F.: Mining user dwell time for personalised web search re-ranking. In: Proceedings of the 22nd International Joint Conference on Artificial Intelligence, vol. 3, pp. 2367–2372, AAAI Press (2011)
7. Khodaei, A., Shahabi, C., Li, C.: SKIF-P: a point-based indexing and ranking of web documents for spatial-keyword search. GeoInformatica **16**(3), 563–596 (2012). GeoInformatica
8. Zobel, J., Moffat, A.: Inverted files for text search engines. Comput. Surv. (CSUR) **38**(2), 6 (2006). ACM
9. Jiang, D., Pei, J., Li, H.: Mining search and browse logs for web search: a survey. Trans. Intell. Syst. Technol. (TIST) **4**(4), 57 (2013). ACM
10. Guo, Q., Agichtein, E.: Beyond dwell time: estimating document relevance from cursor movements and other post-click searcher behaviour. In: Proceedings of the 21st International Conference on World Wide Web, pp. 569–578, ACM (2012)
11. Silverstein, C., Marais, H., Henzinger, M., Moricz, M.: Analysis of a very large web search engine query log. SIGIR Forum **33**(1), 6–12 (1999). ACM
12. Manning, C.D., Raghavan, P., Schütze, H.: Introduction to Information Retrieval. Cambridge University Press, Cambridge (2008)
13. Hassan, A., Jones, R., Klinkner, K.L.: Beyond DCG: user behaviour as a predictor of a successful search. In: Proceedings of the 3rd International Conference on Web Search and Data Mining, pp. 221–230, ACM (2010)
14. Salton, G.: Automatic text processing: the transformation, analysis, and retrieval of information by computer. Addison-Wesley (1989)
15. Hu, Y., Qian, Y., Li, H., Jiang, D., Pei, J., Zheng, Q.: Mining query subtopics from search log data. In: Proceedings of the 35th International SIGIR Conference on Research and Development in Information Retrieval, pp. 305–314, ACM (2012)
16. Ageev, M., Guo, Q., Lagun, D., Agichtein, E.: Find it if you can: a game for modeling different types of web search success using interaction data. In: Proceedings of the 34th International SIGIR Conference on Research and Development in Information Retrieval, pp. 345–354, ACM (2011)

17. Elsweiler, D., Ruthven, I.: Towards task-based personal information management evaluations. In: Proceedings of the 30th Annual International SIGIR Conference on Research and Development in Information Retrieval, pp. 23–30, ACM (2007)
18. Chia-Jung, L., Teevan, J., De La Chica, S.: Characterising multi-click search behavior and the risks and opportunities of changing results during use. In: Proceedings of the 37th International SIGIR Conference on Research and Development in Information Retrieval, pp. 515–524, ACM (2014)
19. Bengio, Y., Grandvalet, Y.: No unbiased estimator of the variance of k-fold cross-validation. J. Mach. Learn. Res. **5**, 1089–1105 (2004)

An Evaluation of Diversification Techniques

Duong Chi Thang, Nguyen Thanh Tam,
Nguyen Quoc Viet Hung[(✉)], and Karl Aberer

École Polytechnique Fédérale de Lausanne, Lausanne, Switzerland
{thang.duong,tam.nguyenthanh,quocviethung.nguyen,karl.aberer}@epfl.ch

Abstract. Diversification is a method of improving user satisfaction by increasing the variety of information shown to user. Due to the lack of a precise definition of information variety, many diversification techniques have been proposed. These techniques, however, have been rarely compared and analyzed under the same setting, rendering a 'right' choice for a particular application very difficult. Addressing this problem, this paper presents a benchmark that offers a comprehensive empirical study on the performance comparison of diversification. Specifically, we integrate several state-of-the-art diversification algorithms in a comparable manner, and measure distinct characteristics of these algorithms with various settings. We then provide in-depth analysis of the benchmark results, obtained by using both real data and synthetic data. We believe that the findings from the benchmark will serve as a practical guideline for potential applications.

1 Introduction

The diversification problem has been long acknowledged in information retrieval [8,10]. Differently from traditional information retrieval techniques which focus on the relevance of search results, diversification is a method of spreading a variety of data shown to user [14,23]. Improving diversity would increase not only the amount of information displayed with a limited number of data items but also the probability of delivering at least one piece of information that truly matches user intent. Moreover, focusing only on relevance might lead to redundancy and biases in the retrieval results due to similarity in structure and content of data items and coverage limitation of search engines [21]. For these reasons, diversity has become a crucial property to enhance user satisfaction through providing a general view that covers different aspects (i.e. subtopics) of data. Moreover, diversification is not only applicable to information retrieval but also numerous other domains, including web search [2,18], large-scale visualization [19], recommender systems [17,28], and novelty detection [12,22].

Diversification implies a trade-off between selecting data of relevance to user intent and filtering data having similar characteristics. As such, diversification is often characterized as a bi-criteria optimization problem, in which the twin objectives of being relevant and being dissimilar compete with each other [9]. To tackle this problem, a rich body of research has proposed different diversification

© Springer International Publishing Switzerland 2015
Q. Chen et al. (Eds.): DEXA 2015, Part II, LNCS 9262, pp. 215–231, 2015.
DOI: 10.1007/978-3-319-22852-5_19

techniques, ranging from *threshold-based* approach to *function-based* approach and *graph-based* approach. In general, the threshold-based approach defines a threshold on one criterion (i.e. either relevance or diversity), and then selects the data that both satisfy this threshold and optimize the other criterion. The function-based approach combines both relevance and diversity in a unified function, and then finds a set of data that maximizes this function. The graph-based approach models data items and their relationships as a graph, and then ranks the data according to the collective information inferred from the graph.

While many diversification techniques have been developed over the last decades, there has been little work on the evaluation of their performance altogether. To this end, various IR test collections have been created for diversity track such as TREC [7], NTCIR [16], and CLEF [4]. In that, participants are able to test their proposed methods using common real-world datasets and metrics [8]. Although these real-world datasets provide a pragmatic view, they do not allow participants to compare their methods in different settings such as the number of subtopics in input data or the number of displayed items. As a result, an algorithm may perform poorly within the common settings but may be able to achieve good performance under another different setting. In addition, the proposed methods may not be evaluated in a comparable manner as the evaluation is not conducted by a third party and under the same system. To this end, we present an evaluation of diversification techniques within a common benchmarking framework that offers the following salient features:

- We selected and implemented the most representative diversification techniques, including threshold-based approach: Swap [24] and Motley [15]; function-based approach: MMR [5] and MSD [13]; and graph-based approach: Affinity Graph [26] and GrassHopper [27]. In addition to comparison purposes, our framework provides reusable components to reduce the development time.
- We designed a generic, extensible benchmarking framework to assist in the evaluation of different diversification techniques, so that subsequent studies are able to easily compare their proposals with the state-of-the-art techniques.
- We compare the above diversification techniques in a fair manner using the same system and settings. Moreover, our experimental results are reliable, reproducible and extensible as the source code of the benchmarking framework as well as the datasets are publicly available.
- We simulated different settings of relevance and diversity measures. In particular, the framework allows users to vary configurable parameters and visualize their effects. Through empirical observations, user is given more insights and better understandings of the behavior of evaluated techniques.
- We offer extensive as well as intensive performance analyses. We believe that these analyses can serve as a practical guideline for how to select a well-suited diversification technique on particular application scenarios.

The remainder of this paper is organized as follows. Section 2 reviews state-of-the-art diversification techniques. We then discuss the benchmarking methodology in Sect. 3. Section 4 describes the experiments on both real and synthetic data. Section 5 summarizes and concludes the benchmarking study, where we provide important suggestions for applications that need diversification.

2   Diversification Techniques

The problem of diversification can be formulated as follows. It takes as input a triple $\langle D, R, M \rangle$. In that, $D = \{d_1, \ldots, d_n\}$ is a set of data items. $R = \{r_1, \ldots, r_n\}$ is a set of relevance scores, in which each r_i is associated with each item d_i. $M = [m_{ij}]_{n \times n}$ is a matrix in which each element m_{ij} measures the dissimilarity between item d_i and item d_j. Given a limited-budget number of displayed items k ($k \ll n$), the problem output is a subset of items $D_k^* \subseteq D$ such that $|D_k^*| = k$ and the selected items are simultaneously having high relevance scores and being highly dissimilar among themselves. When the limited-budget number of displayed items k is clear from the context, we denote the result set D_k^* as D^*. In this section, we offer a description of the six diversification techniques studied in the benchmark. These diversification techniques are carefully selected based on the following criteria: (1) they are the representatives for their respective approach and (2) they are often referred in the research community.

Swap. Swap [24] is a *threshold-based* algorithm that firstly focuses on relevance and then gradually improves diversity. Technically, it initializes the output D^* with the top-k most relevant items and iteratively traverses the remaining items $D \setminus D^*$ in the decreasing order of relevance scores. In each iteration, the currently traversed item is swapped with the item in D^* that is the least dissimilar to the others in D^* if the diversity increases and the relevance drop is not below a pre-defined threshold. The process stops when there is no remaining item left to traverse.

Motley. Motley [15] is also a *threshold-based* algorithm. Unlike Swap, it starts from an empty set and constructs the output by incrementally adding data items in the decreasing order of relevance scores. An item d_i in turn is added to the output D^* if for every item $d_j \in D^*$, the dissimilarity m_{ij} is higher than a predefined threshold. The procedure stops when all items are considered or k items are already included to the output. In the case where all items are considered but there is still available budget, $k - |D^*|$ items from $D \setminus D^*$ are selected randomly and added to D^*.

Maximal Marginal Relevance (MMR). This algorithm belongs to the *function-based* approach in which the twin aspects of relevance and diversity are combined in a comprehensive objective function. More precisely, MMR [5] defines this objective function as $f(D^*) = (1 - \lambda) \sum_{d_i \in D^*} r_i + 2\lambda \sum_{d_i, d_j \in D^*} m_{ij}$, where λ is a tunable parameter that specifies the preference between relevance and diversity. Since maximizing this function is NP-hard, MMR takes a greedy method that builds the output D^* incrementally with k iterations, in each of which an item d_i is selected if the value $\phi(d_i) = (1 - \lambda)r_i + \lambda \sum_{d_j \in D^*} m_{ij}$ is maximal. The core idea is that the objective function can be rewritten as

$f(D^*) = \sum_{d_i \in D^*} \phi(d_i)$ and maximizing the subterm $\phi(d_i)$ in each iteration is expected to approximately maximizing the whole function.

Max-Sum Dispersion (MSD). This is a *function-based* algorithm [13] like MMR that also takes a greedy method to maximize the above objective function. However, MSD takes another rewritten form: $f(D^*) = \sum_{d_i,d_j \in D^*} \varphi(d_i, d_j)$, where $\varphi(d_i, d_j) = \frac{1}{2}(1 - \lambda)(r_i + r_j) + 2\lambda m_{ij}$. The difference between MSD and MMR is that instead of selecting one item at a time, MSD picks two items d_i and d_j that have maximal $\varphi(d_i, d_j)$ value. When k is odd, the last item is selected randomly to add to the result set.

GrassHopper. This technique [27] belongs to the *graph-based* approach. It constructs a graph, which models the diversity and relevance of data items, from the dissimilarity matrix M and the relevance scores R. The graph serves as the representation of underlying states and transitions of an absorbing Markov chain which is used to rank the data items. Based on the Markov chain, the algorithm iterates between two routines: (1) turn the currently selected data items into absorbing states, and (2) add the new item with the highest expected number of visits before absorption to the output. As a consequence, the item set is diversified since the higher the expected number of visits, the more dissimilar between the new item and the previously selected items.

Affinity Graph (AG). AG [26] is also a *graph-based* algorithm, which ranks each data item by taking into account not only its relevance and its diversity but also its importance. The importance of an item is computed from the stationary distribution of a Markov chain whose transition matrix is constructed from the dissimilarity matrix M. AG in turn combines the relevance, the diversity, and the importance into a unified ranking score. The output D^* is then constructed by selecting the top-k items with the highest scores.

Summary: we have implemented representative algorithms in each approach. The implemented algorithms are AG, GrassHopper, MSD, MMR, Swap and Motley. Each algorithm exhibits various diversification characteristics. In fact, often these characteristics are not exclusive; a technique might have multiple ones. Table 1 features each implemented technique with the following key characteristics.

- **Balance:** algorithms that provide a tunable parameter to balance between diversity and relevance.
- **Data processing:** the ability to perform (online or offline) in response to the new arrival of data items. An online technique can process item-by-item in a serial fashion, whereas offline ones have to re-compute the whole result set.
- **#Parameters:** this characteristic is important as the higher number of parameters an algorithm requires, the harder for the users to configure the algorithm correctly.

It is noteworthy that all the diversification techniques require some parameters and the implementation of these techniques requires searching for a suitable parameter settings, which is a critical issue to evaluate the techniques. In addition to the tradeoff parameter λ, some of these methods need other tunable parameters such as the damping factor in graph-based approach which might significantly affect the performance of the algorithms. However, finding the best parameters for each technique would be difficult; and even if we could find them, it would be unfair to the techniques with fewer parameters. Therefore, for fair comparison, all the parameter settings are decided based on the original authors' recommendation and fixed for all runs. In addition, although the pseudo code for most of the algorithms are available, there are many implementation details that we have to decide. For each decision, we strive for the option which would be fair for all algorithms and report the decision for future references on our website[1].

3   Benchmark Methodology

This section describes the setup used in the benchmark. We first present the system architecture of our benchmarking framework. Then, we provide the detail of two real-world datasets used in the benchmark. Next, we describe the methodology and procedure to generate synthetic data used in the experiments. Furthermore, we offer the descriptions of the measures used to assess the diversification techniques. Finally, we discuss the functionality of our benchmarking tool in analyzing the diversification results.

3.1   Framework

A primary goal of this study is to provide a flexible yet powerful tool to support the comparison and analysis of diversification techniques. To this end, we have developed a framework that employs the original performance study of these techniques. Figure 1 illustrates the component-based architecture of the framework, which is built upon three layers: *data access layer, computing layer*, and *application layer*. The data access layer abstracts the underlying data items (synthetic or real data) and feeds them to upper layers. The application layer interacts with users to receive configurable parameters and visualize outputs from the computing layer. The computing layer consists of two main components: diversification module and simulation module. While the former is plugged with diversification techniques that take data from the data access layer and deliver evaluation measures to the application layer, the latter is responsible for generating synthetic data that is designed to show how well the techniques perform in general.

We believe that subsequent studies are able to easily compare their proposals with the state-of-the-art techniques by using our framework. As presented, it is flexible and extensible, since a new technique as well as a new measurement can be easily plugged in. The framework is available for download from our website[2].

---

[1] https://code.google.com/p/diversity-benchmark.
[2] https://code.google.com/p/diversity-benchmark.

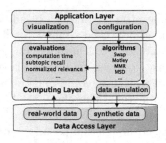

Fig. 1. Benchmarking Framework

Table 1. Characteristics of the algorithm

	Balance	Data processing	#Parameters
Swap	No	Offline	1
Motley	No	Online	1
MMR	Yes	Online	1
MSD	Yes	Online	1
AG	Yes	Offline	3
GrassHopper	No	Offline	1

3.2 Datasets

In this section, we show how to create the data input for the benchmark, which can adapt to both real-world data and synthetic data. While the former provides a pragmatic view, the latter offers different settings for deep examination of the algorithms.

Real-World Dataset. Our framework is adaptable to different real-world datasets. One can import real datasets from his application into our system by converting them into the standard format, which is a set of feature vectors for the data items and their relevance scores. From the provided feature vectors, the dissimilarity between the data items can be computed using existing distance measures (e.g. Euclid, Jaccard, Hamming, fuzzy, and categorical [3]). The dissimilarities are then fed along with the relevance scores to the diversification algorithms in the computing layer. In the following, we discuss two real-world datasets and the procedure to convert them to the standard format.

– *Camera dataset* [11]. The Camera dataset contains information about 497 cameras with 8 attributes per camera such as price, #megapixels and brand.These attributes are used to calculate the dissimilarity between the cameras (using Hamming distance) while the relevance of a camera in the dataset is computed from the price attribute. The referenced diversity for evaluating an algorithm output is measured by the number of brands it contains (see subtopic recall metric).
– *TREC dataset.* We also evaluate the algorithms on the ClueWeb12 dataset which is used in various TREC tracks [7]. More specifically, we leverage the freely-accessible portion of the dataset provided at [1]. It contains 17796 web pages for 50 queries. In order to use the TREC dataset in our framework, a preprocessing step is required to build the feature vectors and relevance scores for the web pages. First, we model the textual information of the web pages by tf-idf feature vectors. As the number of features is high, we also perform dimensionality reduction using latent semantic analysis to keep 50 most significant features. Then, we calculate the relevance scores of the web pages w.r.t a query using the baseline run of the TREC web track [7]. Finally, since the subtopics of the web pages are not available, we perform clustering based

on the feature vectors to group the web pages into 20 clusters (subtopics). It is worth noting that our evaluation results on other test collections (e.g. NCTIR [16], CLEF [4]) are similar and omitted for brevity sake.

Synthetic Data. Synthetic data is generated from the simulation module to help benchmark users to study unbiased evaluations of diversification techniques in a wide range of scenarios. To this extent, we vary the five parameters $\langle n, m, \sigma, \delta, \theta \rangle$: (i) n – the number of data items, (ii) m – the number of subtopics, reflecting the possible characteristics of original data, (iii) σ – the relevance difference between subtopics, reflecting the spectrum of relevance scores of data items, (iv) δ – the subtopic distance, reflecting the dissimilarity between items in different subtopics (the higher the distance between subtopics, the more dissimilar between their items), and (v) θ – the subtopic density difference, reflecting the spectrum of number of data items in the subtopics. The importance of these parameters to the diversification problem is described in Sect. 3.3.

Technically, we model data items as data points. The dissimilarity between items is measured by the Euclid distance between the corresponding points. They are partitioned into various clusters, where each cluster represents a subtopic that the associated items belong to. From this modeling, our simulation module generates synthetic data using the above parameters in the following steps:

1. First, we generate m cluster centroids such that the distance between two centroids is δ, which is varied in $[0, 1]$.
2. We calculate a set of density ratios α: $\{\frac{1}{m} - \lfloor \frac{m}{2} \rfloor \theta, \frac{1}{m} - (\lfloor \frac{m}{2} \rfloor + 1)\theta, ..., \frac{1}{m}, ..., \frac{1}{m} + (\lfloor \frac{m}{2} \rfloor - 1)\theta, \frac{1}{m} + \lfloor \frac{m}{2} \rfloor \theta\}$. The density ratios allow us to calculate the size of the clusters where the first density ratio $\alpha_1 = \frac{1}{m} - \lfloor \frac{m}{2} \rfloor \theta$ is associated with the first cluster and so on.
3. In the x-th cluster, generate $\approx n \times \alpha_x$ points gathering around the centroids (the total number of points is n and α_x is the density ratio of the x-th cluster)
4. Generate relevance scores such that items in the same cluster follow a normal distribution and the difference between the distribution means of two clusters is σ.
5. All relevance and dissimilarity values are then normalized into $[0, 1]$.

It is worth noting that the above simulation process can be customized as there are various parameters that can be changed in addition to the above parameters $\langle n, m, \sigma, \delta, \theta \rangle$. However, for the sake of simplicity, we limit ourselves to the above parameters. There are various design choices that we made regarding the simulation process. First, as each pair of data items has a dissimilarity value and the larger this value is, the more dissimilar between them, it is intuitive that we model the data items as data points and the distance between them as their dissimilarity values. Second, documents or data items that belong to the same subtopics are more similar to ones from different subtopics as they reflect the same aspects of the search keyword. As a result, data items that belong to a subtopic have smaller distance between them, hence, they form a cluster. This means each cluster represents a subtopic. Third, the number of documents varies among subtopics as some subtopics are more popular than the others. This motivates us to vary the size of the subtopics using the subtopic density difference parameter θ. Lastly, among all the subtopics, at most one can express the user intention when searching for the keyword. As a result, some subtopics are more relevant than the others, which means the relevance values vary among the subtopics. We simulate this observation using the relevance difference parameter σ.

3.3   Evaluation Procedure

For comparative evaluation, we use two well-known metrics: *normalized relevance* [20] and *subtopic recall* [25].

- **Normalized relevance:** indicates how well an algorithm preserves relevance when diversity is taken into account. The normalized relevance of a subset $D^* \subset D$ of k items is defined as the sum of their relevance scores over the sum of k highest relevance scores:

$$nRev(D^*) = \frac{\sum_{d_i \in D^*} r_i}{\max_{D_k \subseteq D, |D^k| = k} \sum_{d_i \in D^k} r_i}$$

- **Subtopic recall:** reflects the actual degree of diversity of resulting items. Formally, the subtopic recall of an item set D^* is calculated as the proportion of the number of subtopics it covers over the total number of subtopics in the original item set D:

$$tRec(D^*) = \frac{num\_of\_subtopics(D^*)}{num\_of\_subtopics(D)}$$

For deep understanding, we characterize the diversification methods implemented in the benchmark using six different measures:

- **Computation time:** is an important measure, as every system has limited resources. It helps to choose the right techniques for particular applications under time constraints.
- **Effect of #displayed items:** in practice, it is common that users want to refine the result set by changing the number of displayed items they want to see. As the displayed budget increases, we expect that algorithms are able to cover more subtopics and include more relevant items into the result set. To validate this hypothesis, we need to examine the effect of number of displayed items to the diversification result.
- **Stability:** this property is important to support incremental data exploration. If a user is first presented with the top-10 data items, but then extends the result to the top-20, the expectation is clearly that the top-10 remain unchanged. In other words, the algorithms have to be stable in the sense that the result set can be extended in size to support a user in drilling down into data items. Therefore, there is a need to analyze the stability of the algorithms as the number of displayed items changes.

- **Effect of #subtopics:** the number of subtopics is a common measure to capture the degree of diversity. Since a dataset might cover a large number of subtopics, including all of them in a subset of output data items is challenging. To understand the behavior of diversification techniques, we need to study the effects of varying the number of subtopics in data.
- **Effects of relevance distribution:** beside diversity, relevance is another aspect that determines the course of the diversification problem. For example, if the relevance scores of data items are very similar to each other, relevance becomes a less significant aspect; and thus, algorithms that select data items regardless of their relevance scores might become the winners. To validate this intuition systematically, we examine wide-ranging configurations of the relevance difference parameter σ when generating the synthetic data.
- **Effects of dissimilarity distribution:** another factor that affects diversification is the distribution of dissimilarity values between data items. Algorithms that mainly focus on relevance might become the preferred ones if these dissimilarity values are small. Therefore, it is necessary to analyze the sensitivity of diversification algorithms to the dissimilarity distribution. To this end, we vary the distance parameter δ to obtain various distributions and then compare the algorithms case by case.
- **Effects of subtopic density difference:** the density of the subtopics is an important factor that has a significant effect on the diversification results. For instance, if the difference between subtopic density is high, an algorithm that does not focus on diversifying may include most data items from the densest subtopics while ignoring ones from sparser subtopics. As a result, there is a need to examine the effect of subtopic density difference θ on the performance of diversification techniques.

For the purposes of simplifying understanding and easing comparison, we intentionally restrict the study here to the highlighted metrics and measures. But recall that our benchmarking framework is extensible. In that, various metrics such as Expected Reciprocal Rank [6], α-nDCG [8], and harmonic mean can easily be added. Interested readers can find further details, potential extensions and updates on our website[3].

3.4   Benchmarking Tool

Our benchmarking tool is designed with three main user-interface panels, namely *configuration, quantitative evaluation,* and *qualitative evaluation.*

- *Configuration.* This panel offers benchmark users the flexibility to configure our framework to their own needs. Users can import their own datasets or customize simulation factors to generate synthetic data. Users can select the diversification algorithms and evaluation metrics they want to compare.

---

[3] https://code.google.com/p/diversity-benchmark.

- *Quantitative Evaluation.* The quantitative evaluation panel shows the numeric results plotted through the evaluation metrics. In that, user is able to observe the effects of configured parameters on the performance of diversification algorithms. More importantly, we offer a cross-metric comparison between diversification algorithms. By simultaneously displaying the evaluation results of different metrics, our system allows user to select a well-suited algorithm for his particular applications.
- *Qualitative Evaluation.* We also provide a data visualization in 3D view (when the synthetic data is two-dimensional), in which the XY-axes capture the distance (dissimilarity) between data items whereas the Z-axis represents the relevance score. Users can rotate, zoom in, zoom out the view to analyze different aspects of the results. Through the visualization, users can check the distribution of the output data items and qualitatively evaluate their coverage over the subtopics of original data.

Interested users can download the tool from our website[4] to test the above features.

4   Experimental Evaluation

Now we proceed to report benchmarking results, which ran on a CPU 2.8 GHz - 4GB RAM system. The main goal of the experiments is not only to compare the diversifying performances, but also to analyze the effects of configurable parameters on the performance behavior. Due to space limitations, further details can be found on our website[5].

4.1   Computation Time

We study computation time on different configurations of the number of input items n and the displayed budget number k. Table 2 shows the computation time of each technique, averaged over 100 runs, when varying $n \times k$ from 100×10 to 500×30.

In general, each algorithm category has its own winner on this concern (less than 100ms for all settings). This result is straightforward to understand – these techniques only need one iteration to select an item and do not execute expensive routines. In contrast, GrassHopper and MSD have the highest running time. For example, with $n = 500$ and $k = 20$, GrassHopper runs nearly 6s, MSD needs 272ms while the others require less than 100ms. This is because MSD finds two data items in each iteration while GrassHopper needs multiple iterations that involve complex matrix operations to compute the stationary distribution and absorption probabilities.

---

[4] https://code.google.com/p/diversity-benchmark.
[5] https://code.google.com/p/diversity-benchmark.

Table 2. Average computation time (ms) over 100 runs (the lower, the better)

$n \times k$ *	Motley	Swap	MSD	MMR	AG	GrassHopper
100×10	2	4	8	3	1	29
100×20	2	7	11	4	1	50
100×30	2	12	13	8	1	66
200×10	7	14	25	8	6	213
200×20	7	21	41	13	7	393
200×30	7	29	56	22	7	564
500×10	42	56	160	48	43	3029
500×20	42	68	272	63	45	6062
500×30	42	84	380	86	46	8943

* $n \times k$: n data items and k results

4.2   Effect of Number of Displayed Items

Figure 2 illustrates the effects of varying the displayed budget number from $k = 2$ to $k = 20$ on the performance of the diversification algorithms. The experiment is conducted on both real-world and synthetic data and due to space constraint, we only report the results on the real-world datasets. Regarding the TREC dataset, since it contains 50 different queries, we calculate the subtopic recall and normalized relevance values for each query and report the result as the average over 50 queries.

A key finding is that as the number of displayed items increases, the subtopic recall values also increase. For example, when k increases from 2 to 20, the subtopic recall of GrassHopper increases from 0.1 to 0.4 for the TREC dataset

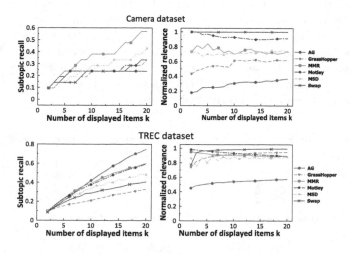

Fig. 2. Effects of number of displayed items k

and from 0.1 to 0.3 for the Camera dataset. This is expected and can be explained as follows. Since the number of displayed items increases, the algorithms are able to include more items into the result set. These items may belong to existing or new subtopics, which makes the subtopic recall values increase. Another interesting observation is the performance of MSD w.r.t both datasets as its subtopic recall and normalized relevance follow a zigzag pattern. The reason is that MSD picks two data items as a time. When the number of displayed items is odd, MSD needs to select the last data item randomly, which incurs in the decrease of both subtopic recall and normalized relevance.

4.3 Stability of the Algorithms

In order to check the stability of the algorithms, for both real and synthetic dataset, we verify whether the output set $D^*_{k_1}$ is a subset of $D^*_{k_2}$ if $k_1 < k_2$ as we increase the number of displayed items k. If this proposition is not satisfied for any dataset or any value of k, we conclude that the algorithm is not stable.

An important finding is that only Swap, Motley and MSD are not stable. For Swap algorithm, since it initializes the result set by the top-k relevant data items, the initial result set is different for different values of k. This affects subsequent swaps as the items to be swapped are selected based on the current result set. Since the swap sequences are different w.r.t k, Swap algorithm is not stable. Regarding Motley and MSD, as they may add items to the result set randomly, there are cases that $D^*_{k_1} \not\subset D^*_{k_2}$. Other algorithms such as MMR, AG and GrassHopper, as they build the result set by adding one element at a time and do not involve any randomness or depend on current result set, are stable.

4.4 Effects of #subtopics

Figure 3 depicts the result of this experiment, which was conducted by varying the number of subtopics m from 2 to 8. The number of input items n increases from 200 to 800 as the number of subtopics increases. The displayed budget number k is fixed to 10, which is often the default number of many well-known applications (e.g. google.com, bing.com). As this experiment requires changing the number of subtopics, which can only be achieved on synthetic data, this experiment is not conducted on the real datasets.

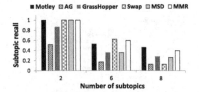 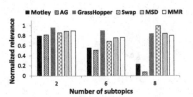

Fig. 3. Effects of the number of subtopics

In general, we observe a reduction in subtopic recall as m increases. For example, the subtopic recall of GrassHopper decreases from 0.8 to 0.2 while that of Swap decreases from 1 to 0.1 as m increases from 2 to 8. The reason is that the number of subtopics covered by the algorithms does not increase linearly with the number of subtopics in the dataset. Another important remark is that MMR can balance between diversity and relevance despite the changes in the number of subtopics. More specifically, MMR ranks second regarding subtopic recall and third with respect to normalized relevance. This can be explained as follows: in the beginning, when the result set D^* contains few items, the relevance term in the objective function of MMR is the dominant term, which makes MMR to select items with high relevant scores. In contrast, when D^* has many items, the diversity term becomes dominant, which forces MMR to find diverse items.

4.5   Effects of Dissimilarity Distribution

The empirical analysis is illustrated in Fig. 4, in which the subtopic distance δ varies from 0.05 to 0.25. We fix the number of subtopics $m = 5$, the displayed budget number $k = 10$, and the number of input items $n = 500$. This experiment is only conducted on synthetic data as it requires changing the subtopic distance δ.

An interesting observation is that the normalized relevance values of all algorithms tend to decrease while their subtopic recall values remain the same as the subtopic distance δ increases. For instance, the normalized relevance value of AG decreases from 0.4 to 0.1 while its subtopic recall stays at 0.2. This can be explained as follows: since the dissimilarity between items increases as we increase δ, the algorithms are able to select items from low relevant subtopics as the loss in relevance is offset by the gain in dissimilarity. Another interesting observation is the performance of Motley. While its subtopic recall values are the highest (over 0.9) and increase with the subtopic distance, its normalized relevance values are the second lowest among the algorithms. The reason lies in its computing model. Motley are able to cover most subtopics since it explicitly aims to maximize diversity by selecting an item only if it is dissimilar with other items in the result set. In addition, since it traverses the candidate items in descending order of relevance, many highly relevant items are not selected if they belong to the same subtopic with the highest relevant item, thus low normalized relevance values.

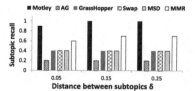
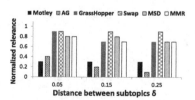

Fig. 4. Effects of dissimilarity distribution

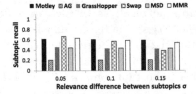 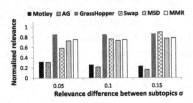

Fig. 5. Effects of relevance variance between domain topics

4.6 Effects of Relevance Distribution

Figure 5 illustrates the effects of varying the relevance difference between the subtopics from $\sigma = 0.05$ to $\sigma = 0.15$ and fixing the number of subtopics $m = 5$, the displayed budget number $k = 10$, and the number of input items $n = 500$. Since the relevance difference between the subtopics can not be changed in the real-world dataset, we only perform this experiment on the synthetic dataset.

A noticeable observation is the tradeoff between diversity and relevance of the algorithms. This can be seen clearly through the performance of Swap. It has high subtopic recall and low normalized relevance when σ is small but the situation is reversed when σ is high. The reason is that Swap is highly sensitive to the relevance drop threshold that a swap can occur. When σ is lower than this threshold (i.e. $\sigma < 0.1$), Swap is able to diversify the result set by exchanging high relevant items in the current result set with low relevant ones as these swaps do not violate the constraint. However, when the relevance difference σ is high, Swap tries to retain highly relevant items since swapping these items out would violate the constraint. Another interesting observation is that the subtopic recall of MSD remains the same at 0.4 as σ increases. Since MSD picks two items at a time, these items tend to be from a pair of subtopics with the highest dissimilarity, which explains the constant subtopic recall.

4.7 Effects of Subtopic Density Difference

The experimental results are shown in Fig. 6, in which the subtopic density difference θ varies from 0.05 to 0.08. We fix the number of subtopics $m = 5$, the displayed budget number $k = 10$, and the number of input items $n = 500$.

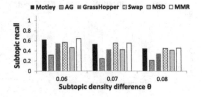 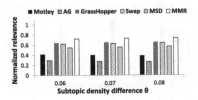

Fig. 6. Effects of subtopic density difference

Since this experiment requires changing the subtopic density difference θ, it is conducted on the synthetic data.

In general, we observe that as θ increases, the subtopic recall decreases while normalized relevance values change marginally. For example, the subtopic recall of Motley decreases from 0.62 to 0.44 while its relevance values remain around 0.4 as we vary θ. The reason behind this observation is that with higher values of θ, some subtopics contain a larger number of data items in comparison with other subtopics. As a result, the data items in smaller subtopics may not be included to the final result. On the other hand, the algorithms are still able to select highly relevant data items from large subtopics which makes the relevance values change very slightly.

5   Conclusions and Future Work

This paper presented a thorough evaluation and comparison of diversification techniques widely used in various domains. We offered an overview of three major classes (threshold-based, function-based and graph-based) of diversification techniques, while discussing about the characteristics of their underlying models. We then introduced the component based benchmarking framework, in which a new diversification technique as well as a new measurement or a dataset can be easily plugged. During the framework development, we made the best effort to re-implement and integrate the most representative diversification techniques, and evaluated them in a fair manner. We also analyzed various performance factors for each technique, including computation time, effects of number of subtopics, number of displayed items, dissimilarity distribution, relevance distribution and stability of the algorithms.

We here summarize the principal findings as a set of guidelines for how to choose appropriate diversification techniques on the following scenarios:

– Overall, MMR performs best in terms of multiple criteria. It has low computation time while it can find relevant items belonging to different subtopics.
– For applications that require fast computation, Motley and AG are the winners. But the two other runner-ups, Swap and MMR, are not significantly slower.
– For applications that focus on diversity, Motley should be used as it can return the highest number of subtopics.
– For datasets whose dissimilarity between subtopics is high, Motley is the best choice in terms of diversity while MMR is the best in terms of balance between diversity and relevance.
– For datasets that have high variance of relevance distribution, we suggest using MMR. In contrast, Swap is the best if the relevance difference is low and diversity is the main concern.
– For datasets in which the number of data items between subtopics vary widely, MMR is the best choice in terms of diversity and relevance.

– For applications that the number of displayed items can be changed, we suggest not to use Swap, Motley and MSD as these algorithms are not stable. Among stable algorithms, MMR and AG are both good choices in terms of diversity.

Category	Winner	2nd best	Worst
Computation time	**Motley**	AG	GrassHopper
Number of subtopics	**MMR**	Swap	AG
Dissimilarity distribution	**MMR**	Motley	AG
Relevance distribution	**MMR**	Swap	AG
Subtopic density difference	**MMR**	Swap	AG
Number of results^a	**MMR**	AG	GrassHopper

^{a} Only stable algorithms are considered.

As a concluding remark, we recommend potential applications to use our benchmarking framework as a tool to find out the well-suited diversification techniques accordingly. As the benchmark source code as well as the datasets used in the benchmark are publicly available, we expect that the experimental results presented in this paper will be refined and improved by the research community, in particular when more data become available, more experiments are performed, and more techniques are integrated into the framework.

Acknowledgment. The research has received funding from the ScienceWise project.

References

1. http://boston.lti.cs.cmu.edu/clueweb12/TRECcrowdsourcing2013/
2. Agrawal, R., et al.: Diversifying search results. In: WSDM, pp. 5–14 (2009)
3. Boriah, S., et al.: Similarity measures for categorical data: a comparative evaluation. In: SIAM, pp. 243–254 (2008)
4. Braschler, M.: CLEF 2001 - overview of results. In: Peters, C., Braschler, M., Gonzalo, J., Kluck, M. (eds.) CLEF 2001. LNCS, vol. 2406, p. 9. Springer, Heidelberg (2002)
5. Carbonell, J., et al.: The use of MMR, diversity-based reranking for reordering documents and producing summaries. In: SIGIR, pp. 335–336 (1998)
6. Chandar, P., et al.: Preference based evaluation measures for novelty and diversity. In: SIGIR, pp. 413–422 (2013)
7. Clarke, C.L., et al.: Overview of the trec 2009 web track. Technical report, DTIC Document (2009)
8. Clarke, C.L., et al.: Novelty and diversity in information retrieval evaluation. In: SIGIR, pp. 659–666 (2008)
9. Deng, T., et al.: On the complexity of query result diversification. Proc. VLDB Endowment **6**, 577–588 (2013)
10. Drosou, M., et al.: Search result diversification. ACM SIGMOD Rec. **39**, 41–47 (2010)

11. Drosou, M., et al.: Disc diversity: result diversification based on dissimilarity and coverage. Proc. VLDB Endowment **6**, 13–24 (2012)
12. Gabrilovich, E., et al.: Newsjunkie: providing personalized newsfeeds via analysis of information novelty. In: WWW, pp. 482–490 (2004)
13. Gollapudi, S., et al.: An axiomatic approach for result diversification. In: WWW, pp. 381–390 (2009)
14. Hasan, M., et al.: User effort minimization through adaptive diversification. In: KDD, pp. 203–212 (2014)
15. Jain, A., Sarda, P., Haritsa, J.R.: Providing diversity in k-nearest neighbor query results. In: Dai, H., Srikant, R., Zhang, C. (eds.) PAKDD 2004. LNCS (LNAI), vol. 3056, pp. 404–413. Springer, Heidelberg (2004)
16. Kando, N., et al.: Overview of IR tasks at the first NTCIR workshop. In: NTCIR, pp. 11–44 (1999)
17. Küçüktunç, O., et al.: Diversified recommendation on graphs: pitfalls, measures, and algorithms. In: WWW, pp. 715–726 (2013)
18. Rafiei, D., et al.: Diversifying web search results, pp. 781–790. In: WWW 2010 (2010)
19. Skoutas, D., et al.: Tag clouds revisited. In: CIKM, pp. 221–230 (2011)
20. Tong, H., et al.: Diversified ranking on large graphs: an optimization viewpoint. In: KDD, pp. 1028–1036 (2011)
21. Vaughan, L., et al.: Search engine coverage bias: evidence and possible causes. Inf. Process. Manag. **40**, 693–707 (2004)
22. Verheij, A., et al.: A comparison study for novelty control mechanisms applied to web news stories. In: WI, pp. 431–436 (2012)
23. Vieira, M.R., et al.: On query result diversification. In: ICDE, pp. 1163–1174 (2011)
24. Yu, C., et al.: It takes variety to make a world: diversification in recommender systems. In: EDBT, pp. 368–378 (2009)
25. Zhai, C.X., et al.: Beyond independent relevance: methods and evaluation metrics for subtopic retrieval. In: SIGIR, pp. 10–17 (2003)
26. Zhang, B., Li, H., Liu, Y., Ji, L., Xi, W., Fan, W., Chen, Z., Ma, W.Y.: Improving web search results using affinity graph. In: SIGIR, pp. 504–511 (2005)
27. Zhu, X., et al.: Improving diversity in ranking using absorbing random walks. In: NAACL, pp. 97–104 (2007)
28. Ziegler, C.N., et al.: Improving recommendation lists through topic diversification. In: WWW, pp. 22–32 (2005)

XML and Semi-structured Data

TOIX: Temporal Object Indexing for XML Documents

Rasha Bin-Thalab[1] and Neamat El-Tazi[2]([⊠])

[1] Department of Computer Engineering, Faculty of Engineering and Petrol,
Hadhramout University, Al Mukalla, Yemen
aziz30@yahoo.com
[2] Department of Information System, Faculty of Computers and Information,
Cairo University, Giza, Egypt
n.eltazi@fci-cu.edu.eg

Abstract. Managing temporal data gets increased demand for many significant areas such as scientific and financial applications. This paper proposes a new index (TOIX), which is designed in order to provide more efficient evaluation of temporal queries on XML documents. The index lies on mapping the twigs into temporal objects and then using these objects instead for answering the query. An improvement of the naive algorithm using a B-tree as well. Finally, a set of conducted experiments were performed and showed that our proposed algorithm outperforms the state of the art indexing algorithms in certain cases.

Keywords: Temporal XML · Indexing · Query processing · Semi-structured data · Summary schema · Twig query

1 Introduction

The study of storage and management of temporal data has emerged due to the increasing importance of time dimension in many applications. As time evolves, a significant amount of temporal data needs to be managed. Typically, indexing plays a key role in managing data efficiently. However, a few work have been addressed on temporal indexes compared with conventional XML indexes.

In temporal XML document literature [12], the temporal XML model defines two attributes (*start* and *end*) for each element to store the recorded (validity) time of the element. Two types of time dimensions have been considered; valid time and transaction time. The literature focuses on transaction time since it is amenable in application systems. The transaction time depends on the committed transaction time of operations. Consequently, physical deletion is not allowed. On the other hand, valid time allows the nodes to be inserted, deleted, or modified according to their own time events in the real world.

Figure 1 shows an example of NBA DB portion in a tree representation using transaction time dimension. Edges, in the temporal XML tree, store the recorded time of their ending nodes. In case of an edge without a recorded time, the ending

© Springer International Publishing Switzerland 2015
Q. Chen et al. (Eds.): DEXA 2015, Part II, LNCS 9262, pp. 235–249, 2015.
DOI: 10.1007/978-3-319-22852-5_20

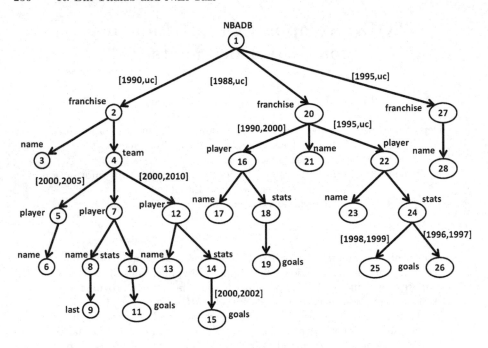

Fig. 1. Temporal XML for NBA DB portion

node has the same recorded time of its parent node. The interval is represented by using two values $[t_s, t_e]$ where t_s represents the start time when the data is inserted and t_e represents the time when the data is modified or deleted. We use the variable uc to indicate that the end time of the interval is valid until changed. Each node is represented by a pair $<n_{id}, t>$ where n_{id} is the unique identifier of the node and t represents its recorded time interval in the DB system.

The general case of retrieving nodes in XML document is performed by using twig queries. A twig pattern is defined as a subtree of nodes (tags, attributes, or text value) connected by either a parent-child or an ancestor-descendant relationship. The purpose of processing these twig queries is to find all occurrences of such patterns in the XML document with an efficient query evaluation. Several traditional XML indexes have been proposed to enhance twig queries evaluation such as BPI-Twig [8] and enhanced range encoding [13]. Since twig queries can also be found in temporal XML, in this paper we describe a variation of XML temporal indexing based on twigs. The index extends the BPI-Twig [8] index since it provided an efficient evaluation of twig queries without the need for node range encoding. Such range encoding is not suitable in the dynamic environment of temporal XML documents. We add time dimension in the proposed indices of BPI-twig to evaluate the temporal twig queries accordingly.

However, in a previous work, we proposed a temporal index [2], called TMIX, which summarized the XML document based on edges available inside the tree. However, the performance is limited by multiple joins of intermediate results in the case of complex twig queries. To avoid such limitations, this paper uses a summary of XML document based on subtrees rather than edges. Furthermore, we present an extra index to enhance the performance retrieval of temporal data. The work depends on Rizzolo temporal XML model [12] where a single document is maintained during the elements evolution over time and each node is associated by a transaction time interval.

The contributions in this paper can be summarized as follows:

1 Building an abstract model for temporal XML documents based on temporal objects, classes, and paths formed from these objects and classes as presented in Sect. 3.
2 Propose two algorithms TOIX and TOIX-V to evaluate temporal twig queries based on the proposed model as will be shown in Sect. 4
3 Evaluate the performance of the proposed algorithms against TMIX [2] as presented in Sect. 5.

2   Related Work

Temporal XML models are divided into multi-version [14] and key-based [3, 11]. In this paper, our proposed algorithms focus on key-based model where a single document is maintained to keep track of all changes of elements over time. TempIndex [11] is an early example of temporal XML indices. The index was extended later to support graph model [12]. The authors created two summaries based on the document structure and level using path continuous concept. TempIndex had a big overhead of both complexity processing and large storage space. Later, Gao [6] proposed an interval B-tree index based on valid time dimension. Each node is represented by (node_code, Interval and depth). Two main limitations of such index existed. They were mainly the space overhead and update costs. Next, a time dimension was added to a suffix tree [16] to build a temporal XML tree. Similarly, another approach was introduced called TFIX [17] by adding a time feature vector to characterize the temporal XML nodes. Another indexing technique TXDIM [15] presented a mathematical framework for a tree XML document model. The model grouped periods into equivalence classes based on their temporal connection and inclusion relations. The work neither considered the effects of updates on the index nor considered handling complex queries such as twig queries. A pattern-based [10] was proposed with a set-based to support temporal XML query language instead of XPath or XQuery.

In 2011, Dyresen and Mekala [7] proposed a new prefix-numbering scheme for temporal XML nodes. Combi [4] also in the same year proposed a new model for temporal dimension for a graph-based in semi structured data. Baazizi [1] also presented an efficient storage encoding for multi-version documents using two methods; the first was based on the knowledge of abstract of the document and the second supposed that the abstract was generated from the initial document by a sequence of updates

For the purpose of achieving consistency, Currim [5] and his colleague described how to interpret integrity constraint in XML schema as time evolves.

Signature	Label
S1	NBA
S2	franchise
S3	team
S4	player
S5	name
S6	stats
S7	goals
S8	last

Fig. 2. Labels with signatures

Recently, Faisal and Sarwar [9] provided a comparative analysis on various schemes of temporal XML documents.

In this paper, we describe a Twig-based approach to index temporal XML which avoid the overhead of intermediate results used in joining nodes to evaluate temporal queries. The work is based on BPI-TWIG [8] which provided a labels-independent index for XML documents with efficient twig query processing engine.

3   Index Structure Definitions

In this section, we define the basic structures needed to construct our proposed temporal XML index. A comprehensive analysis for space utilization is presented after the illustration of each index structure.

First, to simplify the representation, distinct labels (tag names of nodes) are encoded by a unique code called *signature*. For example we have eight distinct signatures in our running example NBA DB of Fig. 1. The signature labels with their encoding is shown in Fig. 2. We define six index structures to preserve the structural relationship between objects.

Definition 1 (Temporal Object). *An XML object, in our proposed algorithm, is a subtree with one level and is represented by a root node with its children $O(<nid_{root},t_{root}>, <nid_1,t_1>, .., <nid_n,t_n>, s_{root}, s_1, s_2,..,s_n)$. where $<nid_{root},t_{root}>$ is the root node of object followed by a set of children nodes $<nid_i,t_i>$. Next, s_{root} is the signature of root node followed by a list of signatures s_i for its consequence nodes.*

From Definition 1, we can say that a *temporal object* corresponds to a twig pattern in XML tree context. Figure 3 shows the temporal objects with their corresponding signatures and nodes that are derived from Fig. 1. The required space for object index is based on the maximum number of objects denoted

Object Id	Signatures	Nodes
O1	S1 S2 S2 S2	n1[1970,uc] n2[1990,uc] n20[1988,uc] n27[1995,uc]
O2	S2 S5 S3	n2[1990,uc] n3[1990,uc] n4[1990,uc]
O3	S3 S4 S4 S4	n4[1990,uc] n5[2000,2005] n7[1990,uc] n12[2000,2010]
O4	S4 S5	n5[2000,2005] n6[2000,2005]
O5	S4 S5 S6	n7[1990,uc] n8[1990,uc] n10[1990,uc]
O6	S5 S8	n8[1990,uc] n9[1990,uc]
O7	S6 S7	n10[1990,uc] n11[1990,uc]
O8	S4 S5 S6	n12[2000, 2010] n13[2000, 2010]n14[2000, 2010]
O9	S6 S7	n14[2000, 2010] n15[2000, 2002]
O10	S2 S4 S5 S4	n20[1988, uc] n16[1990,2000] n21[1988,uc] n22[1995, uc]
O11	S4 S5 S6	n16[1990,2000] n17[1990,2000] n18[1990,2000]
O12	S6 S7	n18[1990,2000] n19[1990,2000]
O13	S4 S5 S6	n22[1995,uc] n23[1995,uc] n24[1995,uc]
O14	S6 S7 S7	n24[1995,uc] n25[1998,1999] n26[1996,1997]
O15	S2 S5	n27[1995,uc] n28[1995,uc]

Fig. 3. Temporal object index

as O_{num}. Since each object represents a subtree, each node can be a candidate to a new temporal object root in its subsequent level until reaching the leaves of the tree. The maximum number of objects can be computed as: $(fanout^{depth\_tree-1} - 1)/(fanout - 1)$. That is, it's exponential to $depth\_tree$. The whole object index size is then computed as $O_{num} \times (2 \times fanout)$. The $fanout$ is duplicated since we present both node numbers and corresponding signatures for each object. To provide quick access of objects, a B-tree is built on the top of object ids.

Definition 2 (Class Signature). *An XML class is a set of common signatures between objects represented as $C(s_{root}, s_1, s_2, .., s_n)$ where s_{root} represents the signature of class root. The remainder signatures s_i represent the set of distinct children signatures of s_{root}.*

Generally, a class describes a set of data having common attributes along with the functions that operate on that data. For example, a player is a class that represents all players in NBA having attributes like: name, goals, colleague, etc. Class index in our proposed indexing algorithm keeps all classes in the temporal XML document. These classes are represented by a unique identifier (C_i) attached by a set of its attributes' signatures. For example, Fig. 4 shows the set of classes that exist in NBA DB.

To provide fast recognition of class id given an input signature, we create two auxiliary structures: parent and child signatures. Each signature is assigned as

Class id	Labels	Signatures
C1	NBA(franchise)	S1(S2)
C2	franchise(name, team, player)	S2(S5, S3, S4)
C3	team(name, player)	S3(S5, S4)
C4	player(name, stats)	S4(S5, S6)
C5	stats(goals)	S6(S7)
C6	name(last)	S5(S8)

Fig. 4. Class signature

Signature	Class
S1	C1
S2	C2
S3	C3
S4	C4
S5	C6
S6	C5

Fig. 5. Parent signatures

Signature	Class
S2	C1
S3	C2
S4	C2, C3
S5	C2, C3, C4
S6	C4
S7	C5
S8	C6

Fig. 6. Child signatures

parent or child in its corresponding class as shown in Figs. 5 and 6. Here, the size of class index depends on the max number of classes (C_{num}) in the document which is less than number of objects O_{num}. Thus, the space requirement equals to $C_{num} \times fanout$. On the other hand, the parent and child signatures indices have the space $C_{num} + S_{num}$ where S_{num} is the total number of signatures.

Definition 3 (Path Object Index). *A sequence of objects $<o_1 \; o_2 \; .. \; o_h>$ starting from root XML object and ending with an object that has at least one leaf child in the path sequence. h refers to the path object length.*

Figure 7 shows a list of path objects of the XML document in Fig. 1. Size of path object index depends on number of paths of objects denoted as PO_{num} and bounded by $\frac{O_{num}}{(depth-1)}$. Consequently, the required space for index is equal to $PO_{num} \times (depth - 1)$. Removing depth tree $(depth)$ as a constant, we can say that the space complexity for this index is in O_{num}.

Definition 4 (Path Class). *A sequence of subsequent classes $<c_1 \; c_2 \; .. \; c_h>$ corresponding to a path object*

The index size equals to $O(PC_{num} \times depth\_tree-1)$ where PC_{num} is the number of path classes in the document. The corresponding path class index for our running example is shown in Fig. 8.

PC	P#	Path of Objects
PC1	Po1	O1[1970, uc] O2[1990,uc]
	Po2	O1[1970, uc] O10[1988,uc]
	Po3	O1[1970, uc] O15[1995, uc]
PC2	Po4	O1[1970, uc] O2[1990,uc] O3[1990, uc] O4[2000,2005]
	Po5	O1[1970, uc] O2[1990,uc] O3[1990, uc] O5[1990, uc]
	Po6	O1[1970, uc] O2[1990,uc] O3[1990, uc] O8[2000,2010]
PC3	Po7	O1[1970, uc] O2[1990,uc] O3[1990, uc] O5[1990, uc] O7[1990,uc]
	Po8	O1[1970, uc] O2[1990,uc] O3[1990, uc] O8[2000,2010] O9[2000,2010]
PC4	Po9	O1[1970, uc] O2[1990,uc] O3[1990, uc] O8[2000,2010] O6[2000,2010]
PC5	Po10	O1[1970, uc] O10[1988,uc] O11[1990,2000]
	Po11	O1[1970, uc] O10[1988,uc] O13[1995,uc]
PC6	Po12	O1[1970, uc] O10[1988,uc] O11[1990,2000] O12[1990,2000]
	Po13	O1[1970, uc] O10[1988,uc] O13[1995,uc] O14[1995,uc]

Fig. 7. Path object index

To provide an efficient access for the path class id we create a path class matrix $PCM(p \times c)$ where p represents the path classes ids and c represents the classes. The intersection of any row and column represents the position of class in the path. The matrix serves as an efficient structure to determine the path class id for a given query path class. For example the corresponding PCM of Fig. 8 is shown in Fig. 9. Clearly, size of PCM matrix is $PC_{num} \times C_{num}$.

Definition 5 (Subtree Path Class(STPC)). *A tree of path classes rooted at second level of the XML document tree.*

STPC is used to determine the intersection of path of classes that form the twig query. This will reduce I/O cost of number of fetched path classes. For

Path Class Id	Classes
PC1	C1 C2
PC2	C1 C2 C3 C4
PC3	C1 C2 C3 C4 C5
PC4	C1 C2 C3 C4 C6
PC5	C1 C2 C4
PC6	C1 C2 C4 C5

PC#/C#	C1	C2	C3	C4	C5	C6
PC1	1	2	0	0	0	0
PC2	1	2	3	4	0	0
PC3	1	2	3	4	5	0
PC4	1	2	3	4	0	5
PC5	1	2	0	3	0	0
PC6	1	2	0	3	4	0

Fig. 8. Path of classes for NBADB   **Fig. 9.** Path class matrix (PCM)

Path Class	Subtree
PC1	STPC1, STPC3
PC2	STPC1
PC3	STPC1
PC4	STPC1
PC5	STPC2
PC6	STPC2

PC/ST	STPC1	STPC2	STPC3
PC1	1	0	1
PC2	1	0	0
PC3	1	0	0
PC4	1	0	0
PC5	0	1	0
PC6	0	1	0

Fig. 10. Index for path class with subtrees

Fig. 11. STPC bitmap index

example we have only three subtrees for Fig. 1 that group all path classes shown in Fig. 10.

To efficiently determine the intersection of path classes, the authors of BPI-Twig [8] created a bitmap matrix for subtrees and path classes. Figure 11 shows an example of bitmap matrix of Figure 1, each row represents the path class and each column represents the subtrees. The intersection is then performed using a bitwise AND of the relevant rows. The space complexity for the STPC bitmap index equals to $PC_{num} \times ST_{num}$ where ST_{num} is the max number of subtrees in the XML tree.

4   Temporal Twig Query Evaluation

In this section, we propose a new algorithm (TOIX) that evaluates temporal queries using the aforementioned basic definitions used in the previous section. We used TXPath [12] in expressing the temporal XML queries.

4.1   TOIX Algorithm

Algorithm 1 shows the utilization of the different structures that we created to perform query evaluation over Temporal XML documents. The algorithm starts by parsing an input temporal twig query to find all objects that match the input query tree and it results in outputting all node numbers with their time intervals that belong to the matched objects.

The input temporal twig query is decomposed into a set of objects to find the query path objects (line 2). Next, these path objects are converted into their corresponding path classes using parent, child, and PCM indices (lines 3–4). The positions of join points of path objects are determined in $JPos_i$ (line 5). Positions of temporal predicates on each query object are also determined (line 6). Next, the algorithm searches for path classes that contain the query classes. Although there are more than one path class that can be found, these path classes are joined to find the intersected paths of classes R_{pc} (line 7). Each path class id in R_{pc} is used to retrieve the corresponded list of path of objects that are

Algorithm 1. Temporal Twig Query Processing (TOIX)

Input: TWQ: Temporal Twig Query

Output: Q_R: $n_1[t_1]$, $n_2[t_2]$,..,$n_q[t_q]$, where q number of twig query nodes

    Internal Variables

      – SI: Signatures index

      – PT: Parent table signatures

      – CT: Child table signatures

      – PCM: Path class matrix

      – STI: Subtree intersection bitmap

      – R_{pc}: List of path classes ids

      – PO_{idx}: Path of objects index

      – Obj_{idx}: Temporal object index

1: $TWQ_s \leftarrow$ Convert TWQ into signatures using SI
2: $TWQ_{obj} \leftarrow$ Decompose TWQ_s into objects and build paths of objects
3: $QPath \leftarrow$ Build path of classes from TWQ_{obj} using PT and CT
4: $Path_{Class} \leftarrow$ Determine ids of path of classes using PCM
5: $JPos_i \leftarrow$ Determine join points between path classes
6: $TPos_i \leftarrow$ Determine position of temporal predicates
7: $R_{pc} \leftarrow$ Find the intersected path classes in $Path_{Class}$ using STI
8: **for** each pc in R_{pc} **do**
9:    $PO_{list} \leftarrow$ Retrieve list of paths from PO_{idx} given pc
10:    **for** each pid in PO_{list} **do**
11:       **for** each pos in $JPos_i$ **do**
12:          $o_{id} \leftarrow$ Retrieve object at pos
13:          **if** pos inside $time\_pred_{TPos}$ **then**
14:             **if** timespan(o_{id}) $\subset time\_pred_{TPos_{pos}}$ **then**
15:                $Q_R \leftarrow$ Retrieve queried nodes ids in o_{id} from Obj_{idx}
16:             **end if**
17:          **else**
18:             $Q_R \leftarrow$ Retrieve queried nodes ids in o_{id} from Obj_{idx}
19:          **end if**
20:       **end for**
21:    **end for**
22: **end for**
23: Output nodes in Q_R

stored in the path objects index PO_{idx} (line 9). The desired objects are selected in each list of retrieved path objects based on its position in $JPos_i$. The selected objects are checked whether they satisfy temporal predicates according to their timestamps (lines 13–19). Finally, the algorithm outputs the queried node ids for each resulted object of each intersected path (line 23).

The complexity of TOIX algorithm depends on query nodes, classes, path classes, path objects, and result size. The third line has the complexity $O(|Q|)$ where $|Q|$ is the query size. The fourth line uses the PCM index which is $O(|PC| \times |C_q|)$ where $|PC|$ and $|C_q|$ represent the number of path classes and classes in the query respectively. Intersection of path classes step (line 7) is performed in $O(|PC_q| \times |ST|)$ where $|PC_q|$ is the number of path classes in the query and $|ST|$ is the number of subtrees in the XML document. I/O cost starts from line 9 by retrieving the requested nodes. Here the algorithm has to trace all paths objects of the corresponding path classes to check the temporal predicates in the corresponding position. The worst case of this step is $O(|PC_q| \times |PO|)$ where $|PC_q|$ is the number of path classes in the query and $|PO|$ is the number of path objects within the path class id (line 8). Additional loop is required (line 11) and its complexity is $O(|join_{pos}|)$ where $|join_{pos}|$ is the number of join points in the query. Retrieving nodes within the objects (line 15 or 18) has the complexity $O(\log O_{idx})$ since the indices are built using B-tree. We can compute the total complexity of TOIX algorithm as follows:

$$O(|Q|) + O(|PC| \times |C_q|) + O(|PC_q| \times |ST|) + O(|PC_q| \times |PO|) + O(|join_{pos}|) \times O(\log |O_{idx}|).$$ Since $|Q|$, $|join_{pos}|$, $|PC|$, and $|C_q|$ are constant expressions, the complexity can be reduced to: $O(|PC_q| \times |ST|) + O(|PC_q| \times |PO|) + O(\log |O_{idx}|)$

Example: Consider the following temporal query that searches for the stats in the period 2000 to 2005 of players after 2000:

$$team \; [//player[ts \geq 2000]][//stats[([ts,te] \supseteq [2000,2005])]]$$

The query tree representation is shown in Fig. 12(a). First, the labels *team*, *player*, and *stats* are mapped into signatures: S3, S4, and S6 respectively. There is no need to decompose the query tree since it has one level subtree. We have only one join point at the root *team*(S3). Additionally, we have two temporal predicates attached with S4 and S6 signatures. We keep the predicates' positions within the twig query path in a separate table. We search for the corresponding classes for the query signatures. The root of the query (*team*(S3)) is the parent signature of class C3. Next, we search for the signature S4 in the parent signatures which is found in the class C4. Similarly, S6 is found in class C5. Now we have two paths of classes in the query: C3//C4[T1] and C3//C5[T2]. Using PCM matrix, we can see that path C3//C4[T1] is found in three path classes in the document: PC2, PC3, and PC4. The second path C3//C5[T2] has only one path class (PC3). The intersection function between paths of classes return PC3 as the common path between the two paths as shown in Figure 12(b). Meanwhile, a temporal predicate table is created to keep path class id, position class, and temporal predicate that is attached in the query path classes as shown in Fig. 12(c). PC3 is checked in the path object index (POI) to search for the corresponding paths of objects. There are two paths of objects as shown in Fig. 12(d). For each returned path, the algorithm checks the two temporal predicates in the corresponding positions in the PC3. Since objects O7 does not satisfy the queries temporal predicates, the first path is discarded from the results. Finally, the objects in the second path objects are retrieved from Object index (OI) and its corresponding nodes based on their signatures are returned.

4.2 Optimization

It is clear that the proposed algorithm TOIX is heavily dependent on the structure of path of objects retrieved for the intersected path classes. However, some unrelated paths have to be traversed to check the temporal predicates. We create a B-tree index *POI-V* on the top of path objects index (POI) to provide a quick access to the target objects. The enhanced index is built vertically based on a triple key $<pcid, d, t>$, where $pcid$ is the path class id, d is the depth of the object, and t is the timestamp of the object. The value of index key is a pair of $<pid, oid>$ where pid is the path object id and oid is the object id. Algorithm 1 is modified starting from line 8 according to the new index as shown in Algorithm 2 named as TOIX-V. Thus, the complexity of TOIX-V is reduced to: $O(|PC_q| \times |ST|) + O(|PC_q| \times \log |PO|) + O(\log |O_{idx}|)$

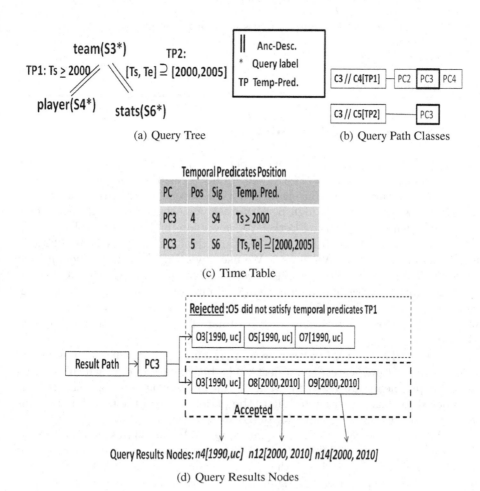

Fig. 12. An illustration of a temporal twig query evaluation

Algorithm 2. Enhanced Temporal Twig Query Processing (TOIX-V)

1: **for** each pc in R_{pc} **do**
2:     **for** each pos in $JPos_i$ **do**
3:         **if** pos inside $time\_pred_{TPos}$ **then**
4:             $P\_O \leftarrow$ Retrieve pair $<Pid,oid>$ from POI-V based on pc, pos, and $time\_pred_{TPos_{pos}}$
5:         **else**
6:             $P\_O \leftarrow$ Retrieve pair $<Pid,oid>$ from POI-V based on pc, pos
7:         **end if**
8:         $L_{obj} \leftarrow$ join objects in P_O based on pid
9:         **for** each o_{id} in L_{obj} **do**
10:             $Q_R \leftarrow$ Retrieve queried nodes ids in o_{id} from Obj_{idx}
11:         **end for**
12:     **end for**
13: **end for**
14: Output nodes in Q_R

5   Experiments and Results

In this section, we perform different experiments to evaluate the performance of the proposed algorithms; TOIX and TOIX-V against the temporal index TMIX [2]. We are comparing against TMIX as one of the state of the art algorithms which showed a high performance against its peers. All algorithms were implemented in C# using Berkeley C# version implantation for B-tree. All experiments were run on a PC with Pentium dual core CPU T4400 @2.2GHz and 6 GB RAM running Windows 7. We used two XML benchmark documents: XBench (9 fan-out and 7 depth) and XMark (9 fan-out and 9 depth) where each element is attached with a random temporal intervals within range [2000,2012] using *year* as granularity. Furthermore, the XML documents size were divided into 25, 50, 75, and 100 Megabytes to evaluate the scalability of the proposed algorithms.

We used different types of temporal twig queries with different number of levels, temporal predicates, and objects. The benchmark queries which are used in our experiments are listed in Table 1. The processing time in cold cache is considered the basic metric to evaluate the performance of the tested algorithms as shown in Figs. 13(a)–(c) and 14(a)–(d) with log-scale.

In XMark, we have three tested queries: Q1, Q2, and Q3 as shown in Table 1. As shown in Fig. 13(a), TOIX-V algorithm showed the best performance over TMIX and TOIX. This is due to the complex structure of query Q1 with respect to TMIX which depends on edge-summary. Additionally, there is only one existing time predicate at the end of the query. TOIX and TOIX-V achieved better results since they depend on the internal structures before fetching the physical data nodes.

The performance of TMIX is different in query Q2 compared with TOIX and TOIX-V as shown in Fig. 13(b). We can see that TMIX is better than TOIX due to its ability to prune retrieved nodes according to temporal conditions

Table 1. Query Benchmarks

Data Set	Q#	Expression
XMark	Q1	//item[[description[//parlist]][shipping="internationally"][//mail[from][to][text[ts>2002]]]]
	Q2	//item[ts ≥ 2002][[description[ts ≥ 2002][//parlist]][shipping="internationally"] [mail[ts >=2002][to[ts ≥ 2002]][from[ts ≥ 2002]][text[ts ≥ 2002]]]]
	Q3	//asia[item[te > 2005]][name][payment="Creditcard"][//mailbox[ts ≥ 2001]]
XBench	Q1	//item[//subject ="BIOGRAPHIES"][ts > 2005]
	Q2	//item[ts ≥ 2003][//author[ts ≥ 2003][//phone_number]][//ISBN][//publisher[ts ≥ 2003]]
	Q3	//item[te ≥ 2008][//author[//name_of_country="New Zealand"]][publisher[ts ≥ 2003] [name]]
	Q4	//item//authors//author//contact_information//mailing_address[name_of_city="Dubai"]

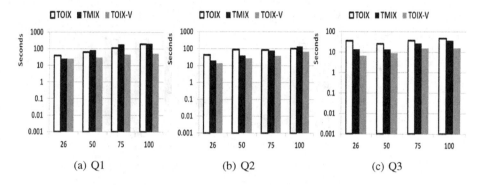

Fig. 13. XMark: query performance with log-scale

predicates compared with TOIX. However, TOIX-V kept its better performance against the TOIX and TMIX. The same occurs for query Q3 which asks for two time predicates, one value predicate, and one object. Whereas TMIX is better than TOIX as shown in Fig. 13(c) for the same reason as in query Q2, TOIX-V is twice faster than TMIX and five times faster than TOIX. In XBench data set, we have four tested queries as shown in Table 1. Query Q1 has only one level and one predicate. The results are shown in Fig. 14(a). Since query Q1 has only one structure join, TMIX recorded best results compared with other competitors. TMIX is faster ranging from 7 to 17 times over TOIX-V and from 27 to 50 times over TOIX.

On the other hand, query Q2 has more complex structure and temporal predicates; we have three branches, two levels, and three temporal predicates. Figure 14(b) shows that the performance of TOIX-V is better ranging from 2 to 4 times over TMIX and from 7 to 32 times over TOIX.

Similarly, query Q3 includes two branches, two levels, two temporal predicates, and one value predicate. The existence of value predicate reduced the number of retrieved nodes and consequently the number of joined nodes. This can be noticed in Fig. 14(c). Whereas the performance of TOIX-V is close to TMIX, TOIX-V is better compared with TOIX ranging from 12 to 21 times. TMIX outperforms TOIX ranging from 11 to 24 times.

248   R. Bin-Thalab and N.El-Tazi

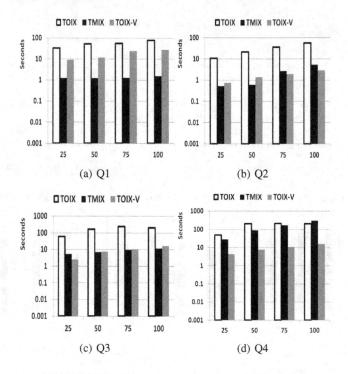

Fig. 14. XBench: query evaluation with log-scale

Query Q4 has more depth which reaches to five levels. Additionally, the query does not contain neither time nor value predicates. However, not having temporal predicate does not ignore the temporal semantics of returned nodes since we asked for the historical information of a given object in the system. The results are shown in Fig. 14(d). The performance of TOIX-V is the fastest compared with TMIX and TOIX. TOIX-V is better ranging from 6 to 18 times over TMIX and from 12 to 26 times over TOIX.

In summary, the results showed that TOIX-V index is better than TMIX in the case where temporal twig queries has rich structure (multiple twigs). This is due to its ability to fetch group of subtrees rather than edges.

6 Conclusion

This paper introduced a new index named TOIX that provides an efficient evaluation of temporal twig queries over temporal XML documents. The new index mapped the twigs into temporal objects and conceptualized those objects into classes for summarization. An optimization of the proposed algorithm was also introduced where B-tree was used on top of the classes and their lifetime to maximize the proposed algorithm performance. Both indexes were evaluated against the state of the art TMIX [2] where they showed a better performance in most cases.

References

1. Baazizi, M.A., Bidoit-Tollu, N., Colazzo, D.: Efficient encoding of temporal XML documents. In: Combi, C., Leucker, M., Wolter, F. (eds.) TIME, pp. 15–22. IEEE (2011)
2. Bin-Thalab, R., El-Tazi, N., El-Sharkawi, M.E.: TMIX: temporal model for indexing XML documents. In: ACS International Conference on Computer Systems and Applications (AICCSA), pp. 1–8 (2013)
3. Buneman, P., Khanna, S., Tajima, K., Tan, W.-C.: Archiving scientific data. ACM Trans. Database Syst. **29**(1), 2–42 (2004)
4. Combi, C., Oliboni, B., Quintarelli, E.: Modeling temporal dimensions of semi-structured data. J. Intell. Inf. Syst. **38**, 1–44 (2011)
5. Currim, F., Currim, S., Dyreson, C.E., Snodgrass, R.T., Thomas, S.W., Zhang, R.: Adding temporal constraints to XML schema. IEEE Trans. Knowl. Data Eng. **24**(8), 1361–1377 (2012)
6. Gao, D., Wang, X., Deng, L.: Indexing temporal XML using interval-tree index. In: Proceedings of the 2008 International Conference on Computer Science and Software Engineering, CSSE 2008, Washington, DC, USA, vol. 04, pp. 689–691. IEEE Computer Society (2008)
7. Dyreson, C.E., Mekala, K.G.: Prefix-based node numbering for temporal XML. In: Bouguettaya, A., Hauswirth, M., Liu, L. (eds.) WISE 2011. LNCS, vol. 6997, pp. 172–184. Springer, Heidelberg (2011)
8. El-Tazi, N., Jagadish, H.V.: BPI-TWIG: XML twig query evaluation. In: Bellahsène, Z., Hunt, E., Rys, M., Unland, R. (eds.) XSym 2009. LNCS, vol. 5679, pp. 17–24. Springer, Heidelberg (2009)
9. Faisal, S., Sarwar, M.: Temporal and multi-versioned XML documents: a survey. Inf. Process. Manag. **50**(1), 113–131 (2014)
10. Li, X., Liu, M., Ghafoor, A., Sheu, P.C.-Y.: A pattern-based temporal XML query language. In: Chen, L., Triantafillou, P., Suel, T. (eds.) WISE 2010. LNCS, vol. 6488, pp. 428–441. Springer, Heidelberg (2010)
11. Mendelzon, A.O., Rizzolo, F., Vaisman, A.: Indexing temporal XML documents. In: Proceedings of the Thirtieth international conference on Very large data bases, VLDB 2004, vol. 30, pp. 216–227 (2004). VLDB Endowment
12. Rizzolo, F., Vaisman, A.A.: Temporal XML: modeling, indexing, and query processing. VLDB J. **17**, 1179–1212 (2008)
13. Song, R., Hong, X., Yang, S.: A new twig query evaluation for XML based on region encoding. In: 2010 3rd International Conference on Biomedical Engineering and Informatics (BMEI), vol. 7, pp. 2682–2686. IEEE (2010)
14. Wang, F., Zaniolo, C.: Temporal queries and version management in XML-based document archives. Data Knowl. Eng. **65**(2), 304–324 (2008)
15. Ye, X., Luo, J., Zhong, G.: Temporal XML index schema. In: Tang, Y., Ye, X., Tang, N. (eds.) Temporal Information Processing Technology and Its Applications, pp. 203–223. Springer, Heidelberg (2010)
16. Zhang, F., Wang, X., Ma, S.: Temporal XML indexing based on suffix tree. In: Proceedings of the 2009 Seventh ACIS International Conference on Software Engineering Research, Management and Applications, SERA 2009, Washington, DC, USA, pp. 140–144. IEEE Computer Society (2009)
17. Zheng, T., Wang, X., Zhou, Y.: Indexing temporal XML using FIX. In: Liu, W., Luo, X., Wang, F.L., Lei, J. (eds.) WISM 2009. LNCS, vol. 5854, pp. 224–231. Springer, Heidelberg (2009)

Expressing and Processing Path-Centric XML Queries

Huayu Wu[1]($\boxtimes$), Dongxu Shao[1], Ruiming Tang[2], Tok Wang Ling[3],
and Stéphane Bressan[3]

[1] Institute for Infocomm Research, A*STAR, Singapore, Singapore
{huwu,shaod}@i2r.a-star.edu.sg
[2] Huawei Noah's Ark Lab, Hong Kong, China
tangruiming@huawei.com
[3] School of Computing, National University of Singapore, Singapore, Singapore
{lingtw,steph}@comp.nus.edu.sg

Abstract. A family of practical queries, which aim to return or manipulate paths as first class objects, cannot be expressed by XPath or XQuery FLWOR expressions. In this paper, we propose a seamless extension to XQuery FLWOR to elegantly express path-centric queries. We further investigate the expression and processing of intra-path aggregation, an analytical operation in path-centric queries.

1 Introduction

A family of practical queries cannot be conveniently expressed in XPath or XQuery. These queries intend to return or manipulate paths as first class objects. We call them *path-centric* queries. Consider an XML document containing Tree-Bank [7], a parsed corpus of English texts annotated with their syntactic structure using XML tags, in Fig. 1(a). One user, a linguist, may legitimately be interested in finding the nesting of prepositional phrases ("PP") enclosing the word "front" in verb phrases ("VP") to study grammatical patterns. He/She could naturally think of expressing this query in XPath as:

$$//\mathtt{VP}[.//\mathtt{PP}//*/\mathtt{text}() = "\mathtt{front}"]$$

However such a query in XPath, or corresponding queries in XQuery, would return a collection of "VP" elements, that is entire sub-trees rooted at a "VP" node. Yet the user is only interested in the paths between "VP" and "front". Without being able to project an interesting path, many path-centric query purposes cannot be easily met. For example, the user could not straightforwardly express a query counting, for every word "front" contained in a verb phrase, the number of prepositional phrases nesting it in the same verb phrase using the FLWOR constructs in XQuery.

Further, consider the XML document outlined in Fig. 1(b), which contains information about flights. One user may be interested in finding the routes from

© Springer International Publishing Switzerland 2015
Q. Chen et al. (Eds.): DEXA 2015, Part II, LNCS 9262, pp. 250–257, 2015.
DOI: 10.1007/978-3-319-22852-5_21

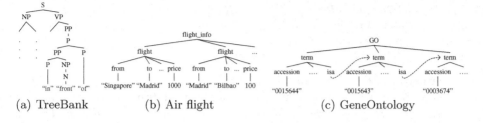

Fig. 1. Three documents to illustrate structural path and semantic path

Singapore to Bilbao. There is no simple XQuery FLWOR query that can produce this result, except writing an XQuery program (user-defined function) that explicitly iterates over the different "flight" elements and tests the connections. Similar queries are natural in the presence of XML ID references. In the GeneOntology data [3], as illustrated in Fig. 1(c), where genes reference their super-class genes. A query to find all other family genes between two known genes cannot be expressed by the current XQuery FLWOR constructs.

Actually XQuery is Turing complete only with the companion of user-defined functions to make it an XQuery program. By doing this, we actually expect an XML database user to be equipped with programming skill on recursive functions, for such a simple query purpose. It will be appreciative if the basic form of XQuery, i.e., FLWOR, can be extended so that such path-centric queries can be easily expressed in a more declarative way.

In this paper we propose seamless extensions to XQuery that can elegantly express queries that return paths and perform intra-path aggregation.

2   SPath: Extending XQuery for Path-Centric Queries

We name our extension SPath, with two functions, StrucPath and SemPath. The function output is a set of chains of document nodes that are structurally or semantically connected. Further, each node may have additional descendent nodes to describe it. We simply call the chains of nodes outputted by the two functions *paths*. More detailed description about *path* can be found in [1].

Definition 1. *(Structural Path, Semantic Path) In a returned path, if the condition to link adjacent nodes is the structural constraint of parent-child relationship in the document tree, the path is called a structural path; if the condition is semantics-based join, the path is called a semantic path.*

2.1   SPath Expressions

Definition 2. *(Definition of SPath functions)*
StrucPath function $::= StrucPath[\![tp \times tp \rightarrow sp\_set]\!]$
SemPath function $::= SemPath[\![tp \times tp \times tp \times cond \rightarrow sp\_set]\!]$
$tp ::= XPathExpr \ (`/' \mid `//') \ tpRel$

$tpRel ::= nodeName \mid tpRel$ ('[' ('//')? $tpRel \mid XPathExpr$ op value ']')*
$cond ::=$ 'parent' $XPathExpr$ op 'child' $XPathExpr$
$op ::=$ '=' | '$\neq$' | '>' |'<' | '$\geq$' | '$\leq$'
$sp\_set ::= \{path^*\}$

XPathExpr *is the simplified XPath expression with only axis of* '/' *and* '//', nodeName *is the XML tag label,* **parent** *and* **child** *are two reserved keywords to represent the parent node and child node in a pair of adjacent nodes, and* value *denotes constant string or numeric values.*

Definition 3. *(Semantics of SPath functions) Let* sp *be a structural path or a semantic path, in terms of a list of nodes, where* sp[first] *(*sp[last]*) denotes the first (last) node in* sp. sp_set *is the set of paths to be returned. For a node* n, n.next *denotes the next adjacent node of* n *in the path. Let the predicate* sat(X, Y) *return a Boolean value meaning whether* X *satisfies the constraint specified in* Y *and the* rel(X, Y) *stands for the linking relationship between node* X *and* Y. diff(X, Y) *represents the set difference between* X *and* Y. PC *denotes the constraint of parent-child relationship in an XML tree:*

$StrucPath(start, end) ::= \{sp \mid sat(sp[first], start) \land sat(sp[last], end) \land$
    $(\forall\ n \in diff(sp, \{sp[last]\}), sat(rel(n, n.next), PC)\}$
$SemPath(nodePattern, start, end, linkCond) ::=$
    $\{sp \mid sat(sp[first], start) \land sat(sp[last], end) \land$
    $(\forall\ n \in diff(sp, \{sp[last]\}), sat(n, nodePattern) \land$
    $sat(rel(n, n.next), linkCond))\}$

As mentioned, each path node in the output paths may have additional descendant nodes, i.e., it is similar to element. Thus the existing XPath functions can work on the `SPath` output in the same way as they do on an element.

2.2   XQuery Extension

In an XQuery query, a `SPath` function can be appended to an XPath expression, to find all paths under the element selected by the XPath expression.

Definition 4. *(Extended Path) Let* PathExpr *be the original XPath expression (or some built-in function returning element) used in XQuery, whose syntax and semantics is defined in XQuery specification [8] and will not be repeated here. Let* ExtPathExpr *be the extended path expression, and* tp *and* cond *are defined in Definition 2:*

$ExtPathExpr ::= PathExpr \mid PathExpr$ '.StrucPath(' tp ',' tp ')' |
    $PathExpr$ '.SemPath(' tp ',' tp ',' tp ',' cond ')'

The following example illustrates how desired structural path and semantic path are found by the two `SPath` functions under an element.

Example 1. **Structural path.** In the TreeBank document in Fig. 1(a), all structural paths starting at "VP" and ending at some POS tag with child value of "front" are specified by

```
doc("TreeBank.xml").StrucPath(//VP, //*[text()="front"])
```

Semantic Path. In the Flight document in Fig. 1(c), all routes from Singapore to Bilbao (the price of each flight must be outputted for analytical purpose) can be found by

```
doc("Flight.xml").SemPath(//flight[from][to][price], //flight
  [from ="Singapore"], //flight[to="Bilbao"], parent/to=child/from)
```

Because each path is an ordered list of nodes with additional descendants, in the extended XQuery FOR and LET can be used to iterate over a path and bind variables to nodes.

Example 2. The XQuery expression and result for the query to find the structural paths starting at "VP" and ending at some tag with child value of "front" in the document in Fig. 1(a) are:

```
FOR $p IN doc("TreeBank.xml").StrucPath(//VP, //*[text()="front"])
RETURN
    <structural_path>{FOR $e IN $p RETURN $e}</structural_path>
```

3   Intra-path Aggregation: A Practical Case

In this section, we discuss a practical query purpose, intra-path aggregation, as an example to illustrate the importance of the proposed SPath functions in expressing path-centric queries with the basic form (FLWOR) of XQuery.

3.1   Expression

Many query purposes aim to aggregate nodes along a single path, which is referred as *intra-path* aggregation. The existing XQuery language is inconvenient in expressing a single path, and so is it for intra-path aggregation.

Example 3. (**Tag Aggregation**) Consider the TreeBank data in Fig. 1(a), and the query to find the total number of *PP* in each path that starts at *VP* and ends at some tag with value of "front". The XQuery expression with our extended functions is shown as:

```
FOR $p IN doc("TreeBank.xml").StrucPath(//VP, //*[text()="front"])
RETURN {count(FOR $e IN $p WHERE $e.name()="PP" RETURN $e)}
```

In the outer FOR clause, StrucPath returns a set of paths and $p iterates over these paths. In other words, in each iteration, the variable is bound to a *path*, rather than the element returned by the doc() function. In the nested FOR clause, $e iterate over the nodes of a certain path, and is bound to each *node*. Only with these new output units, intra-path aggregation for path nodes is expressible under FLWOR constructs.

The new GROUP BY construct introduced in XQuery 3.0 can be used in intra-path aggregation. If the query asks for the total number of different syntactic tags in each path starting at VP and ending at "front", it will be:

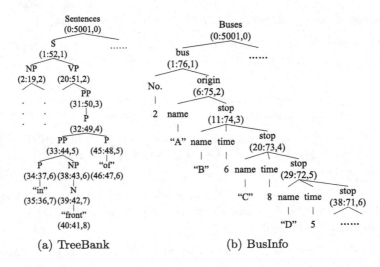

Fig. 2. XML data with containment labeling (partial)

```
FOR $p IN doc("TreeBank.xml").StrucPath(//VP, //*[text()="front"])
RETURN { FOR $e IN $p
         GROUP BY $e.name()
         RETURN <{$e.name()}>{count($e)}</{$e.name()}> }
```

Example 4. (**Value Aggregation**) Now we consider the intra-path aggregation for values in semantic paths. Consider the query to find the quotation of all direct and transit flight from Singapore to Bilbao in the document in Fig. 1(b). The XQuery expression is:

```
FOR $p IN doc("Flight.xml").SemPath(
  //flight[from][to][price], //flight[from="Singapore"],
         //flight[to="Bilbao"],parent/to =child/from)
RETURN {sum(FOR $e IN $p RETURN $e/price)}
```

3.2   Execution

In this section, we study how to extend the existing query processing algorithms to support intra-path aggregation. Structural path and semantic path have different characteristics, so the query execution for the two types of paths should adopt different approaches. Basically, the query processing over semantic paths can reduce to graph search problem, in which problems such as recursion handling can be solved accordingly. In this paper, we put the focus on structural path. More discussion about the query processing over the two types of paths can be found in our technical report [1].

We use two examples to illustrate our algorithm for tag-based and value-based intra-path aggregation.

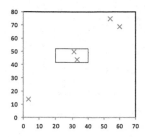

Fig. 3. Example R-tree encoding with range query rectangle

Example 5. Consider the query in Example 3. The TreeBank data with labels is shown in Fig. 2(a). We first find all structural paths by issuing a structural join query `//VP//"front"`. Then many pairs of labels corresponding to *VP* and *front* are returned, each of which represents a structural path. (20:51,2) and (40:41,8) is one such pair. Then we perform an outer structural join between all tuples of labels representing paths and the inverted list of *PP*. For the path starting at (20:51,2) and ending at (40:41,8), there are two PP labels (31:50,3), (33:44,5) are inside the path. Then for this path, the aggregation result is 2.

Example 6. Consider the BusInfo XML document shown in Fig. 2(b). Consider a query to find the total traveling time from stop B to stop D:

```
FOR $p IN doc("BusInfo.xml").StrucPath(
        //stop[name="B"], //stop[name="D"])
RETURN {sum(FOR $e IN $p WHERE $e/name/text() != "B"
        RETURN $e/time/text())  }
```

In this query, the node, the aggregate attribute and the predicates form a twig, `//stop[name!="B"]/time`. By matching this twig, we get the labels of each satisfied *stop* and the corresponding time values, i.e., {(20:73,4), 8} and {(29:72,5), 5}. The desired structural path can be found by another pattern matching for `//stop[name="B"]//stop[name="D"]`. The path shown in Fig. 2(b) is one of the answers, i.e., the path from (11:74,3) to (29:72,5). Last, we perform an outer structural join between the path and the tuples of answers found by first twig matching. Since both (20:73,4) and (29:72,5) are within the path from (11:74,3) to (29:72,5), the time value 8 and 5 are summed up as the result.

The pseudo code and more details on the algorithm can be found in [1].

We further proposed an R-tree based optimization to address all nodes contained in a given path.

Example 7. Consider the query in Example 5. The R-tree for the inverted list of PP is shown in Fig. 3, in which each label in the inverted list will be matched by a point in the two-dimensional R-tree space. Each path instance is encoded as a range-search rectangle in R-tree. For example, the path starting at (20:51,2) and ending at (40:41,8) corresponds to the rectangle in Fig. 3. The PP nodes within the path will be the points enclosed by the rectangle.

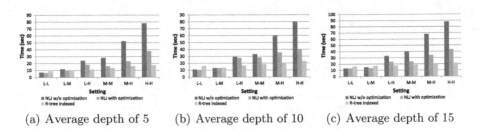

(a) Average depth of 5   (b) Average depth of 10   (c) Average depth of 15

Fig. 4. Efficiency test on synthetic data (Color figure online)

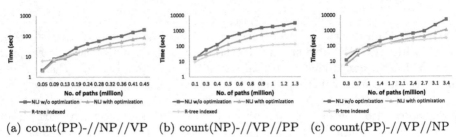

(a) count(PP)-//NP//VP   (b) count(NP)-//VP//PP   (c) count(PP)-//VP//NP

Fig. 5. Efficiency test on TreeBank data

4   Experiments

We assess the usability of our SPath extension compared to the XQuery user-defined function to express path-centric queries and conclude that our extension is more user friendly. The details can be found in [1]. In this section, we report the query performance evaluations for intra- structural path aggregation.

We implement three algorithms for intra- structural path aggregation. The first one is the baseline algorithm we proposed. In the second algorithm, we optimize the outer structural join. For each path, we only start structurally joining it with an inverted list when a label in the inverted list falls behind the starting node of the path, and skip the rest of the labels in the inverted list once we find one label falls behind the end node of the path. We name it *NLJ with optimization*. Finally, we implement the R-tree index to avoid structural joins.

We first randomly generate a 50MB XML data and queries to test. The detailed settings can be found in [1]. The experimental result is shown in Fig. 4. We can see that the result is similar for all the three tested data with different depth, i.e., the optimization can improve the performance, and R-tree index has obvious advantage when the frequency of path nodes increases.

In the second part, we use a real-life data set, TreeBank for evaluation. We choose simple path predicates that generate a large number of paths. The experimental results are shown in Fig. 5, in which x-axis stands for the number of path instances obtained by executing the corresponding path query for different sizes of documents, and y-axis is the running time in log-scale. We can see that for all queries, as the document size increases, the running time of the naive nested loop join increases very fast. After optimizing the algorithm, the performance is better. The best one is the R-tree indexed algorithm, in which the increasing rate is rather low. However, for the first few smaller documents, the R-tree indexed algorithm is not always better than others due to the overhead.

5 Related Work

There are a number of extensions to XPath and XQuery in recent years. [9] extends XPath by introducing a *Related Axis*, to specify the related relationship between query nodes. [2] introduces functions to express fuzzy search in XPath expression, which relaxes the strict requirement on the knowledge of XML structure for query issuers. SXPath [5] allows spatial navigation for Web documents.

Although some existing extensions enhance the expressivity of XPath or XQuery, they are still built on element selection and cannot project interesting path from the element subtree. In [6], a new XML query language XSQuirrel is proposed, which projects a sub-document from an XML document. However, it requires the user to input the detailed pattern of the path, which is similar to the "hard coding" in the XQuery RETURN clause. [4] presents a projection technique to filter unnecessary elements for in-memory XQuery processors, but it is not for expressing path-centric queries discussed in this paper.

6 Conclusion

We argue in this paper that the support for path-centric queries in XPath and XQuery is not sufficient. We propose a simple and seamless extension to XQuery that can express such queries: the **SPath** extension with two functions for path expressions. We show how XQuery with this extension can be used to effectively and elegantly express path-centric queries of interest. We use intra-path aggregation as a practical example to illustrate the importance of our extension, and also discuss query processing issues when the number of path instances is large.

References

1. http://www1.i2r.a-star.edu.sg/~huwu/report.pdf
2. Campi, A., Damiani, E., Guinea, S., Marrara, S., Pasi, G., Spoletini, P.: A fuzzy extension of the XPath query language. J. Intell. Inf. Syst. **33**(3), 285–305 (2009)
3. GeneOntology. http://www.geneontology.org/
4. Marian, A., Siméon, J.: Projecting XML documents. In: VLDB (2003)
5. Oro, E., Ruffolo, M., Staab, S.: Sxpath - extending xpath towards spatial querying on web documents. PVLDB **4**(2), 129–140 (2010)
6. Sahuguet, A., Alexe, B.: Sub-document queries over XML with XSQirrel. In: WWW, pp. 268–277 (2005)
7. The Penn Treebank Project. http://www.cis.upenn.edu/~treebank/
8. W3C Consortium: XQuery 1.0: An XML query language. http://www.w3.org/TR/xquery/ (2007)
9. Zhou, J., Ling, T.W., Bao, Z., Meng, X.: Related axis: the extension to XPath towards effective XML search. J. Comput. Sci. Technol. **27**(1), 195–212 (2012)

A Logical Framework for XML Reference Specification

C. Combi, A. Masini, B. Oliboni, and M. Zorzi[✉]

Department of Computer Science – University of Verona,
Cà Vignal 2, Strada le Grazie 15, 37134 Verona, Italy
{carlo.combi,andrea.masini,barbara.oliboni,margherita.zorzi}@univr.it

Abstract. In this paper we focus on a (as much as possible) simple logic, called XHyb, expressive enough to allow the specification of the most common integrity constraints in XML documents. In particular we will deal with constraints on ID and IDREF(S) attributes, which are the common way of logically connecting parts of XML documents, besides the usual containment relation of XML elements.

1 Introduction

XML (eXtensible Markup Language) is the main mark up language used for representing data to exchange on the Web and for data integration. XML allows one to represent structured and semistructured data through a hierarchical organization of mark up elements. An XML document is typically endowed with a DTD (Data TypeDefinition). DTDs allow the specification in a simple and compact way of the main structural features of XML documents. DTDs easily express hierarchies, order between elements, and several types of element attributes. In particular, the ID/IDREF mechanism of DTDs describes identifiers and references in a similar (but not equivalent) way to keys and foreign keys in a relational setting. The value of an attribute of type ID uniquely identifies an element among all the elements of the entire document; the value of an attribute of type IDREF(S) allows the reference to element(s) on the base of their ID values. DTD simplicity is paid in terms of expressiveness: a DTD efficiently models the structure of XML documents (it is able to provide a "syntactical" control such as context-free grammar), but it is not powerful enough for capturing subtle, semantic features. As an example, (unique) values of ID attributes have the overall document as a scope. Consequently, attributes of type IDREF(S) cannot be constrained to refer to only a subset of elements. Complex specification languages such as XML Schema [9] represent a powerful alternative to DTD: XML Schema supports the specification of a very rich set of constraints (in terms of XPath expressions) and seems to overcome DTD issues and limitations. Unfortunately, as observed in [3,4], XML Schema is too complicate and not compact at all in the specification of even simple integrity constraints.

In this paper we focus on the issue of retaining in a logical framework the simplicity of DTDs with the capability of expressing meaningful integrity

© Springer International Publishing Switzerland 2015
Q. Chen et al. (Eds.): DEXA 2015, Part II, LNCS 9262, pp. 258–267, 2015.
DOI: 10.1007/978-3-319-22852-5_22

constraints. In this context, some interesting theoretical solutions have been proposed [3,4]. With respect to previous proposals, the novelty of our work is that we look specifically for a very simple formal language which is able to model constraints with respect to XML reference specification. In this direction, we propose a logical language, called XHyb, able to express *in a direct and explicit way* constraints on XML documents.

2   Motivating Example

In this paper we will use the DTD shown in Fig. 1 as a running example. The considered DTD describes a subset of information related to the university domain. It represents the fact that a university is composed of many students, professors, courses, and examinations; a student may have a supervisor, when she starts her thesis work; a professor may act as both thesis supervisor and thesis reviewer.

```
<!ELEMENT university (student*,professor*,course*,examination*)>
<!ELEMENT student (name,surname,supervisor?)>
<!ELEMENT professor (name,surname,thesis_stud?,thesis_reviewer?)>
<!ELEMENT course (title)>
<!ELEMENT examination (mark)>

<!ELEMENT name      (#PCDATA)>
<!ELEMENT surname   (#PCDATA)>
<!ELEMENT title     (#PCDATA)>
<!ELEMENT mark      (#PCDATA)>
<!ELEMENT supervisor       EMPTY>
<!ELEMENT thesis_stud      EMPTY>
<!ELEMENT thesis_reviewer  EMPTY>

<!ATTLIST student          stud_id   ID      #REQUIRED>
<!ATTLIST professor        prof_id   ID      #REQUIRED>
<!ATTLIST course           cour_id   ID      #REQUIRED
                           prof_ref  IDREF   #REQUIRED>
<!ATTLIST examination      stud_ref  IDREF   #REQUIRED
                           cour_ref  IDREF   #REQUIRED>
<!ATTLIST supervisor       prof_ref  IDREF   #REQUIRED>
<!ATTLIST thesis_stud      stud_refs IDREFS  #REQUIRED>
<!ATTLIST thesis_reviewer  prof_ref  IDREFS  #REQUIRED>
```

Fig. 1. An example of DTD for XML documents.

The link between a student and her supervisor is modeled by using attribute prof_ref of type IDREF within element supervisor (which is contained in element student). On the other side, the corresponding link between a professor and her thesis students is modeled by means of attribute stud_refs of type IDREFS within element thesis_stud. Both attributes supervisor and stud_refs refer to elements identified by a suitable attribute of type ID. It is worth noting that the DTD grammar does not allow us, for example, to constrain the value of a prof_ref to correspond to the value of attribute prof_id of some element professor.

In general, DTD grammar allows us only to validate containment relations (restriction on the element structure of the document [5]) and links between

IDREF/IDREFS values and ID values within the whole document. Thus, many domain-related constraints cannot be explicitly modeled and some XML documents could be valid according to the given DTD but provide meaningless information (such as, for example, that a thesis reviewer is a course).

In Fig. 2 we report an example of XML document valid against DTD in Fig. 1. Let us consider in the following some examples of requirements we would like to represent and verify in XML documents related to the university domain.

- The supervisor of a student must be a professor.
- A professor may be the supervisor of one or more students.
- A student can be evaluated only once for a given course.

These constraints are clearly not expressible by DTD, as well as more complex constraints such us the following ones, which require to linguistically express the interplay between containment relation and reference constraints specification:

- A professor cannot be both supervisor and reviewer of the same student.
- A professor can be supervisor only for students that attended and passed a course she taught.

In the following section we will introduce the *referential logic* XHyb. In XHyb it is possible to encode constraints in terms of (as much as possible) simple modal formulas. Moreover, the Kripke-style XHyb models naturally fit the shape of XML documents, representing explicitly and distinctly both containment relation and reference specification. The formal description of the relationship XML-documents/XHyb and the encoding in XHyb of the constraints above are in Sects. 4 and 5 respectively.

3   XHyb: Hybrid Logic for XML Reference Constraints

Logic XHyb is an extension of a fragment of *hybrid logic* [1,2], obtained by adding to the syntax the operator $@_a$ (where a ranges over a particular set of variables, called nominal variables): $@_a$ is the hybrid *at* operator and it provides a direct access to the state (uniquely) named by a. The peculiar feature of XHyb is the extension of quantified hybrid logic by means of a new modal operator $*_c$, which explicitly captures the presence of ID/IDREF(S) relation between elements of XML documents.

3.1   Syntax

The alphabet of XHyb is built out of some sets of symbols for constants and variables. We define three distinct sets of constants:
$E = e_0, \ldots, e_k$ is a finite set of *element names*; $R = r_0, \ldots, r_p$: a finite set of *reference names*; $C = c_0, \ldots, c_m$ is a finite set of *colors*.

In the following we will use symbols e, f for element names, r, s for reference names, and c, d for color names, possibly in their indexed version. We assume C, E, R are pairwise disjoint.

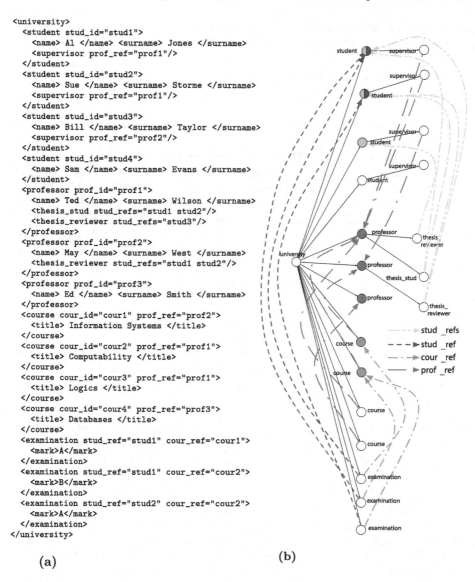

```
<university>
  <student stud_id="stud1">
    <name> Al </name> <surname> Jones </surname>
    <supervisor prof_ref="prof1"/>
  </student>
  <student stud_id="stud2">
    <name> Sue </name> <surname> Storme </surname>
    <supervisor prof_ref="prof1"/>
  </student>
  <student stud_id="stud3">
    <name> Bill </name> <surname> Taylor </surname>
    <supervisor prof_ref="prof2"/>
  </student>
  <student stud_id="stud4">
    <name> Sam </name> <surname> Evans </surname>
  </student>
  <professor prof_id="prof1">
    <name> Ted </name> <surname> Wilson </surname>
    <thesis_stud stud_refs="stud1 stud2"/>
    <thesis_reviewer stud_refs="stud3"/>
  </professor>
  <professor prof_id="prof2">
    <name> May </name> <surname> West </surname>
    <thesis_reviewer stud_refs="stud1 stud2"/>
  </professor>
  <professor prof_id="prof3">
    <name> Ed </name> <surname> Smith </surname>
  </professor>
  <course cour_id="cour1" prof_ref="prof2">
    <title> Information Systems </title>
  </course>
  <course cour_id="cour2" prof_ref="prof1">
    <title> Computability </title>
  </course>
  <course cour_id="cour3" prof_ref="prof1">
    <title> Logics </title>
  </course>
  <course cour_id="cour4" prof_ref="prof3">
    <title> Databases </title>
  </course>
  <examination stud_ref="stud1" cour_ref="cour1">
    <mark>A</mark>
  </examination>
  <examination stud_ref="stud1" cour_ref="cour2">
    <mark>B</mark>
  </examination>
  <examination stud_ref="stud2" cour_ref="cour2">
    <mark>A</mark>
  </examination>
</university>
```

(a)

(b)

Fig. 2. An XML document valid against the DTD in Fig. 1 and its graphical representation (reference colors are represented through lines with both colors and different dashes) (Color figure online).

Set PROP of propositional symbols is the union of the above sets, i.e. PROP $= C \cup E \cup R$.

Moreover, we define the following sets of variables for *nominals* and *sequences of nominals*:

$N = i_0, i_1, \ldots$ is a denumerable set of nominals; $\varGamma = \gamma_0, \gamma_1, \ldots$ is a denumerable set of variables for finite sequences of nominals. We assume that $\varGamma \cap N = \emptyset$ and $\varDelta = N \cup \varGamma$.

In the following we will use symbols i, j, l for nominals, γ, δ for sequences of nominals, and x, y for nominals/nominal sequences, possibly in their indexed version.

Set $\varTheta$ of *terms* τ is the smallest set Y defined by stipulating that: $N \subseteq Y$; $\varGamma \subseteq Y$; if $\tau', \tau'' \in Y$ then $\tau'\tau'' \in Y$.

We equip the language of XHyb with logical connectives $\rightarrow, \perp, \forall, \in, *_c, @_a, \Box$ and $\bigcirc^\forall$. Formulas are built out of the set of terms by means of logical connectives. Formally, the set Z of *well-formed formulas* (only *formulas* in the following, ranged by A, B, C possibly indexed), is the smallest set Y such that: $N \subseteq Y$; PROP $\subseteq Y$; **if** $i \in N$ **and** $A \in Y$ **then** $(@_i.A) \in Y$; **if** $i \in N$ **and** $\tau \in \varTheta$ **then** $(i \in \tau) \in Y$; **if** $\tau \in \varTheta$ **and** $c \in C$ **then** $*_c(\tau) \in Y$; **if** $i \in N$ **and** $A \in Y$ **then** $(\forall i.A) \in Y$; **if** $\gamma \in \varGamma$ **and** $A \in Y$ **then** $(\forall \gamma.A) \in Y$; **if** $A, B \in Y$ **then** $(A \rightarrow B) \in Y$; $\perp \in Y$; **if** $A \in Y$ **then** $\Box A, \bigcirc^\forall A \in Y$.

Connectives $\rightarrow, \perp, \forall$ are defined in the usual way. The intuition about the other connectives is as follows:

– $@_i A$ means that formula A holds at state i. Following hybrid logic tradition, equality between two worlds i and j is represented as $@_i j$;
– $\Box A$ means that A holds at the current state and at all the descendant states;
– $\bigcirc^\forall A$ means that A holds in each children of the current state;
– $*_c$ is the *reference operator*: if $*_c(i)$ holds in a given state, then there exists a reference, labelled by c, to state i.

Notation 1. *In the rest of the paper, we will use the following (quite standard) abbreviations: $\neg A$ stands for $A \rightarrow \perp$; $A \vee B$ stands for $(\neg A) \rightarrow B$); $A \wedge B$ stands for $\neg(A \vee B)$; $\bigwedge_k A(k)$ stands for $A(c_0) \wedge (A(c_1) \wedge (\cdots \wedge A(c_k))))$; $\bigvee_k A(k)$ stands for $A(c_0) \vee (A(c_1) \vee (\cdots \vee A(c_k))))$; $\exists i.A$ stands for $\neg(\forall i.(\neg A))$; $\exists \gamma.A$ stands for $\neg(\forall \gamma.(\neg A))$; $\bigcirc^\exists A$ stands for $\neg \bigcirc^\forall \neg A$; $\Diamond A$ stands for $\neg \Box \neg A$.*

In the following, given γ and γ' sequences of nominals, we will write $\gamma \subseteq \gamma'$ for $\forall i.(i \in \gamma \rightarrow i \in \gamma')$. We will always omit the most external parentheses in formulas. Moreover we will adopt useful precedence between operators in order to simplify the readings of formulas, in particular we stipulate that $\neg, \forall, @, \Box, \bigcirc^\forall$ have the higher priority.

The only binder for variables is $\forall$. Therefore, the definition of the set of free variables in terms and formulas is standard.

Definition 1 (Free and Bound Variables). *The set FV of names of free variables for terms and formulas is inductively defined as follows:*
$FV[i] = \{i\}$; $FV[\tau'\tau''] = FV[\tau'] \cup FV[\tau'']$; $FV[@_i.A] = \{i\} \cup FV[A]$; $FV[i \in \tau] = \{i\} \cup FV[\tau]$; $FV[*_c(\tau)] = FV[\tau]$; $FV[\gamma] = \{\gamma\}$; $FV[\forall i.A] = FV[A] - \{i\}$; $FV[\forall \gamma.A] = FV[A] - \{\gamma\}$; $FV[A \rightarrow B] = FV[A] \cup FV[B]$; $FV[\perp] = \emptyset$; $FV[p] = \emptyset$ *for* $p \in$ PROP; $FV[\Box A] = FV[A]$; $FV[\bigcirc^\forall A] = FV[A]$.

An occurrence of i (of γ) in a formula A is bound iff there is a sub-formula of A of the kind $C = \forall i.B$ ($C = \forall \gamma.B$). In this case we say also that B is the *scope of i (of γ). We say that an occurrence of i (of γ) in a formula A is free iff it is not bound.*

3.2 Semantics

Definition 2 (Frames). *A **structure** is a tuple $S = \langle W, V_E, V_C, V_R, Y, \mathcal{N} \rangle$ where:$|W| < \aleph_0$ is a set of worlds; $V_E : E \to 2^W$, $V_C : C \to 2^W$, $V_R : R \to 2^W$ and $V : \mathsf{PROP} \to 2^W$ is defined as $V = V_E \cup V_C \cup V_R$; $Y: W \to 2^W$ is the reference relation. $\mathcal{N} \subseteq W \times W$ is the relation father-son.*

*An **interpretation** is a tuple $\mathcal{I} = \langle S, g, h, w \rangle$ where S is a structure, $g : N \to W$, $h : \Gamma \to 2^W$, and $w \in W$.*

Informally, the reference relation maps a world w into the sets of worlds the state w "points to". As usual, we will denote by $\mathcal{N}^*$ the transitive and reflexive closure of $\mathcal{N}$.

Definition 3 (Satisfaction). *The satisfiability relation $\mathcal{I} \models A$ is defined in the following way:*

1. $S, g, h, w \not\models \bot$
2. $S, g, h, w \models p \Leftrightarrow w \in V(p)$ *with* $p \in \mathsf{PROP}$
3. $S, g, h, w \models i \Leftrightarrow w = g(i)$
4. $S, g, h, w \models *_c(x_1 \dots x_n) \Leftrightarrow$
   $V = \{v | v \in (g \cup h)(x_1) \cup \dots \cup (g \cup h)(x_n)\} \subseteq Y(w),$
   $\forall v \in V, S, g, h, v \models c;$
5. $S, g, h, w \models @_i.A \Leftrightarrow S, g, h, g(i) \models A$
6. $S, g, h, w \models \forall i.A \Leftrightarrow \forall v \in W, S, g[i \mapsto v], h, w \models A$
7. $S, g, h, w \models \forall \gamma.A \Leftrightarrow \forall M \in 2^W, S, g, h[\gamma \mapsto M], w \models A$
8. $S, g, h, w \models A \to B \Leftrightarrow S, g, h, w \not\models A$ *or* $S, g, h, w \models B$
9. $S, g, h, w \models \Box A \Leftrightarrow \forall v \in W(w\mathcal{N}^*v \Rightarrow S, g, h, v \models A);$
10. $S, g, h, w \models \bigcirc^\forall A \Leftrightarrow \forall v \in W(w\mathcal{N}v \Rightarrow S, g, h, v \models A);$

If $S, g, w \models A$ we say that $\langle S, g, h, w \rangle$ satisfies A.

We say that: *A **is satisfiable*** if there exists $\mathcal{I}$ s.t. $\mathcal{I} \models A$; *S **is a model** of A]($S \models A$)* if for each g, h, w, $S, g, h, w \models A$; *A **is valid** ($\models A$)* if for each S, $S \models A$; *A **is semantical consequence of a finite set** Σ **of formulas** ($\Sigma \models A$)* if $\forall \mathcal{I}((\forall B \in \Sigma.\mathcal{I} \models B) \Rightarrow \mathcal{I} \models A)$.

Let us now briefly focus on the semantics of XHyb particular connectives. The meaning of a formula $@_i A$ is defined by stipulating that A holds in a world w if and only if $w = g(i)$, i.e. the interpretation by g of the nominal i is exactly w. The meaning of a formula $*_c(x_1 \dots x_n)$ is defined upon the relation Y. $*_c(x_1 \dots x_n)$ holds in a world w if and only if the interpretation by g or h of variable x_i ($i = 1, \dots, n$) (for nominals or sequences of nominals) belongs to the set of worlds w points to according to Y. Moreover, the proposition $c \in C$ holds in each $v = (g \cup h)(x_i)$ for some $i = 1, \dots, n$.

Table 1. From XHyb to XML

XHyb constructs	XML interpretation
C (Colors)	IDREF(S) attribute declared in the DTD
E (Element names)	Tag names declared in the DTD
R (Identifier Names)	ID attributes declared in the DTD
W (Worlds)	Values of ID attributes in the XML document
$V_E : E \rightarrow 2^W$	Each element name e is mapped to the set of ID values identifying occurrences of e
$V_C : C \rightarrow 2^W$	Each attribute name of type IDREF(S) is mapped to the set of ID values referenced by values of the given attribute
$V_R : R \rightarrow 2^W$	Each attribute name of type ID is mapped to the set of corresponding ID values in the document
$\mathcal{N} : W \rightarrow 2^W$	Containment relation (parent-child relation)
$\curlyvee : W \rightarrow 2^W$	Each attribute name of type ID is mapped to the set of corresponding ID values in the document

4 From XML to **XHyb**

In this section we describe the relationship between the XHyb logic and XML documents. In Table 1 we summarize the XML interpretation of XHyb, by providing a simple mapping between XHyb syntactical and semantic objects and the corresponding meaning in the XML document.

It is mandatory to say that the tree-like structure of XML documents naturally fits the shape of (most) modal/temporal logic Kripke models. This has been observed and exploited in [6,7]. In this paper we start from the same observation, maintaining a slightly different viewpoint. Given an XML document, we will adopt the (quite) standard graph-representation (see e.g. [3]), but we choose a bit more informative graphical depiction:

- we represent XML elements as nodes, labeled with the element name and, when explicitly required, the ID attribute;
- black edges represent the containment relation;
- colored edges represent the presence of an ID/IDREF(S) link;
- nodes pointed by colored edges are colored accordingly.

More formally, the overall structure of an XML document may be represented as in the following.

Definition 4 (Colored XGraph, Xtree and colored Xstructure). *A colored XGraph is a tuple* $C\mathcal{G} = \{P, E, r, Col, E_{Col}, l_v\}$ *such that:* P *is a set of nodes and* r *is a particular node called* root; E *is a set of ordered pairs of nodes where, for all* $v \in P - \{r\}$, *there exists a node* $u \in P$ *such that* $(u, v) \in E$ *and if* $(u_1, v) \in E$ *and* $(u_2, v) \in E$ *then* $u_1 = u_2$; Col *is a set of color labels;* l_e *is a labeling function* $l_e : P \rightarrow Col$. E_{Col} *is a set of pairs* $((u, v), c)$ *where* (u, w) *is an ordered pair of nodes,* $c \in Col$ *and if* $((u, w), c) \in E_{Col}$ *then* $l_v(w) = c$.

Table 2. XHyb overall picture

Connectives	$*_c$	$\square, \bigcirc^{\forall}$	$*_c + \square, \bigcirc^{\forall}$
Relations/constraints	References	Containment	References + Containment
Shape of the models	colored Xstructure	Xtree	colored Xgraph

– Xtree *is the substructure* $\{P, E\}$*;*
– colored Xstructure *is the substructure* $\{P, E_{Col}, Col, l\}$*.*

The introduction of colored Xgraphs allows us to represent at the same time both the containment relation and the accessibility relation (through references) between nodes. This is possible since in XHyb IDREF(s) attributes are explicitly denotable (thanks to the reference operator $*_c$) and their linguistic treatment is completely independent from the denotation of the containment relation (Table 2). Our graphical representation reflects the way the syntax and the semantics of XHyb are defined. In particular, we can stipulate a bijection between the set of color labels *Col* and the IDREF(s) declaration in the DTD and so with the set of constant C.

Let us now sketch the translation of the DTD University Record and the XML documents proposed in Fig. 2 into the Referential Logic XHyb. We will actually build a concrete alphabet for the XHyb language and a related semantical model by processing the content of the DTD and the XML instance. Intuitively, this can be achieved by reading right-left Table 1 and building step-by-step propositional symbols (the constants of the logic) and a semantical structure (actually a colored Xgraph: a set of nodes equipped with two distinct accessibility relations). Notice that we need both the DTD and the XML instances, since names of elements and attributes (in particular ID and IDREF(S) attributes) can be "statically" determined from the DTD, whereas element occurrences, ID values, and IDREF(s) values can be only "dynamically" extracted from to the XML instance.

In Fig. 2.(b) we propose the graphical representation of the XML document reported in Fig. 2.(a), which is valid against the DTD in Fig. 1. We assume that red, blue, green and pink represent the attributes prof_ref, stud_ref, cour_ref and stud_refs respectively. As an example, consider a node professor. It is red (i.e., it has the same color of link prof_ref), since it is pointed by a node supervisor through a (red) IDREF prof_ref. Any attribute IDREF corresponds, in XHyb, to a propositional symbol: in the example, prof_ref belongs to set C of colors and thus to set PROP. By Definition 3, it is possible to see where propositional symbol/color prof_ref holds. The presence of the IDREF relation between supervisor and professor can be easily encoded as $*_{prof\_ref}(\text{professor})$. This formula clearly holds in a node (a world) supervisor, i.e., we can state (forgetting about interpretation) supervisor $\models *_{prof\_ref}(\text{professor})$. Following Definition 3, Case 4, clearly professor $\models$ prof_ref.

Summing up, the way the logic has been defined allows: (i) to express reference constraints in terms of (simple) XHyb formulae, overcoming DTDs expressive limitations. Some interesting examples related to the university domain are provided in Sect. 5, and (ii) to map an XML document into an XHyb (Kripke-like) model. This does not only confirm that XHyb is a suitable formalism to reason about XML, but it also represents the first step toward the static automated verification of XML constraints.

5   Expressing XML Constraints by XHyb

We provide now an XHyb encoding of some interesting constraints (non-expressible by DTD) that must hold for the XML document reported in Fig. 2. In the following, i, j, k, m, n are variables for nominals and γ is a variable for a finite sequence of nominals.

1. The supervisor of a student must be a professor.
   Attribute prof_ref of element supervisor must refer to an element professor.

$$\forall i((\text{supervisor} \wedge *_{\text{prof\_ref}}(i)) \rightarrow @_i\text{professor})$$

2. A professor may be the supervisor of one or more students.
   When thesis_stud appears, its attribute stud_refs must refer to at least one student element.

$$\text{thesis\_stud} \rightarrow \exists\gamma.(*_{\text{stud\_refs}}(\gamma) \wedge \forall i.(i \in \gamma \rightarrow @_i\text{student}))$$

3. A course must be taught by a professor.
   Attribute prof_ref of element course must refer to an element professor.

$$\text{course} \rightarrow \exists k(*_{\text{proof\_ref}}(k) \wedge @_k\text{professor})$$

4. An examination must be related to a student.
   Attribute stud_ref of element examination must refer to an element student.

$$\text{examination} \rightarrow \exists k(*_{\text{stud\_ref}}(k) \wedge @_k\text{student})$$

5. A student can be evaluated only once for a given course.
   Attributes stud_ref and cour_ref of an element examination cannot have the same values (couple of values) in different examination elements.

$$\forall i.\forall j.((@_i\text{student} \wedge @_j\text{course}) \rightarrow$$
$$\forall m.\forall n(@_m(\text{examination} \wedge *_{\text{stud\_ref}}(i) \wedge *_{\text{cour\_ref}}(j))$$
$$\wedge @_n(\text{examination} \wedge *_{\text{stud\_ref}}(i) \wedge *_{\text{cour\_ref}}(j)) \rightarrow @_m n)$$

6. A professor cannot be both supervisor and reviewer of the same student.
   Attribute stud_refs of element thesis_stud and attribute stud_refs of element thesis_reviewer, when thesis_stud thesis_reviewer are in the same element professor, refer to two different and disjoint sets of elements student.

$$\neg\exists k.j.\gamma(@_k(\text{professor} \wedge \bigcirc^\exists(\text{thesis\_reviewer} \wedge *_{\text{stud\_refs}}(\gamma) \wedge j \in \gamma)) \wedge$$
$$@_j(\text{student} \wedge \bigcirc^\exists(\text{supervisor} \wedge *_{\text{proof\_ref}}(k))))$$

7. A professor can be supervisor only for students that attended and passed a course she taught. *Attribute* stud_refs *of a given element* thesis_stud *must have values among those of attribute* stud_ref *of an element* examination, *where its attribute* cour_ref *refers to an element* course *having attribute* prof_ref *referring to the element* professor *containing the given element* thesis_stud.

$$\forall i(@_i \texttt{professor} \rightarrow \forall k(@_k(\texttt{student} \wedge \bigcirc^{\exists}(\texttt{supervisor} \wedge *_{\texttt{proof\_ref}}(i))) \rightarrow$$
$$\exists m(@_m(\texttt{course} \wedge *_{\texttt{proof\_ref}}(i) \wedge$$
$$\exists n(@_n(\texttt{examination} \wedge *_{\texttt{cour\_ref}}(m) \wedge *_{\texttt{stud\_ref}}(k)))))))$$

6 Conclusions

In this paper we proposed a simple extension of hybrid logic with a reference operator $*_c$. We show how this logic, called XHyb, is suitable to express references specification, overcoming, in an feasible way, some limitations of DTD expressiveness.

References

1. Blackburn, P.: Representation, reasoning, and relational structures: a hybrid logic manifesto. Logic J. IGPL **8**(3), 339–365 (2000)
2. Blackburn, P., Tzakova, M.: Hybridizing concept languages. Ann. Math. Artif. Intell. **24**(1–4), 23–49 (1998)
3. Fan, W., Libkin, L.: On XML integrity constraints in the presence of DTDs. J. ACM **49**(3), 368–406 (2002)
4. Fan, W., Siméon, J.: Integrity constraints for XML. J. Comput. Syst. Sci. **66**(1), 254–291 (2003). Special Issue on PODS 2000
5. Fan, W.: Xml constraints: specification, analysis, and applications. In: Proceedings, DEXA, pp. 805–809 (2005)
6. Franceschet, M., de Rijke, M.: Model checking hybrid logics (with an application to semistructured data). J. Appl. Logic **4**(3), 279–304 (2006)
7. Marx, M.: Xpath and modal logics of finite dag's. In: Proceedings of Automated Reasoning with Analytic Tableaux and Related Methods, International Conference, TABLEAUX 2003, Rome, Italy, 9–12 September 2003, pp. 150–164 (2003)
8. Rodrigues, K.R., dos Santos Mello, R.: A faceted taxonomy of semantic integrity constraints for the XML data model. In: Wagner, R., Revell, N., Pernul, G. (eds.) DEXA 2007. LNCS, vol. 4653, pp. 65–74. Springer, Heidelberg (2007)
9. van der Vlist, E.: XML Schema - The W3C's Object-oriented Descriptions for XML. O'Reilly, Sebastopol (2002)

XQuery Testing from XML Schema Based Random Test Cases

Jesús M. Almendros-Jiménez$^{(\boxtimes)}$ and Antonio Becerra-Terón

Department of Informatics, University of Almería, 04120 Almería, Spain
{jalmen,abecerra}@ual.es

Abstract. In this paper we present the elements of an XQuery testing tool which makes possible to automatically test XQuery programs.
The tool is able to systematically generate XML instances (i.e., test
cases) from a given XML schema. The number and type of instances is
defined by the human tester. These instances are used to execute the
given XQuery program. In addition, the tool makes possible to provide
an user defined property to be tested against the output of the XQuery
program. The property can be specified with a Boolean XQuery function.
The tool is implemented as an oracle able to report whether the XQuery
program passes the test, that is, all the test cases satisfy the property,
as well as the number of test cases used for testing. In the case of the
XQuery program fails the testing, the tool shows counterexamples found
in the test cases. The tool has been implemented as an XQuery library
which makes possible to be used from any XQuery interpreter.

1 Introduction

Testing [21] is essential for ensuring software quality. The automation of testing enables the programmer to reduce time of testing and also makes possible
to repeat testing after each modification to a program. A testing tool should
determine whether a test is passed or failed. When failed, the testing tool should
provide evidences of failures, that is, counterexamples of the properties to be
checked. Additionally, a testing tool should generate test cases automatically
[1]. Fully random generation could not be suitable for an effective and efficient
tool. Distribution of test data should be controlled, by providing user-defined test
cases, that is, data distribution should be put under the human tester's control.
For testing XML based applications some benchmarks datasets are available
(for instance, *XMark* [20], *Michigan Benchmark* [19] and *XBench* [22]). However, they are not always suitable for testing applications. There are some cases
of automatic data generators for XML: *ToXgene* [2] using XML Schemas with
annotations of data distribution functions, *VeXGene* [16] using DTDs, and *XBe-
Gene* [14] based on examples. XML test case generation can find an application
field in *Web Services* [3,12] and *Access Control Policies* [7].

This work was supported by the EU (FEDER) and the Spanish MINECO Ministry
(*Ministerio de Economía y Competitividad*) under grant TIN2013-44742-C4-4-R, as
well as by the Andalusian Regional Government under Project P10-TIC-6114.

Q. Chen et al. (Eds.): DEXA 2015, Part II, LNCS 9262, pp. 268–282, 2015.
DOI: 10.1007/978-3-319-22852-5_23

XQuery has evolved into a widely accepted query language for XML processing and many XQuery engines have been developed. Even though some tools
provide mechanisms for debugging (for instance, XMLSpy[1], Oxigen[2] and Stylus
Studio[3], among others), users should be equipped with a large number of mechanisms for detecting failures in their applications. Among them testing would
facilitate the detection of bugs due to mistakes when using XQuery expressions. Most of programming errors come from wrong XPath expressions (i.e.,
requesting paths/nodes of XML trees that do not exist), unsatisfiable Boolean
conditions and incompatible XPath expressions. Validation of XML documents
against XML Schemas mitigates some of these drawbacks. When an XQuery
expression against a given document returns an empty value/wrong answer,
stepwise/breakpoint/trace based debuggers can help to detect failures but it is
only useful for ensuring the correct execution for a single XML input instance. In
order to have a stronger confidence, XQuery programs should be tested against
a test suite covering a large range of test cases.

In this paper we present the elements of an XQuery testing tool which makes
possible to automatically test XQuery programs. The tool is able to systematically generate XML instances (i.e., test cases) from a given XML schema. The
number and type of instances is defined by the human tester. These instances
are used to execute the given XQuery program. In addition, the tool makes possible to provide an user defined property to be tested against the output of the
XQuery program. The property can be specified with a Boolean XQuery function. Our proposal can be seen as a black-block approach to XQuery testing.
The tool takes as input an XML Schema, an XQuery program and a property to
be checked against the output of the XQuery program. The tool automatically
generates a test suite from the XML Schema, it executes the XQuery program
and for each output of the program it checks the given property.

The tool is implemented as an oracle able to report whether the XQuery
expression passes the test, that is, all the test cases satisfy the property, as well
as the number of test cases used for testing. In case of success (i.e., all the outputs
of the XQuery program satisfy the property), the tool reports "Ok". Otherwise
the tool reports counterexamples (i.e., inputs of the XQuery program for which
the output does not satisfy the given property). The tool is customizable in the
following sense. The human tester can define the structure and content of the
test cases from the XML schema. Additionally, the human tester can control
the number of test cases. He or she can tune the number and size of XML trees
generated from the XML schema. Thus, although the tool generates random
test cases, the tester can control the size of the test suite. The property to be
checked is defined by a Boolean XQuery function. Usually, the property expresses
a constraint in terms of XPath expressions and logical connectors: for all, exists,
and, or, etc., possibly making use of XQuery functions. It makes possible to
express a rich repertory of properties against output documents: the occurrence

---

[1] http://www.altova.com/xmlspy/xquery-debugger.html.

[2] http://www.oxygenxml.com/xml_editor/xquery_debugger.html.

[3] http://www.stylusstudio.com/xquery_debugger.html.

of a certain value, the range of a certain attribute, the number of nodes, etc. The code of the XQuery program is not used to generate the test suite. The adequacy of the test suite to a given XQuery program is determined by the human tester, that is, the human tester has to select an XML Schema suitable to generate a test suite covering as many cases as possible. In essence he or she gives with the schema the required paths as well as the values of tags and attributes of the input documents. The tool randomly generates test cases as combinations of paths and values. In summary, the tool enables a partial validation of the XQuery program: for a subset of the input documents and a given property. Since the testing is automatic and customizable, the human tester can easily change values and paths as well as he or she can increase the number of test cases and play with properties to have an stronger confidence about the soundness of the program.

Our approach is inspired by similar tools in functional languages. This is the case of the *Quickcheck* tool [9] for Haskell, and the *PropEr* tool [18] for Erlang. Properties in these approaches are specified by functions, and are automatically tested from random test cases. Since XQuery is also a functional language, there seems natural to provide a similar tool for XQuery programs. Nevertheless, the context is different. XQuery is a functional language handling XML documents, which are in essence ordered labeled trees, and XQuery has a main element XPath expressions. Thus, here random test case generation is focused on trees and thus specific algorithms can be defined to automatically generate trees of a certain size. Properties in Quickcheck and PropEr are also defined by functions. Thanks to the Higher Order capabilities of XQuery, the property can be passed as argument to the tester, as well as the XQuery program, enabling the implementation of the tool in a similar way to Haskell and Erlang, that is, the tool is an XQuery program implemented as an XQuery library and thus can be used from any interpreter.

Our work is also inspired by some previous works about testing of XML applications. In [4] the *XPT (XML-based Partition Testing)* approach is presented which makes possible the automatic generation of XML instances from a given XML Schema. The *TAXI* tool[4] has been developed in this framework [5]. XPT is an adaptation of the well-known *Category Partition Method* [17], used to generate instances with all the possible combinations of elements. In this approach, a test selection strategy is studied including user defined weight assignments for *choice* statements in XML Schemas and derivation of XML Schemas from an initial XML Schema according to choice statements. XPT is able to generate XML instances according to a fixed number of instances, a fixed functional coverage (in percentage terms) and also a mixed criterium (fixed number of instances and functional coverage). TAXI populates instances by specifying a source (i.e., a URL), by manual insertion of values, or by taking values from the schema (in the *enumeration* section). In [6] they use TAXI to test *XSLT* stylesheets. The tool is able to report the result of the testing, using XML Schemas as model to which output XML instances must conform. In our approach, we take the XML Schema as input for test case generation similarly to TAXI. But there

---

[4] http://labsewiki.isti.cnr.it/labsedc/tools/taxi/public/main.

are some differences. In our case, the XML Schema rather than representing all the possible inputs of the program, is specifically used to generate test cases. Thus, even when an XML Schema for the program exists, the human tester has to select from the XML Schema those relevant elements for the program: paths and values to be tested. It does not mean that the original schema does not serve for test case generation, but usually the number of instances can be greater than necessary, affecting the performance of the tool. In particular, the human tester has to select relevant values and to incorporate them to the XML Schema in the *enumeration* section, and *choice* statements have to explicitly be selected by the human tester. It does not mean that the human tester can test more than one combination of them. Additionally, our work can be seen as an extension of the TAXI approach enabling property-based testing of XQuery programs.

In [10] they propose the automatic generation of XML documents from a DTD and an example. The framework called *GxBE (Generate-XML-By-Example)* uses a declarative syntax based on XPath to describe properties of the output documents. They claim that datasets are useful for testing when they conform a certain schema (an DTD schema extended with cardinality constraints), when they have some specific characteristics (i.e., datasets returning empty answers are not very useful) and the data values match an expected data distribution. They are able to express global properties on the document, in particular, to express the so-called count constraints making use of XPath and the count function. With regard to GxBE, our approach randomly generates test cases from the XML Schema in which values have been incorporated, while properties are not considered to test case generation. Rather than properties are used for testing after test case generation and query execution. In GxBE they propose to use properties in test case generation as pre-conditions, in order to generate a more suitable test suite. We believe that this is an interesting idea and we will incorporate it to our tool in the future. In GxBE they are concerned about test case generation efficiency. In our approach, test case generation efficiency cannot be ensured but it is under control. Firstly, the human tester can customize the number of test cases by modifying the input XML Schema, as well as by selecting the size of the trees. Additionally, test case generation is dynamic. It means that when testing a given program, test cases are incrementally generated and the tool stops when a failure (i.e., the property is not satisfied) has been found. It drastically reduces the time required for testing.

In [15] they describe how to test output documents of XML queries. They permit the specification of properties on XML documents via an XML template in which expected nodes, unexpected nodes, expected ordering and expected cardinality can be specified. In this work rather than test case generation, they propose the specifications of some properties on the output (i.e., post-conditions) to be checked against the output document. We adopt a similar approach but enabling the specification of a richer repertoire of properties. In fact, we do not restrict the type of property, making possible to use any XQuery expression to test the program. Nevertheless, we implicitly assume that the property to be checked is considerably simple. Testing should be focused on simple properties, otherwise the testing process would be as complex as the programming process. Moreover, complex properties affect the performance of the tool.

There are some works about white-box testing in this context. A partition-based approach for XPath testing has been proposed in [11]. They propose the construction of constraints from categorization and choice selection in order to generate test cases for XPath expressions. In [8] they study test case generation based on the category partition method and make use of *SMT* and *Z3* solvers for input data generation. Test cases are specified in XML. Our approach is still a black-box technique, but we will consider as future work to include white-box test case generation. In particular, we believe that the XML schema input of the test case generation can be filtered/generated from the program.

1.1 Example

Now, we would like to show an example to illustrate the approach. Let us suppose the following XQuery program:

```
for $book in $file/book return
    if ($book/title="UML" and $book/price<100) then
        <book_UML>
        {$book/@year}{$book/author}{$book/price}
        </book_UML>
    else if ($book/title="XML" and $book/@year>2000 and $book/price<100) then
        <book_XML>
        {$book/@year}{$book/author}{$book/price}
        </book_XML>
    else ()
```

And let us suppose that we would like to know whether the output of the program satisfies the following properties: *"Prices of books are smaller than 100"* and *"Book are after 2000"*. Our tool works as follows. Firstly, we have to define an XML Schema for generating test cases. For instance, the schema shown in Fig. 1. There, we intentionally select values (defined in the enumeration statement) for titles, authors, prices and years, oriented to the selected properties (i.e., years after and before 2000, and prices greater and smaller than 100). Titles are selected according to the intended behavior of the program (i.e., selection of UML and XML books), and one author is provided for completing book information. Next, we convert the query into a function as follows:

```
declare function tc:books_query($file)
{for $book in $file/book return
    if ($book/title="UML" and $book/price<100) then
        <book_UML>
        {$book/@year}{$book/author}{$book/price}
        </book_UML>
    else if ($book/title="XML" and $book/@year>2000 and $book/price<100) then
        <book_XML>
        {$book/@year}{$book/author}{$book/price}
        </book_XML>
    else ()
};
```

After, we can define the following Boolean XQuery functions that act on the output document.

```
<xs:schema xmlns:xs="http://www.w3.org/2001/XMLSchema">
<xs:simpleType name="authorType">
  <xs:restriction base="xs:string">
    <xs:enumeration value="Buneman"/>
  </xs:restriction>
</xs:simpleType>
<xs:simpleType name="yearType">
  <xs:restriction base="xs:integer">
    <xs:enumeration value="1995"/>
    <xs:enumeration value="2005"/>
  </xs:restriction>
</xs:simpleType>
<xs:simpleType name="priceType">
  <xs:restriction base="xs:integer">
    <xs:enumeration value="80"/>
    <xs:enumeration value="150"/>
  </xs:restriction>
</xs:simpleType>
<xs:simpleType name="titleType">
  <xs:restriction base="xs:string">
    <xs:enumeration value="UML"/>
    <xs:enumeration value="XML"/>
  </xs:restriction>
</xs:simpleType>
<xs:element name="bib">
<xs:complexType>
<xs:sequence>
<xs:element name="book" minOccurs="1" maxOccurs="unbounded">
  <xs:complexType>
    <xs:sequence>
      <xs:element name="author" type="authorType" minOccurs="1"
                   maxOccurs="unbounded"/>
      <xs:element name="title" type="titleType"/>
      <xs:element name="price" type="priceType"/>
    </xs:sequence>
    <xs:attribute name="year" type="yearType" use="required"/>
  </xs:complexType>
</xs:element>
</xs:sequence>
</xs:complexType>
</xs:element>
</xs:schema>
```

Fig. 1. XML Schema example

```
declare function tc:books_price_100($book){
  $book/price<100
};
```

```
declare function tc:books_year_2000($book){
  $book/@year>2000
};
```

Now, we can call the tool with the first property as follows, reporting the answer below:

```
tc:tester("schema.xsd","tc:books_query", "tc:books_price_100",1)
Ok: passed 80 tests.
Trivial: 35 tests.
```

and also, with the second property, reporting the answer below:

```
tc:tester("schema.xsd","tc:books_query", "tc:books_year_2000",1)
Falsifiable after 8 tests.
Counterexamples:
<bib>
  <book year="1995">
    <author>Buneman</author>
    <title>UML</title>
```

```
    <price>80</price>
  </book>
</bib>
....
```

The first reported answer means that all the books of the output have price smaller than 100 for the given 80 test cases. From them, 35 produce an empty answer, and thus the property cannot be checked. Empty answers come from test cases in which prices are greater than 100. The second reported answer means that the tool finds at least one counterexample (shown bellow) from eight test cases. Thus, the second property is not satisfied. We can inspect now the program revealing that UML books are not filtered by year. In the first case, we are not sure whether the property is satisfied, but 80 test cases passed it. We have incorporated a parameter to the tool (the "1" occurring in the example) to limit the number and size of generated test cases. Increasing this number, we can have a stronger confidence about this property. More details will be given in Sect. 2. Let us remark that for the examples shown the tool spends 101 and 20 ms, respectively. We have evaluated our tool with several examples (see Sect. 4).

The rest of the paper is structured as follows. Section 2 will describe the algorithm for test case generation. Section 3 will present how XQuery programs are tested by our tool. Section 4 will show examples of evaluation. Finally Sect. 5 will conclude and present future work.

2   Test Case Generation

In this section we will describe how test cases are randomly generated from an XML Schema. As was commented in the introduction, the XML schema used for test case generation is not necessarily the XML Schema given to the input program. Rather than, the human tester has to define, in most cases, a new XML Schema based on the original one, in which he or she selects the elements that are sufficient to test the program. The original schema leads to a too wide range of test cases that could not be required to test the program, affecting the performance of testing. Although our approach is a black box testing mechanism, the human tester has to select from the space of input data, those relevant to the program. Basically, the selected number of tags/paths of the document as well as values should be relevant.

The XML Schema offers a wide range of mechanisms to express input data structure. However, from testing process point of view, only a small subset of mechanisms is actually required. For instance, *choice* and *all* statements for declaring content of complex types are used in the XML Schema to produce variants of XML trees in which some tags can occur or not in the trees. We only consider the case of *sequence*, in which the human tester can play with *minOccurs* and *maxOccurs* elements, to force the occurrences of tags. Allowing zero in *minOccurs*, *choice* is a particular case of *sequence* in the test case generation. The current implementation does not consider the case of *all* (which permits any ordering of tags) given that it produces a huge number of cases. There are other mechanisms explicitly avoided like *any*, *anyAttribute* and *list*. Additionally, *attributeGroup, Group, redefine, extension, union, import* and *include* enabling reuse of definitions are not considered. Keys are also not considered. *ref* can be used to define recursive definitions in which a recursive *complexType* has to be defined. Finally, *enumeration* is used for defining the values of a certain tag. When a certain tag does not have the corresponding enumeration statement, the tool assigns a default value. In summary, an XML Schema for test case generation includes: *attribute* (*use*), *element* (*minOccurs/maxOccurs*), *sequence* (*minOccurs/maxOccurs*), *complexType*, *ref* and *enumeration*. Neither *use* nor *minOccurs/maxOccurs* are required assuming default values: optional and 1, respectively.

Test case generation can be controlled by the human tester. XML trees are generated in a certain order of increasing size. Basically, the random test case generator starts from the smallest XML tree conforming the given schema, and in each step it increases the size of previously generated trees by adding a new branch up to *maxOccurs* value. Recursive definitions are also unfolded in each step. In case *maxOccurs* is unbounded or with recursive definitions, the random test case generator is able to produce an infinite number of trees, but it never happens because the number of steps is a parameter of the test case generator. When the testing process is carried out, the test cases are dynamically generated up to the required number of steps, but whenever the program fails to satisfy the given Boolean property the tester stops and no more cases are generated. Thus, there will happen that only the required cases to fail the program are computed. In the case of success, they will be fully computed. From a practical point of view, the human tester should request a small number of steps in the beginning and increase in case of success. In Sect. 4 we will show experiments made for different XML Schemas. Figure 2 shows an example of test case generation for an XML Schema. Starting from the XML Schema shown in the left corner (schematically represented), in each step, the test case generator produces new schemas, and schemas are populated with values. In Fig. 2, step 1 generates 1, step 2 generates 1.1, step 3 generates 1.1.1, 1.1.2, and 1.1.3, and step 4 generates 1.1.1.2,

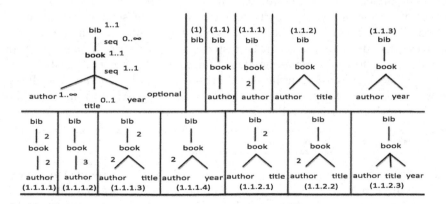

Fig. 2. Test Case Generation Example

1.1.1.3, 1.1.1.3, 1.1.2.1,1.1.2.2,1.1.2.3, and so on. In each step the schemas previously generated are used to compute new ones. Basically, the idea is to increase minOccurs values, to add optional attributes and to unfold recursive definitions in each step. The order in which the trees increase is top-down.

3   Property-Based Testing

Now, we would like to explain how testing process is carried out. Firstly, we will define the meaning of a certain program passing a test, and secondly we will describe the implementation of the tool. We only consider the case of unary Boolean properties, and programs with single input. The following definitions can be generalized to the case of n-ary properties and multiple inputs.

Assuming a finite set of test cases $t_1, \ldots, t_n$ of an XML Schema Σ, a program $\mathcal{P}$ and a Boolean property p, we say that $\mathcal{P}$ passes the test p in $t_1, \ldots, t_n$ whenever for each t_i, $i = 1, \ldots, n$ such that $\mathcal{P}(t_i) \neq \emptyset$ then $p(\mathcal{P}(t_i))$. We say that $\mathcal{P}$ fails the test p whenever there exists t_i, $i = 1, \ldots, n$ such that $\mathcal{P}(t_i) \neq \emptyset$ and $\neg p(\mathcal{P}(t_i))$. Finally, we say that p cannot be checked for $\mathcal{P}$ in $t_1, \ldots, t_n$, whenever for all t_i, $i = 1, \ldots, n$, $\mathcal{P}(t_i) = \emptyset$.

Given a program $\mathcal{P}$ and an XML Schema Σ, we say that $\mathcal{P}$ conforms to Σ whenever there exists t which conforms to Σ such that $\mathcal{P}(t) \neq \emptyset$. In other words, $\mathcal{P}$ conforms to Σ whenever at least an answer of $\mathcal{P}$ is not empty, and thus it means that Σ is correct for $\mathcal{P}$.

Obviously, if p cannot be checked for $\mathcal{P}$ then the test cases $t_1, \ldots, t_n$ are not sufficient and thus we cannot say anything about the program $\mathcal{P}$. It can happen in the following cases: (a) $\mathcal{P}$ does not conform to Σ and (b) $t_1, \ldots, t_n$ are not enough. To solve (a) the human tester has to modify Σ. In case (b), a more complete set of test cases has to be used. The implemented tester is able to detect (for a finite set of test cases) when a program $\mathcal{P}$ passes or fails a test t_i as well as when p cannot be checked for $\mathcal{P}$.

Now, we present the main elements of the implementation. The testing tool has been implemented as an XQuery function that takes as arguments the schema, the program, the property and the number of steps of test case generation. The program and property have to be represented as functions. It does not mean that only functions can be tested. In case of large XQuery programs the main query has to be represented as a function taking as argument an input XML document. In the current implementation only unary functions are allowed. We will consider as future work the extension of the tester to n-ary functions. The code of the tester is as follows:

```
declare function tc:tester($schema as node()*,$query as xs:string, $property
    as xs:string, $i as xs:integer){
 tc:tester_loop($schema,$query,$property,0,$i,0,0)};
```

Basically, the tester is a loop of i steps (according to the provided argument), in which in each step a set of schemas is generated. The schema instances are populated with values, and they are taken as input of the given program, and the given property is checked for each output. Whenever all the instances pass the test the loop continues. Otherwise, no more schemas and instances are generated and the tester reports "Falsifiable", showing the number of tested cases and the counterexamples. In case of loop continues, it ends when the number of steps is reached. In such case, the tester reports "Ok" whenever all the instances pass the test, showing the number of test cases and empty answers. When the number of empty answers is equal to the number of test cases, the tool reports "Unable to check the property". The tester loop is as follows:

```
declare function tc:tester_loop($schema as node()*,$query as xs:string,
    $property as xs:string,$k as xs:integer, $i as xs:integer,$tests as xs:
    integer,$empties as xs:integer){
if ($k>$i) then if ($tests=$empties) then tc:show_unable()
                        else tc:show_passed($tests,$empties)
else tc:tester_schema($schema,$schema,$query,$property,$k,$i,$tests,$empties)
};
```

and each step is implemented as follows:

```
declare function tc:tester_schema($schemas as node()*,$all as node()*,
    $query as xs:string,$property as xs:string,$k as xs:integer,
    $i as xs:integer,$tests as xs:integer,$empties as xs:integer){
if (empty($schemas))
        then let $new := tc:new_schemas($all)
            return tc:tester_loop($new,$query,$property,$k + 1,$i,
                                $tests,$empties)
        else
        let $sc := head($schemas)
        let $structure := tc:skeleton($sc/xs:schema/xs:element)
        let $examples := tc:populate($structure,tc:getTypes($structure),
                        tc:getVal($sc/xs:schema,
                        tc:getTypesName($structure)))
        let $total := count($examples) return
        if (not($total=0))
            then
            let $fquery := function-lookup(xs:QName($query),1)
            let $fproperty:=function-lookup(xs:QName($property),1)
            let $no := (for $example in $examples
                let $result := $fquery($example)
                where not($fproperty($result)) return
                if (empty($result)) then <empty/> else $example)
            let $noempty := $no[not(name(.)="empty")]
            let $falsifiable := count($noempty)
            let $newempties :=  count($no[name(.)="empty"])+$empties
            let $newtests := $tests + count($examples)
            return
            if ($falsifiable=0)
            then tc:tester_schema(tail($schemas),$all,$query,
                            $property,$k,$i,$newtests,$newempties)
            else tc:show_falsifiable($newtests,$noempty)
            else if ($tests=$empties) then tc:show_unable()
                            else tc:show_passed($tests,$empties)
};
```

where tc:new_schemas is responsible to the generation of schemas, sc:skeleton produces the skeleton of the instance and tc:populate fills the skeleton with elements.

3.1 Examples

Now, we would like to show a batch of examples, in order to prove the benefits of the tester.

Example 1. Assuming the schema of Fig. 1, we can test the following program:

```
declare function tc:yearofUMLbooks($file){
for $x in $file/book
where $x/title="UML" and $x/@year>2000
return $x/@year};
```

with regard to the following property:

```
declare function tc:after2000($year){
  $year>2000};
```

obtaining the following answer:

```
Ok: passed 80 tests.
Trivial: 48 tests.
```

48 trivial cases (i.e., empty answers) are checked, corresponding to the case of non "UML" and before 2000 books, which are generated according to the given schema. In the case of the property:

```
declare function tc:before2000($year){
  $year<2000};
```

the following answer is reported:

```
Falsifiable after 8 tests.
Counterexamples:
<bib>
  <book year="2005">
    <author>Buneman</author>
    <title>UML</title>
    <price>80</price>
  </book>
</bib>
<bib>
...
```

The counterexamples shown are non-empty solutions which do not satisfy the given property. The human tester has to include enough values for obtaining a suitable answer. In this example, the schema should include at least a value after 2000 (i.e., 2005), and "UML" as value for title. In case the human tester omits the following line in the XML Schema:

```
<xs:attribute name="year" type="yearType" use="required"/>
```

Now, the tester reports *"Unable to test the property"*. The same happens when the title is omitted. In both cases, it is due to empty answers. In the opposite case, author and price are not required to test this program. The human tester can remove from the schema unnecessary tags and attributes for improving performance.

Example 2. In the case of the following program:

```
declare function tc:books_query($file){
for $book in $file/book return
if ($book/title="UML" and $book/price<100) then
    <book_UML>
    {$book/@year}{$book/author}{$book/price}
    </book_UML>
else if ($book/title="XML" and $book/@year>2000 and $book/price<100) then
    <book_XML>
    {$book/@year}{$book/author}{$book/price}
    </book_XML> else ()
};
```

we can ask about:

```
declare function tc:books_price_100($book){
  $book/price<100};
```

obtaining the following answer with respect to the schema of Fig. 1:

```
Ok: passed 80 tests.
Trivial: 35 tests.
```

35 trivial cases corresponding to tests in which prices are greater than 100, as well as XML books published before 2000. But we can express more complex properties like:

```
declare function tc:allbooksofBuneman($book){
  every $b in $book satisfies $b/author="Buneman"};
```

In this case the tester also answers "Ok" because "Buneman" is the only value for author. Also we can check a more complex property like the following, obtaining the same answer:

```
declare function tc:price_and_year($book){
  every $b in $book satisfies
  if (name($b)="book_UML") then $b/price<100
  else $b/@year>2000 and $b/price<100};
```

Example 3. The adequacy of the selected XML Schema does not only depend on the selected values for each tag and attribute, but also on the required number of branches. Here the minOccurs and maxOccurs values play a key role. For instance, let us suppose we test the following program:

```
declare function tc:third_book($bib){
  let $third := $bib/book[3]
  where $third/title="UML"
  return <third>{$third/title}{$third/author[3]}</third>};
```

with regard to the following property:

```
declare function tc:UML($book){
  $book/title="UML"};
```

In this case, if we call the tester with:

```
tc:tester(.,"tc:second_book", "tc:UML",1)
```

then we obtain *"Unable to test the property"*. It happens because the value one has been selected as number of steps for test case generation. It means that according to the XML Schema only bibliographies with at most one book have been created. The same happens with two steps. Only for three steps the following answer is reported:

```
Ok: passed 656 tests.
Trivial: 400 tests.
```

In this example the human tester should modify the XML Schema and declare minOccurs as three for the sequence of books. It forces the generation of useful test cases in the first step. It also happens for optional attributes. The same happens for branches. For instance, let us suppose that in the schema of Figure 1, minOccurs is set to zero in all the cases. The same previous query and property cannot be checked after eight steps. In the case of long paths and recursive definitions, the number of steps has to be larger, otherwise useful test cases are not generated in early stages.

Table 1. Benchmarks of testing

Query	Time	Passed	Tests Passed	Falsifiable	Counterexamples	Property
Q1	419 ms	✓	584	✗	✗	every $book in $bib/bib satisfies $book/@year>1991
Q1	121 ms	✗	✗	✓	8	some $book in $bib/bib satisfies $book/@year<1991
Q2	1.401 ms	✓	306	✗	✗	every $result in $results/result satisfies $result/title and $result/author
Q2	31 ms	✗	✗	✓	2	some $result in $results/result satisfies not($result/title) or not($result/author)
Q3	1.428 ms	✓	306	✗	✗	every $result in $results/result satisfies $result/title and $result/author
Q3	24 ms	✗	✗	✓	2	some $result in $results/result satisfies not($result/title) or not($result/author)
Q6	1.424 ms	✓	306	✗	✗	every $book in $bib/book satisfies count($book/author)<=2
Q6	23 ms	✗	✗	✓	2	some $book in $bib/book satisfies count($book/author)>2
Q7	1.348 ms	✓	1.884	✗	✗	let $count := count($bib/book) return every $i in 1 to $count - 1 satisfies $bib/book[$i]/title<=$bib/book[$i+1]/title) satisfies $b=true()
Q7	27 ms	✗	✗	✓	12	let $count := count($bib/book) return some $i in 1 to $count - 1 satisfies $bib/book[$i]/title>$bib/book[$i+1]/title
Q8	438 ms	✓	696	✗	✗	every $book in $books satisfies every $item in $book/* satisfies contains(string($item),"Suciu") or name($item)="title"
Q8	26 ms	✗	✗	✓	24	some $book in $books satisfies some $item in $book/* satisfies not(contains(string($item),"Suciu")) and not(name($item)="title")
Q9	1.500 ms	✓	1.518	✗	✗	every $result in $results/results satisfies contains($result/text(),"XML")
Q9	17 ms	✗	✗	✓	2	some $result in $results/results satisfies not(contains($result/text(),"XML"))
Q10	1.989 ms	✓	1.365	✗	✗	count($results/minprice) = count(distinct-values($results/minprice/@title))
Q10	25 ms	✗	✗	✓	4	not(count($results/minprice) = count(distinct-values($results/minprice/@title)))

4 Evaluation

We have evaluated our tool with the XQuery use cases available in the W3C page[5]. We have checked properties on the queries of Sect. 1.1.9.1 of the cited repository (only those queries with a single input parameter). We have analyzed the time required for each query and the number of tests. The properties, response times (in milliseconds) as well as number of tests are shown in Table 1. Benchmarks have been made on a 2.66 GHz Inter Core 2 Duo MAC OS machine, with 4 GB of memory. We have used the BaseX XQuery interpreter [13]. Properties have been selected to be representative to the computation of each query. We have tested each property and its negation in order to measure time required to pass and fail the property. From the table, we can conclude that for a reasonable number of tests (from three hundred to thousand test cases in some examples), the tester is able to answer in a short time. The selected schema is crucial to have a short time. When the number of values for each type, as well the number of selected paths is high the performance is worst. The tester implementation as well as the examples shown in the paper can be downloaded from http:// indalog.ual.es/TEXTUAL.

---

[5] http://www.w3.org/TR/xquery-use-cases/.

5 Conclusions and Future Work

In this paper we have presented a tool for testing XQuery programs. We have shown how the proposed tool is able to automatically generate a test suite from the XML Schema of the input documents of the program in order to check a given property on the output of the program. It makes possible to have an stronger confidence about the correct behavior of the program. As future work, we would like to extend our work as follows. Firstly, there is a natural extension to cover programs with more than one input document. The current implementation of the tool works with any kind of XQuery expression but having a single input document. Secondly, we would like to extend the tool by adding more information about test cases. In the current implementation, only the number of empty solutions are shown, but other than empty solutions are relevant. For instance, non-empty solutions returning empty values for parts of the query (i.e., paths, subexpressions, etc.) can be useful for the human tester. Thirdly, we will consider how to filter test case generation by considering properties on the input document. The XML Schema imposes restrictions in terms of cardinality, but input programs can require to satisfy more complex properties. Input-output properties (i.e., properties relating input and output) will be also subject of study in the future. Fourthly, we would like to extend our work to the case of white box testing. In particular, XML Schemas can be automatically filtered/generated from the program in order to generate useful test cases. Finally, a richer repertoire of XML Schema statements will be included in the implementation.

References

1. Anand, S., Burke, E.K., Chen, T.Y., Clark, J., Cohen, M.B., Grieskamp, W., Harman, M., Harrold, M.J., McMinn, P., et al.: An orchestrated survey of methodologies for automated software test case generation. J. Syst. Softw. **86**(8), 1978–2001 (2013)
2. Barbosa, D., Mendelzon, A., Keenleyside, J., Lyons, K.: ToXgene: a template-based data generator for XML. In: Proceedings of the 2002 ACM SIGMOD, pp. 616–616. ACM (2002)
3. Bartolini, C., Bertolino, A., Marchetti, E., Polini, A.: WS-TAXI: a WSDL-based testing tool for web services. In: International Conference on Software Testing Verification and Validation, 2009, pp. 326–335. IEEE (2009)
4. Bertolino, A., Gao, J., Marchetti, E., Polini, A.: Automatic test data generation for XML schema-based partition testing. In: Proceedings of the Second International Workshop on Automation of Software Test (AST), p. 4. IEEE Computer Society (2007)
5. Bertolino, A., Gao, J., Marchetti, E., Polini, A.: TAXI-a tool for XML-based testing. In: Companion to the Proceedings of the 29th International Conference on Software Engineering, pp. 53–54. IEEE Computer Society (2007)
6. Bertolino, A., Gao, J., Marchetti, E., Polini, A.: XModel-based testing of XSLT applications. WEBIST **2**, 282–288 (2007)
7. Bertolino, A., Lonetti, F., Marchetti, E.: Systematic XACML request generation for testing purposes. In: 36th EUROMICRO Conference on Software Engineering and Advanced Applications, pp. 3–11. IEEE (2010)

8. Chimisliu, V., Wotawa, F.: Category partition method and satisfiability modulo theories for test case generation. In: 2012 7th International Workshop on Automation of Software Test (AST), pp. 64–70. IEEE (2012)
9. Claessen, K., Hughes, J.: QuickCheck: a lightweight tool for random testing of Haskell programs. ACM SIGPLAN Not. **46**(4), 53–64 (2011)
10. Cohen, S.: Generating XML structure using examples and constraints. Proc. VLDB Endow. **1**(1), 490–501 (2008)
11. De La Riva, C., Garcia-Fanjul, J., Tuya, J.: A partition-based approach for XPath testing. In: International Conference on Software Engineering Advances, p. 17. IEEE (2006)
12. Fisher, M., Elbaum, S., Rothermel, G.: An automated analysis methodology to detect inconsistencies in web services with WSDL interfaces. Softw. Test. Verification Reliab. **23**(1), 27–51 (2013)
13. Grün, C.: BaseX. The XML Database (2015). http://basex.org
14. Harazaki, M., Tekli, J., Yokoyama, S., Fukuta, N., Chbeir, R., Ishikawa, H.: XBeGene: scalable XML documents generator by example based on real data. In: Gaol, F.L. (ed.) Recent Progress in DEIT, Vol. 1. LNEE, vol. 156, pp. 449–460. Springer, Heidelberg (2013)
15. Kim-Park, D.S., de la Riva, C., Tuya, J.: An automated test oracle for XML processing programs. In: Proceedings of the First International Workshop on Software Test Output Validation, pp. 5–12. ACM (2010)
16. Jeong, H.J., Lee, S.H.: A versatile XML data generator. Int. J. Softw. Effectiveness Effi. **1**, 21–24 (2006)
17. Ostrand, T.J., Balcer, M.J.: The category-partition method for specifying and generating functional tests. Commun. ACM **31**(6), 676–686 (1988)
18. Papadakis, M., Sagonas, K.: A PropEr integration of types and function specifications with property-based testing. In: Proceedings of the 2011 ACM SIGPLAN Erlang Workshop, pp. 39–50. ACM Press, New York, September 2011
19. Runapongsa, K., Patel, J.M., Jagadish, H.V., Chen, Y., Al-Khalifa, S.: The Michigan benchmark: towards XML query performance diagnostics. Inf. Syst. **31**(2), 73–97 (2006)
20. Schmidt, A., Waas, F., Kersten, M., Carey, M.J., Manolescu, I., Busse, R.: XMark: a benchmark for XML data management. In: Proceedings of the 28th International Conference on Very Large Data Bases, VLDB Endowment, pp. 974–985 (2002)
21. Utting, M., Pretschner, A., Legeard, B.: A taxonomy of model-based testing approaches. Softw. Test. Verification Reliab. **22**(5), 297–312 (2012)
22. Yao, B.B., Ozsu, M.T., Khandelwal, N.: XBench benchmark and performance testing of XML DBMSs. In: Proceedings of the 20th International Conference on Data Engineering, 2004, pp. 621–632. IEEE (2004)

Data Partitioning, Indexing

Grid-File: Towards to a Flash Efficient Multi-dimensional Index

Athanasios Fevgas[(✉)] and Panayiotis Bozanis

Electrical and Computer Engineering Department,
University of Thessaly, Volos, Greece
{fevgas,pbozanis}@inf.uth.gr

Abstract. Spatial indexes are of great importance for multidimensional query processing. Traditional data structures have been optimized for magnetic disks in the storage layer. In the recent years flash solid disks are widely utilized, as a result of their exceptional features. However, the specifics of flash memory (asymmetric read/write speeds erase before update, wear out) introduce new challenges. Algorithms and data structures designed for magnetic disks experience reduced performance in flash. Most research efforts for flash-aware spatial indexes concern R-tree and its variants. Distinguishing from previous works we investigate the performance of Grid File in flash and enlighten constrains and opportunities towards a flash efficient Grid File. We conducted experiments on mainstream and high performance SSD devices and Grid File outperforms R*-tree in all cases.

1 Introduction

Flash memory technology has enable the development of a new storage medium with unique features compared to the traditional magnetic disks. High read and write speeds, low power consumption, shock resistance, small size and lighter weight contributed to its popularity. However, flash have some special characteristics, writes are slower than reads and erases are even slower. Reads and writes are performed at page level (e.g. 4 KB) while erases at block level (e.g. 512 KB). Details for page size and erase block size are not publicly available by most vendors. In place updates are not feasible, therefore write operations require free valid pages. Invalided pages are not reclaimed without cost, as continuous erasures damage flash cells and limit block endurance by a certain number of erase cycles (wear out). Flash-based SSDs (Solid State Disks) are widely adopted in consumer as well as in enterprise computer systems. SSDs are composed by several flash memory chips and one or more controllers. The SSD controller utilizes FTL (Flash Translation Layer) to emulate a block interface compatible to HDD using an out-of-place updates mechanism. It also incorporates algorithms for wear leveling, garbage collection, error correction and data encryption. Higher class models comprise more than one controllers tied up under a raid controller. Usually SSD manufacturers reserve a portion of the total drive capacity as over provision space for enhancing its performance and endurance. The size of over

© Springer International Publishing Switzerland 2015
Q. Chen et al. (Eds.): DEXA 2015, Part II, LNCS 9262, pp. 285–294, 2015.
DOI: 10.1007/978-3-319-22852-5_24

provision space depends on the total drive capacity and the demands of the applications that SSD is intended for. Generally, there are significant differences across various SSD models in terms of performance, reliability and cost.

The distinct features of flash based storage motivate researchers to review data structures and algorithms designed for magnetic disks. Database indexing is a research area that several flash efficient data structures have been proposed recently. Most efforts aim to reduce expensive random writes and seek to exploit the efficiency of sequential reads and writes [1]. A straightforward method to achieve this, is to postpone updates and perform them in batches. Thus, some proposals buffer updates into main memory and persist them in flash at once [2–5]. Another study uses a in-memory head B+-tree to convert random writes to sequential ones, as merging into lower levels in flash. The authors in [6] propose the lazy updating of the leaf nodes by accumulating several updates into flash-resident cascaded buffers. In [7], the researchers experimentally found that reads and writes can benefit by SSD 's internal parallelism and demonstrate an efficient method to perform parallel I/O. Overflow techniques are used in [8,9] to delay node splitting, and logging in [10,11] to amortize the cost of random writes. From a different perspective the authors in [12] aim to optimize index size instead of performance using probabilistic data structures.

Spatial index optimizations for flash based solid disks are discussed in the present work. R-tree [13] (and its variants) is the most popular spatial index and it is implemented in several DBMSes (MySQL, PostgreSQL, Oracle, etc.). Thus, the vast majority of the published work about flash efficient spatial indexes concern R-Tree. The balanced structure of the R-tree introduce small random writes as necessitates splits and merges at the leaf nodes that are propagated up to the root. In order to overcome this, most previous studies aggregate updates into buffers and perform them in batches.

Few published work exist for, other than R-tree, flash optimized spatial indexes. To the best of out knowledge there is not any study for the performance of grid file [14] in flash based storage. Grid file is a multidimensional index that uses a grid to partition the space, each cell of the grid points to one page in the disk. It guarantees single point retrieval with two disk accesses, and it executes range and partially defined queries efficiently as well [14,15]. Distributed variants of the grid file have been utilized recently to support, spatial queries [16] and multidimensional range queries for massive smart meter data [17], in the Hadoop MapReduce environment. Moreover, a distributed grid index has been proposed in [18] to enhance spatial query processing in wireless broadcast environments. All these motivated us to study the performance of Grid file in flash based solid state disks. Our contributions can be summarized as following:

- We implemented Grid file and a page buffering scheme for performing batch writes.
- We conducted extensive experiments on Grid file and R-tree using both mainstream and high performance SSD devices and present a comparative analysis of the results.

The rest of the document is organized as follows. In Sect. 2 we review related work in spatial indexes. In Sect. 3 we present background information on grid-file and R-tree. In Sect. 4 we describe testing environment and present the results. Our work summarizes in Sect. 5 and presents our ideas for future work.

2   Related Work

In the recent years, researchers have perform intensive work to exploit the full potentialities of flash storage and anticipate its idiosyncrasies. A considerable number of research papers have been presented in spatial indexing and most of them concern R-Tree and its variants. Log-based data structures are utilized to avoid expensive page rewrites. Authors in [2] (RFTL) and [4] propose the usage of an in memory buffer that gathers incoming update requests, aiming to enhance the performance of R-tree and Aggregated R-tree respectively. Updates are sequentially persisted in flash when the buffer is full. A node translation table is employed to enable node retrieval by mapping nodes with corresponding flash pages. LCR-tree [10] stores all deltas for a particular R-tree node to the same flash page, this guarantees only one additional read for each node access. Log records are merged in the original tree when log space in flash is exceeded. LCR-tree can be easily adapted to work with any R-tree variant as underlying structure. F-KDB [11] is another multidimensional index that relies to logging in order to improve update performance of K-D-B-tree [19]. It uses an in-memory buffer to cache log entries organized into sets. Log records, unlike [2,4], are gradually evicted into flash according to a proposed replacement policy. An online algorithm is introduced to increase read performance by merging scattered records. Authors in [8,9] aim to reduce random writes using an unbalanced R-tree structure. Therefore, splitting of full nodes is avoided and new records are inserted to overflow nodes. New search and update operations are realized and a merging policy of overflow nodes is defined. A study for a buffer replacement algorithm is conducted in [9] as well. FAST [20] provides a high level framework for flash efficient indexes. It encapsulates the original tree algorithms providing optimized search and update functionalities along with a crash recovery mechanism. FAST-RTree overcomes RFTL's performance in all experiments.

Authors in [21] spotlight the implication of buffering in the performance of spatial indexes and propose some ideas towards efficient buffer managers. Traditional replacement algorithms aim to maximize hit ratio neglecting writes, since reads and writes have equal cost in HDDs. Therefore, new replacement algorithms have been proposed that pursue to reduce writes in SSD resident data storages. Authors in [22] propose eviction of clean pages from a window of the w least recently used in the buffer. A newer algorithm [23], additionally to recency, takes into account the frequency of references.

3   Background

3.1   Grid File

Grid File [14] is a multidimensional index structure for secondary storage, that uses a k-dimensional grid to partition the data space. It consist of a k-dimensional

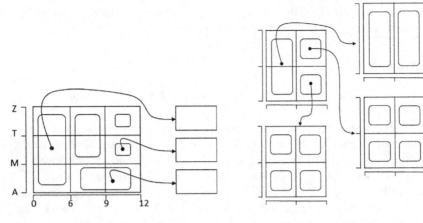

Fig. 1. Grid file   **Fig. 2.** Root directory and subdirectories

array, the *grid array*, and k 1-dimensional arrays, the *linear scales*. The linear scales define the boundaries of each partition of the data space. Figure 1 depicts a 2-d Grid File, each grid cell contains a pointer to a page in the disk (bucket), which stores the actual data. Two or more cells can point to the same page forming a bucket region. Bucket regions are disjoint and their size expresses the density of data. The scales are always kept in the main memory, but the grid or parts of it are stored to the disk. During a single point query, first the grid cell is located by searching the linear scales, then the pointer is followed to retrieve the bucket is retrieved from disk. Grid File accomplishes a single record retrieval with at most two disk accesses (one for the part of the Grid File and one for the data bucket). It supports also range and partial queries achieving good performance too [14,15]. Insertion of new records may cause bucket overflows, triggering reorganization of the entire directory and scales by adding new rows or columns. Similarly, deletions lead to underflow buckets, launching merge operations. Since grid is disk resident, splits and merges increase disk accesses.

Looking for an efficient implementation of the Grid File, authors in [24] propose a two level grid. They utilize an in-memory scaled-down root directory (Fig. 2) which provides a coarse representation of the data space. Each cell of the root directory points to a subdirectory. Subdirectory can be thought as a small Grid File and occupies only one page in the disk. The query operations described before, apply to the root directory as well as for the subdirectories, as is. But, the number of pages are touched by splits and merges is minimized. The two disk access principle in not violated, since single record retrieval is still fulfilled by accessing the appropriate subdirectory and the data bucket.

3.2   R-tree

R-tree can be thought as an extension of B+-tree towards indexing of multi-dimensional data. The leaves of R-tree store minimum bounding rectangles

(MBR) of geometric objects and pointers to the actual data. An internal node consists of a pointer to a child subtree and the corresponding MBR of the sub-tree. A tree $\mathcal{T}$ MBR encloses all the MBRs contained in $\mathcal{T}$. Each R-tree node stores a minimum m and a maximum M number of records, where $m \leq M/2$. A search query can involve several paths starting from the root node to the leaves. Thus, the cost to retrieve even a few objects can be linear to the size of data (worst case). Figure 3 presents the corresponding R-tree instance for a set S of rectangles.

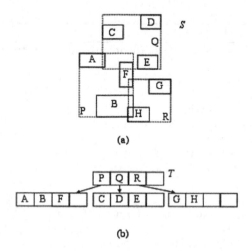

Fig. 3. An R-tree instance.

 Several improvements for the R-tree have been introduced and all of them pursue to improve its performance by adjusting various parameters. The R*-tree [25] performs the best among them. It utilizes several heuristics, like re-insertions during insertions, buffering, MBR adjusting and optimizations for node splitting/merging. Therefore, R*-tree is recognized as the best performing from R-tree variants. However, the construction of any R-tree with repeated insertions impedes the query efficiency; actually, queries can not avoid linear worst-case time complexity.

4   Grid-File Implementation

Aiming to evaluate Grid File for flash storage and identify its performance char-acteristics, we implemented the two level Grid File following [24]. Inserts, deletes and exact match queries are currently supported, while LRU is used as buffer replacement policy. A new buffering scheme is introduced aiming to exploit SSD unique characteristics.

4.1 Bucket Buffering and Replacement Policy

Most of the previous efforts for flash efficient data structures postpone random writes and perform them as batches. Moreover, authors in [7] show that SSD performance increases by the number of outstanding I/Os. Motivated by previous research, we present a simple buffering scheme for Grid File pursuing to accelerate write performance in flash storage. We introduce a small *Cold Dirty Buffer* (CDB) for storing buckets that are evicted from the *Main Buffer* (MB). The pages in the CDB are persisted to the SSD as batch.

Algorithm 1. Fetch Page

Data: the *id* of the requested page, the main buffer MB, the cold dirty buffer
      CDB
Result: a reference to the requested page p
if *p is in MB* **then**
    move p to the MRU position of MB;
    return a reference to the requested page p;
else if *p is in CDB* **then**
    move p to the MRU position of MB;
    return a reference to the requested page p;
else
    if *MB is full* **then**
        select victim v from the LRU position of MB;
        if *v is dirty* **then**
            if *CDB is full* **then**
                flash CDB to the disk;
            end
            evict v from MB to CDB;
        end
    end
    fetch p;
    set p to the MRU position of MB;
    return a reference to the requested page p
end

The process of fetching a page is described in the Algorithm 1. A lookup is performed to the MB and on hit, the page is moved to the most recent unit (MRU) position. On miss, the CDB is checked and if the page is found, then it is moved to the MRU position of the MB (Fig. 4). Otherwise, the page is retrieved from the disk. When the CDB is fulfilled it is flashed to the SSD drive as batch.

5 Performance Evaluation

In this section, we present the experimental results for the Grid File. We compare our implementation with R*-tree, since R*-tree is considered among the most

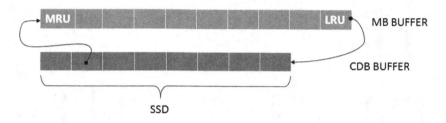

Fig. 4. Replacement policy

popular spatial data structures. The R*-tree implementation is derived by the libspatialindex [26]. Two different solid drives were used for the experiments, a mainstream and a high performance one, in order to identify possible impact of their internal architecture to the index performance.

5.1 Methodology

The experiments were conducted on a DELL Precision T3500 workstation running Centos 6.6 64-bit linux with kernel 2.6.32. It is equipped with 8 GB of DDR3 RAM and a 4-core Intel Xeon W3550 3.06 GHz CPU. The operating system is hosted to an Intel 520 240 GB connected to SATA-III interface. Two additional drives an Intel 530 SSD 240GB SATA-III and an OCZ Revodrive 350 480 GB PCIe are, exclusively, utilized for the experiments.

Direct I/O (O_DIRECT) is used, in order to bypass file system's buffer cache and measure the real performance. Therefore, both data structures (Grid File and R*-tree) were modified accordingly. The experiments were conducted using mixed synthetic workloads with varying operation ratio with the page size to be fixed to 4 KB.

5.2 Results

We, first, compare the performance of Grid File against of that of R*-tree. During this series of experiments, a simple LRU was utilized for the Grid File, thus, the number of outstanding write operations is 1 for both indexes. Figure 5 presents the wall-clock time for inserting 1 million records with buffer size ranging from 128 KB to 4 MB. In all cases, Grid File outperforms R*-tree by a factor of 1.35 to 2.60 for the Intel SSD and from 1.49 to 2.80 for the OCZ one. The elapsed time is decreased as buffer size increases for both structures. Grid File speedup is bigger than those of R*-tree (2.85 times against 1.5 times from 128 KB to 4 M).

A set of experiments using various synthetic workloads is presented in Fig. 6. The search/insert ratio varies from 100 % searches to 100 % inserts by a 20 % step. Grid File performs better and for the varying workloads e.g. for a search oriented workload (60 % searches, 40 % inserts) is faster by a factor of 1.5 for Intel SSD and 1.75 for OCZ one.

The final set of experiments considers the CDB buffer and presents the elapsed time for increasing number of write operations. The total size of main

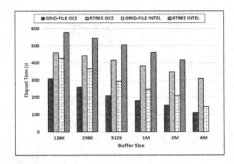
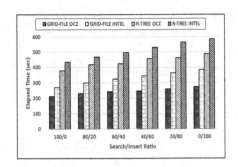

Fig. 5. Elapsed time for 1 M insertions   **Fig. 6.** Elapsed time for mix workload

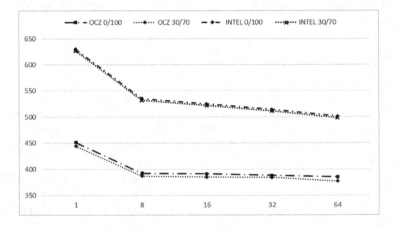

Fig. 7. Elapsed time by increasing outstanding write operations

memory buffers $(MB_{size} + CDR_{size})$ is set to 512 KB in all cases. We used two write sensitive workloads, each containing 2 million queries. The workloads were selected to be write sensitive since the number of the outstanding read requests is 1. Figure 7 presents the elapsed wall clock time while the number of outstanding writes gradually increasing from 1 to 64. The elapsed time for the OCZ drive is improved by 17 % and 15 % for the (0/100) and (30/70) loads respectively. Similarly, the elapsed time is improved by 12 % for both workloads in the INTEL drive. In all cases the greatest improvement is observed between 1 to 8 requests.

6   Conclusions

Flash is a new type of non-volatile memory with outstanding efficiency compared to traditional magnetic drives. Its specifics, motivated many researchers in the area of spatial indexing to study R-tree and propose several optimizations. We, first, study the performance of Grid file in flash solid disks and we compare it with R*-tree. Experimental results show that Grid file outperforms R*-tree in all cases. A buffering scheme for batch updates towards to a flash efficient Grid file is also demonstrated. Our future work deals with optimization of Grid file for non-volatile based storages and enrich available query types implementing region and KNN search queries.

References

1. Bouganim, L., Jónsson, B., Bonnet, P.: uFLIP: understanding flash IO patterns (2009). arXiv preprint arXiv:0909.1780
2. Wu, C.H., Chang, L.P., Kuo, T.W.: An efficient R-tree implementation over flash-memory storage systems. In: Proceedings of the 11th ACM International Symposium on Advances in Geographic Information Systems, GIS 2003, pp. 17–24. ACM, New York (2003)
3. Wu, C.H., Kuo, T.W., Chang, L.P.: An efficient B-tree layer implementation for flash-memory storage systems. ACM Trans. Embed. Comput. Syst. (TECS) 6(3), 19 (2007)
4. Pawlik, M., Macyna, W.: Implementation of the aggregated R-tree over flash memory. In: Yu, H., Yu, G., Hsu, W., Moon, Y.-S., Unland, R., Yoo, J. (eds.) DASFAA Workshops 2012. LNCS, vol. 7240, pp. 65–72. Springer, Heidelberg (2012)
5. Roh, H., Kim, S., Lee, D., Park, S.: As B-tree: a study of an efficient B+-tree for SSDs. J. Inf. Sci. Eng. 30(1), 85–106 (2014)
6. Agrawal, D., Ganesan, D., Sitaraman, R., Diao, Y., Singh, S.: Lazy-adaptive tree: an optimized index structure for flash devices. Proc. VLDB Endow. 2(1), 361–372 (2009)
7. Roh, H., Park, S., Kim, S., Shin, M., Lee, S.W.: B+-tree index optimization by exploiting internal parallelism of flash-based solid state drives. Proc. VLDB Endow. 5(4), 286–297 (2011)
8. Wang, N., Jin, P., Wan, S., Zhang, Y., Yue, L.: OR-tree: an optimized spatial tree index for flash-memory storage systems. In: Xiang, Y., Pathan, M., Tao, X., Wang, H. (eds.) ICDKE 2012. LNCS, vol. 7696, pp. 1–14. Springer, Heidelberg (2012)
9. Jin, P., Xie, X., Wang, N., Yue, L.: Optimizing R-tree for flash memory. Expert Syst. Appl. 42, 4676–4686 (2015)
10. Lv, Y., Li, J., Cui, B., Chen, X.: Log-compact R-tree: an efficient spatial index for SSD. In: Xu, J., Yu, G., Zhou, S., Unland, R. (eds.) DASFAA Workshops 2011. LNCS, vol. 6637, pp. 202–213. Springer, Heidelberg (2011)
11. Li, G., Zhao, P., Yuan, L., Gao, S.: Efficient implementation of a multi-dimensional index structure over flash memory storage systems. J. Supercomput. 64(3), 1055–1074 (2013)
12. Athanassoulis, M., Ailamaki, A.: BF-tree: approximate tree indexing. In: Proceedings of the 40th International Conference on Very Large Databases. Number EPFL-CONF-201942 (2014)
13. Guttman, A.: R-trees: a dynamic index structure for spatial searching, vol. 14. ACM (1984)
14. Nievergelt, J., Hinterberger, H., Sevcik, K.C.: The grid file: an adaptable, symmetric multikey file structure. ACM Trans. Database Syst. (TODS) 9(1), 38–71 (1984)
15. Papadopoulos, A.N., Manolopoulos, Y., Theodoridis, Y., Tsotras, V.: Grid file (and family). In: Liu, L., Özsu, M.T. (eds.) Encyclopedia of Database Systems, pp. 1279–1282. Springer, New York (2009)

16. Eldawy, A., Mokbel, M.F.: A demonstration of spatialhadoop: an efficient mapreduce framework for spatial data. Proc. VLDB Endow. **6**(12), 1230–1233 (2013)
17. Liu, Y., Hu, S., Rabl, T., Liu, W., Jacobsen, H.A., Wu, K., Chen, J.: DGFindex for smart grid: enhancing hive with a cost-effective multidimensional range index (2014). arXiv preprint arXiv:1404.5686
18. Park, K.: Location-based grid-index for spatial query processing. Expert Syst. Appl. **41**(4), 1294–1300 (2014)
19. Robinson, J.T.: The KDB-tree: a search structure for large multidimensional dynamic indexes. In: Proceedings of the 1981 ACM SIGMOD INTERNATIONAL CONFERENCE on Management of DATA, pp. 10–18. ACM (1981)
20. Sarwat, M., Mokbel, M.F., Zhou, X., Nath, S.: FAST: a generic framework for flash-aware spatial trees. In: Pfoser, D., Tao, Y., Mouratidis, K., Nascimento, M.A., Mokbel, M., Shekhar, S., Huang, Y. (eds.) SSTD 2011. LNCS, vol. 6849, pp. 149–167. Springer, Heidelberg (2011)
21. Koltsidas, I., Viglas, S.D.: Spatial Data management over flash memory. In: Pfoser, D., Tao, Y., Mouratidis, K., Nascimento, M.A., Mokbel, M., Shekhar, S., Huang, Y. (eds.) SSTD 2011. LNCS, vol. 6849, pp. 449–453. Springer, Heidelberg (2011)
22. Park, S.y., Jung, D., Kang, J.u., Kim, J.s., Lee, J.: CFLRU: a replacement algorithm for flash memory. In: Proceedings of the 2006 International Conference on Compilers, Architecture and Synthesis for Embedded Systems, pp. 234–241. ACM (2006)
23. Jin, P., Ou, Y., Harder, T., Li, Z.: AD-LRU: an efficient buffer replacement algorithm for flash-based databases. Data Knowl. Eng. **72**, 83–102 (2012)
24. Hinrichs, K.: Implementation of the grid file: design concepts and experience. BIT Numer. Math. **25**(4), 569–592 (1985)
25. Beckmann, N., Kriegel, H.P., Schneider, R., Seeger, B.: The R*-tree: an efficient and robust access method for points and rectangles, vol. 19. ACM (1990)
26. Hadjieleftheriou, M.: libspatialindex 1.8.5 (2015). http://libspatialindex.github.io/. Accessed 20 Feb 2015

Supporting Fluctuating Transactional Workload

Ibrahima Gueye[1]($\boxtimes$), Idrissa Sarr[1], Hubert Naacke[2,3], and Joseph Ndong[1]

[1] LID Laboratory, University Cheikh Anta Diop, Dakar, Senegal
{ibrahima82.gueye,idrissa.sarr,joseph.ndong}@ucad.edu.sn
[2] Sorbonne Universités, Paris, France
[3] LIP6, UPMC Univ Paris 06, Paris, France
hubert.naacke@lip6.fr

Abstract. This work deals with a fluctuating workload as in social applications where users interact each other in a temporary fashion. The data on which a user group focuses form a *bundle* and can cause a peak if the frequency of interactions as well as the number of users is high. To manage such a situation, one solution is to partition data and/or to move them to a more powerful machine while ensuring consistency and effectiveness. However, two problems may be raised such as how to partition data in a efficient way and how to determine which part of data to move in such a way that data are located on one single site. To achieve this goal, we track the bundles formation and their evolution and measure their related load for two reasons: (1) to be able to partition data based on how they are required by user interactions; and (2) to assess whether a machine is still able of executing transactions linked to a bundle with a bounded latency. The main gain of our approach is to minimize the number of machines used while maintaining low latency at a low cost.

Keywords: Transaction · Data placement · Elasticity · Load balancing

1 Introduction

Many popular applications are designed to run on top of a network of users and to allow them interacting each other. Therefore, users' data involved in an interaction may be updated through a transaction. In order to handle efficiently such transactions while ensuring scalability and good performances, data of all users may be partitioned. However, a user is free to be in touch with anyone without restriction by creating new social links. Therefore, it is a tough game creating data partitions to ensure that interactions of a tied users fit perfectly into one single partition [1,2]. Meanwhile, temporary interactions may lead to affinities and group behavior such as in Facebook or Google+. Data related to the interactions of a given group may be gathered in a *bundle*. A data item belongs to a *bundle* if it is involved in a group interactions during a time window. Since the number of interactions may increase and decrease at any moment, thus, the bundle as well as the generated load vary over time. Hence, facing properly the variation of such *bundles* and their related workload may help increasing the scalability of the system.

© Springer International Publishing Switzerland 2015
Q. Chen et al. (Eds.): DEXA 2015, Part II, LNCS 9262, pp. 295–303, 2015.
DOI: 10.1007/978-3-319-22852-5_25

The overall objective of this work is to deal with a fluctuating workload by exploiting the features of social applications and precisely the social graph that portrays the interactions between users. The goal is to partition data based on group interactions and to finally keep all related data in one single node and avoid multi-partitions transactions. Actually, we rely on the social graph to figure out characteristics of a given workload and/or to predict its variation in order to face it as earlier as possible in an efficient manner.

The key novelties of our contributions can be portrayed as follows:

(i) A formal mechanism to identify and track bundles and their evolution based on social features.

(ii) A dynamic data placement mechanism taking into account the variation of bundles over time, which guarantees to process any transaction locally.

The rest of the paper is organized as follows. We describe in Sect. 2 a few works related to ours in order to highlight the importance of the targeted problem as well as our solution. In Sect. 3, we present the formal description of a bundle and its identification. In Sect. 4 we describe the components of our system, and we present the grouping process, which relies on bundle-based dynamic data placement in order to ensure transaction execution on single node. We validate our approach through experimentation in Sect. 5 before concluding in Sect. 6.

2   Related Works

Most of the studies related to our work are in two fields: data placement and distributed transaction processing.

Data placement. The problem of data placement has been widely studied in the literature [3–9], etc. However, these studies were conducted for load balancing and are suited to an environment where the number of machine (nodes) is fixed. SCADS Director is another data placement solution which addresses the problem in a context of simple key-value storage systems using a greedy heuristic [10]. And more recently, Accordion [11] is solution that deals with dynamic data placement for DBMS that support the ACID transaction. It performs dynamic placement while explicitly taking into account the *affinity between partitions*. It is designed for in memory data base systems, where distributed transactions represent a major source of overload [12,13].

Even though there is plenty of works handling dynamic data placement, to the best of our knowledge, none of them are using social features to split and adapt partitioning according to user behaviors of an application.

Distributed transaction processing. In addition to the dynamic data placement, we deal with transaction processing and particularly with distributed transactions. Schism [8], Sword [9] or Calvin [1] are among the well-known solutions dealing with distributed transactions. Schism and Sword are very closed to our approach and rely on bundles. The authors propose to build the bundle by excluding some transactions to restrict the content of a bundle, which leads transactions to require data belonging to several bundles. Thus, transactions are more expensive and may require a distributed transaction validation. However, the overhead of distributed transactions can become prohibitive at a large scale [1]. To reduce this overhead, the authors of Schism and Sword propose an algrithm to construct the bundle by minimizing the number of distributed transactions. Our approach differs from those solutions by avoiding distributed transactions running on two bundles.

3   Bundle Identification

The *bundle* is tightly dependent of the number of simultaneous users that access to the linked data. That is, a *bundle*'s workload evolves over time and proportionally to the users interactions.

A crucial question is how may we manage efficiently transactions while taking into account the bundle variation in terms of size and load. Our solution to this question is to partition data based on bundle while ensuring that a bundle fits into a single machine. That is, the partitioning process is guided by the bundle size and its variations over time, and we guarantee with a "best effort" policy that each bundle fits in a machine with respect to its capacites. The motivation of doing so is avoiding distributing the execution of transactions over several nodes that require additional costs in terms of overhead.

We represent the kth bundle of the time window w by the pair $\beta_w^k = \{D; I_w\}$, where I_w is the set of interactions that require the set of data items D. In other words, a *bundle* is defined by the data items D on which the interactions of a group users rely during a time window. An interaction involving two data items a and b within w is noted as $i_{w_{a,b}}$. Bundles can be depicted as a undirected and weighted graph where vertices are the data items and edges are the interactions. The frequency of an interaction is represented as the weight of the edge and is used to keep a data item into a *bundle* or not.

To formally define a bundle, we rely on the k-core concept of graph theory. A k-core is a maximal group of actors where each actor is connected to at least k other members of the group. With this in mind, a *bundle* is formed by the set of data items involved at least in k interactions during a time window with a given frequency. We mention that the higher the k value, the denser the bundle. It is worth-noting that *bundles* may evolve over time since only interactions of a given time window is considered. That is, if a data item d is required less times in a time window than in a previous one, then d may be excluded from the *bundle* based on a threshold value. In this respect, *bundles* may grow, merge, split and so on over time.

The examples in Fig. 1 illustrate the formation and evolution of bundles based on a 2-core, which means that a bundle should satisfy a 2-core at least. We suppose a set of seven data items {a,b,c,d,e,f,g} belonging to seven users that interact during a period of time. Since each interaction takes place between two users, hence, it ties at least two data items, each for a user. In Fig. 1(a),

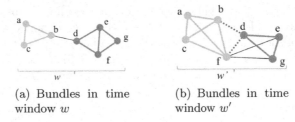

(a) Bundles in time window w

(b) Bundles in time window w'

Fig. 1. Bundles formation and evolution

we observe two bundles at w.

$$\beta_w^1 = \{(a, b, c); (i_{w_{a,b}}, i_{w_{b,c}}, i_{w_{a,c}})\}$$

$$\beta_w^2 = \{(d, e, f, g); (i_{w_{d,e}}, i_{w_{e,f}}, i_{w_{d,f}}, i_{w_{e,g}}, i_{w_{g,f}})\}.$$

Based on recent interactions happened in a subsequent window w', the bundles have changed as depicted in Fig. 1(b).

$$\beta_{w'}^3 = \{(d, e, g); (i_{w'_{d,e}}, i_{w'_{e,g}}, i_{w'_{d,g}})\}$$

$$\beta_{w'}^4 = \{(a, b, f, c); (i_{w'_{a,b}}, i_{w'_{a,f}}, i_{w'_{a,c}}, i_{w'_{f,b}}, i_{w'_{f,c}}, i_{w'_{c,b}})\}.$$

The bundle $\beta_{w'}^4$ is a 3-core. That is why the data item d is excluded and as one can see, the bundle is more homogeneous and dense.

4 System Overview and Data Grouping Process

4.1 System Overview and Notations

Our solution relies on an architecture divided in three layers as shown on Fig. 2: (i) the client applications nodes (CN), which produce the workload through the transactions they submit; (ii) the routing layer that is responsible of routing incoming transactions from client applications; and (iii) the database layer composed of DB nodes, which are responsible of executing transaction requests. The routing layer is composed of Router nodes (RN), Bundle monitor node (BM) and Resources monitor node (RM). The RN is the main component of the routing layer since it receives and forward transactions according to the execution policy. The BM is responsible of tracking the bundles and participates to their placement while the RM ensures the monitoring of the DB load and can provide detailed informations.

4.2 Grouping Data for Transaction Processing

The router knows (either explicitly or deduced from the transaction call) if the transaction T requests for accessing a single bundle or a set of bundles. In case a

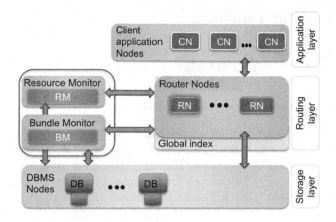

Fig. 2. Global architecture

transaction requires more than one bundle, the overall bundles form a temporary graph connecting all bundle graphs $(\Gamma_{\beta_w}(T))$. We distinguish two cases whether the bundles are in the same DB or not. If all the bundles are in the same DB, then, the transaction is routed to that DB. Otherwise, the bundles are first grouped at one single DB. This grouping operation is handled by one dedicated Router node called coordinator and proceeds as follows.

1. The coordinator asks each involved DB_i to send its bundles description and afterwards performs the merging operations to get the temporary graph $\Gamma_{\beta_w}(T)$ covering the entire required data.

$$\Gamma_{\beta_w}(T) = \{\beta_w^1, \beta_w^2, ..., \beta_w^n\} \tag{1}$$

   Then, the coordinator ensures that the related load of $\Gamma_{\beta_w}(T)$ will not exceed the maximum capacity of a machine. To this end, we adjust the size of w in order to remove the least frequently used data items.
2. Based on the size of the temporary graph $\Gamma_{\beta_w}(T)$, the router chooses the appropriate DB_i that minimizes the bundles transfer from others while ensuring enough resources capacity to handle the transactions. If there is no DB that has enough capacity to manage $\Gamma_{\beta_w}(T)$, then the router instantiates a new one to this respect.

The bundle transfer among DB nodes is done through live data migration. To mitigate the additional load caused by this migration, the bundle is moved as late as possible and precisely when a transaction accesses it. The migration is achieved by pulling on-demand, one by one, the data items of that bundle. In fact, when a bundle is about to be migrated, the source DB node marks on it's local index this bundle as migrated and the destination DB node marks it as being migrated to it. Then the bundles' items are pulled on-demand as they are requested by transactions.

This process has the edge to smooth the whole migration cost over time through several transactions.

5   Experimental Validation

In order to support fluctuation, our solution relies on a strategy to decide which data to group (into a bundle) and where to locate it.

5.1   Evaluating the Data Grouping Strategy

The grouping strategy aims to gather related data at a common location, which allows for fast local access. Ideally, a perfect grouping strategy would maximize the occurrence of local transactions. However, since we assume a fluctuating workload, we continuously face the case where the data do need grouping. Consequently, we can "split" the workload over time, and focus on a time period during which the workload is regular (*i.e.*, almost no fluctuation). This allows us to study the transition from an initial data placement which is not optimal with respect to the workload, to a final data placement which is near optimal. We consider that the relative performance of a solution over one period is generalizable to quantify the relative performance over a long run with fluctuating workload.

Our prototype implementation of the data grouping strategy extends an open source key-value database Redis [14]. Redis is an in memory key-value store. We use Storm [15], a distributed system for real-time processing, to implement all the entities of our system. Storm is fault tolerant and ensures delivery of messages. We generate a workload composed of transactions accessing between 3 and 5 users' data, randomly choosen following a standard normal distribution. This aims to reflect the workload pattern of our targeted application class, as surveyed in [16].

Impact of the Bundle Grouping Strategy. We investigate the impact of grouping the bundles on transaction processing (especially the distributed ones). For comparison purpose, we conduct two experiments with or without grouping. The first experiment aims to be a reference for comparison with the distributed case. We switch off the bundle grouping strategy. All data are stored at a single DB. Thus, every transaction is local. Figure 3(a) shows the baseline results plotted in red. In the second experiment, the data are uniformly partitioned over four DB nodes. We activate the partitions grouping strategy. The workload remains the same as in the first experiment. The bundle grouping algorithm controls the size and placement of partition bundles. Figure 3(a) and (b) show the results plotted in blue.

These two experiments demonstrate that our system is able to process transactions overs a distributed system while managing distributed accesses. Moreover, the second experiment (using 4 DB) performs with higher latencies than the first one (using 1 DB) because of the grouping overhead. However the difference in terms of transaction latency remains relatively small. This illustrates the effectiveness of our grouping strategy.

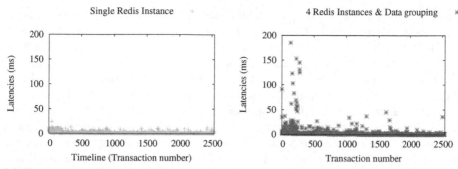

(a) Comparison: Single Redis Instance vs   (b) Latencies: With grouping algorithm
Our distributed deployment

Fig. 3. Local instance vs distributed processing with grouping

Impact of the Bundle Fluctuation. We now consider that bundles may
fluctuate (occuring and vanishing), and investigate the consequences of bundle
fluctuation on latency and data transfer.

Impact of the bundle fluctuation on latency. We characterize the fluctuation as
the bundle lifetime that is the average period during which a bundle occurs.
When the lifetime expires, the bundle breaks down and its parts may form new
bundles together with other data. Consequently the overall load remains almost
constant in this experiment.

We run the experiment for 10 rounds. We vary the lifetime of a bundle for
each round, ranging from 100 to 1000 s respectively for the rounds 1 to 10. We
report the 95th percentile of the transaction latency on Fig. 4(a). We observe that
the shorter the lifetime of the bundle is, the greater the value of the latency is.
Below 300 s lifetime the frequent bundle grouping operations make our solution
unusable in practice. Hopefully, beyond 300 s lifetime our solution achieves rather
stable latency.

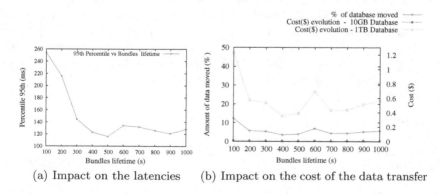

(a) Impact on the latencies   (b) Impact on the cost of the data transfer

Fig. 4. Impact of the bundle fluctuation

Impact of the bundle fluctuation on the data transfer's cost. In this experiment, we aim to quantify the monetary cost (expressed in $) of our bundle grouping strategy. We consider the case where our solution is deployed over a cloud computing infrastructure. In such *pay-per-use* environment, the billing of the resources (machines, storage, network) is based on how much we use them. Since our solution results in moving data across machines, we focus on the cost of data exchange while we fix the number of machines we use. We vary the bundle lifetime by assigning to it the values $100s, 200s, ..., 1000s$ successively. For each lifetime value, we run the same workload and measure the amount of transferred data. On Fig. 4(b) we report the amount of transferred data as a ratio of the total database size.

Moreover, we take Amazon Web Services (EC2 and S3) [17] as a reference for pricing. Such services charge 0.01 USD to transfer 1GB of data between two machines located in the same region (i.e. the same data center). Based on the actual database size, we extrapolate the costs for larger size values (10 GB and 1 TB), and report them on Fig. 4(b). The figure shows that the cost of data transfer is negligible for lifetimes bundles above 200s. This is suitable with the application context which assumes an average bundle time longer than one minute (e.g. the lifetime of a chat session). Therefore our solution for bundle migration should not imply significant extra cost when deployed on a cloud platform.

6 Conclusion

In this work, we deal with a fluctuating workload by exploiting the features of social applications. Actually, we propose a formal mechanism to identify and track bundles and their evolution based on social features in order to dynamically adjust the partitioning schema. Moreover, we propose a data placement scheme that permits to entirely store a bundle on a single machine while bounding the response time of transactions. In this respect, we monitor the bundle's size and the linked load to asses whether they still fit on a single node or not. In case the size of a bundle exceeds a machine capacity, then additional resources are used to migrate the bundle and the load. Finally, we propose an execution model with the principle of executing one transaction at one single node. We propose a grouping protocol that gather all required data on a single node whenever a transaction has to access data distributed over several nodes. Ongoing works are conducted to study the scalability of our solution [18].

References

1. Thomson, A., Diamond, T., Weng, S.C., Ren, K., Shao, P., Abadi, D.J.: Calvin: fast distributed transactions for partitioned database systems. In: SIGMOD, pp. 1–12 (2012)
2. Liu, B., Tatemura, J., Po, O., Hsiung, W.P., Hacigumus, H.: Automatic entity-grouping for oltp workloads. In: IEEE ICDE, pp. 712–723 (2014)

3. Apers, P.M.G.: Data allocation in distributed database systems. ACM TODS **13**(3), 263–304 (1988)
4. Madathil, D., Thota, R., Paul, P., Xie, T.: A static data placement strategy towards perfect load-balancing for distributed storage clusters. In: IEEE IPDPS, pp. 1–8 (2008)
5. Copeland, G., Alexander, W., Boughter, E., Keller, T.: Data placement in bubba. SIGMOD Rec. **17**(3), 99–108 (1988)
6. Mehta, M., DeWitt, D.J.: Data placement in shared-nothing parallel database systems. VLDB J. **6**(1), 53–72 (1997)
7. Sacca, D., Wiederhold, G.: Database partitioning in a cluster of processors. ACM TODS **10**, 29–56 (1985)
8. Curino, C., Jones, E., Zhang, Y., Madden, S.: Schism: a workload-driven approach to database replication and partitioning. VLDB Endow. **3**(1–2), 48–57 (2010)
9. Abdul, Q., Kumar, K., Deshpande, A.: Sword: Scalable workload-aware data placement for transactional workloads. In: EDBT, pp. 430–441 (2013)
10. Trushkowsky, B., Bodík, P., Fox, A., Franklin, M.J., Jordan, M.I., Patterson, D.A.: The scads director: Scaling a distributed storage system under stringent performance requirements. In: 9th USENIX, FAST, pp. 12–12 (2011)
11. Serafini, M., Mansour, E., Aboulnaga, A., Salem, K., Rafiq, T., Minhas, U.F.: Accordion: elastic scalability for database systems supporting distributed transactions. PVLDB **7**(12), 1035–1046 (2014)
12. Lee, J., Kwon, Y.S., Frber, F., Muehle, M., Lee, C., Bensberg, C., Lee, J.Y., Lee, A.H., Lehner, W.: Sap hana distributed in-memory database system: transaction, session, and metadata management. In: ICDE, IEEE Computer Society, pp. 1165–1173 (2013)
13. Pavlo, A., Curino, C., Zdonik, S.: Skew-aware automatic database partitioning in shared-nothing, parallel oltp systems. In: SIGMOD, pp. 61–72 (2012)
14. Redis Inc.: http://redis.io/. Online Retrieved on Aug 2014
15. Apache Storm: http://storm.incubator.apache.org/. Online Retrieved on Aug. 2014
16. Amstrong, T.G., Ponnekanti, V., Borthakur, D., Callaghan, M.: Linkbench: a database benchmark based on the facebook social graph. In: SIGMOD, PP. 1185–1196 (2013)
17. Amazon Web Services Pricing: http://aws.amazon.com/fr/ec2/pricing/. Online Retrieved on Nov 2014
18. Gueye, I.: Large scale web 2.0 transaction processing with on-demand dynamic resources adjustment: toward a transactional engine with energy saving. PhD thesis, University Cheikh Anta Diop (2015)

Indexing Multi-dimensional Data in Modular Data Centers

Libo Gao, Yatao Zhang, Xiaofeng Gao$^{(\boxtimes)}$, and Guihai Chen

Shanghai Key Laboratory of Scalable Computing and Systems,
Department of Computer Science and Engineering,
Shanghai Jiao Tong University, Shanghai 200240, China
{nosrepus,confidentao}@gmail.com, {gao-xf,gchen}@cs.sjtu.edu.cn

Abstract. Providing efficient multi-dimensional indexing is critically important to improve the overall performance of the cloud storage system. To achieve efficient querying service, the indexing scheme should guarantee lower routing cost and less false positive. In this paper, we propose RB-Index, a distributed multi-dimensional indexing scheme in modular data centers with Bcube topology. RB-Index is a two-layer indexing scheme, which integrates Bcube-based routing protocol and R-tree-based indexing technology. In its lower layer, each server in the network indexes the local data with R-tree, while in the upper layer the global index is distributed across different servers in the network. Based on the characteristics of Bcube, we build several indexing spaces and propose the way to map servers into the indexing spaces. The dimension of these indexing spaces are dynamically selected according to both the data distribution and the query habit. Index construction and query algorithms are also introduced. We simulate a three-level Bcube to evaluate the efficiency of our indexing scheme and compare the performance of RB-Index with RT-CAN, a similar design in P2P network.

Keywords: Multi-dimensional data · Distributed index · Modular data center

1   Introduction

Recent years have witnessed an increasing need of cloud storage systems due to the emergence of modern data-intensive applications. Various cloud storage systems are put forward to meet requirements like scalability, manageability and low latency. Examples of such systems include BigTable [4], DynamoDB [6],

This work has been supported in part by the National Natural Science Foundation of China (Grant number 61202024, 61472252, 61133006, 61422208), China 973 project (2012CB316200), the Natural Science Foundation of Shanghai (Grant No. 12ZR1445000), Shanghai Educational Development Foundation (Chenguang Grant No. 12CG09), Shanghai Pujiang Program 13PJ1403900, and in part by Jiangsu Future Network Research Project No. BY2013095-1-10 and CCF-Tencent Open Fund.

Q. Chen et al. (Eds.): DEXA 2015, Part II, LNCS 9262, pp. 304–319, 2015.
DOI: 10.1007/978-3-319-22852-5_26

Cassandra [9], HyperDex [11], etc. One of the requirements of these systems is to support large-scale analytical jobs and high concurrent OLTP queries. To achieve this, many works [1,5,7,12–14,16] have been devoted to designing a new indexing scheme and data management system. A typical indexing scheme is RT-CAN [7]. RT-CAN is a two-layer indexing scheme integrating R-tree structure and CAN-based routing protocol. Its higher-layer index, called global index, is built upon the local index. In RT-CAN, each server in the network builds an R-tree as the local index, then selects a set of R-tree nodes and publishes them into the global index. At the same time, a multi-dimensional indexing space is constructed. Each server in the network maintains a zone and stores the global index in its responsible zone. Consequently, the global index, composed of R-tree nodes from different servers, works as an overview index and is distributed across servers. When a query is given, we first check the global index to determine which servers may contain the required data and then search the local R-trees on the related servers to get the result.

The design of the distributed two-layer index makes RT-CAN efficient and robust, however, like most of other works, RT-CAN is conducted in P2P network, rather than in data centers. Unlike the P2P network, of which nodes may scatter widely in the real world and the latency among nodes may vary greatly, the data center interconnects a great number of servers via a specific Data Center Network (DCN) and reaches high-reliability, scalability and regularity in its structure. A specific type of data center is Modular Data Center (MDC) [8,18,19], in which thousands of servers are interconnected via switches and then packed into a 20- or 40-feet shipping container. It can be rapidly employed anywhere to meet different requirements of applications. As the MDC gains its popularity, it brings new challenges for researchers to design a new efficient indexing scheme to support query processing for it. There are two main challenges. First, the indexing scheme should utilize the MDC's architecture topology to improve the performance. Second, since in modern data-intensive applications, multi-dimensional data are commonly processed, like photos, videos, etc., the indexing scheme should support multi-dimensional indexing. Although RT-CAN supports multi-dimensional data indexing, it requires that the dimension of stored data cannot surpass the dimension of the overlay network, which means that RT-CAN is not scalable in terms of data dimension. Therefore, to index higher-dimensional data, the original overlay network should be expanded and more servers should be added to build the index, which costs a lot. Moreover, the high-dimensional space is usually sparse. Directly building a high-dimensional indexing space is not efficient.

In this paper, we try to transplant the two-layer indexing scheme on MDC and the new indexing scheme should be scalable in terms of dimension. We present our RB-Index, a distributed indexing scheme for multi-dimensional query processing in MDC with Bcube [2] topology. Bcube is a server-centric architecture for MDC. Lots of mini-switches and links are used to form its hyperspace-liked network structure, which results in some attracting features like low-diameter, high-bandwidth and fault tolerance. In RB-Index, we adopt

R-tree as the local index. For the global index, instead of setting up one indexing space, we set up several distinct indexing spaces based on the feature of Bcube topology and build different global index for each space. The dimension of these global index is selected according to both data distribution and query habit. We design the rule to map servers into indexing spaces and publish R-tree nodes into the global index. A cost model is proposed to select the least-cost set of R-tree nodes to publish. We also discuss the query processing algorithms. In the end, we simulate a three-level Bcube and compare the performance of RB-Index with RT-CAN.

The contribution of this paper is threefold: (1) A multi-dimensional indexing scheme is proposed for MDC with Bcube topology. We design a strategy to build several indexing spaces, on which different global indexes with different dimension are set up. (2) We propose the rule to select the indexed dimension according to both the data distribution and query habit. The indexed dimension can be replaced along with the change of query habit, so theoretically RB-Index can support any-dimensional data indexing. (3) We simulate a three-level Bcube and evaluate the efficiency of RB-Index.

The rest of paper is organized as follows. Section 2 discusses the related work. Section 3 gives an overview of RB-Index. Section 4 introduces the global index construction, including the dimension selection, mapping scheme and the cost model. In Sect. 5, we present the query algorithms including point query, range query and KNN query. Section 6 illustrates the simulation and proves the efficiency of our indexing scheme. Finally, Sect. 7 gives the conclusion.

2   Related Work

Efficient indexing scheme is crucial for fast data retrieval in cloud infrastructures. Due to the unprecedented scale of data, extending the traditional indexing technologies, like B-tree and R-tree, has drawn the attention in distributed environment. One part of these efforts can be classified as the primary index. The primary index usually adopts key-value based indexing strategies. Given a key, the index will efficiently locate the objects linked with the key by utilizing a range index like a distributed B^+-tree [10] or by a multidimensional index like SD-Rtree [3]. However, these indexes do not support query on other data attributes. In real world, data like videos and photos usually have more than one keys. Therefore, supporting efficient secondary index becomes a very useful feature for many applications.

To handle these application cases, a new indexing scheme with two-layer structure, called RT-CAN, was proposed in [7]. In RT-CAN, servers are organized into an overlay network and the overlay network forms the global indexing space. The dimension of the overlay network is determined by data dimension, so after the data being mapped into the indexing space, every data attribute can be queried with the index. There are also some similar designs in [12,13]. All of these works are implemented in P2P network, until recently, several research work [5,14,16] implemented such a design in data centers. One problem of these

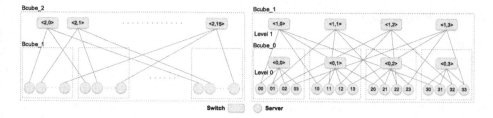

Fig. 1. A $Bcube_2$ with n=4 (left) is constructed from 4 $Bcube_1$ (right) and 16 4-port switches.

works is that the index is not scalable with respect to dimension. Take RT-CAN as an example, the scale of the overlay network determines the query efficiency under different number of dimension. When the dimension of stored data increases, the overlay network should also be extended and more server nodes should be added to guarantee the efficiency. Thus, this introduces great cost, especially in data centers where the physical layout and scale is usually fixed.

To address this problem, we want to build a scalable two-layer indexing scheme. Our index is designed for MDC. The scale of MDC is usually limited with the size of a shipping container. This property pushes such problem of dimension-scalability into an extreme condition and solving this problem becomes urgent and significant.

The Modular Data Center is proposed to meet the growing flexibility of customers. Many companies has already deployed the MDCs. Sun first presented an MDC in 2006 with up to 2240 servers and 3PB storage. Later, HP and IBM employed their own MDCs. The design of the MDC architecture is driven by application needs. A typical architecture is called Bcube [2].

Bcube is a server-oriented network for container-based, MDCs. By using much mini-switches and links, Bcube accelerates one-to-x traffic and provides high network capacity for all-to-all traffic. The construction is recursively-defined. $Bcube_0$ is the smallest module that consists of n servers connecting to an n-port switch. A high-level $Bcube_k$ employs $Bcube_{k-1}$ as a unit cluster and connects n such clusters with n^k n-port switches. Figure 1 illustrates an example of $Bcube_2$ with $n = 4$, which consists of 64 servers. The servers and switches are labeled in a string: $a_k a_{k-1} \ldots a_0 (a_i \in [0, n-1], i \in [0, k])$, where a_i represents port number of the switch in the ith level to which the server connected.

3   System Overview

In this section, we first define the indexing space for $Bcube_k$. Then we design the rule to mapping servers in the $Bcube_k$ into its corresponding indexing space and assign the potential index range. Finally, we give an overview of the RB-Index.

3.1 Indexing Space

As is mentioned, Bcube is a recursively defined structure and a $Bcube_k$ is constructed by n $Bcube_{k-1}$s. This property guarantees that a $Bcube_k$ contains n^k $Bcube_0$s and n^{k+1} servers. To efficiently index multi-dimensional data, we construct a $(k+1)$-dimension indexing space for a $Bcube_k$. All the data mapped into the space is normalized, so the range on each dimension is the same and the indexing space is actually a hyperspace. We set the dimension of the indexing space to the level of $Bcube_k$, hoping that the servers in the network can be mapped into the indexing space uniformly and each one can take the responsibility for the indexes in a specific zone. We will talk about the mapping scheme later and here we want to give a closer view of the indexing space by sharing an example.

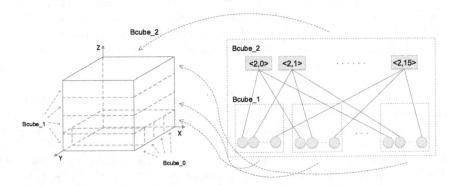

Fig. 2. The structure of indexing space for $Bcube_2$ with n=4.

Figure 2 gives an example of the indexing space for $Bcube_2$. Since the level of $Bcube_2$ is three, the indexing space is a cube. Here $n = 4$, so four $Bcube_1$s are arranged along with z-coordinate to form the cube, while each $Bcube_1$ contains four $Bcube_0$s. The servers in $Bcube_0$ are arranged along with x-coordinate. The construction of a higher dimension indexing space takes the similar way. First, n servers in each $Bcube_0$ are placed uniformly along with the first dimension, then the process continues iteratively until finally n $Bcube_{k-1}$ are arranged uniformly along with the $(k+1)$th dimension.

Through this construction, the indexing space has some features. Referring to Bcube's label pattern, we know if two servers' labels have exactly one digit difference, then these two servers are adjacent and the path length is one. More generally, the path length between any two servers is not longer than $k+1$. Taking the indexing space in Fig. 2 as an example, the path length of any two servers, which are in line with the same coordinate, is one; the path length of any two servers on a plane, which is parallel to a coordinate plane, is not longer than two.

3.2   Mapping Scheme and Potential Index Range

Previously, we introduced the indexing space for $Bcube_k$. We set the dimension of the indexing space to the level of $Bcube_k$, so all n^{k+1} servers in the network can be mapped uniformly into the indexing space. Each server will hold a zone in the indexing space and take the responsibility for the indexes in that zone. We call that zone the potential index range. When building the global index, each server selects a set of R-tree nodes from its local R-tree, then these R-tree nodes will be published to the servers of which potential index range intersects with the node range. Here, we introduce the rule to allocate the potential index range to each server.

Suppose the indexing space is bounded by $\mathbf{B}=(B_0, B_1 \ldots B_k)$, where B_i is $[l_i, l_i + \gamma], i \in [0, k], \gamma \in \mathbb{R}^+$, the potential range of server t is $pir(t)$. Here, we denote a server t with a label string $a_k a_{k-1} \ldots a_0$ ($a_i \in [0, n-1], i \in [0, k]$). Equivalently, t equals to $\sum_{i=0}^{k} a_i n^i$. Then $pir(t)$ can be calculated by the following function.

$$
\begin{aligned}
pir(t) &= pir(a_k a_{k-1} \ldots a_0) \\
&= ([l_0 + a_0 \frac{\gamma}{n}, l_0 + (a_0 + 1)\frac{\gamma}{n}], \ldots, [l_k + a_k \frac{\gamma}{n}, l_k + (a_k + 1)\frac{\gamma}{n}]).
\end{aligned} \tag{1}
$$

A close observation reveals that in $Bcube_k$, the labels of servers in the same $Bcube_0$ only have the first bit, a_0, different. According to Eq. (1), these servers will be arranged along with the first coordinate. Figure 3 presents the arrangement of servers in the $3D$ space. We assume that l_0, l_1 and l_2 all equal to zero. Each of four $Bcube_1$s has 16 servers which are placed in a plain parallel to the xy-coordinate plain. Every server occupies a $3D$ zone. For example, $pir(000) = ([0, \frac{1}{4}], [0, \frac{1}{4}], [0, \frac{1}{4}])$, while $pir(133) = ([\frac{3}{4}, 1], [\frac{3}{4}, 1], [0, \frac{1}{4}])$

3.3   Two-Layer Index Architecture

In a distributed storage system, data are randomly distributed over servers. Each server builds a local R-tree for local data retrieval. When given a query, instead of searching all the local R-trees, the global index is given to guide the query to the related servers, which significantly reduces the query region. One

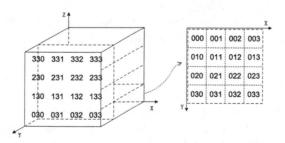

Fig. 3. Potential index range of servers

way to achieve this goal is to build a centralized global index. Queries are first sent to several master servers which possess the global index, then, according to the results, they are broadcast to the related servers for local queries. However, this strategy may face problems when the number of queries increases. Several master servers will be the bottleneck of the query performance. An alternative is to distribute global indexes to all servers and each server is responsible for a portion of global index in its potential index range. By balancing the workload, the performance of query is guaranteed. However, both of the indexing schemes require that the dimension of the indexed data cannot surpass the dimension of the indexing space. Therefore, to index the high-dimensional data, we need to build a high-dimensional indexing space. In this paper, we propose RB-Index, which is a distributed two-layer indexing scheme supporting scalability in terms of data dimension.

Previously, we have introduced the indexing space formed by $Bcube_k$ and the dimension of the space is limited to $k+1$. Usually, k is not a big number and the dimension of data may be far more than $k+1$. If we try to reduce the dimension of data and map them to a low-dimensional indexing space, when users query on some unindexed dimension, all the servers need to be searched to get the result. In RB-index, rather than building only one global indexing space, we build $n+1$ indexing spaces, containing one $(k+1)$-dimension indexing space for $Bcube_k$ and n k-dimension indexing spaces for n $Bcube_{k-1}s$. The global indexes built in these indexing spaces are classified into two types, **the main global index** and **the subsidiary global index**. The main global index is the index in the $(k+1)$-indexing space. Among all the global indexes, the main global index has the highest dimension, $k+1$, and is distributed across all the servers in the network. The subsidiary index is built in the k-dimension indexing space. In each of n $Bcube_{k-1}s$ in $Bcube_k$, we build a subsidiary index, which is distributed only across servers in that $Bcube_{k-1}$. Therefore, there are n subsidiary global indexes in all. Each server t in the network is mapped into exactly two indexing spaces and responsible for two potential index ranges, $pir(t)$ for main global index and $pir'(t)$ for subsidiary global index. Theoretically, each time RB-Index can support $((n+1)k+1)$-dimension indexing. If the data dimension is less than $((n+1)k+1)$, extra zero vector can be appended to the data. If the data dimension exceeds $((n+1)k+1)$, we first pick $((n+1)k+1)$ dimensions from data and gradually rebuild the subsidiary indexes according to the query habit by replacing the indexed dimension.

For clarity, we summarize symbols with their meaning in Table 1. Some of them will be used in the description of the rest of this paper.

4   Index Construction

In this section, we first introduce the adaptive rule to determine the dimension of the indexing space, then we present the way to publish the R-tree nodes into the indexing space. Finally, a cost model is introduced to select a proper set of R-tree nodes from a local R-tree.

Table 1. Symbol description

Sym	Description	Sym	Description
t	Code for servers	k	The level of Bcube
N_t	t's R-tree node set for publishing	n	Number of switch ports
$pir(t)$	t's potential index range in main global index	$pir'(t)$	t's potential index range in subsidiary global index

4.1 Dimension Selection

RB-Index builds several indexing spaces to support high-dimensional data indexing. The dimension of these indexing spaces need to be selected properly to face all kinds of query. An adaptive method is to select the dimension according to the query habit. It is common that during a certain period of time, users may tend to query on some certain combinations of the dimension. We build the index for them. Once the query habit changes, we first judge whether the new query habit exists, then we rebuild the indexing space based on the new query habit. There is no denying that rebuilding the indexing space will introduce a great cost, however, the primary aspect we consider here is that the gained efficiency should surpass the cost. In RB-Index, the dimension of the main global index is fixed once being established, while the dimension of the subsidiary global index will be replaced along with the change of query habit.

As is mentioned, the main global index has $k+1$ dimension and is distributed across all the servers in the network. We want to distribute the main global indexes as uniformly as possible to avoid the hot spots during query, so principle component analysis (PCA) [17] is adopted to select dimension for the main global index. PCA is a traditional dimension-reduction method. By analyzing the distribution of data, PCA always picks up top x dimension on which data are most uniformly distributed. Since the data are randomly distributed across the servers in the network, we use distributed PCA [15] to get the first $k + 1$ dimension.

In $Bcube_k$ with n, there are n $Bcube_{k-1}$s, which means that there are n subsidiary global indexes. For each subsidiary global index, we select k dimension according to the query habit. There are some restrictions on selecting dimension. First, the dimension of the subsidiary global index cannot totally be contained in the global index. Second, the dimension of any two subsidiary global indexes cannot be exactly the same. These two requirements are raised to avoid redundancy among the global indexes. Initially, we list the top n frequently occurred query patterns. Here, the query pattern is defined as a group of k dimensions that are usually queried together. Each subsidiary global index is initialized with one of these query patterns. As more and more queries happen, we use LRU-algorithm, a caching algorithm, to choose one subsidiary index to be rebuilt. The subsidiary index that is least recently hit will be required to be rebuilt. Here *a hit* means the global index contains some dimension of a given query. We also assume that only when a query does not hit any global index, including one main global index and n subsidiary global indexes, the system will trigger rebuilding the least recently hit subsidiary index. In practice, to avoid frequent index rebuilding, the frequence of a query pattern should exceed a threshold in a time quantum before triggering the rebuilding of indexing space. Otherwise, no rebuilding will happen.

4.2   Publishing Scheme

As is discussed above, every server builds a local R-tree to accelerate local data retrieval. To build the global indexes, each server adaptively selects a set of R-tree nodes, $\mathbf{N_t} = \{N_t^1, \ldots N_t^n\}$, from its local R-tree and publishes them into the global index.

We use the same mapping scheme as in [7]. Since the publishing scheme of both the main global index and the subsidiary global index is similar, we take the main global index as an example. The format of the published R-tree node is (ip, mbr), where ip records the physical address of the server which publishes the node and mbr gives the range of the R-tree node. For each selected R-tree node, the center and radius are two criteria used for mapping. The center decides the position where the R-tree node is mapped, while the radius decides whether the node is mapped to one server or several servers. A threshold, R_{max}, is set to be compared with the radius. The detailed process of the mapping scheme is as follows: given an R-tree node, we calculate its center and radius. First, the node is directly mapped to the server whose potential index range contains the center. Then the radius is compared with R_{max}. If the radius is larger than R_{max}, then the node will also be mapped to the servers whose potential index range intersects with the R-tree node range.

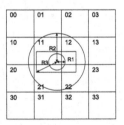

Fig. 4. Mapping scheme in $2D$ place

Figure 4 shows an example in $2D$ space. Suppose that the radius of the indexed R-tree node N is R_3 and R_{max} equals R_2, then N will only be mapped to $\{11\}$. If we set R_{max} to R_1, then since R_3 is larger than R_1, N will be mapped to servers $\{11, 12, 21, 22\}$. In higher-dimensional space, the mapping scheme works in the same way.

4.3 Cost Model

In RB-Index, each server chooses a set of nodes from its local R-tree and publishes them into the global index. [7] raises two properties as the basic requirements in selecting the local R-tree node set: index completeness and unique index. These two criteria guarantee that the selected set of R-tree nodes covers the whole local data range with the least redundancy. Based on these requirements, a cost model is introduced here to optimize the index selection further. We combine the cost model raised in [7,13] and take advantage of both the cost models.

In the rest of the paper, we use hop number as the metric to evaluate the routing cost. A hop refers to the trip from one server to its neighbor. Since servers in Bcube are connected in a quite close physical distance, the communication delay between any two servers is approximately equal. Thus, using hop number to evaluate the real routing cost is reasonable. The cost model considers two aspects to value the cost of publishing a local R-tree node $C(N)$: index maintenance cost $C_m(N)$ and query process cost $C_q(N)$,

$$C(N) = C_m(N) + C_q(N) \tag{2}$$

The maintenance cost refers to the cost of essential node update operations: split and merge. Once the published node is split or merged, additional routing cost is caused by republishing the node. Given that the average routing cost between any server in $Bcube_k$ is $k + 1$, the routing cost of splitting includes deleting the original node and publishing two new nodes. Similarly, the routing cost of merging includes deleting two old nodes and publishing a new node. Both of the two operations cost $3(k + 1)$ approximately. Assume that the probabilities of splitting and merging for node N are $p_{split}(N)$ and $p_{merge}(N)$, we can get Eq. (3) as follows:

$$C_m(N) = 3(k + 1)(p_{split}(N) + p_{merge}(N)) \tag{3}$$

Now the problem becomes how to estimate the value of $p_{split}(N)$ and $p_{merge}(N)$. In [7], a two-state markov chain model is applied on each R-tree node to calculate its $p_{split}(N)$ and $p_{merge}(N)$. However, this takes a great amount of computation. Here, we only use this method to calculate $p_{split}(N)$ and $p_{merge}(N)$ of the leaf node. For non-leaf nodes, we use the method in [13]. Suppose $p_1(N)$ is the probability of insertion in node N and $p_2(N)$ is the probability of deletion in node N. The relationship between p_1, p_2, p_{split} and p_{merge} can be formulated as:

$$p_{split} = \frac{(\frac{p_2}{p_1})^{\frac{3m}{2}} - (\frac{p_2}{p_1})^m}{(\frac{p_2}{p_1})^{2m} - (\frac{p_2}{p_1})^m}, \quad p_{merge} = \frac{(\frac{p_2}{p_1})^{2m} - (\frac{p_2}{p_1})^{\frac{3m}{2}}}{(\frac{p_2}{p_1})^{2m} - (\frac{p_2}{p_1})^m} \tag{4}$$

For any non-leaf node N, suppose its children are $c_1, c_2, \ldots, c_i$, then p_1 and p_2 can be calculated by the Eq. (5):

$$p_1(N) = \prod_{j=1}^{i}(1 - p_{split}(c_j)), \quad p_2(N) = \prod_{j=1}^{i}(1 - p_{merge}(c_j)) \tag{5}$$

Therefore, once we get $p_{split}(N)$ and $p_{merge}(N)$ of leaf nodes, we can get the update probabilities of the internal node iteratively.

The second cost is query processing cost. We mainly consider false positive. Due to the overlay between R-tree nodes, it is common for the global index to guide the query to servers which actually does not contain the needed datum. Suppose $R(N)$ represents the range of node N and $D(N)$ represents the range of data in node N, then the probability of false positives p_{fp} can be simply defined in Eq. (6). And the average routing cost of query processing can be calculated in Eq. (7).

$$p_{fp}(N) = \frac{R(N) \cap D(N)}{R(N)} \tag{6}$$

$$C_q(N) = (k+1)p_{fp}(N) \tag{7}$$

After getting the maintenance cost (Eq. 3) and query processing cost (Eq. 7), we can get the cost of a node N:

$$C(N) = (k+1)(3p_{split}(N) + 3p_{merge}(N) + p_{fp}(N)) \tag{8}$$

5   Query Processing

In this section, we show how the global index can be applied to efficiently process high-dimensional data queries.

5.1   Point Query

Before we talk about point query, we make some explanations: here, point query refers to full-dimensional point query, rather than partial-dimension point query, since partial point query can be regarded as range query, which will be introduced later. Point query can be processed in either one of the global indexes. Usually it is processed in the main global index since the dimension selection of the main global index determines that data are most uniformly distributed in that index and the main indexing space is better grained with more responsible servers. Given a point $p=(x_0, \ldots, x_d)$, the process of point query $Q(p)$ can be divided into two phases:

In the first phase, we forward the query to server t whose potential index range contains the point. Then we generate a circle centered at point p with radius R_{max}. All the servers whose potential index range intersects with the circle should be searched. We take advantage of multi-pathes between servers in Bcube and forward the query in parallel. We check the main global index buffered in these servers and get related R-tree nodes whose range contains the point. In the second phase, according to the result R-tree nodes, we continue the query on local R-trees in the corresponding physical servers. For point query processing, suppose the circle covers M servers and the average path length between any two server is $(k+1)$, then the routing cost of forwarding the query in the global index is $O((k+1) * M)$. Besides, forwarding query to the related servers for local search brings additional cost. If the overlay of R-tree is properly controlled, this process will have a tiny cost.

5.2   Range Query

Suppose the dimension of the stored data is d. The range query can be either the d-dimension range query or partial-specific range query. We denote a range query as $Q(range)$, where $range=([l_{d_1}, u_{d_1}] \ldots [l_{d_t}, u_{d_t}])$, $\{d_1, d_2 \ldots d_t\}$ is a subset of $\{1 \ldots d\}$. The range query processing can be divided into three phases.

In the first phase, we choose one global index which matches the query most. Here, matching means having the most number of the same dimensions with the query dimension. We prefer to choose the global index which guides query to lest number of servers. If the dimension of the range query is not indexed, then we will check whether this query happens frequently in the past time. If so, one subsidiary global index which is the least recently used will be rebuilt. Otherwise, we will omit the following phases and directly broadcast the range query to all the servers and query on local R-trees. In the second phase, first, we send the query to servers whose potential index range intersects with the query range. Then due to the publishing scheme, the query should also be forwarded to some other servers to get the full result. We calculate the center of the range query and its radius. In the global index space, if only part of dimension of the query is indexed, then the center maybe a line or a plain. The server t is searched if and only if $|t.pir.center - range.center| < radius + R_{max}$. The buffered R-tree nodes in these servers will be searched. In the third phase, according to result R-tree nodes, the query is forwarded to the related physical servers to continue searching on their local R-trees.

The cost of range query processing is related to the radius of the range. As we can see from the process, if the radius is larger, more servers will be searched to get the buffered R-tree nodes. The worst case is broadcasting the query to all the servers. However, this case will not always happen. Thus, compared with the cost of broadcasting the query to all servers, the cost of range query processing with the help of RB-Index is reduced significantly.

5.3   KNN Query

We denote the KNN query $Q(p, K)$, which requires the K nearest neighbors for the point p. Again, we choose the main global index. The reason is the same as point query. We first generate a circle C centered at p with a given radius R_{init}. R_{init} is set according to the data distribution and the value K. During the process of the query, if the result of range query $Q(C)$ contains K nearest data, then the KNN query complete. Otherwise, we extend the circle radius with δ until the result of range query returns enough nearest data. The cost of KNN query is related to the value K. With higher K value, bigger range needs to be searched, which reduces the query efficiency.

6   Performance Evaluation

In this section, we simulate a $Bcube_2$ with $n = 4$ and evaluate the performance of RB-Index on it. The data are generated and distributed randomly across the

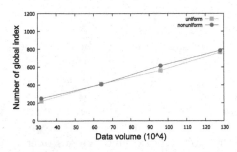 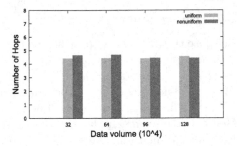

Fig. 5. Average number of global index in one server

Fig. 6. Point query

servers in the Bcube. We generate two different datasets to evaluate the performance of RB-Index. In the uniform dataset, we generate 320000 to 1280000 data. These data have 11 attributes and values of attributes are uniformly distributed between $(0, 1)$. In the nonuniform dataset, data are generated following the 80/20 rule, which means 80 percentage of data are concentrated in 20 percentage of the space. For both of the datasets, we disseminate data to servers and keep every server roughly maintain the same number of data. A $Bcube_2$ with $n = 4$ contains 64 servers and each server builds a local R-tree indexing local data. For each internal R-tree node, the maximum number of entries is set to 10. Initially, we choose the last but one level from R-trees to publish into global indexes, since these nodes are not frequently updated and have the modest false positive.

In the following simulation, we calculate the number of hops to evaluate the query performance of RB-Index. As is mentioned, in MDC the communication delay between any two servers is approximately equal. Thus, using hop number to evaluate the real routing cost is reasonable. We mainly focus on point query and range query since in RB-Index, KNN query can be achieved by several range queries. We also build the RT-CAN [7] in $Bcube_2$ as a reference object. Due to the algorithm of point query, the point query processing of RT-CAN and RB-Index is actually same, so we only compare the range query performance of RT-CAN with RB-Index.

Before evaluating query performance, we first estimate the space cost of the global index under different data volume. We record the number of buffered R-tree nodes in each server and calculate the average number of buffered R-tree nodes in a server. The result is shown in Fig. 5. As the data volume increases, the average number of buffered R-tree nodes increases as well. In addition, the nonuniform dataset causes slightly more published R-tree nodes than the uniform data. This may be resulted from the property of R-tree and the publishing scheme. Although holding two portions of different global indexes in each server introduces more space cost, we still think it is a tolerant cost and the gained efficiency overweighs the space cost.

For the point query, to evaluate the performance, we adopt the single-path strategy, which means when we get the result R-tree nodes through the global index, we continue local searching from one server to another, rather than in parallel. Once we find the data, the query terminated. For the range query, we

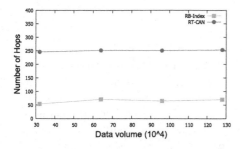

Fig. 7. Range query (uniform dataset)

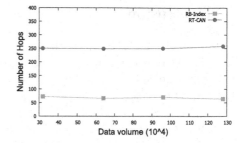

Fig. 8. Range query (nonuniform dataset)

limit the size of query range. Each time the search range will cover approximately 10 percentage of the range in each dimension. We respectively conduct 1000 randomly generated queries and take the average result. Figure 6 shows the average routing cost of point query. The result shows that even with increasing data volume, the routing cost still maintains stable. Figures 7 and 8 show the average routing cost of range query. Since RT-CAN only indexes three dimension of data, the average routing cost is almost the cost of broadcasting query in the network. In contrast, RB-Index improves the performance by more than 50 % in both uniform dataset and nonuniform dataset.

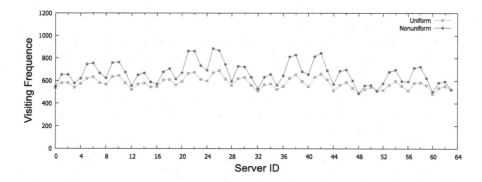

Fig. 9. Visiting frequence

Finally, we record the visiting frequence of each server in the network. As in real systems, both the point query and range query are processed, we conduct 2000 random queries with both point query and range query. The result is showed in Fig. 9. Under uniform dataset, the visiting frequence of each server is approximately same, while some servers mapped in the center of the indexing space are a bit more frequently visited since they possess more published R-tree nodes. Under nonuniform dataset, the overall visiting frequence decreases and due to the skewed distribution, some servers are particularly frequent-visited.

7   Conclusion

In this paper, we propose RB-Index, a multi-dimensional indexing scheme for Modular Data Center with Bcube topology. RB-Index adopts the two-layer indexing scheme, of which the global index is built upon local R-tree and is distributed over servers. Based on the characteristics of Bcube, we set up several indexing spaces to build several global indexes. We select the dimension of these indexing spaces adaptively to meet query habits. As a result, theoretically RB-Index can support any-dimensional data indexing. We propose the index construction rule and query algorithms to guarantee efficient data management in the network. In the evaluation, we simulate a 64-node Bcube and compare the query performance of RB-Index with RT-CAN, a most related previous work. Although building several global indexes increases space cost, the result verifies the query efficiency of RB-Index.

References

1. Chen, G., Vo, H.T., Wu, S., Ooi, B.C., Özsu, M.T.: A framework for supporting DBMSlike indexes in the cloud. VLDB 4(11), 702–713 (2011)
2. Guo, C., Lu, G., Li, D., Wu, H., Zhang, X., Shi, Y., Tian, C., Zhang, Y., Lu, S.: BCube: a high performance, server-centric network architecture for modular data centers. In: SIGCOMM (2009)
3. du Mouza, C., Litwin, W., Rigaux, P.: SD-Rtree: a scalable distributed Rtree. In: ICDE pp. 296–305 (2007)
4. Chang, F., Dean, J., Ghemawat, S., Hsieh, W.C., Wallach, D.A., Burrows, M., Chandra, T., Fikes, A., Gruber, R.E.: Bigtable: a distributed storage system for structured data. In: OSDI, pp. 205–218 (2006)
5. Li, F., Liang, W., Gao, X., Yao, B., Chen, G.: Efficient R-tree based indexing for cloud storage system with dual-port servers. In: Decker, H., Lhotská, L., Link, S., Spies, M., Wagner, R.R. (eds.) DEXA 2014, Part II. LNCS, vol. 8645, pp. 375–391. Springer, Heidelberg (2014)
6. DeCandia, G., Hastorun, D., Jampani, M., Kakulapati, G., Lakshman, A., Pilchin, A., Sivasubramanian, S., Vosshall, P., Vogels, W.: Dynamo: amazons highly available key-value store. SIGOPS 41(6), 205–220 (2007)
7. Wang, J., Wu, S., Gao, H., Li, J., Ooi, B.C.: Indexing multi-dimensional data in a cloud system. In: SIGMOD, pp. 591–602 (2010)
8. Hamilton, J.: An architecture for modular data centers. In: CIDR (2007)
9. Avinash, L., Prashant, M.: Cassandra: a decentralized structured storage system. ACM Spec. Interest Group Oper. Syst. (SIGOPS) 44(2), 35–40 (2010)
10. Aguilera, M.K., Golab, W., Shah, M.A.: A practical scalable distributed B-tree. PVLDB 1, 598–609 (2008)
11. Escriva, R., Wong, B., Sirer, E.G.: HyperDex: a distributed, searchable key-value store. In: SIGCOMM, pp. 25–36 (2012)
12. Sai, W., Wu, K.-L.: An indexing framework for efficient retrieval on the cloud. ICDE 32(1), 75–82 (2009)
13. Wu, S., Jiang, D., Ooi, B.C., Wu, K.-L.: Efficient B-tree based indexing for cloud data processing. VLDB 3(1–2), 1207–1218 (2010)

14. Gao, X., Li, B., Chen, Z., Yin, M., Chen, G., Jin, Y.: FT-INDEX: a distributed indexing scheme for switch-centric cloud storage system. In: ICC (2015)
15. Liang, Y., Balcan, M.-F., Kanchanapally, V.: Distributed PCA and k-means clustering. In: NIPS (2012)
16. Liu, Y., Gao, X., Chen, G.: Design and optimization for distributed indexing scheme in switch-centric cloud storage system. In: ISCC, pp. 804–809 (2015)
17. Jolliffe, I.T.: Principal Component Analysis, 2nd edn. Springer, New York (2002)
18. IBM: Scalable Modular Data Center. http://www-935.ibm.com/services/us/its/pdf/smdc-eb-sfe03001-usen-00-022708.pdf
19. Rackable Systems: ICE Cube Modular Data Center. http://www.rackable.com/products/icecube.aspx

Data Mining IV, Applications

A Modified Tripartite Model for Document Representation in Internet Sociology

Mikhail Alexandrov[1,2], Vera Danilova[1,2]([✉]), and Xavier Blanco[1]

[1] Autonomous University of Barcelona, Barcelona, Spain
{MAlexandrov.UAB,XblancoE}@gmail.com
[2] Russian Presidential Academy of National Economy and Public Administration,
Moscow, Russian Federation
vera.danilova@e-campus.uab.cat

Abstract. Seven years ago Peter Mika (Yahoo! Research) proposed a tripartite model of actors, concepts and instances for document representation in the study of social networks. We propose a modified model, where instead of document authors we consider textual mentions of persons and institutions as actors. This representation proves to be more appropriate for the solution of a range of Internet Sociology tasks. In the paper we describe experiments with the modified model and provide some background on the tools that can be used to build it. The model is tested on the experimental corpora of Russian news (educational domain). The research reflects the pilot study findings.

Keywords: Tripartite model · Document representation · Ontology · Social networks analysis · Internet sociology

1 Introduction

The automatic detection of document and corpus topic is one of the most popular tasks within the natural language processing. The traditional model, where each document is represented by a vector describing the distribution of keywords in a document or corpus, works quite well. On the basis of this representation, it becomes possible to group similar documents and reveal the corpus structure, as well as to group keywords and build topic vocabularies [6,11]. Considering each document as an ontology instance and a set of keywords as a concept, we obtain a two-layer ontology model. The appearance of social networks allowed the researchers to bring in the social aspect, thus creating a tripartite model of ontology for Social Networks Analysis containing document authors (actors), document content (instances), and tags (concepts) [13]. Three-layer model is a tool for studying the emergence and dynamics of communities on the basis of user-generated content. Author profiling gave an opportunity to analyze the

Under partial support of the Catholic University of San Pablo (grant FINCyT-PERU).

© Springer International Publishing Switzerland 2015
Q. Chen et al. (Eds.): DEXA 2015, Part II, LNCS 9262, pp. 323–330, 2015.
DOI: 10.1007/978-3-319-22852-5_27

attitude of different user categories towards particular events and topics. However, it is not only the author's personality that matters for solving Internet Sociology tasks, but also the awareness of the connection between the text mentions of personalities and organizations and concepts and instances within text. This knowledge is important, when dealing with traditional mass media - e-publications created by analysts and journalists. In this case, the Actor in the Actor-Concept-Instance triple is a Named Entity (NE) mention (person and organization names), the Concept is an automatically generated topic description, not user-generated tags, and the Instance is a set of phrases with a high level of specificity that frequently denote events (olympiad, demonstration, etc.). The present paper considers the use of this modified tripartite model of ontologies for the purposes of Internet Sociology studies.

The rest of the article is organized as follows. Section 2 provides a description of the three-layer model. Section 3 presents the experiments on real-world data that include model construction, opinion mining and clustering. Section 4 concludes the paper. This paper outlines our pilot-study experiments performed on the basis of Russian newspaper articles. The listed examples are translated into English.

2   Multipartite Models

2.1   Tripartite Model

The tripartite model uses three layers of descriptors (keywords) for document representation as follows. The first layer is formed from a list of actors or person and organization names that are mentioned in a document; the second layer is a list of concepts that characterize the domain a given document belongs to; the third layer is represented by the instances or document-specific collocations. Each element of each list contains a set of tokens, which number varies from 1 to 5. As for a single document, the tripartite model can be created for a corpus of documents. The corpus representation is a list of actors, concepts and instances that are mentioned throughout a corpus. Table 1 presents an excerpt from a model for a corpus of 250 documents related to the educational domain.

Here, the NE "Dr. Livanov" is a mention of the Russian Minister of Education and Science. "Dr. Rukshin" is one of the leading Russian school teachers. Parliament means a discussion at the State Duma. President means here a discussion at the Public Council under the President's administration. Moscow means the Department of Education of the Moscow Regional Government. It should be

Table 1. An excerpt from our experimental corpus model (educational domain)

No.	Actor	Concept	Instance
1	Ministry of Education	Young Talent	Law of Education, Parliament, Mar. 2013
2	Cabinet of Ministers	Primary School	Law of Education, President, Jul. 2014
3	Higher School of Economics	Secondary School	College Education, Parliament, Dec. 2013

noted that the frequency of occurrence of key terms in a document is not taken into account when building the model. We focus on the presence or absence in a document of the keywords from the corpus-based model.

2.2 Bipartite Model

The notation for the dimensionality of each layer (list) for the corpus of documents is as follows. Let K_a be the number of actors, K_i - the number of instances, and K_c - the number of concepts.

Actors-Concepts model (AC model) shows what keywords each person or organization is connected with in a text. It can be presented as a table (matrix) of size M (K_a, K_c). Matrix transposition MT of size (K_c, K_a) will show the actors that are related to a given concept (CA model).

The connection between an actor and a concept can be evaluated using the Jaccard distance [6]:

$$J(A_i, C_j) = 2 \times \frac{|D_{A_i} \cap D_{C_j}|}{|D_{A_i} \cup D_{C_j}|}, \tag{1}$$

where A_i and C_j represent a specific actor and a concept respectively, D_{A_i} and D_{C_j} correspond to the collections of documents, where the mentioned concept and actor occur. The number of documents, where the given concept and actor co-occur, in the numerator is doubled in order to normalize the connection degree from $[0.0, 1.0]$.

A weight W_A for each actor on the concepts layer, and a weight W_C for each concept on the actors layer can be introduced as follows: $W_A = \sum_{i=1}^{K_{cj}} J(A_i, C_j)$ (K_{cj} is the number of concepts related to a specific actor, and $J(A_i, C_j)$ is the Jaccard distance between the selected actor and one of the related concepts) and $W_C = \sum_{i=1}^{K_{aj}} J(C_i, A_j)$ (K_{aj} is the number of actors related to a specific concept, and $J(C_i, A_j)$ is the Jaccard distance between the selected concept and one of the related actors).

Concepts-Instances model (CI model) shows the specific collocations each concept (keyword set) is related to and vice versa. The weights are introduced in the same way as for the previous models. CI and IC models are commonly used for natural language processing purposes in two-layer model-based document parametrization: CI matrix shows the distribution of keywords throughout the documents, and IC - specific expressions that are "spotted" in each of the documents. The *Instances-Actors* (IA) and *Actors-Instances* (AI) models can be built by transforming the previously obtained matrices: $(AI) = (AC) \times (CI)$ and $(IA) = (IC) \times (CA)$.

2.3 Unipartite Model

Two-layer models can be reduced to single-layer models, which further allows to perform automatic grouping. A one-layer model uses only one of the layers or dimensions (actors, concepts or instances) for document and corpus representation. In order to properly build the unipartite model, let us first consider the (AC) model, where we can find the connections between two actors on the concept layer. To this end, the calculation of the number of shared concepts for each pair of actors should be performed. If all of the concepts are the same, the connection will have its maximum weight, and if no concepts are shared, the connection is insignificant. The connection can be evaluated using the Jaccard distance as follows:

$$J(A_i, A_j) = 2 \times \frac{|K_{ci} \cap K_{cj}|}{|K_{ci} \cup K_{cj}|}, \tag{2}$$

where A_i and A_j are the corresponding actors, K_{ci} and K_{cj} are sets of concepts related to the actors A_i and A_j correspondingly. The number of shared concepts in the numerator is doubled in order to normalize the connection degree from $[0.0, 1.0]$.

Having performed a pairwise evaluation of actors connection on the concepts layer, we can build a matrix Actors - Actors (AA), which is the target unipartite model. The resulting proximity matrix is the input for the cluster analysis. As a result, we obtain the groups of interconnected actors. It is a meaningful result: each cluster contains person and organization names discussed within the same topics. In a similar way, the reversed (CA) matrix provides the Concepts-Concepts model (CC). In this case, clustering also produces a meaningful result: the connections between the concepts that share the same actors can be seen. In a similar fashion, we obtain the matrices Concepts-Concepts (CC) and Instances-Instances (II) from the bipartite (CI) model, and Actors-Actors (AA) and Instances-Instances (II) from the bipartite Concepts-Instances model. As it can be seen, (AA), (CC) and (II) matrices have been obtained twice. The difference is that (AA) in one case reflects the connection between the actors on the concept layer, and in the other case - on the instance layer, and similarly for the matrices (CC) and (II).

3   Experiments

A corpus of 250 articles on education has been compiled for testing. We consider this amount enough for a pilot study, because we can easily check the performance of clustering and opinion measurement algorithms. An excerpt from an article on education (translated) is given in Fig. 1. In this text we can find the elements belonging to the three layers: S. Rukshin, A. Kiryanov, D. Medvedyev (Actors); Education (Concepts); gifted children, lyceums (Instances).

We consider two main operations on the obtained models: clustering (unipartite model) and opinion polarity analysis (tripartite and bipartite models).

3.1   Model Construction

Document models are built based on a corpus model, which includes 3 keyword lists forming the actor, concept and instance dimensions. The corpus is created

LYCEUM NETWORKS

In 2009, a famous Russian teacher S. Rukshin from Saint-Petersburg and his colleague A. Kiryanov developed a strategy for the education of gifted children and presented it to the President D. Medvedyev. Firstly, they proposed to create a network of federal lyceums for generally gifted children. Secondly, for the children with specific talents in mathematics, physics, etc., they suggested opening a network of lyceums under the biggest universities of the country ...

Fig. 1. A sample document (educational domain)

using keyword-based queries, where the keywords cannot be randomly selected and should represent a description of a certain topic. The three-layer model construction includes two steps: (i) identify the descriptors (keywords and keyphases denoting actors, concepts and instances within the given corpus) for each layer; (ii) parametrize texts in the three dimensions.

Identification of Descriptors

The extraction of actors and instances is a named entity recognition (NER) task. As a solution, we extract single terms on the basis of their specificity criterium using LexisTerm-I [2]. Term frequencies can be counted either in "Corpus" mode for the whole corpus or in "Document" mode for each particular document. In this implementation, LexisTerm-I proposes some candidates for the construction of actor, instance and concept lists, and an expert does the manual correction of the obtained list. At a later stage, we plan to make the procedure totally automatic using approaches described in [8–10].

Actors List. The frequency of actors in the domain-specific documents is assumed to be higher than in the general lexis, therefore, they are detected in the "Document" mode. Our education-related corpus contains $n = 50$ actor names, where there are 30 person names and 20 organization names. Each list element includes 1–5 tokens, which allows to record full names with titles (e.g., Dr. Dmitry Livanov), and full organization names (e.g., Ministry of Science and Education).

Concepts List. The terms have been extracted by LexisTerm-I for $K = 10$ in the "Corpus" mode. Each concept in the resulting list includes one or several words. The sample main concepts are as follows: education accessibility, young talents, Unified State Exam, etc.. Within the "Education" domain we distinguish $M = 10$ subtopics, for which keyword lists are manually created. The number of descriptors cannot be higher than 20, which, from our experience, is enough. The number of tokens per concept together with descriptors varies from 1 to 5.

Instances List. The instance list is constructed using the results of LexisTerm-I text processing in the "Document" mode. Each instance is represented by 3 components: (i) an object or event; (ii) object/event location; (iii) object/event date. The sample instances are the Law of Education, the decree on the creation of a lyceum network, Moscow school merging, etc. There is also a fixed list of locations: Parliament, President's Council, Moscow Government, etc. The date includes the related month and year. The resulting list includes thus the following components: $[object_1, object_2, ...object_n]$, $[location_1, location_2, ...location_n]$, and *(month, year)*. For our corpus, the list of objects and events contains $P = 20$ elements, the list of locations - $Q = 15$ elements. Each element's length is of 1–5 tokens.

Document Parametrization

Document parametrization is the representation of a document in the three dimensions of actors, concepts, and instances on the basis of the corpus model performed with the *ParamDoc-3D* program. Operations on two dimensions are performed in *ParamDoc-2D*, and on one dimension - in *ParamDoc-1D* [1]. The base tripartite model is parametrized as follows.

Actor Dimension. All actor mentions from the N-actor list of the corpus model in a document are collected. The actor mention is selected if it corresponds to the expert-established settings. For unigrams, a complete match is obviously required. In case of 2–5 token keywords, a complete or partial match (2-3 tokens) can be set.

Concept Dimension. The weight of each topic in a document is measured using a vocabulary of the corresponding descriptors. A topic with the highest weight, or topics with the weights exceeding certain threshold are selected. The calculation uses the following method [5]: (a) measure the text coverage by the vocabulary (total density of the descriptors), (b) measure the vocabulary coverage by the text (percentage of vocabulary entries' mentions), (c) measure the total coverage. Both coverages can take on a value $[0.0, 5.1]$ using a non-linear scale. The total value can range from 0 to 2. By establishing a threshold, we can sort out a set of topics present in a document.

Instance Dimension. Instance extraction is done in 3 steps: (1) search for an event or object from the list of P-instances of the corpus model, (2) search for location triggers from the list of Q-locations of the corpus model, (3) search for the mentions of month and year in the document body. Objects, events, and locations are extracted similarly to actors: total and partial matches of the 1–5 token sets are considered. As for the date, texts often contain several mentions, in which case at most 5 of them are extracted.

3.2   Opinion Mining

At the end of 2013 the Russian Federation authorities approved the Law of Education. This event was preceded by a long evaluation of the contents of the Law by the educational community, the Parliament and the Government. It was a tough debate, due to the conflict between the market-driven and the traditional approach to state education financing in the country. We have performed the opinion mining using GMDH [1,3] for the pairs from actor and instance dimensions (Table 2).

It can be seen that the negative attitude was expressed mostly towards a public organization (the Civic Chamber), and not towards the governmental

Table 2. Distribution of opinions on the Law of education

Actor	Instance	Neg	Pos.
The Civic Chamber	Law of Education	72%	18%
Ministry of Education	Law of Education	55%	45%

Table 3. Contents of the main actor and instance clusters

No.	Actor cluster contents	Instance cluster contents
1	abramov, rukshin, kovaldzhi, kuzminov, higher school of economics, international monetary fund, world bank, russian ministry of education, civic chamber of the russian federation	Strategy, law, standard, lyceums
2	malinetsky, efimov, moscow institute of physics and technology, institute of international programs	Korea, China, Singapore
3	...	...

structures (Ministry of Education). It can be easily explained, because there are many followers of unpopular reforms in the mentioned public organization.

3.3 Clustering

Clustering experiments have been carried out using the *MajorClust* method [7, 14] to group actors and instances in the concept dimension. Earlier, MajorClust proved its efficiency in short texts processing [4,12]. The results are shown in the Table 3.

The first Actor cluster includes person and organization names related to the topics "secondary school", "gifted children", "reforms financing". The second cluster contains person and organization names, related to the topics "high school", "science", "modeling". The first Instance cluster spans instances, related to lawmaking, which are in turn related to the topics "secondary school" and "gifted children". The second cluster includes all the countries, where the authorities have been actively supporting young talents in the recent years (key phrases: "gifted children", "innovations", "Asia").

4 Conclusion and Future Work

The present paper describes a document representation model, based on the tripartite model of ontologies by Peter Mika (Yahoo! Research), which can be useful for the solution of Internet Sociology tasks. The results of a pilot experiment are presented. Information on the deployment of the developed tools is provided. As the future work, we plan to do the following:

- Investigate the behaviour of the proposed model in more detail. To this end:
  - Evaluate the proposed model on a larger corpus;
  - Perform clustering in the actor and instance dimensions;
- Develop the necessary tools that will allow to:
  - Perform document parametrization automatically;
  - Visualize the results so that they can be easily interpreted.

References

1. Alexandrov, M.: Development of general methodology for analysis of public opinion of Internet-community and its application to given topics (authority, economy, corruption, etc.) on the basis of Data/Text Mining tools. Report on State project 84, RPANEPA [rus] (2013)
2. Alexandrov, M., Beresneva, D., Makarov, A.: Dynamic vocabularies as a tool for studying social processes. In: Proceedings of the 6th International Conference on Intelligent Information and Engineering Systems, ITHEA Publishing, vol. 27, pp. 88–92 (2014)
3. Alexandrov, M., Danilova, V., Koshulko, A., Tejada, J.: Models for opinion classification of blogs taken from Peruvian Facebook. In: Proceedings of the 4th International Conference on Inductive Modeling (ICIM-2013), Kyiv, Ukraine Publishing House ITRC-NASU (Ukraine) & Czech Technical University pp. 241–246 (2013)
4. Alexandrov, M., Gelbukh, A., Rosso, P.: An approach to clustering abstracts. In: Montoyo, A., Muñoz, R., Métais, E. (eds.) NLDB 2005. LNCS, vol. 3513, pp. 275–285. Springer, Heidelberg (2005)
5. Alexandrov, M., Gelbukh, A., Makagonov, P.: Evaluation of thematic structure of multidisciplinary documents and document flows. In: Proceedings of the 11th International DEXA Workshop (Database and Expert System Applications), pp. 125–129 (2000)
6. Baeza-Yates, R., Ribero-Neto, B.: Modern Information Retrieval. Addison Wesley, Boston (1999)
7. Bishop, C.: Pattern Recognition and Machine Learning. Springer, Berkeley (2006)
8. Danilova, V., Alexandrov, M., Blanco, X.: A survey of multilingual event extraction from text. In: Métais, E., Roche, M., Teisseire, M. (eds.) NLDB 2014. LNCS, vol. 8455, pp. 85–88. Springer, Heidelberg (2014)
9. Danilova V., Popova S.: Socio-political event extraction using a rule-based approach. In: Meersman, R., Panetto, H., Mishra, A., Valencia-Garcia, R., Soares, A.L., Ciuciu, I., Ferri, F., Weichhart, G., Moser, T., Bezzi, M., Chan, H. (eds.) Proceedings of the 13th International Conference on Ontologies, DataBases and Applications of Semantics (ODBASE'2014), vol 8842, pp. 537–546, Springer (2014)
10. Gelbukh, A., Sidorov, G., Guzman-Arenas, A.: Use of a weighted topic hierarchy for document classification. In: Matoušek, V., et al. (eds.) TSD 1999. LNCS (LNAI), vol. 1692, pp. 133–138. Springer, Heidelberg (1999)
11. Manning, C., Raghavan, P., Schutze, H.: Introduction to Information Retrieval. Cambridge University Press, Cambridge (2008)
12. Eissen, S.M., Stein, B.: Analysis of clustering algorithms for web-based search. In: Karagiannis, D., Reimer, U. (eds.) PAKM 2002. LNCS (LNAI), vol. 2569, pp. 168–178. Springer, Heidelberg (2002)
13. Mika, P.: Ontologies are us: a unified model of social networks and semantics. J. Web Semant. Sci. Serv. Agents World Wide Web $5(1)$, 5–15 (2007)
14. Stein, B., Niggemann, O.: On the nature of structure and its identification. In: Widmayer, P., Neyer, G., Eidenbenz, S. (eds.) WG 1999. LNCS, vol. 1665, p. 122. Springer, Heidelberg (1999)

Improving Financial Time Series Prediction Through Output Classification by a Neural Network Ensemble

Felipe Giacomel[1] (✉), Adriano C.M. Pereira[2], and Renata Galante[1]

[1] Department of Computer Science, Federal University of Rio Grande do Sul
(UFRGS), Porto Alegre, Brazil
{fsgiacomel,galante}@inf.ufrgs.br
[2] Department of Computer Science, Federal University of Minas Gerais (UFMG),
Belo Horizonte, Brazil
adrianoc@dcc.ufmg.br

Abstract. One topic of great interest in the literature is time series prediction. This kind of prediction, however, does not have to provide the exact future values every time: in some cases, knowing only time series future tendency is enough. In this paper, we propose a neural network ensemble that receives as input the last values from a time series and returns not its future values, but a prediction that indicates whether the next value will raise or fall down. We perform exhaustive experiments to analyze our method by using time series extracted from the North American stock market, and evaluate the hit rate and amount of profit that could be obtained by performing the operations recommended by the system. Evaluation results show capital increases up to 56 %.

Keywords: Artificial neural networks · Classification · Prediction · Stock markets · Time series

1 Introduction

Multilayer perceptron neural networks outperform other techniques when it comes to financial time series prediction [2], hence they are among the most used techniques to predict future values and tendencies of stock markets [1]. Nevertheless, while methodologies based on single artificial neural networks (ANNs) have been largely used for financial time series prediction, neural network ensembles are still little used in this area. [4] showed that simple ensembles can perform better generalizations than a singular ANN, and [12] that a subset of all possible ANNs can achieve better results than a single ANN.

Although ANNs can perform useful predictions of future values in financial time series, studies that focus on this subject generally also focus on minimizing the ANN error and lack a performance test, which would show how much capital would be gained if the operations recommended by the method would be followed. In fact, these studies do not recommend how to operate at all: they

© Springer International Publishing Switzerland 2015
Q. Chen et al. (Eds.): DEXA 2015, Part II, LNCS 9262, pp. 331–338, 2015.
DOI: 10.1007/978-3-319-22852-5_28

show the predicted values for a stock with a certain margin of error, but do not tell us what to do with these results. In practical terms, this kind of prediction is not enough for use in real life. Finally, methods that try to predict the exact value may not behave properly in periods of high volatility [6], so is not safe to trust them all the time. Auxiliary inputs should be added to these ANNs to help them dealing with these periods of uncertainty.

In this paper we propose the resolution of the movement prediction in financial time series by developing a neural network ensemble that classifies a set of inputs into an actual operation order. It receives as input the time series past values and returns a prediction that indicates what is going to happen to the time series in the next period: if it will raise or fall. With this result, one can be told exactly what to do: to buy or to sell a stock. We present as results the hit rate of the method, the variation of capital that one investor could have obtained if he had followed the recommended operations, and a comparison of this variation to what would have been achieved using both the buy-and-hold and naive approaches. To validate our method, we test it in a set of North American stock market time series with a daily granularity. Results in this datasets are promising, with hit rates around 60 % and a capital gain up to 56 % in 166 days.

The remainder of this paper is structured as follows: Sect. 2 gives a background of concepts and review related works. Section 3 introduces our proposed method. Section 4 presents the results obtained with the neural network ensemble. Finally, conclusions and future work are presented in Sect. 5.

2   Related Work

For stock movement prediction, [10] proposes an ANN that predicts stocks movements five days in the future and an algorithm that simulates buy and sell operations based on these predictions. The performance of the method is measured by the mean error of the network and by the capital gain when executing the recommended operations. [8] makes a very interesting study, where the focus is the prediction of the market directional movement for different combinations of inputs, and not the minimization of the error in predicted values. The best results are obtained when external time series (like the exchange rate of foreign currencies) are added to the ANN inputs. Also [9] proposes a ANN where the output tells if a stock is going to raise or fall in the next minutes, enabling the implementation of a high frequency trading system, which executes several trades on the same day. These studies have satisfactory results, but our premise is that they could be improved by a neural network ensemble.

In the ensembles subject, [3] uses a Flexible Neural Tree ensemble to predict the next values of three big stock markets and obtains results with a very low error. [11] combines market news and older prices into a set of Support Vector Machines to predict market movements, and shows that both the prediction performance and profitability of the system increase. Recently, [7] made a comparison among three types of ensembles (mean, median and mode) and a single ANN, with better results obtained by the ensembles. Although these studies improved time series prediction with ensembles, they generally do not try to predict the movements of the time series and, when they do that, the transformation of the predictions in actual trades is not their main focus.

We identified a gap in this area of research, since these two kinds of studies could complement each other. If we improved a classification system by adding it to an ensemble, a higher number of correct classifications would be made and, therefore, a higher capital gain would be achieved. Thus, this study proposes to fill this gap by implementing a neural network ensemble to classify time series output values and use them in a real trade system, showing the results of a set of simulated trades based on the ensemble's recommendations.

3 Proposed Ensemble

We propose a prediction method that ensembles two ANNs to predict the same thing — financial time series movements. The proposed ensemble is viewed by the user as a black box that receives as inputs the last opening, maximum, minimum and closing values for a stock and classifies this set of inputs into one of three possible values, which refer to the next unit of time: *raise*, *fall* or *do not know*. With this information, one trader can decide how she/he will operate in the next time period.

Internally, the ensemble is composed by two ANNs, hence called "neural networks A and B", and a merge module that combines their outputs. Both ANNs are trained to classify their inputs into one of two possible outputs: *raise* or *fall*. Figure 1 shows how the ANNs and the merge module are organized to provide a final classification. The ensemble's inputs x_1, x_2, x_n are passed directly to both A and B ANNs. These ANNs, using an activation function f, are trained to return different kinds of classifications in their outputs y_1 and y_2. Finally, the merge module is responsible to combine the y_1 and y_2 classifications and give a trade advice z, which has to be one of the ensemble's possible outputs.

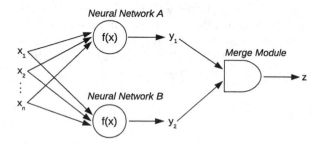

Fig. 1. Architecture of the proposed ensemble

Both A and B ANNs can have similar or different architectures (i.e., number of hidden layers and nodes, training algorithm, etc.), but must receive the same inputs and differ in the kind of output each one returns. These outputs have similar meanings, but are obtained in different ways that vary according to the type of training the ANN received. ANN A is trained to predict the expected opening and closing values for the next time period. With these values, we are able to predict if one stock will raise or fall during a specific time period. If the opening prediction is lower than the closing prediction, we interpret that the ANN is predicting a raise in the prices during the time period. The opposite comparison is made to interpret a prediction of a fall. ANN B, additionally, is trained to predict if a stock is going to raise or fall, no matter how much, returning 1 in the first output node and 0 in the second one when a raise in the prices is predicted for the next time period, or the opposite when a fall is predicted. Due to the nature of ANNs, a decimal value between zero and one is usually the output for both outputs, and never the exact values 1 and 0. What we do is consider the node with the highest value as the winner. The merge module basically works as a logical AND, since both ANNs must agree in the prediction in order to allow the ensemble to give a proper classification. When both networks predict a raise (or a fall) in the prices, the ensemble's recommends to operate according to these predictions. On the other hand, if they disagree, the ensemble gives no advice at all. In this case, we interpret this as an uncertainty period on the market, where any prediction attempt is more likely to be a bet. A stock trader should not perform any operation in this moment.

For the implementation of the networks, we used the Encog Framework [5], where we implemented two feed forward multilayer perceptrons. We chose to use as a training algorithm the resilient propagation, which differs from backpropagation as it does not use fixed values for learning rate and momentum. In fact, in resilient propagation the update on weights and bias are determined by the sign from the derivative of the weight change on each step of the training step, so these values do not have to be initially defined by the user.

In our tests, the generalization of the ANNs did not improve when we added exogenous time series, like foreign currencies and other stock markets indexes, to the input dataset. Additionally, adding financial indicators as inputs also did not impact in the ensemble's performance. This happened because the inherent nature of ANNs makes the use of these indicators unnecessary: during its training period, an ANN already learns the relation between the entries, so a redundancy is created when we input the ANN with technical indicators that are, generally, also an interpretation of the past values of one stock.

4   Evaluation

In order to evaluate our method, we first defined a methodology. Then, a group of 9 time series was assembled into one dataset. Finally, we defined the metrics used to analyze the results. The method will be considered satisfactory if it can give profit to a trader who acts according to its outputs. The higher the profit, the better the method is. Also, it is desirable that the hit rate of the method surpasses 50 %, otherwise it would not be as good as a coin toss.

4.1   Experimental Setup

To test the capability of our method to generate profit, we adopted the following methodology: if the ensemble predicts a raise in the next time period, we perform a buy operation at the beginning of this period and a sell operation at the end of this same period; Conversely, if the ensemble predicts a fall in the next time period, we perform a sell operation at the beginning of this period and buy at the end of this same period. If the ensemble cannot predict what is going to happen, we do not execute any operation and wait for the next time period.

For the data used in the experiment, we have chosen daily time series from the North American stock market. Table 1 contains the complete list of the used time series and a comparative of the number of periods that a raise or fall occurred in each time series, showing that the data is balanced in all time series. We chose the North American stock market because it is the biggest in the world, and therefore, less willing to be influenced by external factors. We collected daily data for these time series from 05/19/2008 to 01/14/2015, totaling 1,677 days of data for each time series. All sets were divided into two: one for training and one for validation of the ANNs, containing, respectively, 90 % and 10 % of the available data. We collected opening, maximum, minimum and closing prices of each day of data. All these entries used a time window of five days, totaling 20 inputs for each network. Before entering data to the ANN, we normalized the entry dataset using the classic linear interpolation method.

Table 1. Time series used in the ensemble's validation and data distribution

Symbol	Name	Raises/Falls	%
S&P 500	Standard & Poor's 500	925/752	55/45
AA	Alcoa Inc	858/819	51/49
BAC	Bank of America Corp	852/825	50/50
C	Citigroup Inc	843/834	50/50
F	Ford Motor Co	878/799	52/48
FCX	Freeport-McMoRan Inc	843/834	50/50
GE	General Electric Co	868/809	51/49
JPM	JPMorgan Chase and Co	850/827	50/50
SWN	Southwestern Energy Co	823/854	49/51

4.2   Experiments and Results

For the experiments, we implemented each ANN on the ensemble with three layers: an input with 20 entry nodes, a hidden with six nodes, and an output with two nodes. Although both networks return an output that means different things, both of them need this amount of outputs to be interpreted by the ensemble. We trained all the networks for a maximum of 50,000 iterations unless a minimum error was reached or the network stopped converging. We measured our method with two metrics: the relation of the number of hits and errors that are obtained with its recommendations, and the capital gain that would be obtained if one trader would strictly follow all of its recommendations. This last one is also compared to the capital gain of two common strategies: the *Buy and Hold*, which buys the stock at the beginning of the time period and sells it in the end of the same period, and the *Trivial Strategy*, which assumes that if a stock raised (or fell) in time period t, it will also raise (or fall) in $t+1$.

Classification Analysis. Here, we measure the amount of days the ensemble gives a prediction (i.e., an output different from *do not know*) and the distribution of raise and fall predictions. After running the ensemble in each time series, we got the results shown in Table 2. Although all time series had the same amount of inputs, the number of recommendations given by each one were not any close to each other. AA and GE time series, for instance, had above 75 % of days with recommendations. BAC and F, on the other hand, had recommendations only in less than 20 % of the days, which means that most of the times the ensemble's internal ANNs did not agree with the classification results. For time series with low prediction rates, it is preferable that the ensemble does not give a prediction at all than a wrong one. A surprising behavior that is shown in this table is that, even with the parity between raises and falls shown in Table 1, this parity does not reflects in the balance of predictions made by the ensemble.

Table 2. Summary of the classifications made by the proposed ensemble

Symbol	# Days test period	Total advices	% Days W/ advices	# Raise predictions	# Fall predictions
S&P 500	166	39	23 %	34	5
AA	166	128	77 %	13	115
BAC	166	26	15 %	25	1
C	166	101	60 %	101	0
F	166	32	19 %	31	1
FCX	166	89	53 %	33	56
GE	166	127	76 %	10	117
JPM	166	106	63 %	93	13
SWN	166	78	46 %	24	54

Performance Analysis. We measured the performance of the ensemble under real conditions. If it can profit and have a good hit rate when confronted with real financial time series, the method can be used by traders in their operations. We simulated an initial capital of $100,000.00 and compared it to the final capital we would have by always investing 100 % of our money according to the ensemble's recommendations. Moreover, we compare our results with the financial return obtained by buy-and-hold and trivial strategies. Among all results, that can be seen in Table 3, the best ones were obtained by S&P 500 time series, which had the highest hit rate and outperformed both buy-and-hold and trivial strategies. It had the lowest capital gain among all time series, but we credit this to the fact that, as we have seen in Table 2, it made a very small number of predictions. Eight out of the nine used time series had a good hit rate (above 56 %), and all of them profited. If we had used the buy-and-hold strategy in all time series, we would have profited in five of them, giving this strategy a success rate of 55 %. Trivial strategy had a better performance: it profited in six out of nine time series, with a success rate of 66 %. Our ensemble, finally, profited in all time series, with a success rate of 100 %. Also, it outperformed both comparative approaches in four time series, one of the comparative approaches in four time series, and none of the comparative approaches in only one time series (even thought it generated profit). Our ensemble has showed itself not as a tool that always gets the best profits, but instead gives good and consistent results. Investors in general prefer techniques that give them small and constant profits than a roller coaster of yields that can generate higher profits but can also lose capital frequently.

Table 3. Results for time series prediction using our ensemble and other methods

Symbol	Initial $	Final $ buy and hold	Final $ trivial	Final $ ensemble	Ensemble hit rate
S&P 500	100,000.00	100,953.44	103,204.82	106,767.06	64.71 %
AA	100,000.00	118,454.61	92,596.88	131,195.15	50.78 %
BAC	100,000.00	120,390.31	119,740.16	115,855.32	69.23 %
C	100,000.00	116,265.58	105,381.62	109,491.78	56.44 %
F	100,000.00	99,168.27	116,384.78	111,210.01	56.25 %
FCX	100,000.00	68,604.99	98,364.03	106,495.49	56.18 %
GE	100,000.00	96,193.38	103,902.88	112,967.79	58.27 %
JPM	100,000.00	116,862.75	89,417.12	113,833.18	61.32 %
SWN	100,000.00	57,965.17	159,294.49	156,903.98	60.26 %

5   Conclusion and Future Work

In this paper we proposed an ensemble of neural networks to predict the stock market movements. To accomplish that, we turned the prediction problem into a classification problem. We tested our method in the North American stock market with a daily granularity, measuring the market movements hit rate and the amount of capital that our method could profit when compared to both buy-and-hold and trivial approaches, with very satisfactory final results.

Our method could be used in the creation of new strategies for algorithmic trading in stock markets, or to perform stock portfolio management, changing stocks according to model trends prediction. Finally, we foresee several possibilities for future work. We plan to test this method by changing one or both of the neural networks to another type of machine learning techniques (like SVM or RBF neural networks). Also, we want to add a technical indicator to the ensemble, which could vote for the classification results. However, since in our tests increasing the number of items in the ensemble made the number of advices decrease, we would have to focus on the increment of advices given by each item in the ensemble before adding new elements to it.

Acknowledgments. This work is partially supported by InWeb (Brazilian National Institute for Web Research), under the MCT/CNPq grant 45.7488/2014-0 and by the authors individual grants and scholarships from CNPq (National Counsel of Technological and Scientific Development) and CAPES (Coordination for the Improvement of Higher Education Personnel).

References

1. Atsalakis, G.S., Valavanis, K.P.: Surveying stock market forecasting techniques-part ii: soft computing methods. Expert Syst. Appl. **36**(3), 5932–5941 (2009)
2. Cao, Q., Leggio, K.B., Schniederjans, M.J.: A comparison between fama and french's model and artificial neural networks in predicting the chinese stock market. Comput. Oper. Res. **32**(10), 2499–2512 (2005)
3. Chen, Y., Yang, B., Abraham, A.: Flexible neural trees ensemble for stock index modeling. Neurocomputing **70**(4), 697–703 (2007)
4. Hansen, L.K., Liisberg, C., Salamon, P.: Ensemble methods for handwritten digit recognition. In: Proceedings of the 1992 IEEE-SP Workshop on Neural Networks for Signal Processing [1992] II, pp. 333–342. IEEE (1992)
5. Heaton, J., Reasearch, H.: Encog java and dotnet neural network framework. Heaton Research Inc, 20 July 2010
6. Kara, Y., Acar Boyacioglu, M., Baykan, Ö.K.: Predicting direction of stock price index movement using artificial neural networks and support vector machines: The sample of the istanbul stock exchange. Expert syst. Appl. **38**(5), 5311–5319 (2011)
7. Kourentzes, N., Barrow, D.K., Crone, S.F.: Neural network ensemble operators for time series forecasting. Expert Syst. Appl. **41**(9), 4235–4244 (2014)
8. O'Connor, N., Madden, M.G.: A neural network approach to predicting stock exchange movements using external factors. Knowl.-Based Syst. **19**(5), 371–378 (2006)
9. Silva, E., Castilho, D., Pereira, A., Brandao, H.: A neural network based approach to support the market making strategies in high-frequency trading. In: International Joint Conference on Neural Networks (IJCNN). pp. 845–852. IEEE (2014)
10. Taylor, B., Kim, M., Choi, A.: Automated stock trading algorithm using neural networks. In: Juang, J., Chen, C.Y., Yang, C.F (eds.) Proceedings of the 2nd International Conference on Intelligent Technologies and Engineering Systems (ICITES2013), Lecture Notes in Electrical Engineering. vol. 293, pp. 849–857. Springer, Switzerland (2014)
11. Zhai, Y.Z., Hsu, A., Halgamuge, S.K.: Combining news and technical indicators in daily stock price trends prediction. In: Liu, D., Fei, S., Hou, Z., Zhang, H., Sun, Changyin (eds.) ISNN 2007, Part III. LNCS, vol. 4493, pp. 1087–1096. Springer, Heidelberg (2007)
12. Zhou, Z.H., Wu, J., Tang, W.: Ensembling neural networks: many could be better than all. Artif. Intell. **137**(1), 239–263 (2002)

Mining Strongly Correlated Intervals
with Hypergraphs

Hao Wang[1] (✉), Dejing Dou[1], Yue Fang[2], and Yongli Zhang[2]

[1] Computer and Information Science, University of Oregon, Eugene, OR, USA
{csehao,dou}@cs.uoregon.edu
[2] Department of Decision Science, University of Oregon, Eugene, OR, USA
{yfang,yongli}@uoregon.edu

Abstract. Correlation is an important statistical measure for estimating dependencies between numerical attributes in multivariate datasets. Previous correlation discovery algorithms mostly dedicate to find piecewise correlations between the attributes. Other research efforts, such as correlation preserving discretization, can find strongly correlated intervals through a discretization process while preserving correlation. However, discretization based methods suffer from some fundamental problems, such as information loss and crisp boundary. In this paper, we propose a novel method to discover strongly correlated intervals from numerical datasets without using discretization. We propose a hypergraph model to capture the underlying correlation structure in multivariate numerical data and a corresponding algorithm to discover strongly correlated intervals from the hypergraph model. Strongly correlated intervals can be found even when the corresponding attributes are less or not correlated. Experiment results from a health social network dataset show the effectiveness of our algorithm.

1 Introduction

Correlation is a widely used statistic measure for mining dependencies in multivariate data sets. Its value typically reflexes the degree of covariance and contravariance relationships in numerical data. Previous data mining algorithms focus on discovering attribute sets with high piecewise correlations between attributes. Such correlation measure shows a high level picture of the dependency profile in data, nevertheless, they can only reveal the correlations of numerical attributes in the scope of full range. The numerical data themselves are often considered as containing richer information than just the high level attribute-wise correlations. For example, in the meteorology data, the rate of precipitation is more positively (or negatively) correlated with humidity (or air pressure) especially when the humidity is large enough (e.g., when humidity $\geq 80\%$).

In this paper, we address the problem of discovering the intervals with strong correlations from numerical data. The patterns discovered are in the form of interval sets, for example, *"Humidity[20%, 30%], Precipitation[70%, 90%], Correlation 0.81"*. The correlation of the intervals, in this example, 0.81, is

Q. Chen et al. (Eds.): DEXA 2015, Part II, LNCS 9262, pp. 339–348, 2015.
DOI: 10.1007/978-3-319-22852-5_29

calculated by the correlation of all data instances that fall inside of the ranges of intervals. The strongly correlated intervals would provide us valuable insights with more detail dependencies hidden in the data. For example, in the financial market data, the demands of stocks and bonds generally raise as the prices fall. This market principle of the price and demand is the corner stone of a stable financial market. However, such principle might not hold under certain circumstances, such as a potential economic crisis. When the prices of stocks fall below certain thresholds and a crash in the stock market is triggered, the demand would fall along with the decline of the price for a certain price range. In this example, the strong correlated intervals from historical transaction data will provide investors with useful insights on when and how to avoid the risks in the financial investment.

Discretization [6] is one of the intuitive ways to generate correlated intervals on numerical attributes. Mehta et al. [9] proposed an unsupervised discretization method which preserves the underlying correlation information during the discretization process. It can serve as a preprocessing step for further data mining applications such as classification and frequent itemset mining. While it can discover strongly correlated intervals, there are some fundamental problems for discretization. For example, the *crisp boundary problem* [5] forces the discretization boundaries to make trade-offs between adjacent intervals on all attributes. Information in data may lose during the discretization as well. With regarding to the size of intervals, we usually face a dilemma to decide the quantity of segmentations. More segmentations means less information loss during the discretization process, while less segmentations will lead to large intervals that strong correlations between small intervals cannot be discovered.

In this paper, we propose a novel method to discover the strongly correlated intervals without suffering from the problems in the discretization based methods. We propose to model the numerical data with a hypergraph representation and use the average commute time distances to capture the underlying correlations. We propose a corresponding algorithm to discover the intervals with high piecewise correlations. One strength of our algorithm is that the discovery of intervals and their correlation optimization are achieved in a single step. Each boundary of the intervals are optimized independently. Therefore, they would not suffer from the crisp boundary problem or information loss problem.

The correlation measure we use in this paper is the Spearman's rank correlation coefficient [4]. The Spearman's rank correlation coefficient is defined as:

$$\rho = \frac{\sum_i (x_i - \bar{x})(y_i - \bar{y})}{\sqrt{\sum_i (x_i - \bar{x})^2 \sum_i (y_i - \bar{y})^2}},$$

in which $\bar{x}$ and $\bar{y}$ stand for the average order ranks of attribute x and y respectively. Based on the Spearman's correlation coefficient, we also propose the correlation gain and normalized correlation to evaluate our approach from different perspectives.

The correlation gain is defined as the ratio between the correlation of the intervals and the correlation of the related attributes,

$$\rho_{gain} = \frac{\rho}{\rho_{att}},$$

in which ρ_{att} is the correlation between related attributes. High correlation gain implies that even though the two attributes are less correlated in general, strong correlations might still be found between intervals on the attributes. The intervals with high correlation gain are valuable because it can reveal the strongly correlated intervals that hide under the less correlated attributes.

The normalized correlation is an estimation of correlation which depends on both the underlying correlation priori and the number of data instances available. Statistically, the estimation deviation has a negative relationship with the number of instances [4]. For example, for a dataset with only 2 data instances the correlation is always ± 1 with infinite deviation, while for a dataset with infinite number of instances, the estimation converges to the true correlation value with 0 deviation. In this paper we also introduce normalized correlation ρ_{norm} [11] to generate strongly correlated intervals. It is defined as

$$\rho_{norm} = \rho\sqrt{n},$$

in which ρ is the correlation estimation and n is the number of data instances used for the estimation. The normalized correlation makes trade off between the estimated value and the estimation accuracy. The intervals with too less data instances will result in low ρ_{norm} due to the large deviation and so does the intervals with many data instances such as the full range intervals due to lower correlation value. It also relieves the effort to pre-define a threshold for selecting certain intervals as "strongly correlated."

Our main contributions in this paper are:

– We propose a hypergraph random walk model to capture the underlying correlation of the numerical data. This model can capture the correlation relationship at the interval level rather than at the attribute level using a measure based on average commute time distance.
– We propose an algorithm to discover strongly correlated intervals based on the hypergraph random walk model. We are able to discover strongly correlated intervals with high accuracy without suffering from the information loss and crisp boundary problem in the discretization based methods.
– We propose the normalized correlation to generate strongly correlated intervals without a pre-defined threshold. We also propose the correlation gain to find the highly correlated intervals even the corresponding attributes are less correlated.
– We conduct experiments in a health social network dataset and the results show the effectiveness of our algorithm.

The rest of this paper is organized as follows: we give a brief introduction of related works in Sect. 2. We make a detailed description of our method in Sect. 3. We report experiment results in Sect. 4. We conclude the paper in Sect. 5.

2   Related Work

Previous research efforts have proposed various methods for the discovery of correlated intervals. Discretization is the one of the most straight forward way to generate intervals for various optimization goals. Kotsiantis and Kanellopoulos [6] made a thorough survey of the discretization techniques. These discretization methods target different optimization goals, such as minimizing the information loss, maximizing the entropy etc. Mehta et al. [9] proposed a PCA based unsupervised discretization method which can preserve the underlying correlation structure during the discretization process. The discretization process serves as an independent process which can be fed into many other data mining tasks such as association mining and clustering.

Quantitative association mining is another technique related to discovering strongly correlated intervals. The quantitative association mining is an extension of the traditional association mining. It generates intervals on the numerical data instead of the categorical data. Srikant and Agrawal [10] first proposed an algorithm that deals with numerical attributes by discretizing numerical data into categorical data. Fukuda et al. [12,13] proposed several methods that either maximize the support with pre-defined confidence or maximize the confidence with pre-defined support by merging up adjacent instance buckets. However, the support and confidence measures were attested to be inadequate to discover strongly correlated intervals due to the *catch-22* problem [10] and the crisp boundary problem [2] as well.

3   Discovering Strongly Correlated Intervals with Hypergraphs

In this section, we present our hypergraph based method which can efficiently discover strongly correlated intervals. We propose a hypergraph model to represent the correlation structure of numerical data. The correlation measure is captured by the average commute time distance between vertices in the hypergraph model.

3.1   Hypergraph and Average Commute Time Distance

Hypergraph is a generalization of regular graph that each edge is able to incident with more than two vertices. It is usually represented by $G = (V, E, W)$, in which V, E and W are the set of vertices, edges and weights assigned to corresponding edges respectively. The incident matrix of a hypergraph G is defined by H in which

$$H(v, e) = \begin{cases} 1 \; if \; v \in e \\ 0 \; if \; v \notin e \end{cases} \tag{1}$$

Zhou et al. [14] generalized the random walk model on hypergraph and defined the average commute time similarity S_{ct} and the Laplacian similarity S_{L_+}. The

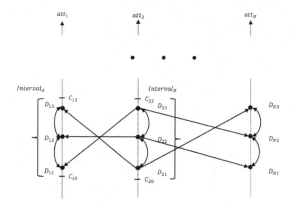

Fig. 1. Hypergraph representation of numerical data

average commute time similarity $n(i,j)$ is defined by

$$n(i,j) = V_G(l_{ii}^+ + l_{jj}^+ - 2l_{ij}^+), \tag{2}$$

where l_{ij}^+ is the ith and jth element of matrix $\mathbf{L}^+$, $\mathbf{L}$ is the hypergraph Laplacian:

$$\mathbf{L} = \mathbf{D}_v - \mathbf{H}\mathbf{W}\mathbf{D}_e^{-1}\mathbf{H}^T, \tag{3}$$

and $\{.\}^+$ stand for Moore-Penrose pseudoinverse. D_v and D_e denote the diagonal matrix containing the degree of vertices and edges respectively. $V_G = tr(\mathbf{D_v})$ is the volume of hypergraph. The average commute time distance is defined by the inversion of normalized average commute time similarity [7]. As mentioned in [8] the commute-time distance $n(i,j)$ between two node i and j has the desirable property of decreasing when the number of paths connecting the two nodes increases. This intuitively satisfies the property of the effective resistance of the equivalent electrical network [3].

3.2 Hypergraph Representation of Numerical Data

As illustrated in Fig. 1, it shows an example of numerical data with values sorted in the ascending order on attributes. Let $A = \{att_1, att_2, ..., att_M\}$ be the set of attributes, in which M is the number of attributes. The jth value of the ith attribute att_i is denoted as D_{ij}. The boundary candidates $c_{i1}, c_{i2}, ... , c_{i(N-1)}$ are the averages of each two adjacent values on the corresponding attribute att_i. The boundaries of intervals are defined on these boundary candidates. The set of attributes, instances and intervals are denoted as S_{att}, S_{inst} and S_{inter} respectively. Based on the sorted numerical data, we further build the data representation with a hypergraph model. Each data value D_{ij} corresponds to a vertex in the hypergrah. Each pair of vertices with adjacent values, D_{ij} and $D_{i(j+1)}$, are connected through a hyperedge with a weight proportion to the

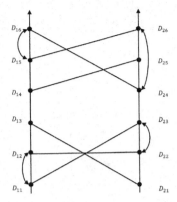

Fig. 2. Relationship between correlation and average commute time distance

inversion of the distance between them. The vertices in the same data instance with value D_{i1}, D_{i2},..., D_{iM} are connected by a hyperedge as well.

As shown in Fig. 2, the data instances below are strongly negative correlated and the data instances above are less strongly correlated. The non direct random walk paths on strongly correlated data instances, for example, $D_{11} \rightarrow D_{23} \rightarrow D_{22} \rightarrow D_{12}$, is shorter than the path on loosely correlated data instances, for example, $D_{15} \rightarrow D_{26} \rightarrow D_{24} \rightarrow D_{16}$, because the corresponding data values on the other correlated attributes are closer as well. In this case, the average commute time distance between the strongly correlated vertices is relatively shorter than the not strongly correlated vertices. For the reason above, the average commute time distance from random walk model is capable of capturing the correlation measures in our problem.

3.3   Algorithm Description

Based on the hypergraph model, we propose an algorithm for discovering strongly correlated intervals in numerical data. The pseudo code of our correlated interval discovery algorithm is shown in Algorithm 1. The algorithm first builds up the hypergraph model as described in Sect. 3.1. Adjacency matrix H is built up for the hypergraph. In Function Interval_Set_Discovery, the Laplacian matrix is computed from Eq. 3. The average commute time distance matrix is generated with entry i, j according to Formula 2. The distance between the adjacent data instances are the inversion of the average commute time similarity. In Function Merge_Interval, a bottom up mining process is applied simultaneously on all the attributes. For each attribute, an interval i_k is initialized with boundary $[c_i, c_{i+1}]$ for node n_k, i.e., exactly one node for each interval. A distance matrix C_i is maintained for each attribute, the distances between intervals are initialized as the average commute time distance for the node corresponding to this interval. In each iteration, for each attribute, the algorithm looks up the minimum distance between the adjacent intervals i_k and i_{k+1}, then merges these

---

Algorithm 1. Correlated Interval Discovery Algorithm

Function: Interval Set Discovery
Initialize: $\mathbf{C} = \mathbb{R}_{M \times M}$
for i, j = 1 to M **do**
$\quad \mathbf{C}_{i,j} = V_G(l_{ii}^+ + l_{jj}^+ - 2l_{ij}^+)$
end for
while mergeable **do**
$\quad$ **for** i, j = 1 to M **do**
$\quad\quad$ Merge Interval
$\quad$ **end for**
end while

Function: Merge Interval
Initialize: $D = C$
while intervals generated **do**
$\quad k = min\_index(D_{k,\ k+1}, \ k \in 1, \ ..., \ M-1)$
$\quad i_k = \text{Merge}(\boldsymbol{i}_k, \ \boldsymbol{i}_{k+1})$
$\quad$ **for** k = 1 to M-1 **do**
$\quad\quad D_{i,k} = D_{i,k} + D_{i+1,k}$
$\quad\quad D_{k,i} = D_{k,1} + D_{k,i+1}$
$\quad$ **end for**
end while

---

two intervals. The distance matrix is updated accordingly. In every few iterations, for each generated interval, we scan the corresponding intervals on the rest attributes. The normalized correlation is calculated for these pairs of intervals. The intervals are used to update the final result, and only the top k correlated interval sets with best normalized correlations are kept in set. If the correlation metrics of certain intervals are above the user pre-defined threshold, such as the $Interval_A$ and $Interval_B$ in Fig. 1, then the two intervals are combined into one attribute/interval set $\{S_{att}, S_{inter}\}$ and an interval set is generated. The mining process continues to generate interval sets till intervals on all attributes merges into full ranges.

4 Experimental Results

We evaluate our method on a real life health social network dataset. SMASH [1] is the abbreviation of Semantic Mining of Activity, Social, and Health Data Project. The dataset collected in this project include social connections and relations, physical activities, and biomarkers from 265 overweight or obese human subjects. After preprocessing, the input data in our experiment contain the following attributes for the physical activities and biomarkers. The physical activity indicator *Ratio No.Steps* is the ratio of steps that a human subject walked through in two consecutive periods of time. Three biomarkers *HDL*, *LDL* and *BMI* are used for the health condition indicators. The *HDL* and *LDL* stand for the high density lipoprotein and low density lipoprotein respectively. The rate of

Table 1. Experiment results from the SMASH data

(a) Top Five Rules from Correlation Preserving Discretization

Attribute Set	Correlation	Correlation Gain	Normalized Correlation
Ratio LDL[0.74, 0.85] Ratio No.Steps[0.32, 0.89]	0.62	13.70	1.68
Ratio BMI[0.93, 1.01] Ratio No.Steps[0.89, 1.21]	0.82	5.13	1.31
Ratio LDL[0.99, 1.09] Ratio BMI[0.93, 1.01]	0.81	2.86	0.73
Ratio BMI[1.01, 1.12] Ratio LDL[0.99, 1.09]	0.78	2.75	1.16
Ratio BMI[1.01, 1.12] Ratio LDL[1.09, 1.31]	0.64	2.25	1.34

(b) Top Five Rules from Hypergraph Based Method Ranked by Correlation Gain

Attribute Set	Correlation	Correlation Gain	Normalized Correlation
Ratio HDL[0.97, 1.26] Ratio BMI[0.89, 1.00]	0.94	45.22	1.67
Ratio HDL[0.79, 0.91] Ratio No.Steps[0.78, 3.94]	0.57	28.15	2.51
Ratio LDL[1.00, 1.02] Ratio No.Steps[0.39, 3.82]	1.00	22.11	2.01
Ratio HDL[0.78, 1.25] Ratio No.Steps[0.27, 3.96]	0.37	18.41	3.85
Ratio LDL[0.88, 1.02] Ratio No.Steps[0.01, 3.82]	0.72	16.12	1.65

(c) Top Five Rules Ranked by Normalized Correlation

Attribute Set	Correlation	Correlation Gain	Normalized Correlation
Ratio LDL[0.70,1.00] Ratio HDL[0.86,1.00]	0.55	2.01	5.75
Ratio LDL[1.00,1.18] Ratio HDL[1.00,1.14]	0.53	1.93	4.99
Ratio LDL[0.69,0.93] Ratio BMI[0.97,0.99]	0.63	21.01	4.14
Ratio LDL[0.92,1.05] Ratio No.Steps[1.17,2.31]	0.56	10.68	3.87
Ratio HDL[1.02,1.08] Ratio No.Steps[1.20,1.91]	0.79	194.41	3.71

HDL usually relates with decreasing rate of heart related disease and the reverse case for *LDL*. The *BMI* stands for body mass index which is a common indicator of the obesity level.

In Table 1 we list our experiment results from the SMASH dataset. The interval sets in Table 1(a) and 1(b) are results from the correlation preserving discretization [9] and our hypergraph based method respectively. Comparing the results in the two tables, our algorithm not only returns intervals with higher correlations, but also higher correlation gains. Note that the interval set discovered by our algorithm has the ability to overlap with each other. For example, the second and third interval sets in Table 1(b) *"Ratio HDL[0.79, 0.91], Ratio No.Steps[0.78, 3.94]"* and *"Ratio LDL[1.00, 1.02], Ratio No.Steps[0.39, 3.82],"* the intervals on *Ratio No.Steps* has a large overlapping between 0.78 and 3.82. On the contrary, the interval set found by correlation preserving discretization clearly suffers from the *crisp boundary problem*. Note the boundaries of intervals on attribute *Ratio BMI* in the last four interval sets in Table 1(a) only have two choices, *Ratio BMI[0.93, 1.01]* and *Ratio BMI[1.01, 1.12]*. The decision of the boundaries on each attribute has to take into consideration of the correlations with all other attributes. The trade-off in the discretization methods makes them hard to make an optimization for every interval set discovered in the data. Therefore, intervals found by the correlation preserving discretization method are suboptimal.

Strongly correlated intervals provide us interesting information with regarding to relationships between the health condition of cardiovascular system and obesity. As we mentioned before, as a good health indicator, *HDL* is usually expected to have a strong correlation with other health condition factors such as *BMI*. However, the correlation between the two attributes, *HDL* and *BMI*, will be mostly ignored because the correlation is not large enough to raise any attention. The interval sets discovered by our algorithm show that, when *Ratio BMI* changes under the moderate range, say close to 1.00, the correlations are much larger than in the rest conditions. It indicates that with regarding to the weight variations, no matter increasing or losing weight, the first few pounds might be ones that affect the health condition most. On the other hand, it also indicates that drastic exercises with a rapid weight losing rate will be likely, on the contrary, result in a deterioration of cardiovascular health condition.

Table 1(c) shows the results when we use normalized correlation as the correlation measure. Note that the last interval set in Table 1(c) which has fairly high correlation gain does not show up in Table 1(a) and 1(b). This interval set is not discovered in the first two experiments because the sizes of intervals are not above the user defined threshold. The normalized correlation renders us the potential to find the intervals that is below the user defined threshold. This indicates the fact that even the situation is rare, *HDL* and *Ration No.Steps* has a positive correlation when the amount of exercise increases drastically. Although generally exercises do not make drastic changes on *HDL*, in the situation when the subject changes the amount of exercises drastically, such as at the beginning of weight reducing program, it will result in a greater change of *HDL*.

5   Conclusions

We present a novel algorithm for discovering strongly correlated intervals from numerical data. Previous research either dedicates to discover the correlated attribute sets from the full ranges of data or uses discretization methods to transform the numerical data before the pattern discovery. These methods, however, suffer from the information loss or crisp boundary problems. The method we proposed in this paper can discover strongly correlated intervals from the less correlated attributes. These discovered intervals are not only strongly correlated but also have independently optimized boundaries with regarding to the correlation measures. Experiment results in a health social network dataset show the effectiveness of our method.

Acknowledgment. This work is supported by the NIH grant R01GM103309. We thank Brigitte Piniewski and David Kil for their input.

References

1. Semantic Mining of Activity, Social, and Health data, AIMLAB, University of Oregon. http://aimlab.cs.uoregon.edu/smash/. Accessed June 2015
2. Aumann, Y., Lindell, Y.: A statistical theory for quantitative association rules. In: ACM International Conference on Knowledge Discovery and Data Mining, pp. 261–270 (1999)
3. Doyle, P., Snell, L.: Random walks and electric networks. Appl. Math. Comput. **10**, 12 (1984)

4. Freedman, D., Pisani, R., Purves, R.: Statistics. W.W. Norton & Company, New York (2007)
5. Ishibuchi, H., Yamamoto, T., Nakashima, T.: Fuzzy data mining: effect of fuzzy discretization. In: IEEE International conference on Data Mining, pp. 241–248 (2001)
6. Kotsiantis, S., Kanellopoulos, D.: Discretization techniques: a recent survey. Int. Trans. Comput. Sci. Eng. 32(1), 47–58 (2006)
7. Liu, H., LePendu, P., Jin, R., Dou, D.: A hypergraph-based method for discovering semantically associated itemsets. In: IEEE International Conference on Data Mining, pp. 398–406 (2011)
8. Lovász, L.: Random walks on graphs: a survey. Combinatorics, Paul erdos is eighty 2(1), 1–46 (1993)
9. Mehta, S., Parthasarathy, S., Yang, H.: Toward unsupervised correlation preserving discretization. IEEE Trans. Knowl. Data Eng. 17(9), 1174–1185 (2005)
10. Srikant, R., Agrawal, R.: Mining quantitative association rules in large relational tables. In: ACM International Conference on Management of Data, pp. 1–12 (1996)
11. Struc, V., Pavesic, N.: The corrected normalized correlation coefficient: a novel way of matching score calculation for lda-based face verification. In: International Conference on Fuzzy Systems and Knowledge Discovery, pp. 110–115 (2008)
12. Takeshi, F., Yasuhido, M., Shinichi, M., Takeshi, T.: Mining optimized association rules for numeric attributes. In: Symposium on Principles of Database Systems, pp. 182–191 (1996)
13. Takeshi, F., Yasukiko, M., Shinichi, M., Takeshi, T.: Data mining using two-dimensional optimized association rules: scheme, algorithms, and visualization. In: ACM International Conference on Management of Data, pp. 13–23 (1996)
14. Zhou, D., Huang, J., Schölkopf, B.: Learning with hypergraphs: clustering, classification, and embedding. In: Advances in Neural Information Processing Systems, pp. 1601–1608 (2007)

WWW and Databases

Diagnosis Model for an Organization Based on Social Network Analysis

Dongwook Park[✉], Soungwoong Yoon, and Sanghoon Lee

Korea National Defense University, 33 Jae 2 Jayu-ro, Deokyang-gu, Goyang-si,
Gyeonggi-do 412-170, Korea
redrumdw@naver.com, ysw1209@empas.com, hoony@kndu.ac.kr
http://www.kndu.ac.kr

Abstract. Human resources are one of the most important elements of an organization. Previous studies focused on analyzing the character of members based on sociological ideas e.g. questionnaires and interviews to diagnose an organization. However, answering detailed reports is a time-consuming process, has potential for concealment, and struggles to display organizational level problems. We propose a *multi-facet diagnosis model* based on both sociological questionnaires and social network analysis to reveal relationships among organization members. We present a light-weighted questionnaire guided by psychological studies and propose SNA-based algorithms to detect specific members, such as those at-risk and leaders, which are meaningful when diagnosing an organization. Experimental results show that the proposed method covers the core of psychological diagnosis results and can observe specific members.

Keywords: Organization diagnosis · SNA · Visualization

1 Introduction

People are networked in society, and naturally they have formed groups to work together and share their daily lives to achieve common goals. We call this type of group an *organization*, and hierarchical structure is common-place for managing an organization. An organization needs components such as resources and rules, and human resources are the most valuable assets from the viewpoint of sociology. Organization members have their own roles in order to achieve common goals, which can be an expression of their relationships. Therefore, managing an organization means checking not only members but also their relationships.

Managing members' relationships to diagnose an organization has been underestimated. Past research to reveal the importance of human resources has analyzed personal attributes [7,19], categorizing their roles [10], or context of one's personal environment [12], which are based on self-diagnostic tests, such as voluntarily filled questionnaires and/or personal interviews. Thus, well-generated questions and sincere answers are considered to be the best means to gather personal data which guides proper organization diagnosis, so studies in this field have focused on generating appropriate questions.

© Springer International Publishing Switzerland 2015
Q. Chen et al. (Eds.): DEXA 2015, Part II, LNCS 9262, pp. 351–358, 2015.
DOI: 10.1007/978-3-319-22852-5_30

However, these methods have several points of concern, such as: (a) Answering questionnaires is an arduous and time-consuming task. Moreover, analyzing them may present difficulties to explain an organizational level state. (b) It may encounter reluctance in revealing one's private concerns, such as mental conditions, even though detecting these are critically important. (c) The potential for concealment. Members who do not want to open themselves to scrutiny could provide intentionally misleading answers.[1]

In this research, we suggest the novel organization diagnosis model in collaboration with sociology and social network analysis (SNA) to solve the problems of an organization through revealing relationships, called the *multi-facet diagnosis model*. Through analyzing members' relationships in an organization, we attempt to construct members' personal link networks and visualize organizational level analysis results. For this, we generate a light-weighted questionnaire guided by psychological studies and propose an algorithm to identify specific members, such as those at-risk and leaders, who make significant contributions in order to locate and solve problems within an organization.

2   Related Work

Detecting problems in an organization through networking has been widely studied in the sociology and physics field [11,15]. In accordance with the concept of an organization, A community means a closed and localized society [18]. On computer science, a community is regarded as a sized network or a neighborhood structure. The key issue is detecting relevant communities to solve their problems. Many studies to detect a community predict link tendency using local neighborhood structure [20], clustering category [14], and a link-based structural analysis method for a community [1].

In Social Network (SN) studies, member relationships of an organization are considered connections among nodes in networks [3]. From the research of friendships within a class [2] and based on graph theory [18] to analyzing sites on the Web, numerous SN studies focused on revealing structures of SN and their problems [4,8,13,16,17]. Even though most past research was about the human community, a few studies were about the connection of a real community and problems [4,9], so detecting real problems through SN is just a starting point.

However, earlier studies have focused on existing networks and strongly relied on huge data sources and the possibility of finding optimal solutions [5]. Few studies paid attention to real networks in which each node's analysis is important. We attempt to demonstrate implicit networks in physically connected organizations, which can be easily expanded toward any type of real networks. We also adapt SNA methodology to estimate members by only their relationships, so we show a way of assuming that problems of real networks can be projected onto virtual networks, such as SN.

---

[1] Reports of criminal acts such as 'A Murder Committed by One Who Seemed so Kind', 'A Bullied Student', or 'Sexual Harassment in the Office' show the difficulty in identifying suspects through the analysis of superficial facts and self-diagnosis.

3   Multi-facet Diagnosis Model

SN theory assumes that members network with one another. However, we can hardly obtain such links because they are invisible. To project extensive physical relationships of an organization onto network structure, such as SN, our model has an assumption.

Assumption 1. *In an organization, members are linked by various strengths.*

We try to connect each member as a node and measure their link strength. We propose a two-step link generation model. First, we survey members with a simplified questionnaire arranged by sociology. Next, we form SNs through analyzing and evaluating their answers by proposed algorithms adapted from SNA.

3.1   Survey

For surveying, we adapt the sociological questions used to investigate bullying in middle schoolers, which are simplified to fit our research objective following three rules: (a) Assess members' contexts and mental state using as few questions as possible. (b) Ask not for oneself, but for other members. This idea is from the assumption that people respond more openly about others than themselves. (c) Do not assign guilt to members.

Following these rules, we carefully chose questions and revised them several times with the assistance of a researcher with expert experience in large-scale surveys and interviews. Then, questions are corrected to conform to the rules and receive additional information regarding members and an organization. Finally, we chose 9 revised questions which are relatively simple[2], and the summary of the questionnaire is shown as Table 1.

Table 1. Questionnaire summary

Feature	# of Q	Example
Friendship	3	People you rarely talk with?
Leadership	3	People whose advice you seek for personal matters?
Knowledge	2	People who can easily be sought for information about one's duties?
Amusement	1	People who lead the group in entertainment performances?

Organization members are required to answer each question by designating a max. of 7 members. To enable ease of response and to qualify their answers,

---

[2] Ordinarily 32 questions (Career guidance test, http://www.career.go.kr/cnet/front/eng/eng_home_new.do/) to 567 items (MMPI, http://psychcentral.com/lib/minnesota-multiphasic-personality-inventory-mmpi/0005959/) are needed to diagnose personal character.

Algorithm 1. Identifying at-risk members and leaders

Require:

1: Member nodes(m): $M = \{m\}$

2: Feature(f): $f1, f2, f3, f4$

3: Question Types(T): Positive(P), Negative(N)

4: $Degree_{ik}^{fT} = \sum_{j=1}^{M} Z_{ijk}^{fT}$
$$\triangleright\, i, j, k \in M,\, i \neq j \neq k,\, |f| = F,\, T = \{P, N\}$$

5: $DegreeCentrality_i^{fT}(C_i^{fT}) =$
$\sum_{j=1}^{M}(Z_{ij}^{fT} + Z_{ji}^{fT}) / \sum_{i=1}^{M}\sum_{j=1}^{M}(Z_{ij}^{fT})$
$$\triangleright\, i, j \in M,\, i \neq j,\, 0 \leq C_i^{fT} \leq 1$$

    procedure FOR AT-RISK MEMBERS
        Degree Detection
        for M **do**
            if $\sum_{f=1}^{F} Degree_{mk}^{fN} \geq T_X$ AND $\sum_{f=1}^{F} Degree_{mk}^{fP} \leq T_y$ **then** choose
            end if
        end for
        Return $\{m\}$
    end procedure

    procedure FOR LEADERS
        Degree Centrality Detection
        $maxC = 0,\, maxM = 0$
        for M **do**
            if $\sum_{f=1}^{F} C_m^{fP} \geq maxC$ AND $\sum_{f=1}^{F} C_m^{fN} < T_z$ **then**
                $maxC = \sum_{f=1}^{F} C_m^{fP}$
                $maxM = m$
            end if
        end for
        Return $maxM$
    end procedure

---

a list of all members is provided on the first page of questionnaire. We do not limit the time available to answer. And though there is the rule of answer up to 7 members, they can leave answers blank if there is no suitable person.

3.2 Generating Social Networks

Using survey results, we attempt to locate real relationships among members in an organization in order to project relationships as virtual *links* and members as *nodes*. Following Assumption 1, we regard the names designated in each question as links from responders to designated members with feature tags. As a result, we can generate SN-like networks among organization members.

Degree and centrality are significant to analyze given SNs. Members who possess many degrees are regarded as important subjects. Concurrently, individuals with few degrees are suspected of being at-risk members [6]. We propose algorithm 1 of functional bi-directional degrees and centrality analysis methods, which has functions as follows.

Identifying at-risk members: We set links among nodes as dependent variables and SN from a survey as independent variables. Following the type of questions, each member's positive and negative SN are generated. Then we identify specific members who have dependent variables less than the given threshold (T_y) in positive SN and more than the given threshold (T_x) in negative SN as highly at-risk members.

Identifying leaders: We attempt to locate a member of abundant in-links through calculating degree centrality of each member on positive and negative SN, and then find one who possesses the topmost centrality score in positive SN under the condition in which the centrality score in negative SN is less than the given threshold (T_z).

4   Experiment

We selected ROK Army units as examples of an organization to test our model. Unit members share their daily lives as well as engage in regular training together, and their roles and duties are separated by military hierarchy toward their unit's goals. The size of test units is about 70 to 200 members. When surveying them, we focus on nothing but relationships among members and do not aggregate any personal data, such as origin, scholastic ability, age, parents' job, and position etc., which could be regarded sensitive and private.

ROK Army has an annual psychological test program to evaluate soldiers' mental state. Our questionnaire is similar to its evaluation forms to alleviate any resistance from testers towards taking the test. Average survey time is around 15 min, which is only 12.5 % of the existing psychological test. After the survey,

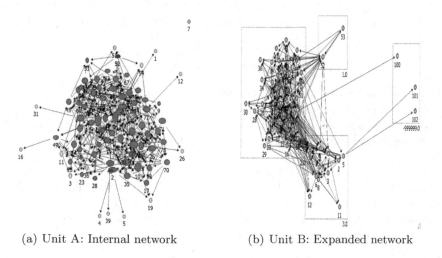

(a) Unit A: Internal network   (b) Unit B: Expanded network

Fig. 1. Visualized networks of test units

we analyze and visualize the organization's networks and all members' relationships as links using a SNA tool, Netminer version 4.2[3]. Figure 1 visualizes SN
of sample organizations. A dot represents a member, and the edges indicate
links. We determine that SN of organizations structurally features the scale-free
network and follows the power-law.

Figure 1(a) shows a number of isolated nodes on the boundary, and we can
state them to be at-risk members. In particular, node 7 has no connection with
others, and node 52 has no in-link, which means they are potentially at-risk and
we need to observe them to manage an organization. Figure 1(b) displays the
visualization of unit B clustered by members' military hierarchy. In this, +100
nodes are in other units. Nodes 7, 19, and 35 appear to be core members, while
node 5 appears to be isolated, and Node 5 is connected with members of another
organization. We can assume that node 5 may offer the possibility of difficulty.

To evaluate our model, we compared our analysis results to the results of the
annual psychological test. The 4-level danger grade (D.Grade) is given (consisting of sub-levels of A, B, C, and nothing) to ROK soldiers based on a personal
psychological survey to prevent possible accidents and find suitable tasks for
them. In particular, the highest D.Grade members (A grade) are judged after
careful observation by both commanders and doctors for 3+ months.

Reversed Novelty (RN) scoring scheme is new measurement criteria to estimate our model's efficiency. The objective of RN is how we can locate the most
at-risk members of given organizations, so we weight D.Grade and service period
because the stress of adapting generally would decrease as service time passed.
Given this reason, in ROK army, it is generally regarded as true that a soldier
would totally adapt to their organizational environment when he is promoted
to a 25 % higher level of the military hierarchy because at that time his service
period must be over 16+ months and he will complete his duty in a short time.
Therefore, if second year soldiers are still at-risk members, we weight them more
since it is more objective evidence of long-term observation and psychological
test. It is as follows

$$RN = \sum_i \beta_i * \frac{Y_i * \alpha_{Y_i}}{X_i * \alpha_{X_i}} \tag{1}$$

X : Members at-risk by psychological survey
Y : Members at-risk by proposed method
α : D.Grade weight (A : 1 / B : 0.5 / C : 0.3)
β : Service period weight (Year 2: 1 / Year 1: 0.5)

Table 2 compares the at-risk member list of the psychological test and the list
of our model. RN score is 0.81 and our model can detect the highest D.Grade
members. The average RN score of all test units is 0.74, and we can detect
all highest D.Grade members, which implies that our model covers the core of
psychological survey.

---

[3] http://www.netminer.com/.

Table 2. Detecting at-risk members: test unit A

Psychological test results			Multi-facet diagnosis model results		
Node ID	Name	D.Grade	Node ID	Name	D.Grade
7	SGT. Woo	A	7	SGT. Woo	A
51	CPL. Yi	B	51	CPL. Yi	B
59	PFC. Jin	B	26	CPL. Jung	-
68	PFC. Yun	B	52	PFC. Yi	-
28	CPL. Baek	C	28	CPL. Baek	C
12	CPL. Song	C	12	CPL. Song	C

5 Conclusion and Future Work

In this paper, we suggest a novel multi-facet diagnosis model collaborated with sociology and SNA to reveal relationships, so we can evaluate the contexts of members in an organization. We attempt to make inter-personal data analyzable links through a simple questionnaire and SNA-based algorithms, so we can locate specific members who are core for managing an organization. Experimental results show that proposed the model can reveal at-risk members in relationship networks, and achieves meaningful RN scores compared to existing psychological analysis methods. And with our model, we can obtain useful data of an organization in which varying manner of incidents occurs. Then we could offer prompt information for managers to prevent such incidents when there are at-risk patterns in organizations.

In many ways, Our study may offer a starting point to understand the real world consisting of people's relationships. The objectivity of our model and the RN score must be improved by unifying other context data, such as members' movement logs or other data. Implementing our model in a real-time diagnosis system is also important because at-risk individuals could cause difficulties whenever they desire. In the future, we seek to improve our model through both sociology and network theory for the problem of how we can virtually connect people who are physically linked. A new concept called *networked community* will be needed to outperform existing solutions.

References

1. Ahn, Y.Y., Bagrow, J.P., Lehmann, S.: Link communities reveal multi-scale complexity in networks. Nature **466**, 761–764 (2010)
2. Almack, J.C.: The influence of intelligence on the selection of associates. School Soc. **16**, 529–530 (1922)
3. Barabsi, A.: Linked: The New Science of Networks. Perseus Pub., New York (2002)
4. Cao, B., Yin, J., Deng, S., Wang, D., Wu, Z.: Graph-based workflow recommendation: on improving business process modeling. In: Proceedings of the 21st ACM International Conference on Information and Knowledge Management, CIKM 2012, pp. 1527–1531. ACM, New York (2012)

5. Devooght, R., Mantrach, A., Kivimäki, I., Bersini, H., Jaimes, A., Saerens, M.: Random walks based modularity: application to semi-supervised learning. In: Proceedings of the 23rd International Conference on World Wide Web, WWW 2014, pp. 213–224. ACM, New York (2014)
6. Freeman, L.C., Borgatti, S.P., White, D.R.: Centrality in valued graphs: a measure of betweenness based on network flow. Soc. Netw. **13**(2), 141–154 (1991)
7. Halim, S., Zeki, A.: Manpower model for human resource planning management. In: 2012 International Conference on Advanced Computer Science Applications and Technologies (ACSAT), pp. 336–339 (2012)
8. Hsu, W.H., King, A.L., Paradesi, M.S.R., Pydimarri, T., Weninger, T.: Collaborative and structural recommendation of friends using weblog-based social network analysis. In: Computational Approaches to Analyzing Weblogs, Papers from the 2006 AAAI Spring Symposium, Technical report SS-06-03, pp. 55–60, Stanford, California, USA, 27–29 March 2006 (2006)
9. Kim, Y., Hau, Y., Song, S., Ghim, G.: Trailing organizational knowledge paths through social network lens: integrating the multiple industry cases. J. Knowl. Manage. **18**(1), 38–51 (2014)
10. Legato, P., Monaco, M.F.: Human resources management at a marine container terminal. Eur. J. Oper. Res. **156**(3), 769–781 (2004)
11. Leszczensky, L., Pink, S.: Ethnic segregation of friendship networks in school: testing a rational-choice argument of differences in ethnic homophily between classroom- and grade-level networks. Soc. Netw. **42**, 18–26 (2015)
12. Ordiz, M., Fernndez, E.: Influence of the sector and the environment on human resource practices' effectiveness. Int. J. Hum. Resour. Manage. **16**(8), 1349–1373 (2005)
13. Park, G., Seo, S., Lee, S., Lee, S.: Influencerank: trust-based influencers identification using social network analysis in q&a sites. IEICE Trans. **95-D**(9), 2343–2346 (2012)
14. Prat-Pérez, A., Dominguez-Sal, D., Larriba-Pey, J.L.: High quality, scalable and parallel community detection for large real graphs. In: Proceedings of the 23rd International Conference on World Wide Web, WWW 2014, pp. 225–236. ACM, New York (2014)
15. Ravasz, E., Barabási, A.L.: Hierarchical organization in complex networks. Phys. Rev. E **67**, 026112 (2003)
16. Ribeiro, B.: Modeling and predicting the growth and death of membership-based websites. In: Proceedings of the 23rd International Conference on World Wide Web, WWW 2014, pp. 653–664. ACM, New York (2014)
17. Vieira, M.V., Fonseca, B.M., Damazio, R., Golgher, P.B., Reis, D.D.C., Ribeiro-Neto, B.: Efficient search ranking in social networks. In: Proceedings of the sixteenth ACM conference on Conference on information and knowledge management, CIKM 2007, pp. 563–572. ACM, New York (2007)
18. Wasserman, S.: Social Network Analysis: Methods and Applications, vol. 8. Cambridge University Press, Cambridge (1994)
19. Xin-ru, L., Haijuan, G.: International human resource management research focus and cutting-edge analysis. In: 2012 International Symposium on Management of Technology (ISMOT), pp. 495–500 (2012)
20. Zhang, J., Wang, C., Wang, J.: Who proposed the relationship?: recovering the hidden directions of undirected social networks. In: Proceedings of the 23rd International Conference on World Wide Web, WWW 2014, pp. 807–818. ACM, New York (2014)

Integration Method for Complex Queries Based on Hyponymy Relations

Daisuke Kitayama[1]([✉]) and Takuma Matsumoto[2]

[1] Faculty of Informatics, Kogakuin University, 1-24-2 Nishishinjuku,
Shinjyuku-ku, Tokyo 163-8677, Japan
kitayama@cc.kogakuin.ac.jp
[2] LAC Co., Ltd., Hirakawacho Mori Tower, 2-16-1, Hirakawacho,
Chiyoda, Tokyo 102-0093, Japan
takuma.matsumoto@lac.co.jp

Abstract. Users often find it difficult to build suitable Web search queries even if they can explain the target of their search. In this situation, users must create complex queries using Boolean search operators such as "AND" and "NOT" by trial and error. However, complex queries often narrow down search results to a great extent. Therefore, we consider that users would like to replace their queries with simple queries that can retrieve the same target as complex queries. In this paper, we propose an integration method for complex queries based on hyponymy relations. We retrieve the same target as a complex query using a keyword that represents a narrower concept of one of the keywords in a complex query. Using this method, users can use understandable, simple queries and obtain many Web search results that have the same target as the complex query.

Keywords: Query translation · Web page clustering · Web search · Hyponymy relation

1 Introduction

In recent years, a huge number of Web pages have been added to the Internet, and this number continues to increase. In this situation, users can use Web search engines such as Google and Bing to retrieve target information. Users can obtain Web pages that contain their target information by simply inputting a search query into a Web search engine. However, users often find it difficult to build suitable Web search queries even if they can explain the target of their search. For example, a user might want to obtain information about certain fruit. If this user does not know or remember the name of fruit, they may not be easily able to build a suitable Web search query. In this case, this user inputs query keywords by trial and error to obtain their target information. The query expansion technique is one solution to this problem. This technique adds and replaces terms related to the user-inputted search query. Using this technique, users can search the Web

© Springer International Publishing Switzerland 2015
Q. Chen et al. (Eds.): DEXA 2015, Part II, LNCS 9262, pp. 359–368, 2015.
DOI: 10.1007/978-3-319-22852-5_31

with a suitable query by only selecting candidate queries, even if they find it difficult to think of additional or replacement keywords.

The number of keywords in Web search queries is increasing year after year [1]. In general, when users add keywords to a Web search query, they can obtain more specific the Web search results, however the number of results decreases. Therefore, we believe that reducing the keywords while retaining the meaning of the input query is the correct approach. However, the conventional methods cannot create an integrated query from a complex Web search query.

In this paper, we propose a method to integrate keywords for user-inputted complex queries based on hyponymy relations. We define a complex query as a query consisting of two or more keywords, and define an integrated query as a query consisting of one keyword that has the same meaning as a complex query. We focus on a keyword that represents a narrower concept of one of the keywords in a complex query. This keyword can retrieve specific Web pages and exclude unrelated Web pages. In addition, we detect integrated queries using the similarity between Web page clusters returned by keyword with a narrower concept and a complex query.

We detect integrated queries using the following procedure.

1. We cluster the Web search results of a user-input complex query.
2. We extract the hyponyms of a keyword in the user-input complex query.
3. We narrow down the list of hyponyms to select integrated query candidates using terms related to the input complex query.
4. We cluster the Web search results of the integrated query candidates.
5. We detect integrated queries by comparing the results of steps 1 and 4.

We illustrate an example when a user inputs "Fruit NOT Sour" as a complex query. First, we cluster the Web search results. Next, we extract the hyponyms of "Fruit." For instance, we may get "Apple" and "Banana" as hyponyms. At this time, many hyponyms are extracted. Therefore, we narrow down the hyponyms using terms related to the hyponyms and input complex query. The reason for this step is that if a hyponym has the same meaning as an input complex query, they will have similar related terms. Finally, we detect integrated queries by comparing the Web clusters formed from the complex query and extracted candidate hyponyms. If a candidate hyponym has the same meaning as an input complex query, it will also have similar search results. In this example, we might detect "Banana," "Melon," and "Loquat" as integrated queries.

The remainder of this paper is organized as follows. We describe our approach and other work related to this topic in Sect. 2. In Sect. 3, we explain a Web page clustering method for identifying integrated queries. We then describe an integration method for complex queries in Sect. 4. We evaluate and discuss our experimental results in Sect. 5. Finally, we summarize our method in Sect. 6.

2   Related Work

Query expansion has been well studied. Query expansion using a conceptual dictionary began in the 1990s, as is well-known. In recent years, it has become more difficult to search for target information on the Web. Therefore, researchers have actively developed query expansion for Web search.

Yoshida et al. [9] proposed a visualization of search result using graphs. Their method generates queries and reranks search results based on the distance between nodes on the graph. Users can generate complex Boolean queries by manipulating the nodes on the graph. Oishi et al. [4] developed an algorithm to extract terms related to query keywords based on the distance of a sentence. They then developed a query expansion system using this algorithm. Otsuka et al. [5] proposed a query extension method based on topic diversity and ambiguity of search criterion using Q&A site information. They developed a Web search system that helped a user to materialize their search criterion. These studies focused on expanding queries rather than specifically on integrated queries and complex queries.

3   Web Page Clustering

We can classify Web pages according to a certain rule using Web page clustering methods. In this research, we suppose that equivalent queries have similar clusters of Web search results. Therefore, we use Web page clustering to extract the integrated query.

We use Nishina's method [3] for Web page clustering. In addition, we customize a part of their method as follows. We use MeCab[1][2] as the morphological analyzer and exclude stop words from a set of nouns. We generate a compound noun using morpheme patterns, and we repeat cluster extraction until we cannot extract any further important terms. We summarize the Web page clustering process as follows.

1. We retrieve the top n Web pages, consisting of a title and a snippet, of an input query.
2. We morphologically analyze the title and snippet, and extract nouns including compound nouns.
3. We exclude stop words from the set of nouns.
4. We sort the set of nouns according to their TF-IDF score.
5. We generate groups of term using the noun cosine similarity, as in Nishina's method.
6. We classify Web pages using the term groups. This clustering is a soft clustering.

We extract nouns from Web page titles and snippets using morphological analysis in step 2. In this step, we create compound nouns when there is a noun surrounded by a noun suffix and noun prefix. However, we do not extract nouns that are not independent or the stem of an adjectival verb.

We exclude stop words from extracted nouns. We define numbers, dates, and words in the stop-word dictionary as stop words. The stop-word dictionary comprises registered words of short-length terms, frequent terms in snippets such as "Information," "Corporation," and so on.

---

[1] http://mecab.sourceforge.net/.

4   Integrated Keywords from Complex Queries

We detect the integrated query using hyponyms of the keywords that constitute the input query. We narrow down the search results using Boolean AND and NOT operators. On the other hand, we can also exclude unrelated Web pages using hyponyms, as the hyponyms represent specific search intentions. We note that we can replace a query that uses AND and NOT operators with hyponyms of the query keyword without changing the search intention. Therefore, we use hyponyms as candidates for an integrated query.

In this paper, we use a dictionary of hyponymy relations to extract keyword hyponyms that constitute an input complex query. Complex input queries consist of two or more keywords. Therefore, we must detect the most effective keyword for extracting hyponyms. We assumed that the first keyword of the input query corresponds to the user's search intention. Therefore, we use this keyword in the complex query for extracting hyponyms.

4.1   Hyponymy Relation Database

We built the hyponymy relation database using the Hyponymy Extraction Tool v1.0[2][6–8]. For this database, we used Wikipedia[3] dump data from June 24, 2014. The Hyponymy Extraction Tool can extract hyponymy relations from three types of source, as Category, Definition and Hierarchy. This tool extracts sets of hypernyms, hyponyms, and validity scores. For example, a term "Anime titles" has 3,916 hyponyms, and the validity score of hyponym "Case Closed" is 1.240835.

We use the hierarchy source because it extracts the largest amount of data. We call this dataset the hyponymy database. Note that we do not use hyponyms that have negative scores, because that data often does not have a hyponymy relationship.

4.2   Extracting Integrated Query Candidates

There are only a few hyponyms when we use exact matches from the hyponymy database. For example, there is no hyponym that is an exact match of "Fruit." This is because only the hypernyms "Kind of fruit" and "Vegetables can be eat like a fruit" exist in the hyponymy database. Therefore, we retrieve hyponyms by prefix[4] search. However, processing time increases according to increase in candidate hyponyms terms because we detect integrated queries for each candidate term by the method described in Sect. 4.3. At this point, we narrow down the list of retrieved hyponyms by determining the relationship between the input query and retrieved hyponyms, as described below.

---

[2] http://alaginrc.nict.go.jp/hyponymy/.

[3] http://ja.wikipedia.org/.

[4] We developed this method using Japanese Wikipedia. In Japanese, target hyponyms often appear at the prefix. We consider that target hyponyms appear at the suffix in English.

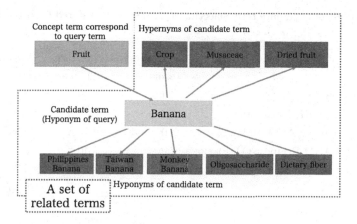

Fig. 1. Set of related terms for the candidate term "Banana"

Related Candidate Term Collection. First, we collect a set of related terms for hyponyms as follows.

1. We retrieve terms using the first term of input query from the hyponymy database using a prefix search.
2. We extract the hyponyms of these terms.
3. We extract hypernyms and hyponyms of the hyponyms from step 2.
4. We collect the terms extracted in steps 2 and 3 as a set of related terms.

We call the hyponyms extracted in step 2 "candidate terms" to distinguish them from the hyponyms of step 3 (that is, the hyponyms of the hyponyms). We show an example of related terms for "Banana" in Fig. 1.

Candidate Selection. The next step of related term collection is to narrow down the candidate terms. In this step, we select the best 10 candidate terms by ranking according to extracted candidate terms. We do this because processing time can be reduced by narrowing down the list of terms. In addition, the correct answer for an integrated query is likely to be included in the top 10 terms.

First, we calculate the score of each term using related terms according to the following formula.

$$candidate(k) = \frac{pm(R_k, C_q)}{|C_q| + |R_k| - pm(R_k, C_q)} \tag{1}$$

where k is a candidate term, q is an complex input query, R_k is the set of terms related to k, and C_q is a set of representative terms of a Web page cluster. The function $pm(R_k, C_q)$ return the number of corresponded related terms with representative terms. Therefore, when there are many partially matched representative and related terms, $candidate(k)$ has a high value. Finally, we extract the top 10 candidate terms as the result.

Table 1. Result of candidate term selection. Queries are translated from the Japanese results.

Complex query	# of candidate terms	Percent included	Processing time
Tea AND Big three	74	60 %	4:48
Measurement AND Radiation	78	40 %	5:30
Official language AND Finland	290	40 %	19:59
Thai food AND Recipe NOT Fiery	50	40 %	3:08
Aquarium AND Japan AND Sea otte	52	20 %	3:36
Bank AND Government	415	0 %	29:13

4.3 Integrated Query Detection

We calculate the equivalence score to detect integrated queries from candidate terms using the following formula.

$$equiv(k) = \frac{|C_k \cap C_q|}{|C_k|} \times \sum_{c \in C_q} w(c, C_k) \qquad (2)$$

$$w(c, C_k) = \begin{cases} |c| & (c \equiv c_x \in C_k) \\ 0 & (\text{other}) \end{cases} \qquad (3)$$

where k is a candidate term, C_k is a set of representative terms of the cluster for candidate term k, and c is a representative term. Furthermore, $|c|$ denotes the number of Web pages in a cluster that have a representative term c. At this point, C_k includes not only representative terms extracted by the Web clustering method, but also candidate term k. This is because k can be a representative term of a Web page cluster formed by query q, however, it cannot be a representative term of Web page cluster formed by query k.

We calculate the percentage of correct representative terms C_q in C_k in Eq. (2), which functions like a precision value. In other words, the equivalence score is high when the Web cluster of a candidate term is similar to the input query cluster.

On the other hand, we also calculate the weight of clusters in Eq. (2). This weight is the sum of the number of Web pages in the input query clusters that match representative terms with a candidate term cluster.

In brief, $equiv(k)$ is the precision times the weight of the correct cluster. We can then extract candidate terms that generate many clusters that have the same representative term and include many Web pages.

5 Experiments

We developed a prototype system, and evaluated candidate term selection and integrated term detection. We collected 100 Web search results for clustering using the Bing search engine.

5.1 Evaluation of Candidate Selection

First, we evaluated the candidate term selection that is described in Sect. 4.2. For this evaluation, we calculated the $equiv(k)$ score for all candidate terms, and then calculated the number of the 10 selected candidate terms that are included in the top 10 results of all candidate terms. We use six complex queries as input queries for this experiment. We show the results of this experiment in Table 1. For example, the complex query "Tea AND Big three" has 74 hyponyms as candidate terms. We ranked these candidate terms using $equiv(k)$ score. In this case, 60 % of the 10 selected candidate terms are included in the top 10 results. These results show that 33.3 % of the terms extracted by $equiv(k)$ are actually in the top 10 results.

This result shows that the method is not very accurate when candidate terms have only a few related terms. The median number of related terms for each complex query is 4.3. We conclude that the number of related terms tends to be low. When the number of related terms is low, the $pm(R_k, C_q)$ score in $equiv(k)$ is low. Therefore, our method may not detect relationships between input query and candidate terms. We conclude that the accuracy of the selection method depends on the number of terms related to the candidate term.

We next evaluated processing time. We compared integrated term detection with and without term selection. Table 1 shows the processing time of integrated term detection without selection. This result shows that the processing time increases according to the number of hyponyms used as candidate terms. On the other hand, the average processing time of integrated term detection with selection was 1 m 26 s. Therefore, we conclude that our proposed selection method is effective at reducing processing time and maintaining accuracy.

5.2 Evaluation of Integrated Query Detection

Setting of Experiment. We investigated whether the integrated query can replace a complex query using a survey of 10 participants. Participants selected integrated queries that could replace a complex query from six candidate queries. If participants decided that there were no correct integrated query options, they did not select any. We show the top three results each for our proposed method and the conventional method. We used 25 complex queries and extracted 15 complex queries at random for each participant. We prepared complex queries such that there was at least one correct integrated query in the hyponymy database. The conventional method is the TF-IDF method. We extracted the top three terms in the Web search results of a complex query. For instance, we first performed steps 1 to 3 in Sect. 3. We next calculated the TF-IDF score for each term. In this method, TF is the term frequency over all of the titles and snippets of the Web search result, and DF is the document frequency, treating a set of one title and snippet as one document.

Experimental Results. We show an example of the output of our method on two complex queries in Table 2. The columns correspond to Eq. (2) in Sect. 4.3.

Table 2. Ranking of candidate integration queries on "Tokyo AND Autumn tints"

Rank	Candidate	# of correct cluster	# of cluster	Precision	Weight	Score
1	Koishikawa Korakuen Garden	14	79	0.177	117	20.734
2	Rikugi-en	11	67	0.164	95	15.597
3	Mukojima-Hyakkaen Garden	10	96	0.104	89	9.270
4	Yakusiike Garden	11	94	0.117	76	8.893
5	Senzoku Pond	9	86	0.104	68	7.116
6	Botanical Gardens at the Univ. of Tokyo	8	85	0.094	63	5.929
7	National Museum of Nature and Science	5	63	0.079	29	2.301
8	Yumenoshima Tropical Greenhouse Dome	6	98	0.061	35	2.142
9	Central Garden	5	74	0.067	23	1.554
10	National Museum of Emerging Science and Innovation	0	5	0	0	0

The # of correct clusters is $|C_k \cap C_q|$ in Eq. (2), # of clusters is $|C_k|$, precision refers to the ratio in Eq. (2), weight refers to the sum in Eq. (2), and score is its result.

Table 2 shows the ranking of the integrated queries when the complex input query is "Tokyo AND Autumn tints." Our method can extract the famous spots for beautiful colored leaves in Tokyo such as "Koishikawa Korakuen Garden" and "Rikugi-en" from 23,965 candidate terms. For this experiment, we used the top three integrated queries.

Discussion. Table 3 shows a part of the experimental results when we detect the correct answer as the term that was selected as an acceptable replacement for a complex query by a majority of the participants. We explain this table using "Dog AND Aiful."[5] We used the conventional method to extract "Chihuahua," "Aiful dog," and "Popularity" as the top three terms. Then, using our method, we extracted "Chihuahua," "Jikkoku dog"[6] and "Japanese Spitz" as the top three terms. The underlined terms are the correct answers of the integrated query that were selected by most participants. The "Accuracy" column shows the percentage of correct answers in the extracted integrated query.

First, we discuss the average accuracy of both methods. In this result, the accuracy of the conventional method is 28.0 %. On the other hand, the accuracy of our method is 22.7 %. Therefore, our method is worse than the conventional method by 5.3 %. However, the conventional method often extracts simple-joined integrated queries. For example, "Aiful dog" is extracted from "Dog AND Aiful," and "Radiation measurement" is extracted from "Measurement AND Radiation." These simple-joined integrated queries are the correct answer because these terms can replace a complex term. However, they may retrieve the same

---

[5] Aiful is a consumer credit company in Japan. The mascot of this company is a Chihuahua that is a kind of dog.

[6] Jikkoku dog is a kind of Japanese dog.

Table 3. Results of experiment for evaluation of integrated query detection

Complex query	# of participants	Conventional method		Proposed method	
		Integrated query (Top 3)	Accuracy	Integrated query (Top 3)	Accuracy
Dog AND Aiful	5	Chihuahua, Aiful dog, Popularity	67%	Chihuahua, Jikkoku dog, Japanese Spitz	33%
Play AND Child NOT Spot	7	Child, Past, Outdoor play	33%	Make-a-face, Game, Play on words	0%
Measurement AND Radiation	6	Radiation measurement, measuring machine for radiation, measurement	33%	Scintillation counter, Semiconductor detector, Direct measurement	0%
Sweetener AND Calorie-free	9	Calorie-free sweetener, Sugar, Artificial sweetener	33%	Natural sweetener, Sucralose, Aspartame	33%
Brazil NOT Soccer	3	Brazil's national bond, World, Japan	33%	Cerrado Protected Areas, Historic Center of the Town of Goias, Historic Centre of Sao Luis	67%
Tokyo AND Autumn tints	3	Noted place, Autumn tint spots, Tokyo metropolitan	0%	Koishikawa Korakuen Garden, Rikugi-en, Mukojima-Hyakkaen Garden	67%
	6	Research, Summer vacation, Plant observation	33%	Row of zelkova at Babadaimon, Rosarium, Morning glory	0%
Theme park AND Harry Potter	7	Harry, Potter, USJ (An abbreviation of Universal Studio Japan)	33%	Universal Studio Florida, Walt Disney Studios Park, Epcot	0%
Bank AND Government	7	Center, Organ, institutional bank	0%	Saitama Resona Bank, Kitanippon Bank, Michinoku Bank	0%
Fish AND Nemo	5	Anemone fish, Movie, Finding Nemo	67%	Piranha, Swordfish, Tetra	0%
Food AND Hot	5	Thai food, Seoul food, Recipe	33%	Mapo doufu, Kimchi, Oden	33%
Asteroid AND Hayabusa	5	Unmanned spacecraft, Itokawa, Earth	33%	Itokawa, Asteroid cross Venus, Quasi-satellite	0%
Snowboard player AND Big name	8	Snowboard, Entertainer, Athlete	0%	Ryo Aono, Ayumu Hirano, Shaun White	100%
Onsen AND Oita AND Famous NOT Inn	4	Oita, Beppu, Beppu Onsen	33%	Beppu Onsen, Higashiyama Onsen, Sujiyu Onsen	33%
		Average	28.0%	Average	22.7%

Table 4. Ranking results

Rank	Conventional method (21 terms)	Proposed method (17 terms)
1	47.6 %	70.6 %
2	19.0 %	17.6 %
3	33.3 %	11.8 %

search result, therefore, they are not suitable for query suggestion. On the other hand, our method can extract integrated queries that have different points of view. Given these considerations, we conclude that our method can extract an integrated query that is suitable for query suggestion with the same accuracy as the conventional method.

Finally, we show the result of ranking at Table 4. Each percentage indicates the percent of correct integrated queries in the target rank out of all the correct integrated queries identified by each method. In this table, for the conventional method, 47.6 % of extracted correct answers were in the top rank, 19.0 % extracted correct answers were in the second rank, and 47.6 % extracted correct answers were in the third rank. For our method, 70.6 % extracted correct answers were in the top rank, 17.6 % extracted correct answers were in the second rank, and 11.8 % extracted correct answers were in the third rank. Therefore, our method can rank integrated terms effectively.

6   Conclusions

We proposed a method to integrate complex queries using hyponymy relations. In particular, we extract a term that is a hyponym of a complex query and generates similar Web clusters to the Web clusters of a complex query. We confirmed its effectiveness for ranking integrated queries and reducing processing time by narrowing down the number of candidate terms using selection. As future work, we will improve the selection method for candidate terms with respect to extraction accuracy. We plan to develop a selection method for the keywords for extracting hyponyms. For example, it is important to select a keyword that has a broad meaning.

Acknowledgments. This work was supported by ISPS KAKENHI of Grant-in-Aid for Young Scientists (B) Grant Number 15K16091.

References

1. Hearst, M.A.: Search User Interfaces, Chap. 4.1.2, 1st edn. Cambridge University Press, New York (2009)
2. Kudo, T., Yamamoto, K., Matsumoto, Y.: Applying conditional random fields to Japanese morphological analysis. In: Proceedings of EMNLP, pp. 230–237 (2004)
3. Nishina, T., Utsumi, A.: Web document clustering based on the clusters of topic words (in Japanese). J. Nat. Lang. Process. **17**(4), 23–41 (2010)
4. Oishi, T., Mine, T., Hasegawa, R., Fujita, H., Koshimura, M.: Related word extraction algorithm for query expansion – an evaluation. In: Bai, Q., Fukuta, N. (eds.) Advances in Practical Multi-Agent Systems. SCI, vol. 325, pp. 33–48. Springer, Heidelberg (2010)
5. Otsuka, A., Seki, Y., Kando, N., Satoh, T.: Qaque: faceted query expansion techniques for exploratory search using community qa resources. In: Proceedings of the 21st International Conference Companion on World Wide Web, WWW 2012 Companion, pp. 799–806. ACM (2012)
6. Sumida, A., Torisawa, K.: Hacking wikipedia for hyponymy relation acquisition. In: Proc. of IJCNLP 2008 (2008)
7. Sumida, A., Yoshinaga, N., Torisawa, K.: Boosting precision and recall of hyponymy relation acquisition from hierarchical layouts in wikipedia. In: Proceedings of the Sixth International Language Resources and Evaluation (2008)
8. Yamada, I., Hashimoto, C., Oh, J.H., Torisawa, K., Kuroda, K., De Saeger, S., Tsuchida, M., Kazama, J.: Generating information-rich taxonomy from wikipedia. In: Universal Communication Symposium (IUCS), pp. 97–104, Oct 2010
9. Yoshida, T., Nakamura, S., Oyama, S., Tanaka, K.: Query transformation by visualizing and utilizing information about what users are or are not searching. In: Buchanan, G., Masoodian, M., Cunningham, S. (eds.) Digital Libraries: Universal and Ubiquitous Access to Information. Lecture Notes in Computer Science, vol. 5362, pp. 124–133. Springer, Berlin Heidelberg (2008)

An Expertise-Based Framework for Supporting Enterprise Applications Development

Devis Bianchini, Valeria De Antonellis, and Michele Melchiori[✉]

Department of Information Engineering, University of Brescia, Via Branze 38,
25123 Brescia, Italy
{bianchin,deantone,melchior}@ing.unibs.it

Abstract. Currently, Web mashups are becoming more and more popular for organizations and enterprises with the aim to implement applications based on third party software components. These components may offer sophisticated functionalities and access to high valuable datasources through Web APIs. However, developing a Web mashup may require a rather specialized knowledge about specific Web APIs, their technological features and how to integrate them. If we consider a large organization, knowledge required to implement a mashup can be available, but distributed among different developers that are not easy to identify and assess. To this purpose, we propose a framework and a software tool for searching experts inside the organization that own valuable knowledge about specific Web APIs and the way to integrate them meaningfully. Retrieved experts are ranked based on: (i) the expertise level on the specific request, and (ii) the social distance with the developer that issued the request. The approach integrates knowledge both internal and external to the organization and represented as a linked data. We include a preliminary evaluation based on an implementation of the framework.

1 Introduction

Currently, implementing Web applications by composing third party software components is becoming more and more popular. This approach is usually known as Web mashup and is supported by Web sites offering large, ever growing, repositories of components accessible through Web APIs (e.g., ProgrammableWeb, Mashape). In these repositories, providers like Google but also single developers, share Web APIs. Web designers can therefore search for the component they need and integrate them into Web mashups. This profusion of ready-to-use components makes them interesting also for developing applications for organizations [8] as parts of enterprise information systems.

The activity of developing mashups requires specialized knowledge on specific Web APIs, their technologies and the way to combine Web APIs in a meaningful way. This kind of knowledge can be available inside an enterprise, but may be distributed among various experts. Recently, collaboration has taken advantage of Web 2.0 social-based paradigm and technologies by specializing them to needs and requirements of enterprise. This conceptual specialization and related tools

© Springer International Publishing Switzerland 2015
Q. Chen et al. (Eds.): DEXA 2015, Part II, LNCS 9262, pp. 369–379, 2015.
DOI: 10.1007/978-3-319-22852-5_32

are commonly referred to as Enterprise 2.0 [9]. Enterprise 2.0 tools organize knowledge workers as a social network with the purpose to facilitate collaboration and joint creation of contents and applications.

In this paper we propose the LINKSMAN (LINKed data Supported MAshup collaboratioN) framework and tool for finding experts in the context of enterprise mashup development. A main difference with general expert finding systems is that they aim at identifying experts on a given subject of interest, and ranking them with respect to their expertise, based on available *sources of evidence*, typically documents [1]. On the contrary, we (i) use in an integrated way non-document sources of evidence that are both internal and external to the organization; (ii) we keep into account a kind of social distance from candidate experts to rank them. In particular, we adopt ProgrammableWeb[1] as external source of evidence because this repository is actually one of the most used and updated for mashup development. The basic ideas of LINKSMAN have been given in [4]. In this paper, we present in a more complete, formalized and mature way the approach. Moreover, a new implementation of the prototype is described and experimentation is provided.

The paper is organized as follows. In Sect. 2, we present some related work. In Sect. 3, we give a motivating example. In Sect. 4, we discuss construction of the RDF models for the internal and external sources. In Sect. 5 we identify types of collaboration requests and provide suitable metrics for implementing them. In Sect. 6, we show the LINKSMAN architecture. In Sect. 7, we present some experimental results for preliminary evaluation of the framework. Section 8 provides final remarks and future work.

2   Related Work

Modeling social aspects for building web mashups has been discussed in [3], where a survey of approaches for Web API search on public repositories is also provided. Related efforts can be found in literature about Linked Web services or Linked Web APIs for discovery purposes, for example [11]. In the context of Enterprise 2.0, social aspects are relevant too. A number of collaboration platforms is currently available to provide enterprises with tools that allow for social linking and tagging of resources, to support the user feedbacks/opinions, and user-produced content management platforms (blogs and Wikis) (e.g., YAMMER[2]). The IBM Connections platform provides also analysis tools and metrics to discover trends in social content and activities. The work [13] states and confirms that cooperation in software teams is more effective when the involved employees have social connections and proposes mining techniques for discovering collaboration patterns inside distributed teams.

Expert search from online data sources is a growing area for both researchers and companies. Traditional expert search approaches and systems use as sources

---

[1] See http://www.programmableweb.com: the last access on March 2015 counts more than 13000 Web APIs and 6000 mashups.

[2] http://www.yammer.com.

of evidence closed sets of document and databases [1]. On the contrary, the work [10] discusses potential benefits and drawbacks of using Linked Open Data (LOD) for expert search. The authors analyze what is currently present and missing in various LOD datasets w.r.t. different kinds of expert search. A more general approach is in [6] that introduces the concept of expertise sphere in order to measure how well the area of expertise of an expert fits a given subject (e.g., the subject of a paper to be reviewed).

3   Motivating Scenario

In this section, we introduce an possible scenario for the considered problem. Let us suppose a developer is working for the business and marketing department of an enterprise, who has to build an application to visualize, on an interactive map, information about current sales and demographic data in order to visualize potential market opportunities. The developer may consider simpler to build this application by composing functionalities and data than developing the application by scratch. Therefore, suitable Web APIs[3] have to be selected (e.g., the GeoData Demographics API, that provides demographic data for a given zone) from public repositories and possibly integrated with enterprise components/ data (e.g., a Web API providing data on enterprise sales). This components selection phase may use advanced search techniques [5]. However, composing the selected Web APIs require to understand their syntax, semantics and data formats. These difficulties may worsen because of limitations and typical heterogeneity in the documentation associated with third party Web APIs. In this context, it is frequent that a developer searches for advices and collaboration from other developers of the enterprise. Moreover, it is reasonable to assume that her/his preference is for experts that can be contacted and involved because some social or organizational contacts exist with them. These considerations are also confirmed by studies, for example [7]. In this perspective, the LINKSMAN framework supports the developer by identifying a list of candidates that might have useful expertise for developing the mashup, assigning a higher rank to the ones closer in the organization to her/him.

4   Linked Data Model for Integrating Public and Enterprise Knowledge

In this section, we describe how the external and the internal sources are modeled as Linked Data (LD) vocabularies. These sources are linked to create an integrated data set. Then, a set of perspectives is identified and associated with key classes of the LD vocabularies in order to identify the information we use for assessing expertize. These perspectives allow for introducing descriptors and similarity measures on them.

---

[3] Following a common practice, from now on in this paper we use the term Web API also to refer to the software component itself.

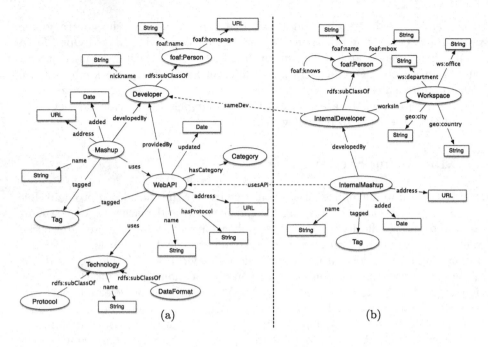

Fig. 1. RDF vocabularies: (a) PW repository; (b) developer organization.

4.1   Vocabularies Definition

The ProgrammableWeb (PW) and Developers Organization (DO) RDF vocabularies are shown in Fig. 1. We maintain them separated and linked so that the PW source can eventually be either extended by integrating it with the content of similar sources (e.g., https://www.mashape.com) or substituted by one of them. In particular, we designed the PW vocabulary by analyzing the content of PW and identifying the classes of main resources, namely Web APIs, mashups/web app, developers, providers and categories. The DO vocabulary representing the knowledge on the enterprise organization and on mashups for internal use, has been designed by taking into account the content of a well known Enterprise 2.0 platform, that is IBM Connections.

Concerning vocabularies definition we follow the LD practices. That is, relying as much as possible on existing vocabularies and ontologies [12]. For instance, we modeled the internal developer class in the DO vocabulary (see Fig. 1(b)) as specialization of foaf:Person described in the FOAF (Friend of a Friend) ontology[4]. Specific ontologies provide concepts as office, department, city, country.

The approach to build the vocabularies is sketched in the following. For each identified class, descriptive properties in the original source are identified. For instance, the PW vocabulary includes Web API properties (e.g., protocols, data formats). Then, URIs are assigned to classes and properties. If for

---

[4] http://xmlns.com/foaf/spec/.

a class/property a corresponding element can be found in an existing vocabulary, the element URI given in the vocabulary is used (e.g., http://xmlns.com/ foaf/spec/#term_knows). For concepts not extracted from an existing vocabulary, URIs in the TBox (ontology model) have the following structure:

$$BaseURI/source\_name/\texttt{ontology}/\{class|property\} \qquad (1)$$

where $BaseURI$ is the URI of the host where the RDF vocabulary is stored and $repository\_name$ is the name of the repository. URIs of the ABox (resources) are defined by exploiting URIs assigned by repositories to resources, when available.

Finally, RDF triples in the PW and DO data sources are linked (see dashed links in Fig. 1) according to the following criteria: (i) a sameDev link is established between a DO InternalDeveloper resource and a PW Developer resource when an developer working for the enterprise is also registered as developer in the PW web site; (ii) a DO InternalMashup, that is an enterprise mashups, is linked through a usesAPI property to each component PW WebAPIs.

4.2   Descriptors for Expert Finding

In the PW and DO vocabularies, we identify a set of key classes that we consider as sources of evidence for establishing developer's expertise. These classes are InternalDeveloper and InternalMashup for DO, and WebAPI and Mashup for PW. For each of these classes we have one or more applicable perspectives. A perspective represents a set of properties that provide knowledge on a specific subject [4]. For example, Mashup has a not empty *WebAPIs perspective* including the property address that is the addresses of its component Web APIs. Similarly, InternalDeveloper class has a not empty *Organization perspective*.

Formally, given a class Cl we introduce a function A_{Cl} where $A_{Cl}(p)$ lists the properties $pr_1, pr_2, \cdots, pr_m$ associated with Cl according to a perspective p. Descriptors are then associated with each object of a class Cl. A descriptor according to the perspective p for an object o_i, instance of Cl, is defined as a list of sets of property values:

$$des_p(o_i) = < st_{i1}, st_{i2}, ..., st_{in} > \qquad (2)$$

where every set st_{ij} is the set of values of the j-th property of $A_{Cl}(p)$.

Example. For an InternalDeveloper object, a descriptor according to the *Organization* perspective is built as union of property values of ws:country, ws:city, ws:department, ws:office of the object workplace associated with the InternalDeveloper object. That is, $A_{InternalDeveloper}(Organization)=$ {ws:country, ws:city, ws:department, ws:office}. A descriptor according to the *Organization* perspective for the *JSmith001* instance of InternalDeveloper, could be:
des (http://.../developers/JSmith001) $= < \{Italy\}, \{Rome, Milan\},$
  $\{Marketing\_Dept, Business\_Dept\}, \{Marketing\_Development\} >.$

We also set some properties as *semantically comparable*. For example, we want to recognize as similar values like *Research and Development, R&D* and *Research projects* for the `ws:office` property. Therefore `ws:office` is set as semantically comparable property. Other properties could not be semantically comparable. For example, concerning values of `hasDataFormat`, it seems reasonable to consider as similar two data formats only if the corresponding names are similar strings (e.g., *JSON, JSON3*). Therefore we associate with the k-th property in $A_{Cl}(p)$ a *comparison criteria* described by the boolean $SemComp(Cl, p, k)$ that is *True* if the property is semantically comparable and *False* if the property is not semantically comparable. Note that this choice has to be done usually only once and at design time.

4.3   Descriptors Similarity

We define a overall similarity measure on pair of descriptors as follows:

$$Sim_O(des_p(o_i), des_p(o_j)) = \frac{\sum_k sim(des_p(o_i)[k], des_p(o_j)[k])}{n} \in [0..1] \qquad (3)$$

where $n = |des_p(o_i)| = |des_p(o_j)|$ and $des_p(o)[k]$ represents the k-th set of property values in the descriptor $des_p(o)$. The term $sim(des_p(o_i)[k], des_p(o_j)[k]) \in [0..1]$ is the maximum similarity between every pair of values, one from the set $des_p(o_i)[k]$ and one from the set $des_p(o_j)[k]$ computed according to the comparison criteria $SemComp(Cl, p, k)$. If $SemComp(Cl, p, k)$=*False* the Jaro-Winkler metrics is used to compare pairs of values. The score is normalized so that 0 means no similarity and 1 is an exact match. If $SemComp(Cl, p, k)$=*True* a semantic similarity score is used basically as defined in [2]. Here we only remark that the evaluation is based on: (i) the WordNet lexical system to deal with semantically related terms, (ii) on specific rules to deal with composed names (e.g., *Research projects*), and (iii) a thesaurus to deal with acronyms like *R&D*.

5   Collaboration Requests

A collaboration request specifies the requested type of expertise. We define in a request R submitted by a developer D_R as the tuple $R =< T_R, D_R, WA_R >$. where T_R is the target of the request and WA_R is a set of one or more Web APIs W_{Ri} of interest specified by D_R. We distinguish among four different targets corresponding to common problems occurring in mashup development:

(T1) search for experts on a specific Web API W_R, where $WA_R = \{W_R\}$;
(T2) search for experts on the technologies of a Web API W_R, where $WA_R = \{W_R\}$;
(T3) search for experts on the functionalities of a Web API W_R, where $WA_R = \{W_R\}$;
(T4) search for experts on a mashup whose component Web APIs are the set $WA_R = \{W_{Ri1}, ..., W_{Rik}\}$.

In the first three cases the request focuses on a single Web API. For example, if in $T3$ is specified a Web API for displaying maps, the developer D_R is searching for experts on APIs for displaying maps (i.e., same functionalities). In the fourth case, D_R is looking for developers that have experience in mashups that are built from the set $\{W_{Ri1}, ..., W_{Rik}\}$ or at least with a not empty subset of it.

Table 1. Definitions of collaboration requests.

Collaboration Request	Precondition for Evaluating $m(R, D_i)$	Expression of $m(R, D_i)$
CR1	Developer D_i has used the Web API W_R in some mashups (that is, $W_R \in des_{pW}(D_i) \neq 0$)	$\alpha \cdot Sim(des_{pO}(D_R), des_{pO}(D_i))$ + $\beta \cdot Sim(des_{pS}(D_R), des_{pS}(D_i))$ $\in [0, 1]$
CR2	Developer D_i has used APIs that share technologies with the Web API W_R (that is, $Sim(des_{pT}(D_i), des_{pT}(W_R)) \neq 0$)	$\alpha \cdot Sim(des_{pO}(D_R), des_{pO}(D_i))$+ $\beta \cdot Sim(des_{pS}(D_R), des_{pS}(D_i))$+ $\gamma \cdot Sim(des_{pT}(D_i), des_{pT}(W_R))$ $\in [0, 1]$
CR3	Developer D_i has used APIs that share some functionalities with the Web API W_R (that is, $Sim(des_{pF}(D_i), des_{pF}(W_R)) \neq 0$)	$\alpha \cdot Sim(des_{pO}(D_R), des_{pO}(D_i))$+ $\beta \cdot Sim(des_{pS}(D_R), des_{pS}(D_i))$+ $\gamma \cdot Sim(des_{pF}(D_R), des_{pF}(W_R))$ $\in [0, 1]$
CR4	Developer D_i has used at least one of the APIs in WA_R specified in the request (that is, $Sim(des_{pW}(D_i), des_{pW}(WA_R)) \neq 0$)	$\alpha \cdot Sim(des_{pO}(D_R), des_{pO}(D_i))$+ $\beta \cdot Sim(des_{pS}(D_R), des_{pS}(D_i))$+ $\gamma \cdot Sim(des_{pW}(D_R), des_{pW}(WA_R))$ $\in [0, 1]$

5.1 Specializing Collaboration Requests

Formally we define a collaboration request pattern in a general way as:

$$CR = < R, m, \delta, \prec > \tag{4}$$

where R is the request for collaboration, m is the metric used to measure the matching between a candidate developer and R, by using the similarity between descriptors as in Eq. (3). Application of a collaboration request to a dataset described according to the model of Sect. 4 produces a list of developers matching the request R. The list is ordered according to a ranking function $\prec$, that is $D_i \prec D_j$ if $m(R, D_i) \leq m(R, D_j)$. A threshold δ is used to filter out the less relevant search results.

The collaboration request pattern CR is specialized for each target identified above producing a collaboration request. In fact, based on suitable expertise assumptions [4], a target determines the structure of R, as seen above, and the expression of m as described in the following and in Table 1. For each specialized pattern, the second column of the table shows the precondition that has to be true in order to evaluate $m(R, D_i)$, otherwise $m(R, D_i)$ is set to 0. The third column describes how $m(R, D_i)$ is evaluated in terms of descriptor similarities. Note

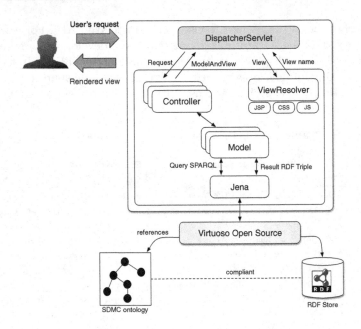

Fig. 2. The LINKSMAN functional architecture

that in the $m(R, D_i)$ expressions, pO, pS, pT, pF and pW denote respectively the perspectives: *Organization, Social, Technologies, Functionalities, WebAPIs*. For example, $CR4$ (searching for experts on a given mashup) is the collaboration request for the target $T4$. In $CR4$, the metrics $m(R, D_i)$ (see Table 1) is based on a weighted sum of three terms: (i) similarity of D_i and D_R with respect the *Organization* perspective, (ii) similarity between D_i and D_R with respect to the *Social* perspective, (iii) similarity between the set of APIs used by D_i in her/his mashups and the set of $\{W_{Ri}\}$ of Web APIs specified in the request R. Note that currently, we set weigths α, β and γ to the same values, with $\alpha, \beta, \gamma \in [0..1]$ and $\alpha + \beta + \gamma = 1.0$.

6   Architecture and Tool for Expert Finding

In this section we briefly present the LINKSMAN architecture and prototype. The functional architecture is illustrated in Fig. 2 and is based on the Virtuoso RDF data store. The LINKSMAN prototype has been developed as Web application and partitioned in three main layers according to a MVC (model-view-controller) approach by adopting for the implementation the *Spring MVC* framework[5]. The base flow of activities is the following. A user request is managed by the *DispatcherServlet* that forwards it to a suitable *Controller* which is able to manage the request. The *Controller*, based on the type of request, selects a suitable processing class as part of the *Model*, which encapsulates the application logic. Once the request has been processed, the *Controller* returns

---

[5] http://spring.io/guides/gs/serving-web-content/.

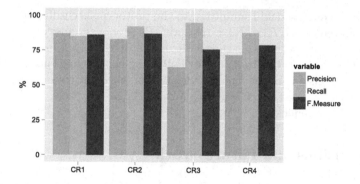

Fig. 3. Precision, Recall and F-measure for each type of request

an answer to the *DispatcherServlet*. The answer includes both data to be shown to the user and a *View* ID. The *DispatcherServlet* loads the *View* corresponding to the ID and the *View* uses HTML, JSP, CSS and Javascript code to present the result to the user.

Concerning the application logic implemented in the *Model* classes, the submitted request is translated by a specific *SPARQL module* into a set of SPARQL queries processed by *ARQ*, a SPARQL query engine to transform the query into Jena API calls. The Jena component (as shown the figure) accesses the Virtuoso RDF Store.

7   Simulation and Preliminary Evaluation

We performed a preliminary evaluation of LINKSMAN based on simulations of expert request.The purpose was to build a test set of data and compare the outputs of the prototype with expected results.

Setting. Initially, we chose ten of the most popular Web APIs from the PW repository (like Google Maps, Twitter and YouTube). Then, we selected four mashups for each API and stored their description in the DO dataset as internal mashups. For every mashup we identified the developer email (`foaf:mbox` field and used it as developer ID. Then we completed the developer's profile by adding randomly chosen information about the organization (country, city, office, department) so that the likelihood that two developers have at least an overlapping according to the *Organization* perspective is about 20 % and we stored the profiles in the PW data set. Furthermore, we added random social connections (instantiation of the DO `foaf:knows` property) to the profiles so that the probability of having a social connection between two profiles is about 10 %.

Simulation. We defined the requests in the following way. One of the developers is chosen as D_R, that is the developer that is searching for collaboration. One of the mashup developed by D_R was selected, we denote it with M_R. The Web APIs in M_R were used to define a request, specifically the W_R part of the request.

Requests were composed and submitted to the prototype according to each of the collaboration requests $CR1, CR2, CR3$ and $CR4$. For every answer, we evaluated the list of developers returned in the result by the LINKSMAN prototype in term of relevance with respect to the request. Following this evaluation, we calculated the average precision, recall and F-measure for each pattern (See Fig. 3). In formulating the request for $CR4$, the Web APIs in M_R were used to set the WA_R part of the request.

8   Conclusions and Future Work

In this paper, we described a Linked Data based framework to support expert search for collaboration in enterprise mashup development. Specifically, LINKS-MAN integrates knowledge that is both internal and external to enterprises. The main contributions of the framework are: (i) internal and external sources are modeled as linked RDF vocabularies; (ii) four types of collaboration request have been defined; (iii) a Web-based architecture and a prototype application based have been designed and implemented. Finally, simple simulations showing encouraging results have been performed. However, we plan to use more Web APIs and a more complex social network in order to provide a complete validation of the approach. Future work includes more extensive experimentation with real developers, identification of additional collaboration queries and use of additional data sources (e.g., Mashape, https://www.mashape.com).

References

1. Balog, K., Azzopardi, L., de Rijke, M.: Formal models for expert finding in enterprise corpora. In: Proceedings of the 29th Annual International ACM SIGIR Conference on Research and Development in Information Retrieval, SIGIR 2006, pp. 43–50. ACM, New York (2006)
2. Bianchini, D., De Antonellis, V., Melchiori, M.: Flexible Semantic-Based Service Matchmaking and Discovery. World Wide Web J. **11**(2), 227–251 (2008)
3. Bianchini, D., De Antonellis, V., Melchiori, M.: Semantic Collaborative Tagging for Web APIs Sharing and Reuse. In: Brambilla, M., Tokuda, T., Tolksdorf, R. (eds.) ICWE 2012. LNCS, vol. 7387, pp. 76–90. Springer, Heidelberg (2012)
4. Bianchini, D., De Antonellis, V., Melchiori, M.: A Linked Data Perspective for Collaboration in Mashup Development. In: Database and Expert Systems Applications (DEXA), 2013 24th International Workshop on Semantic Web (WebS), pp. 128–132, Aug 2013
5. Bianchini, D., De Antonellis, V., Melchiori, M.: Capitalizing the Designers' Experience for Improving Web API Selection. In: Meersman, R., Panetto, H., Dillon, T., Missikoff, M., Liu, L., Pastor, O., Cuzzocrea, A., Sellis, T. (eds.) OTM 2014. LNCS, vol. 8841, pp. 364–381. Springer, Heidelberg (2014)
6. Boeva, V., Krusheva, M., Tsiporkova, E.: Measuring expertise similarity in expert networks. In: 6th IEEE International Conference on Intelligent Systems (IS), 2012, pp. 53–57. IEEE, (2012)

7. Hertzum, M., Pejtersen, A.M.: The information-seeking practices of engineers: searching for documents as well as for people. Inf. Process. Manage. **36**(5), 761–778 (2000)
8. Hoyer, V., Fischer, M.: Market overview of enterprise mashup tools. In: Bouguettaya, A., Krueger, I., Margaria, T. (eds.) ICSOC 2008. LNCS, vol. 5364, pp. 708–721. Springer, Heidelberg (2008)
9. McAfee, A.P.: Enterprise 2.0: The dawn of emergent collaboration. MIT Sloan Manage. Rev. **47**(3), 21–28 (2006)
10. Stankovic, M., Wagner, C., Jovanovic, J., Laublet, P.: Looking for Experts? What can Linked Data do for You? In: Proceedings of Linked Data on the Web (LDOW) at WWW2010 (2010)
11. Taheriyan, M., Knoblock, C.A., Szekely, P., Ambite, J.L.: Semi-Automatically modeling Web APIs to create linked APIs. In: Proceedings of the ESWC 2012 Workshop on Linked APIs (2012)
12. Villazón-Terrazas, B., Vilches, L.M., Corcho, O., Gómez-Pérez, A.: Methodological Guidelines for Publishing Government Linked Data. In: David, W., (ed.) Linking Government Data, pp. 27–49. Springer, New York (2011)
13. Wolf, T., Schröter, A., Damian, D., Panjer, L.D., Thanh Nguyen, H.D.: Mining task-based social networks to explore collaboration in software teams. IEEE Softw. **26**(1), 58–66 (2009)

Data Management Algorithms

A Linear Program for Holistic Matching: Assessment on Schema Matching Benchmark

Alain Berro, Imen Megdiche$^{(\boxtimes)}$, and Olivier Teste

IRIT UMR 5505, INPT, UPS, UT1, UT2J, University of Toulouse CNRS,
31062 Toulouse, Cedex 9, France
{Berro,Megdiche,Teste}@irit.fr

Abstract. Schema matching is a key task in several applications such as data integration and ontology engineering. All application fields require the matching of several schemes also known as "holistic matching", but the difficulty of the problem spawned much more attention to pairwise schema matching rather than the latter. In this paper, we propose a new approach for holistic matching. We suggest modelling the problem with some techniques borrowed from the combinatorial optimization field. We propose a linear program, named LP4HM, which extends the maximum-weighted graph matching problem with different linear constraints. The latter encompass matching setup constraints, especially cardinality and threshold constraints; and schema structural constraints, especially superclass/subclass and coherence constraints. The matching quality of LP4HM is evaluated on a recent benchmark dedicated to assessing schema matching tools. Experimentations show competitive results compared to other tools, in particular for recall and HSR quality measures.

Keywords: Schema matching · Linear programming · Holistic matching · Matching quality assessment

1 Introduction

Schema matching problem is among the most studied problems in the literature. It represents a key task in several application fields such as data integration, ontology engineering, web services composition, query answering in the web, and so on [10]. The schema matching task consists of identifying correspondences, also named mappings or alignments, between schemes [18]. Almost all the approaches transform inputted schemes into an internal data representation, such as graphs or graph-like representations [1,19]. Various forms of schemes cited in [10] such as dictionaries, taxonomies, XML/DTD, relational databases, can be internally transformed into graph-like representation (trees, forests, rooted directed acyclic graphs, etc.), which have hierarchical organization of nodes.

Furthermore, the number of input schemes declines two types of matching approaches as reviewed in [18]: pairwise matching for two input schemes and

© Springer International Publishing Switzerland 2015
Q. Chen et al. (Eds.): DEXA 2015, Part II, LNCS 9262, pp. 383–398, 2015.
DOI: 10.1007/978-3-319-22852-5_33

holistic matching for several input schemes. In the scientific literature, pairwise schema matching gained much more attention than holistic schema matching. This is due to the different challenges [21] and difficulties of the matching task. However, the availability and the rapid evolution of a huge variety of data requires the integration of several data sources at the same time. Even if, some incremental solutions [5] have been proposed to deal with several or large data sources, the incremental process suffers from the closure of the solution.

In this paper, we define a new holistic approach for matching schemes having a graph-like structure. This approach also allows a holistic matching of the schemes of open data tables [4] represented as an hierarchy of concepts in the works of [3].

The approach combines combinatorial optimization techniques and the characteristics of schema matching problem. It consists of an extension of the maximum-weighted graph matching problem [20] adapted with different constraints related to the schema matching problem. The constraints are related to the mapping cardinalities, the threshold setup, super/subclass relations and structural coherence. The main contributions of our approach are as follows:

- it returns a global optimal solution as it extends the maximum-weighted graph matching problem [10];
- it can be used without setting threshold which makes a gap compared to the other tools;
- it is applicable for pairwise and holistic matching with a theoretical polynomial time [20].

Our approach is evaluated on the recent benchmark proposed by [6]. This benchmark is devoted to assess the quality of matching tools. Hence, the matching quality of our approach is confronted to the matching quality of different referenced solutions.

The remainder of this paper is devoted to the description of our approach in Sect. 3, then we present the experimental assessments of our approach in Sect. 4 and we conclude in Sect. 5. In the following section, we briefly describe related works.

2   Related Works

A schema matching task can be summed up by the general workflow described by [18]. This task is composed of a pre-processing step, an execution step for one or several matchers, which can be element-level matchers or structural-level matchers [22] then a combination of matcher(s) results and finally a selection of mappings.

We can notice according to [9,10] that different pairwise approaches, especially in the field of ontology matching, have reduced one or several steps in the workflow described above into a combinatorial optimization problem. S-Match [12] reduces the semantic matching to the propositional validity problem, which is theoretically a co-NP hard problem. The elements of schemes are translated into logical formulas and the semantic matching consists of resolving propositional formula constructed between elements. Similarity Flooding (SF) [16] reduces the selection of the mappings to the stable marriage problem, which returns a local optimal solution [10]. The SF approach proposes a structural-matcher which propagates similarities between neighbourhood nodes until a fixed point computation. OLA [11] reduced the selection of mappings into a weighted bipartite graph matching problem. This approach models structural similarity computation as a set of equations of the different properties of ontologies. The pairwise matcher CODI [13] implements the probabilistic markov logical framework presented in [17]. This approach transforms the matching problem into a maximum-a-posteriori (MAP) optimization problem which is equivalent to Max-Sat problem (NP-hard).

Furthermore, we highlight the existence of some approaches which have been proposed as a platform for the matching task. For instance COMA++ [2] and YAM [6]. COMA++ (COmbination of Matching Algorithms) is a generic matcher offering to users several strategies for matching schemes and several features to combine and reuse the results of matchings. YAM is a schema matcher factory. It uses machine learning techniques to tune thresholds and other parameters to propose a suitable matcher according to the evaluated dataset.

Discussion. Our solution LP4HM and CODI perform, at the same time, the structural matching phase without additional structural similarity computation and the mapping extraction phase. CODI is based on a pairwise approach whilst LP4HM is a holistic approach. For pairwise ontology matching, CODI is more generic than LP4HM because it considers all type of properties. Whereas, LP4HM considers only hierarchical relationships expressed with the subclass property in the ontologies. The integer linear program (ILP) of LP4HM is reduced, in pairwise scenarios, to the maximum-weighted bipartite graph matching problem just like what OLA has done for the extraction phase. Compared to OLA we do not compute structural similarities but we encoded structural properties as linear constraints. The complexity of the maximum-weighted bipartite graph matching problem with ILP is polynomial [20] even with the simplex algorithm [20]. For holistic scenarios, our ILP is reduced to the maximum-weighted non-bipartite graph matching problem [20] which can be also solved in polynomial time [8]. Unlike CODI whose pairwise approach is reduced to a NP-Hard problem, our proposed solution extends a polynomial problem in both pairwise and holistic versions. YAM have to be run several times to learn a threshold whilst LP4HM uses a predefined threshold and it generates an optimal solution with a unique run. Moreover, using a threshold in our approach is an optional feature.

3   A Linear Program for Holistic Matching: LP4HM

The idea of the LP4HM matcher is to extend the maximum-weighted graph matching (MWGM) problem with additional linear constraints. The authors of

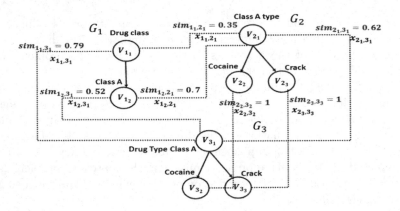

Fig. 1. A running example for holistic matching

[20] define the MWGM problem for a graph G as follows: "to find a matching (= set of disjoint edges) M in G with a weight $w(M)$ as large as possible". It is clear that the definition of the graph matching problem in combinatorial optimization field is different from the definition of the schema matching problem. But a reduction holds between both problems. Indeed, we can consider that G is composed of (i) the nodes representing schema elements and (ii) edges representing virtual connections between these nodes and having as weights the similarities between the concepts of the nodes. From this reduction, searching 1:1 correspondences (i.e. a node matches at most with only another node) for the schema matching problem is the same as finding a set of disjoint edges with a maximum weight in G. Moreover, pairwise and holistic schema matching problems correspond to bipartite and non-bipartite MWGM problem [20]. The latter have been proved to be theoretically polynomial.

To illustrate these reductions and our linear program, we give the running example of Fig. 1. This example represents a part of the schemes of open data tables, which have been transformed into a graph-like structure by the application of the approach of [3]. The data are available on the following links[1]. Each graph in Fig. 1 has continuous edges representing their structure. Between each pair of graph, we have dotted edges having as weight similarities. If we consider a graph G composed of the nodes of G_1 and G_2 and the dotted weighted edges between them, resolving pairwise matching between G_1 and G_2 is equivalent to resolving the MWGM problem in the bipartite graph G. If we consider a graph G composed of the nodes of the three graphs G_1, G_2 and G_3 and the dotted edges, resolving holistic matching between G_1, G_2 and G_3 is equivalent to resolving a MWGM in the non-bipartite graph G.

Given the reductions between the schema matching and the graph matching problem, our contribution consists of using the linear programming technique to inject additional constraints on the graph matching problem to make it more

---

[1] http://www.scotland.gov.uk/Topics/Statistics/Browse/Crime-Justice/TrendData.
http://data.gov.uk/dataset/seizures-drugs-england-wales.

suitable to meet the specificities of the schema matching problem. Our approach is divided into two steps:

- In the first step, we prepare LP4HM input data. According to the work-flow described in the related work section, this step involves a pre-processing step for computing direction matrices, an element-level matching and aggregation phases for computing similarities between the elements of the different schemes;
- In the second step, we construct and execute LP4HM. According to the same workflow, this step ensures structural-level matching in the form of linear constraints as well as mapping extractions.

The following notations will be used in the remainder of this paper:

- N is the number of input graphs;
- i, j are the IDs of the graphs G_i and G_j;
- V_i is the set of vertices in the graph G_i;
- n_i is the cardinality of vertices ($|V_i|$) in the graph G_i;
- E_i is the set of edges in the graph G_i;
- k, l is the vertex order;
- v_{i_k} is the vertex of order k in the graph G_i;
- $e_{i_{k,l}}$ is the edge having as source v_{i_k} and as target v_{i_l} in the graph G_i.

3.1   LP4HM Input Data

Our linear program takes as input the following data: (1) a set of $N \geq 2$ graphs $G_i = (V_i, E_i)$, $i \in [1, N]$, (2) N direction matrices representing the hierarchical relationships between the elements of the graphs, (3) $N(N-1)/2$ similarity matrices representing an aggregated result of different element-level matchers.

Direction Matrices. We compute a set of N direction matrices Dir_i of size $n_i \times n_i$ defined for each graph G_i, $\forall i \in [1, N]$. Each matrix encodes edge directions. It is defined as follows:

$$Dir_i = \{dir_{i_{k,l}}, \forall k \times l \in [1, n_i] \times [1, n_i]\}$$

$$dir_{i_{k,l}} = \begin{cases} 1 & \text{if } e_{i_{k,l}} \in E_i \\ -1 & \text{if } e_{i_{l,k}} \in E_i \\ 0 & \text{otherwise} \end{cases}$$

Similarity Matrices. We compute $N(N-1)/2$ similarity matrices denoted $Sim_{i,j}$ of size $n_i \times n_j$.

$$Sim_{i,j} = \{sim_{i_k,j_l}, \forall k \in [1, n_i], \forall l \in [1, n_j], \forall i \in [1, N-1], j \in [i+1, N]\}$$

For each pairwise graph G_i and G_j, a similarity measure sim_{i_k,j_l} is computed between all combination of labels of vertices v_{i_k} and v_{j_l} belonging respectively to G_i and G_j. These labels are first tokenized and stemmed, second different types of element-level matchers are applied on stemmed tokens, finally we apply the maximum as an aggregation function between the element-level matchers resulting measures.

We have selected different element-level matchers according to their time performance and quality in the recent comparative study of [23]. The selected metrics are as follows: (1) from the category character-based metrics we have chosen Edit distance, Monge-Elkan, Jaro-Winkler, ISUB and 3-gram to compute similarity between tokens and we have applied the generalized Mongue-Elkan [14] method on these metrics to get the similarity between concepts, (2) from the category token-based, we have applied Jaccard, soft TF-IDF with Levenshtein and (3) from the language-based category we have chosen Lin [15] and WUP [24]. WUP was not evaluated in [23] but it was emphasized as a well elaborate metric in [10]. We have used different libraries implementing these metrics: OntoSim[2], SimMetric[3], SecondString[4] and WS4J[5].

In this step, we also compute the value of the predefined threshold. For pairwise matching, this threshold corresponds to the median of all the maximums of rows in the similarity matrix. For the holistic matching, this threshold is the median of all local-threshold of each pairwise graph.

We acknowledge that our choices for the default threshold and the aggregation function are simplistic compared to other works in the literature, which focus more deeply on the problematic of matchers tuning and combination [10]. We point out that the main contribution of this paper resides in the holistic linear program of the following section.

3.2   Description of LP4HM

In this section, we describe the formalization of our linear program for holistic matching. The formalization is generalizable for $N \geq 2$ graphs. We will use the example of Fig. 1 to illustrate some resulting constraints of LP4HM. We emphasize that to solve LP4HM for this example, the complete model is composed of the decision variables and the constraints of the three combinations (G_1, G_2), (G_1, G_3) and (G_2, G_3). Due to space consideration, we illustrate only the decision variables and the constraints of the combination (G_1, G_2).

Decision Variable. We define a single decision variable which expresses the possibility to have or not a matching between two vertices belonging to two different input graphs. For each G_i and G_j, $\forall i \in [1, N-1]$, $j \in [i+1, N]$, x_{i_k,j_l} is a binary decision variable equals to 1 if the vertex v_{i_k} in the graph G_i matches with the vertex v_{j_l} in the graph G_j and 0 otherwise.

Example 1. In Fig. 1, we have 6 decision variables between G_1 and G_2, the shown one are $x_{1_1,2_1}$, $x_{1_2,2_1}$, the not shown one are $x_{1_1,2_2}$, $x_{1_1,2_3}$, $x_{1_2,2_2}$, $x_{1_2,2_3}$.

---

[2] http://ontosim.gforge.inria.fr/.
[3] http://sourceforge.net/projects/simmetrics/.
[4] http://secondstring.sourceforge.net/.
[5] https://code.google.com/p/ws4j/.

Linear Constraints. Our linear program involves four constraints: C1 expresses the matching cardinality, we focus on 1:1 mapping cardinality [19], C2 constraints the selected mappings to be superior than a given threshold, C3 expresses sub/super class relations and C4 expresses coherence between edge directions.

C1 (Matching Cardinality). Resolving 1:1 mapping cardinality is equivalent to resolve a set of disjoint edges in the MWGM problem. To achieve this requirement, each vertex v_{i_k} in the graph G_i could match with at most one vertex v_{j_l} in the graph G_j, $\forall i \times j \in [1, N - 1] \times [i + 1, N]$. The corresponding constraint is as follows:

$$\sum_{l=1}^{n_j} x_{i_k, j_l} \leq 1, \quad \forall k \in [1, n_i]$$

Example 2. Applying C1 between G_1 and G_2 generates the following constraints:
$x_{1_1,2_1} + x_{1_1,2_2} + x_{1_1,2_3} \leq 1$
$x_{1_2,2_1} + x_{1_2,2_2} + x_{1_2,2_3} \leq 1$
$x_{1_1,2_1} + x_{1_2,2_1} \leq 1$
$x_{1_1,2_2} + x_{1_2,2_2} \leq 1$
$x_{1_1,2_3} + x_{1_2,2_3} \leq 1$

C2 (Matching Threshold). Usually matching systems [2, 16] have to use a threshold in the mapping selection phase in order to enhance their matching quality. Our model handles this practice for a predefined threshold *thresh* computed as input data. The following constraint restricts the set of mapping solutions to be greater than a given *thresh*. $\forall i \times j \in [1, N - 1] \times [i + 1, N]$ and $\forall k \times l \in [1, n_i] \times [1, n_j]$

$$sim_{i_k, j_l}\, x_{i_k, j_l} \geq thresh\, x_{i_k, j_l}$$

Example 3. Applying C2 for a *thresh* $= 0.4$ between G_1 and G_2 generates the following constraints:
$0.35 x_{1_1,2_1} \geq 0.4 x_{1_1,2_1}$
$0.7 x_{1_2,2_1} \geq 0.4 x_{1_2,2_1}$

The decision variable $x_{1_2,2_1}$ can be affected to 0 or 1 as its similarity is higher than the value of the thresh. For a threshold equals to zero, our model is functional without this constraint. This is very interesting namely for high heterogeneous datasets.

The second part of our constraints is related to structural matching. The idea is quite simple, it consists of prohibiting some cases of inconsistency by means of linear constraints. Figure 2 illustrates these cases, in the right side we have different possible matching solutions for the graphs given in the left side. The cases (a) and (f) represent incoherence in structural matching. Indeed, case

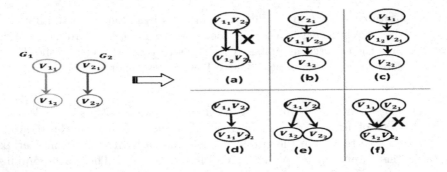

Fig. 2. An example of structural inconsistency

(a) can be generated when parents match with children and vice versa. It gives rise to incoherence in edges directions. Case (f) can be generated when children match and their parents do not match. This case is known in schema matching literature [10] as super/sub class matching. It is particularity interesting for hierarchical structures in graph-like schemes. In the following, we propose the constraint C3 and C4 to resolve respectively the cases (f) and (a).

C3 (Super/Sub Class Matching). This constraint makes parents match when their children match. For each pair of vertex v_{i_k} and v_{j_l} $\forall i \times j \in [1, N-1] \times [i+1, N]$ and $\forall k \times l \in [1, n_i] \times [1, n_j]$, their predecessors $v_{i_{pred(k)}}$ and $v_{j_{pred(l)}}$ have to match.

$$x_{i_k, j_l} \leq x_{i_{pred(k)}, j_{pred(l)}}$$

Example 4. Applying C3 between G_1 and G_2 generates the following constraints:

$x_{1_2, 2_2} \leq x_{1_1, 2_1}$
$x_{1_2, 2_3} \leq x_{1_1, 2_1}$

C4 (Edges Direction Coherence). The purpose of this constraint is to prevent the generation of conflictual edges. By using direction matrices, the product of directions of the pairwise vertices has to be equal to 1 (i.e. both edges have values -1 or 1). The constraint is expresses as follows: $\forall i \times j \in [1, N-1] \times [i+1, N]$ such as $i \neq j$ and $\forall k, k' \in [1, n_i]$ $\forall l, l' \in [1, n_j]$

$$x_{i_k, j_l} + x_{i_{k'}, j_{l'}} + (dir_{i_{k,k'}} dir_{j_{l,l'}}) \leq 0$$

Example 5. Applying C4 between G_1 and G_2 generates the following constraints:

$x_{1_1, 2_2} + x_{1_2, 2_1} + dir_{1_1, 1_2} dir_{2_2, 2_1} \leq 0$
$x_{1_1, 2_3} + x_{1_2, 2_1} + dir_{1_1, 1_2} dir_{2_3, 2_1} \leq 0$
By substituting the direction values $dir_{1_1, 1_2} = 1$, $dir_{2_2, 2_1} = -1$, $dir_{2_3, 2_1} = -1$ these constraints are equivalent to:

$x_{1_1, 2_2} + x_{1_2, 2_1} \leq 1$
$x_{1_1, 2_3} + x_{1_2, 2_1} \leq 1$

These constraints do not allow the matching of (v_{1_1}, v_{2_2}) and (v_{1_2}, v_{2_1}) in the same time (if it is the case we have $1 + 1 \leq 1$ which is impossible). An important aspect in this constraint is that it allows the feasibility of one of these matchings.

LP4HM Model

$$
\left\{
\begin{aligned}
&\max \sum_{i=1}^{N-1} \sum_{j=i+1}^{N} \sum_{k=1}^{n_i} \sum_{l=1}^{n_j} sim_{i_k, j_l}\, x_{i_k, j_l} \\
\\
&s.t. \quad \sum_{l=1}^{n_j} x_{i_k, j_l} \leq 1, \; \forall k \in [1, n_i] \hspace{2.5cm} (C1) \\
&\hspace{3cm} \forall i \in [1, N-1], \, j \in [i+1, N] \\
\\
&\quad sim_{i_k, j_l}\, x_{i_k, j_l} \geq thresh\; x_{i_k, j_l} \hspace{1.5cm} (C2) \\
&\hspace{3cm} \forall i \in [1, N-1], \, j \in [i+1, N] \\
&\hspace{3cm} \forall k \in [1, n_i], \, \forall l \in [1, n_j] \\
\\
&\quad x_{i_k, j_l} \leq x_{i_{pred(k)}, j_{pred(l)}} \hspace{2cm} (C3) \\
&\hspace{3cm} \forall i \in [1, N-1], \, j \in [i+1, N] \\
&\hspace{3cm} \forall k \in [1, n_i], \, \forall l \in [1, n_j] \\
\\
&\quad x_{i_k, j_l} + x_{i_{k'}, j_{l'}} - (dir_{i_{k,k'}} dir_{j_{l,l'}}) \leq 1 \hspace{0.5cm} (C4) \\
&\hspace{3cm} \forall i \in [1, N-1], \, j \in [i+1, N] \\
&\hspace{3cm} \forall k, k' \in [1, n_i], \, \forall l, l' \in [1, n_j] \\
\\
&\quad x_{i_k, j_l} \in \{0, 1\} \;\; \forall i \in [1, N-1], \, j \in [i+1, N] \\
&\hspace{3cm} \forall k \in [1, n_i], \, \forall l \in [1, n_j]
\end{aligned}
\right.
$$

A Relaxation of LP4HM. The LP4HM program focuses on 1:1 mapping cardinalities using binary decision variables. We propose to relax the decision variables in the [0,1] interval. This relaxation enables resolving n:m matching cardinalities. Suppose that we have two vertices "first name" and "last name" both having the same similarity distance to "name". Therefore, we have two binary decision variables with the same similarity value, only one of these decision variables will be chosen. By relaxing variables in the [0,1] interval both variables will be assigned with a 0.5 value. We name LP4HM(relax) a relaxed version of LP4HM with decision variables in [0,1] resolving n:m mapping cardinalities.

4   Experimental Assessments

As far as we know, all existing benchmarks are devoted to pairwise schema matching problem. So, we have experimented our approach on a recent pairwise schema matching benchmark proposed by [6]. This benchmark is composed of ten datasets having different criteria and each one is composed of two XSD

Table 1. Average results of LP4HM, COMA++, SF and YAM

	Precision (%)	Recall (%)	F-Measure (%)	Accuracy (%)	HSR (%)
LP4HM_A	39	50	42	14	53
LP4HM_A(relax)	24	68	34	0	59
LP4HM_B	59	50	52	27	52
LP4HM_B(relax)	50	50	47	25	53
COMA++	66	36	43	28	34
SF	50	41	44	15	39
YAM	61	55	56	27	47

schemes. In this benchmark three tools have been compared: the generic matcher COMA++ [2], Similarity Flooding (SF) [16] and YAM [7]. SF was experimented with a threshold equals to 1. The experimentations of COMA++ have been done on three strategies (AllContext, FilteredContext and NoContext) and the best results have been maintained. The results of YAM correspond to an average of 200 runs per dataset.

Our approach is evaluated on two scenarios denoted "A" and "B". Scenario "A" consists of evaluating LP4HM and LP4HM(Relax) without the threshold constraint i.e. we do not use the constraint C2. For this scenario, we denote our approaches as LP4HM_A and LP4HM_A(Relax). Scenario "B" consists of evaluating LP4HM and LP4HM(Relax) with all the constraints. We have used a default threshold fixed to the median of the maximum similarity of each similarity matrix. This threshold is computed in the pre-processing step and inputted to the model. For this scenario, we denote our approaches as LP4HM_B and LP4HM_B(Relax). Our approach is resolved with an academic version of the CPLEX solver.

In this benchmark, the matching quality is evaluated according to precision, recall, f-measure, overall [16] and HSR (Human Spared Resources) [6]. Precision, recall and f-measure are well known measures issued from the information retrieval field. The overall measure is proposed by [16] and the HSR measure is proposed by [6]. Both measures compute post-match effort gained by the use of the matching tool.

4.1   Average Results

Table 1 and Fig. 3 present average results for precision, recall, f-measure, overall and HSR of LP4HM and LP4HM(Relax) on strategies "A" and "B", COMA++, SF and YAM. Without using a threshold in strategy "A", our approach has low precision results and high recall results particularly for the relaxed version. By using a threshold in strategy "B", the results of precision and recall of the relaxed and non relaxed versions, are more balanced which leads to balanced f-measure results. We can also observe that overall results are correlated to precision results. Indeed, if the precision is lower than or equals to 50 % so the overall results are

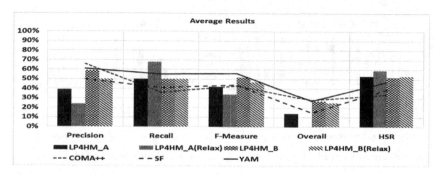

Fig. 3. A representation of the average results of LP4HM, COMA++, SF and YAM

lower than zero as reported in [16] (we note that all negative results have been rounded to 0 for all the approaches). Moreover, we notice that the HSR results are correlated to recall results as reported in [6]. The results of LP4HM_A(Relax) show more clearly these observations: the average precision is equal to 24 %, the average recall is equal to 68 %, the average overall is equal to 0 % and the average HSR is equal to 59 %. Compared to other approaches, LP4HM_A(Relax) has the worst precision and overall but the best recall and HSR. Otherwise, we notice that the results of LP4HM_B are close to the results of YAM, which is the best compared to COMA++ and SF.

The comparison of the different strategies of our approach shows that the non relaxed versions LP4HM_A and LP4HM_B give a better compromise for all the quality measures. Using a pre-defined threshold for LP4HM_B enhances precision results of LP4HM_A. The recall still the same so we have better results on F-Measure and overall. Even if using a threshold improves the results of the different quality measure, we highlight that without using a threshold nor learning the HSR results of both LP4HM_A and LP4HM_A(Relax) are very interesting. These crucial results would be very interesting for holistic scenarios. The matching of $N \geq 2$ schemes is more difficult than matching 2 schemes. Indeed, if precision is better than recall so users have to find the missing mappings for N schemes simultaneously, which is a human difficult task. So when system returns good recall and low precision, users have just to eliminate the not relevant mappings proposed by the matcher.

In the next section, we will detail the results obtained for the different datasets.

4.2 Detailed Results

The authors of [6] have classified the different datasets according to five properties: label heterogeneity, domain specific, average size, structure and number of schemes. In this section, we will present the detailed results by grouping the datasets according to their average size. The average size represents the average number of schema elements. We will briefly report the type of the different properties for each dataset. Please refer to [6] for more details about the descriptions of these datasets.

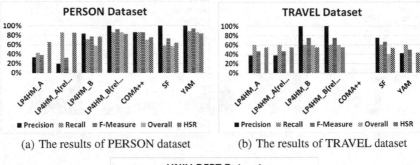

(a) The results of PERSON dataset   (b) The results of TRAVEL dataset

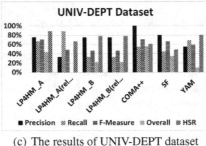

(c) The results of UNIV-DEPT dataset

Fig. 4. Results of small size datasets.

Small Size Datasets (<10 Elements). The datasets PERSON, TRAVEL and UNIVERSITY DEPARTMENT (UNIV-DEPT) are small size datasets. PERSON dataset has low label heterogeneity and a nested structure. TRAVEL dataset has average label heterogeneity and a flat structure. UNIV-DEPT has high label heterogeneity and a flat structure.

The results of PERSON dataset are depicted in Fig. 4(a). We notice that LP4HM_B (Relax) performs, without learning, the same results as YAM. LP4HM_B (Relax) and LP4HM_A (Relax) get the same best results of recall, but LP4HM_B(Relax) outperforms LP4HM_A(Relax) in precision. The recall of relaxed versions is better than the recall of not relaxed versions which is explained by the presence of 1:n expert mappings.

The results of TRAVEL dataset are depicted in Fig. 4(b). We observe that the strategy "B" of our version gives better results than strategy "A". But, it is worth to note that the results of recall are the same for the four versions. This implies that with only the proposed constraints our method is able to find 60 % of expert mappings. For this dataset our approach is better than or equals to other approaches in recall and HSR.

The results of UNIV-DEPT dataset are depicted in Fig. 4(c). Contrarily to the two datasets described above, the strategy "A" seems more efficient than strategy "B". Indeed, we can observe that recall results of strategy "B" are worse than those in strategy "A". So experiments show that fixing a threshold in highly heterogeneous datasets will remove relevant solutions that have low similarity values. In this dataset, our approach is better than other approaches.

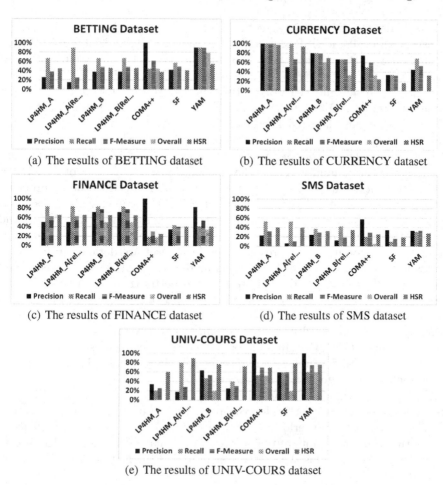

(a) The results of BETTING dataset

(b) The results of CURRENCY dataset

(c) The results of FINANCE dataset

(d) The results of SMS dataset

(e) The results of UNIV-COURS dataset

Fig. 5. The results of average size datasets

For small size datasets, our approach gives competitive results compared to COMA++, SF and YAM. For low and average heterogeneity, nested or flat structures, the two versions of our approach with a prefixed threshold are more effective. For high heterogeneity, the two versions of our approach without using a threshold nor learning on datasets are more effective.

Average Size Datasets (10–100 Elements). The datasets BETTING, CURRENCY, FINANCE, SMS and UNIVERSITY COURSES (UNIV-COURS) are average size datasets. BETTING dataset has average label heterogeneity and a flat structure. CURRENCY dataset has average label heterogeneity and a nested structure. FINANCE is a domain specific dataset. It has average label heterogeneity and a flat structure. SMS dataset has average label heterogeneity and a flat structure.

The results of BETTING dataset are depicted in Fig. 5(a). For this dataset, our results on recall and HSR are roughly the same as YAM. The prefixed threshold for strategy "B" enhances precision but the recall still the same as the not relaxed version of the strategy "A". We can notice that LP4HM_A(Relax) performs 89 % of recall and the worst precision. Contrarily to COMA++ whose precision attempts 100 % but the recall is lower than 50 %. The results of CURRENCY dataset are depicted in Fig. 5(b). The results of our approach in the different versions outperform the results of COMA++, SF and YAM. Strategy "A" of our approach is more effective than strategy "B", which demonstrates that using a threshold eliminates some relevant solutions. The results of FINANCE dataset are depicted in Fig. 5(c). Our approach outperforms the other approaches for all measures except precision measure. Recall is the same for all the versions of our approach. Strategy "B" is little better than strategy "A". We remind that this dataset is a domain specific dataset, using Wordnet[6] as a dictionary is a good external resources. The results of SMS dataset are detailed in Fig. 5(d). For this nested dataset strategy "A" gives more important results on recall and HSR compared to strategy "B". Our approach is better than COMA++, SF and YAM for all measures except precision measure. The results of UNIV-COURS dataset are shown in Fig. 5(e). Users gain at least 60 % and at most 90 % of human spared resources with not relaxed and relaxed versions of strategy "A" of our approach. LP4HM_A(Relax) seems the more efficient strategy for this datasets, because 80 % of relevant mappings have been detected and users have only to remove extra not relevant mappings.

We can notice that all average size datasets have also average label heterogeneity. We have observed that for the nested structure datasets the strategy "A" of our approach outperforms the strategy "B". This shows the efficiency of the structural constraints of our approach. For flat structure, we think that LP4HM_A(Relaxed) seems the best in recall and HSR. But LP4HM_B is more balanced.

Large Size Datasets (>100 Elements). The two large size datasets of this benchmark are BIOLOGY and ORDER. Both datasets have a nested structure. BIOLOGY is a domain specific dataset and has average label heterogeneity, while ORDER has low label heterogeneity.

Figure 6 depicts the results of BIOLOGY and ORDER datasets. In Fig. 6(a), we notice that our approach fails to get relevant matchings for BIOLOGY dataset. We can also observe that the results of the other approaches are very low. As BIOLOGY dataset is domain specific, our approach needs some external sources like specific dictionaries or a list of predefined synonyms. In Fig. 6(b), we can observe that our approach especially LP4HM_A(Relax) successes to find 40 % of the relevant mappings for ORDER dataset. Our results are similar to the results of SF and YAM.

The two large datasets, proposed in this benchmark, show the difficulties to maintain good quality results for large size datasets especially when the dataset

---

[6] https://wordnet.princeton.edu/.

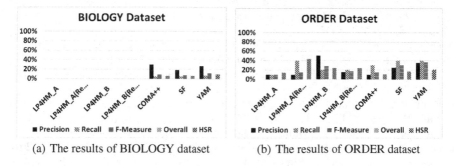

(a) The results of BIOLOGY dataset   (b) The results of ORDER dataset

Fig. 6. The results of large size datasets

is domain specific. We have also noticed that in these datasets an important number of sub-trees is repeated (due to references in XSD schemes). This is another difficulty in these datasets compared to the other datasets.

5 Conclusion

In this paper, we have presented a new approach to resolve the holistic matching of $N \geq 2$ schemes. Our approach is based on the linear programming technique to model the matching problem and its characteristics, such as the mapping cardinality and threshold filtering. Our model also holds constraints for the coherence of structural matching applied on hierarchical structures. We have experimented our approach on a recent benchmark in the literature [6]. The results of our approach are interesting for small and medium size datasets of different types of heterogeneity. We highlight that our approach returns good recall and HSR results without learning techniques nor threshold tuning. Our results are competitive compared to the results of existing approaches. Our future work will be devoted, first to experiment the quality of our model for other large size datasets with different types of aggregation functions and second to extend our model with further constraints to match labelled graphs.

References

1. Agreste, S., Meo, P.D., Ferrara, E., Ursino, D.: XML matchers: approaches and challenges. Knowl.-Based Syst. **66**, 190–209 (2014)
2. Aumueller, D., Do, H.H., Massmann, S., Rahm, E.: Schema and ontology matching with COMA++. In: SIGMOD 2005. pp. 906–908 (2005)
3. Berro, A., Megdiche, I., Teste, O.: A content-driven ETL processes for open data. In: Bassiliades, N., Ivanovic, M., Kon-Popovska, M., Manolopoulos, Y., Palpanas, T., Trajcevski, G., Vakali, A. (eds.) New Trends in Database and Information Systems II. AISC, vol. 312, pp. 29–40. Springer, Heidelberg (2015)
4. Berro, A., Megdiche, I., Teste, O.: Holistic statistical open data integration based on integer linear programming. In: RCIS 2015, pp. 524–535 (2015)

5. Do, H.H., Rahm, E.: Matching large schemas: approaches and evaluation. Inf. Syst. **32**(6), 857–885 (2007)
6. Duchateau, F., Bellahsene, Z.: Designing a benchmark for the assessment of schema matching tools. Open J. Databases (OJDB) **1**(1), 3–25 (2014)
7. Duchateau, F., Coletta, R., Miller, R.J.: Yam: a schema matcher factory. In: CIKM, pp. 2079–2080 (2009)
8. Edmonds, J.: Maximum matching and a polyhedron with 0, 1-vertices. J. Res. Natl. Bur. Stand. B **69**, 125–130 (1965)
9. Euzenat, J., Shvaiko, P.: Ontology Matching. Springer, Heidelberg (2007)
10. Euzenat, J., Shvaiko, P.: Ontology Matching, 2nd edn. Springer, Heidelberg (2013)
11. Euzenat, J., Valtchev, P.: Similarity-based ontology alignment in owl-lite. In: Proceedings of the 16th European Conference on Artificial Intelligence (ECAI), pp. 333–337. IOS press (2004)
12. Giunchiglia, F., Yatskevich, M., Shvaiko, P.: Semantic matching: algorithms and implementation. In: Spaccapietra, S., Atzeni, P., Fages, F., Hacid, M.-S., Kifer, M., Mylopoulos, J., Pernici, B., Shvaiko, P., Trujillo, J., Zaihrayeu, I. (eds.) Journal on Data Semantics IX. LNCS, vol. 4601, pp. 1–38. Springer, Heidelberg (2007)
13. Huber, J., Sztyler, T., Nner, J., Meilicke, C.: CODI: Combinatorial optimization for data integration: results for OAEI 2011. In: CEUR Workshop Proceedings on OM, vol. 814 (2011). http://CEUR-WS.org
14. Jimenez, S., Becerra, C., Gelbukh, A., Gonzalez, F.: Generalized mongue-elkan method for approximate text string comparison. In: Gelbukh, A. (ed.) CICLing 2009. LNCS, vol. 5449, pp. 559–570. Springer, Heidelberg (2009)
15. Lin, D.: An information-theoretic definition of similarity. In. In Proceedings of the 15th International Conference on Machine Learning, pp. 296–304. Morgan Kaufmann (1998)
16. Melnik, S., Garcia-Molina, H., Rahm, E.: Similarity flooding: a versatile graph matching algorithm and its application to schema matching. In: Proceedings of the 18th International Conference on Data Engineering, ICDE 2002, pp. 117–128. IEEE Computer Society (2002)
17. Niepert, M., Meilicke, C., Stuckenschmidt, H.: A probabilistic-logical framework for ontology matching. In: Proceedings of the 24th AAAI Conference on Artificial Intelligence, pp. 1413–1418. AAAI Press (2010)
18. Rahm, E.: Towards large-scale schema and ontology matching. In: Bellahsene, Z., Bonifati, A., Rahm, E. (eds.) Schema Matching and Mapping. Data-Centric Systems and Applications, pp. 3–27. Springer, Heidelberg (2011)
19. Rahm, E., Bernstein, P.A.: A survey of approaches to automatic schema matching. VLDB J. **10**, 334–350 (2001)
20. Schrijver, A.: Combinatorial Optimization: Polyhedra and Efficiency. Springer, Heidelberg (2003)
21. Shvaiko, P., Euzenat, J.: Ontology matching: State of the art and future challenges. IEEE Trans. Knowl. Data Eng. **25**(1), 158–176 (2013)
22. Shvaiko, P., Euzenat, J.: A Survey of Schema-Based Matching Approaches. In: Spaccapietra, S. (ed.) Journal on Data Semantics IV. LNCS, vol. 3730, pp. 146–171. Springer, Heidelberg (2005)
23. Sun, Y., Ma, L., Shuang, W.: A comparative evaluation of string similarity metrics for ontology alignement. J. Inf. Comput. Sci. **12**(3), 957–964 (2015)
24. Wu, Z., Palmer., M.: Verb semantics and lexical selection. In: 32nd Annual Meeting of the Association for Computational Linguistics, New Mexico State University, Las Cruces, New Mexico, pp. 133–138 (1994)

An External Memory Algorithm for All-Pairs Regular Path Problem

Nobutaka Suzuki[1]([✉]), Kosetsu Ikeda[1], and Yeondae Kwon[2]

[1] University of Tsukuba, 1-2 Kasuga, Tsukuba, Ibaraki 305-8550, Japan
{nsuzuki,lumely}@slis.tsukuba.ac.jp
[2] Tokyo University of Science, 2641 Yamazaki, Noda, Chiba 278-8510, Japan
yekwon@rs.noda.tus.ac.jp

Abstract. In this paper, we consider solving the all-pairs regular path problem on large graphs efficiently. Let G be a graph and r be a regular path query, and consider finding the answers of r on G. If G is so small that it fits in main memory, it suffices to load entire G into main memory and traverse G to find paths matching r. However, if G is too large and cannot fit in main memory, we need another approach. In this paper, we propose an external memory algorithm for solving all-pairs regular path problem on large graphs. Our algorithm finds the answers matching r by scanning the node list of G sequentially, which avoids random accesses to disk and thus makes regular path query processing I/O efficient.

1 Introduction

Graph has been used in broad areas such as social network, life science, Semantic Web, and so on. Recently, such graphs are greatly increasing and their sizes are rapidly growing. Regular path query is a popular query language for such graphs and is being studied actively so far. In this paper, we consider processing regular path queries efficiently on large graphs. In particular, we focus on solving the all-pairs regular path problem, which is to find all pairs (n, n') of nodes such that there is a path between n and n' whose sequence of labels matches a given regular path query.

Let us consider finding the answers of a regular path query r on a graph G. If G is so small that it fits in main memory, it suffices to load entire G into main memory, transform r into an NFA A, and find the "answer" paths of G whose sequence of labels matches A. However, if G is too large to fit in main memory, the above approach is not applicable to obtaining the answers of r. Thus we need another approach to solving the problem on large graphs efficiently.

A possible approach to this problem is to store a graph in some graph store and issue regular path queries on the graph store. However, to solve the problem we have to determine, for a large number of pair of nodes (n, n'), whether there exists a path between n and n' matching a given regular path query. This requires

---

Y. Kwon–Currently, the author is with Graduate School of Agricultural and Life Sciences, The University of Tokyo, 1-1-1 Yayoi, Bunkyo-ku, Tokyo 113-8657, Japan.

Q. Chen et al. (Eds.): DEXA 2015, Part II, LNCS 9262, pp. 399–414, 2015.
DOI: 10.1007/978-3-319-22852-5_34

fetching nodes and edges on a disk a large number of times, and thus the approach is not necessarily I/O efficient for solving the problem.

Another possible approach is to use the SPARQL query language. In fact, SPARQL is added support for regular path query processing called "Property Path" since version 1.1, and thus we can use a regular path query on RDF stores supporting SPARQL 1.1. However, Property Path is under "simple walk semantics" (no node appears twice on a path except the start and the end of the path), and thus Property Path evaluation becomes quite inefficient even for very restricted queries [6], e.g., evaluating $(aa)^*$ is NP-complete.

To cope with these problems, we propose another novel approach based on external memory algorithm. Let G be a graph (file) and B be an area allocated in main memory. Firstly, our algorithm scans G sequentially and repeats the following until EOF is found.

1. Load data of size $|B|$ from G into B (thus the loaded data is a subgraph of G) and find answers that can be obtained by traversing B.
2. During the traversal of B, find connections among paths inside and outside B, and store them into another graph called "contracted graph".

After the scan, the algorithm traverses the contracted graph and finds the rest of answers whose path runs across different subgraphs, which is also done by a sequential scan. Our approach highly reduces random accesses to disk, which can make processing regular path queries I/O efficient.

Related Work. There are a number of studies related to regular path queries, e.g., [3,15] are extensive surveys on regular path queries and related query languages. However, studies on regular path query processing on large graphs are unexpectedly not many. [5] proposes an algorithm for solving all-pairs regular path problem efficiently on large graphs. This algorithm is an in-memory algorithm and assumes that a graph fits in main memory. [2,12–14] propose distributed approaches for regular path query processing. These are suitable for graphs inherently distributed over multiple machines or graphs that are too large to be handled in a single machine. On the other hand, for graphs that can be handled in a single machine, we believe that our approach is a reasonable choice for regular path query processing.

A number of graph stores have been proposed, e.g., Pregel [8], Sparksee[1] (formerly known as Dex [9]), Grail [16], and Neo4j[2], and for some of them a powerful query language that can simulate regular path queries is available (Gremlin[3]). However, these are not necessarily suitable for the all-pairs regular path problem on large graphs, since solving the problem on a graph store requires fetching a large number of nodes and edges due to the high complexity of the problem. gStore [19] is an RDF store with an efficient query processor, but this does not support regular path query. On the other hand, SPARQL 1.1 supports regular path query by virtue of Property Path, and thus we can use regular path queries

---

[1] http://www.sparsity-technologies.com/.

[2] http://www.neo4j.org/.

[3] https://github.com/tinkerpop/gremlin.

on RDF stores supporting SPARQL 1.1, e.g., Apache Jena[4] and Sesame[5]. However, Property Path of SPARQL 1.1 is under simple walk semantics, and thus the complexity of regular path query evaluation is NP-hard even for very restricted queries [6]. Therefore, such RDF stores are not necessarily suitable for solving the all-pairs regular path problem on large graphs either.

Several external memory algorithms have been proposed in database research field, e.g., graph triangulation [4], strongly connected components [17], graph reachability [18], and k-bismulation [7]. To the best of our knowledge, however, no external memory algorithm for processing the all-pairs regular path problem has been proposed so far.

2   Definitions

Let Σ be a set of labels. A *labeled directed graph* (*graph* for short) over Σ is denoted $G = (V, E)$, where V is a set of nodes and E is a set of labeled edges. Let $n, n' \in V$ be nodes. An edge from n to n' labeled by $l \in \Sigma$ is denoted $n \xrightarrow{l} n'$. A path p from n to n' is denoted $n \rightsquigarrow_p n'$. By $l(p)$ we mean the sequence of labels on p. For example, if $p = n_1 \xrightarrow{a} n_2 \xrightarrow{b} n_3 \xrightarrow{c} n_4$, then $l(p) = abc$.

A *regular path query* is defined as a regular expression over Σ, as follows.

- ε, $\emptyset$, and $a \in \Sigma$ are regular path queries.
- If $r_1, \cdots, r_n$ are regular path queries, then $r_1 \cdots r_n$ and $r_1| \cdots |r_n$ are regular path queries.
- If r is a regular path query, then r^* is a regular path query.

r^+ and $r?$ are abbreviations for r^*r and $r|\varepsilon$, respectively. By $L(r)$ we mean the *language* of a regular path query r.

A nondeterministic finite automaton (NFA) over Σ is a quintuple $A = (Q, \Sigma, \delta, q_0, F)$, where Q is a set of *states*, $q_0 \in Q$ is the *start state*, $F \subseteq Q$ is a set of *accepting states*, and $\delta : Q \times \Sigma \to 2^Q$ is a *transition function*. The *extension* of δ, denoted $\hat{\delta}$, is defined as follows.

- For any $q \in Q$, $\hat{\delta}(q, \varepsilon) = \{q\}$.
- For any $w \in \Sigma^*$ and any $a \in \Sigma$, $\hat{\delta}(q, wa) = \bigcup_{q' \in \hat{\delta}(q,w)} \delta(q', a)$.

That is, $q' \in \hat{\delta}(q, w)$ iff A enters state q' when A reads w in state q. The *language* accepted by A is defined as $L(A) = \{w \in \Sigma^* \mid q_f \in \hat{\delta}(q_0, w), q_f \in F\}$.

In this paper, we consider finding, for a graph $G = (V, E)$ over Σ and a regular path query r, all pairs $(n, n') \in V \times V$ of nodes such that G contains a path $n \rightsquigarrow_p n'$ such that $l(p) \in L(r)$.

---

[4] https://jena.apache.org/.
[5] http://openrdf.org/.

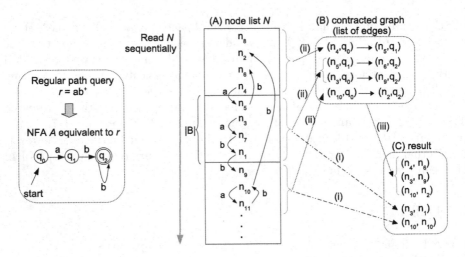

Fig. 1. Outline of our algorithm

3   Overview of Our Algorithm

Let r be a regular path query and $G = (V, E)$ be a graph. The algorithm uses an area B allocated in main memory with $|B| = \epsilon \cdot M$, where $|B|$ denotes the size of B, M is the size of main memory, and $\epsilon < 1$. The input data to our algorithm is a node list N (Fig. 1(A)), which consists of the nodes of V with some information of each node (outgoing edges, etc.). In short, the algorithm works as follows. First, the algorithm reads N sequentially and repeats the following until EOF is found.

1. Load $|B|$ bytes of data from N into B. Thus B contains a subgraph of G.
2. Traverse B and do the following.
   (a) Find the "local" answers (paths matching r) in B that can be obtained by traversing only B (Fig. 1(i)).
   (b) To find answer paths running across boundaries of B, find paths adjacent to edges outside B, and store their connections into another graph file called *contracted graph* (Fig. 1(ii) and (B)).

Then, the algorithm traverses the contracted graph obtained in step (2-b) and outputs the answers not found in step (2-a) (Fig. 1(iii)).

Let us give some definitions related to node list N. Let $n \in V$ be a node. By $order(n)$ we mean the *order* of n in N. For example, in Fig. 1(A) $order(n_8) = 1$, $order(n_2) = 2$, and so on. By $B.min$ we mean the minimum order of the nodes in B. Similarly, by $B.max$ we mean the maximum order of the nodes in B. For example, assuming that the second region of N in Fig. 1(A) is loaded into B, we have $B.min = order(n_5) = 5$ and $B.max = order(n_1) = 8$. By $Out(n)$ we mean the set of outgoing edges of n, that is, $Out(n) = \{n \xrightarrow{l} n' \mid n' \in V, n \xrightarrow{l} n' \in E\}$. By $In(n, l)$ we mean the set of source nodes of incoming edges of n

labeled by l, that is, $In(n, l) = \{n' \in V \mid n' \xrightarrow{l} n \in E\}$. By $inMax(n, l)$, we mean the node having the maximum order in $In(n, l)$, that is, $inMax(n, l) = \operatorname*{argmax}_{n' \in In(n,l)} order(n')$. Similarly, we define that $inMin(n, l) = \operatorname*{argmin}_{n' \in In(n,l)} order(n')$.

Let r be a regular path query and $A = (Q, \Sigma, \delta, q_0, F)$ be an NFA such that $L(A) = L(r)$. Suppose that $|B|$ bytes of data is loaded from N into B. If B contains a path $n \rightsquigarrow_p n'$ such that $l(p) \in L(A)$, then the algorithm outputs (n, n'). For example, in Fig. 1(A) the algorithm outputs (n_3, n_1) as an answer due to $n_3 \xrightarrow{a} n_7 \xrightarrow{b} n_1$, which can be obtained by traversing only B. However, B contains only a subgraph of G, and thus we also have to handle edges running across a boundary of B appropriately, i.e., (i) edges from outside to inside B (e.g., $n_4 \xrightarrow{a} n_5$ in Fig. 1(A)) and (ii) edges from inside to outside B (e.g., $n_1 \xrightarrow{b} n_9$ in Fig. 1(A)). We have to find paths matching r even if the paths contain such "boundary" edges.

To find such paths, our algorithm traverses B and finds paths $n \rightsquigarrow_p n'$ such that $q' \in \hat{\delta}(q, l(p))$ for some $q, q' \in Q$ (i.e., p matches a subexpression of r), then check if p is adjacent to a boundary edge mentioned above (i) and (ii). To do this, for a pair (n, q) of a node n in B and a state $q \in Q$, we define four types T_{in}, T_{out}, T_{start}, and T_{accept}, as follows.

- (n, q) is of type T_{in} if n has an incoming edge from a node outside B.
- (n, q) is of type T_{out} if n has an outgoing edge to a node outside B.
- (n, q) is of type T_{start} if q is the start state of A.
- (n, q) is of type T_{accept} if q is an accepting state of A.

T_{in} and T_{out} are used to check if (n, q) is adjacent to a "boundary" edge, while T_{start} and T_{accept} are used to check if (n, q) is a pair at which a traversal should be started/finished.

Let us formally define the above four types. In the following, if a pair (n, q) is of type T_{in}, then we write $T_{in}(n, q)$ (T_{out}, T_{start}, T_{accept} are denoted similarly).

- $T_{in}(n, q)$ iff for some $l \in \Sigma$,
  - $q \in \delta(q', l)$ for some $q' \in Q$, and
  - n has an incoming edge labeled by l from a node outside B, that is, $order(inMin(n, l)) < B.min$ or $order(inMax(n, l)) > B.max$.
- $T_{out}(n, q)$ iff for some $n \xrightarrow{l} n' \in Out(n)$,
  - $q' \in \delta(q, l)$ for some $q' \in Q$, and
  - n' is a node outside B, that is, $order(n') < B.min$ or $order(n') > B.max$.
- $T_{start}(n, q)$ iff $q = q_0$.
- $T_{accept}(n, q)$ iff $q \in F$.

In the following, we present how the algorithm finds the answers by using the above four types. Suppose that by traversing B we find a path $n \rightsquigarrow_p n'$ in B such that $q' \in \hat{\delta}(q, l(p))$.

1. If $T_{start}(n, q)$ and $T_{accept}(n', q')$, then output (n, n') as an answer. For example, consider path $p' = n_3 \xrightarrow{a} n_7 \xrightarrow{b} n_1$ in Fig. 1(A). Then since $T_{start}(n_3, q_0)$ and $T_{accept}(n_1, q_2)$ (i.e., $q_2 \in \hat{\delta}(q_0, l(p'))$), (n_3, n_1) is outputted as an answer.

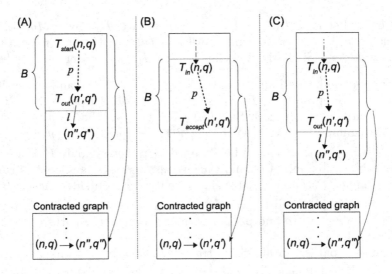

Fig. 2. Construction of contracted graph

2. If $T_{start}(n, q)$ and $T_{out}(n', q')$, then p is a path from n to n' such that $q' \in \hat{\delta}(q, l(p))$ and that n' has an edge leaving B (Fig. 2(A)). By the definition of T_{out}, we have a pair (n'', q'') of a node n'' and a state q'' satisfying the following condition.

(C1) For some edge $n' \xrightarrow{l} n'' \in Out(n')$, n'' is a node outside B and $q'' \in \delta(q', l)$.

This implies that "(n'', q'') is reachable from (n, q)", that is, G contains a path $p' = n \leadsto_p n' \xrightarrow{l} n''$ such that $q'' \in \hat{\delta}(q, l(p'))$. In order to record this fact, we use an extra graph called *contracted graph*, and add an edge $(n, q) \to (n'', q'')$ to the contracted graph. For example, consider a path $n_3 \xrightarrow{a} n_7 \xrightarrow{b} n_1$ in Fig. 1(A). Since we have $T_{start}(n_3, q_0)$, $T_{out}(n_1, q_2)$, and $n_1 \xrightarrow{b} n_9$ satisfies the condition (C1) for state q_2, $(n_3, q_0) \to (n_9, q_2)$ is added to the contracted graph.

The contracted graph is stored as a file, and later it is used to find answer paths containing boundary edges.

3. If $T_{in}(n, q)$ and $T_{accept}(n', q')$, then p is a path from n to n' such that $q' \in \hat{\delta}(q, l(p))$, n has an incoming edge entering B, and that $q' \in F$ (Fig. 2(B)). In other words, "(n', q') is reachable from (n, q)". Therefore, we add an edge $(n, q) \to (n', q')$ to the contracted graph.

4. If $T_{in}(n, q)$ and $T_{out}(n', q')$, then p is a path from n to n' such that n has an incoming edge entering B, n' has an edge leaving B, and that $q' \in \hat{\delta}(q, l(p))$ (Fig. 2(C)). By the definition of T_{out}, there exists a pair (n'', q'') satisfying the condition (C1) above, thus we have that "(n'', q'') is reachable from (n, q)". Therefore, we add an edge $(n, q) \to (n'', q'')$ to the contracted graph. For example, consider node n_5 in Fig. 1(A). Then we have $T_{in}(n_5, q_1)$ and

$T_{out}(n_5, q_1)$. Since $n_5 \overset{b}{\to} n_6 \in Out(n_5)$ satisfies the condition in (C1) for state q_2, an edge $(n_5, q_1) \to (n_6, q_2)$ is added to the contracted graph.

In summary, the algorithm works as follows. First, the algorithm reads N sequentially, and loads $|B|$ bytes of data from N into B repeatedly until EOF is found. For each loaded data in B, by doing (1) to (4) above the algorithm outputs the local answers in B and constructs a contracted graph simultaneously. After the sequential scan of N, the algorithm traverses the contracted graph and finds the paths from (n, q) to (n', q') such that q is the start state and that q' is an accepting state, and outputs (n, n') as an answer. For example, in Fig. 1(B) we have $(n_4, q_0) \to (n_5, q_1) \to (n_6, q_2)$, thus (n_4, n_6) is obtained as an answer from the contracted graph. Similarly, we obtain (n_3, n_9) and (n_{10}, n_2).

4   Details of the Algorithm

In this section, we present the details of our algorithm. Let $G = (V, E)$ be a graph and r be a regular path query. First of all, N is a list consisting of the nodes in V with the following items for each node $n \in V$; $Out(n)$, $inMin(n)$, $inMax(n)$, and $order(n)$.

Let us first present the main part of our algorithm. This procedure reads N sequentially and processes each subgraph loaded into B. For each loaded subgraph, it prunes unnecessary edges (line 7), and then finds the answers obtained by traversing B and constructs the contracted graph $cgraph$ simultaneously (line 8). After the sequential scan of N, the algorithm finds the answers obtained by traversing $cgraph$ (line 10).

<u>Procedure MAIN</u>

Input: Node list N of graph $G = (V, E)$, regular path query r
Output: Set of pairs of nodes (n, n') such that $n \rightsquigarrow_p n'$ and that $l(p) \in L(r)$

1. Create an empty file $cgraph$
2. Construct an ϵ-free NFA $A = (\Sigma, Q, \delta, q_0, F)$ such that $L(A) = L(r)$
3. Allocate an area B in main memory ($|B| = \epsilon \cdot M$)
4. Read N sequentially and do the following until EOF is found
5. **begin**
6.   Load $|B|$ bytes of data from N into B
7.   PRUNE(B, A)
8.   TRAVERSEBUF$(B, A, cgraph)$
9. **end**
10. TRAVERSECGRAPH$(A, cgraph)$

In the following, we present the details of TRAVERSEBUF (line 8), TRAVERSECGRAPH (line 10), and PRUNE (line 7).

4.1   Processing Node List

We present TRAVERSEBUF, which finds the "local" answers in B and constructs $cgraph$. Let $A = (Q, \Sigma, \delta, q_0, F)$ be an NFA such that $L(A) = L(r)$. This procedure works as follows. For each pair (n, q) of a node n in B and a state

q, if $T_{in}(n,q)$ or $T_{start}(n,q)$, then we start a traversal from (n,q)[6] (lines 2 to 4). Let (n',q') be a pair encountered in this traversal. If $T_{start}(n,q)$ and $T_{accept}(n',q')$, then (n,n') is outputted as an answer (lines 6 and 7). If $T_{in}(n,q)$ and $T_{accept}(n',q')$, then $(n,q) \rightarrow (n',q')$ is added to $cgraph$ (lines 8, 9, and 14). If $T_{out}(n',q')$, then for every pair (n'',q'') satisfying the condition (C1), $(n,q) \rightarrow (n'',q'')$ is added to $cgraph$ (lines 10 to 14).

Procedure TRAVERSEBUF

Input: Subsequence of N loaded into B, ε-free NFA $A = (Q, \Sigma, \delta, q_0, F)$, contracted graph $cgraph$ (empty file)

Output: Set of pairs (n,n') of nodes in B such that $n \rightsquigarrow_p n'$ and that $l(p) \in L(r)$, contracted graph $cgraph$

1. $tmp\_cgraph \leftarrow \emptyset$
2.   **for each** pair (n,q) of a node n in B and a state $q \in Q$ **do**
3.   **if** $T_{in}(n,q)$ or $T_{start}(n,q)$ **then**
4.   Traverse B from n and find all pairs (n',q') such that $n \rightsquigarrow_p n'$, $q' \in \hat{\delta}(q, l(p))$ for some $q' \in Q$, and that $T_{out}(n',q')$ or $T_{accept}(n',q')$. Let S be the resulting set.
5.   **for each** $(n',q') \in S$ **do**
6.   **if** $T_{start}(n,q)$ and $T_{accept}(n',q')$ **then**
7.   Output (n,n') as an answer
8.   **if** $T_{in}(n,q)$ and $T_{accept}(n',q')$ **then**
9. $tmp\_cgraph \leftarrow tmp\_cgraph \cup \{(n,q) \rightarrow (n',q')\}$
10.   **if** $T_{out}(n',q')$ **then**
11.   **for each** $n' \xrightarrow{l} n'' \in Out(n')$ and $q'' \in \delta(q', l)$ **do**
12.   **if** $order(n'') < B.min$ or $order(n'') > B.max$ **then**
13. $tmp\_cgraph \leftarrow tmp\_cgraph \cup \{(n,q) \rightarrow (n'',q'')\}$
14. Add each edge in tmp_cgraph to $cgraph$ in ascending order

In line 14, the edges $(n,q) \rightarrow (n',q')$ in tmp_cgraph are added to $cgraph$ in ascending order of $order(n)$, as shown in Fig. 1(B). This ordering is used when traversing $cgraph$ later.

4.2   Processing Contracted Graph

We next present TRAVERSECGRAPH, which traverses $cgraph$ and outputs the rest of answers. If the size of $cgraph$ does not exceed $|B|$, then it suffice to load entire $cgraph$ into B and find the answers by traversing B (lines 1 to 4). In line 3, we write $(n,q_0) \rightsquigarrow (n',q')$ if there exists a path from (n,q_0) to (n',q') in B. In most cases, $cgraph$ is small and the traversal of $cgraph$ is completed here. On the other hand, if the size of $cgraph$ exceeds $|B|$, the procedure repeats a forward scan of $cgraph$ (line 8) and a backward scan of $cgraph$ (line 9) alternatively, until all the answers contained in $cgraph$ are found.

---

[6] We currently adopt breadth first search but any other search strategies can be applicable.

Procedure TRAVERSECGRAPH

Input: ε-free NFA $A = (Q, \Sigma, \delta, q_0, F)$, contracted graph *cgraph*
Output: Set of pairs (n, n') in *cgraph* such that $n \leadsto_p n'$, $q' \in \hat{\delta}(q_0, l(p))$,
    and that $q' \in F$

1. **if** the size of *cgraph* does not exceed $|B|$ **then**
2.   Load entire *cgraph* into B
3.   Traverse B and find all pairs (n, n')
        such that $(n, q_0) \leadsto (n', q')$ and that $q' \in F$
4.   Output the pairs obtained in line 3
5. **else**
6.   Create priority queues *pq_fwd* and *pq_bwd*
        Initially, *pq_fwd* and *pq_bwd* are empty
7.   **do**
8.   SCANFORWARD(A, *cgraph*, *pq_fwd*, *pq_bwd*)
9.   SCANBACKWARD(A, *cgraph*, *pq_fwd*, *pq_bwd*)
10.   **while** *pq_fwd* $\neq \emptyset$ or *pq_bwd* $\neq \emptyset$

In the following, we present SCANFORWARD and SCANBACKWARD in lines 8 and 9 above. To describe these procedures we need some notations. Suppose that a subpart of *cgraph* is loaded into B, and let $(n_f, q_f) \to (n'_f, q'_f)$ be the first edge in B and $(n_l, q_l) \to (n'_l, q'_l)$ be the last edge in B. Then we define that $B.min' = order(n_f)$ and that $B.max' = order(n_l)$. Note that, by the ordering of *cgraph* (line 14 of TRAVERSEBUF), for any edge $(n, q) \to (n', q')$ in B we have $B.min' \leq order(n) \leq B.max'$.

SCANFORWARD and SCANBACKWARD use two priority queues *pq_fwd* and *pq_bwd*. Suppose that, during traversing B, we encounter an edge $(n, q) \to (n', q')$ such that (n', q') is outside B, i.e., $order(n') < B.min'$ or $order(n') > B.max'$. Since (n', q') is outside B, we have to "suspend" the traverse and "restart" it from (n', q') afterwards. *pq_fwd* and *pq_bwd* are used to remember such pairs for suspending and restarting. *pq_fwd* and *pq_bwd* are defined as follows.

– *pq_fwd* is an ascending priority queue w.r.t. $order(n)$, thus pair (n, q) with the least $order(n)$ is at the top of the queue.
– *pq_bwd* is a descending priority queue w.r.t. $order(n)$, thus pair (n, q) with the largest $order(n)$ is at the top of the queue.

Now let us present SCANFORWARD (SCANBACKWARD is defined similarly). The procedure reads *cgraph* sequentially and repeats the following until EOF is found. First, the procedure loads $|B|$ bytes of data from *cgraph* into B, and computes set S of pairs at which a traversal should be (re)started. At the first execution of SCANFORWARD, for any edge $(n, q) \to (n', q')$ in B, $(n, q)_{(n, q)}$ is added to S whenever $q = q_0$ (in lines 5 and 6, the subscription of $(n, q)_{(n, q)}$ is explained later). At the second or later execution, each pair $(n, q)_{(n_s, q_s)}$ in *pq_fwd* is added to S if B contains an edge $(n, q) \to (n', q')$ for some (n', q') (line 8). Then for each (subscripted) pair $(n, q)_{(n_s, q_s)}$ in S, start a traversal on B from (n, q) (line 10). Let (n', q') be a pair encountered by this traversal. If $q' \in F$, then the start and end nodes of the traversal is outputted as an answer (lines 13 and 14). If (n', q') is a pair outside B, then the traversal is suspended and (n', q') is added to pq_fwd or pq_bwd (lines 15 to 18).

Procedure SCANFORWARD
-1.5mm

Input: ε-free NFA $A = (Q, \Sigma, \delta, q_0, F)$, contracted graph $cgraph$,
   priority queues pq_fwd, pq_bwd
Output: Set of pairs (n, n') such that $(n, q_0) \leadsto (n', q')$ and that $q' \in F$
   obtained by scanning $cgraph$ forward

1. Read $cgraph$ sequentially and do the following until EOF is found
2. **begin**
3. $S \leftarrow \emptyset$
4.   Load $|B|$ bytes of data from $cgraph$
5.   **if** this is the first execution of SCANFORWARD **then**
6. $S \leftarrow \{(n, q)_{(n,q)} \mid (n, q) \rightarrow (n', q')$ is an edge in B, $q = q_0\}$
7.   Mark $(n, q)_{(n,q)}$ as "visited" for each $(n, q)_{(n,q)} \in S$
8.   Remove all pairs $(n, q)_{(n_s,q_s)}$ such that $B.min' \leq order(n) \leq B.max'$
   from pq_fwd, and add each removed pair $(n, q)_{(n_s,q_s)}$ to S if B contains
   an edge $(n, q) \rightarrow (n', q')$ for some (n', q')
9.   **for each** $(n, q)_{(n_s,q_s)} \in S$ **do**
10.   Traverse B from (n, q) and find all pairs (n', q') such that
   $(n, q) \leadsto (n', q')$ in B and that $(n', q')_{(n_s,q_s)}$ is unvisited.
   Let T be the resulting set.
11.   Mark $(n', q')_{(n_s,q_s)}$ as "visited" for each $(n', q')_{(n_s,q_s)} \in T$
12.   **for each** $(n', q')_{(n_s,q_s)} \in T$ **do**
13.   **if** $q' \in F$ **then**
14.   Output (n_s, n') as an answer
15.   **if** $order(n') < B.min'$ **then**
16.   Add $(n', q')_{(n_s,q_s)}$ to pq_bwd
17.   **else if** $order(n') > B.max'$ **then**
18.   Add $(n', q')_{(n_s,q_s)}$ to pq_fwd
19. **end**

Let us explain SCANFORWARD and SCANBACKWARD by examples. According to TRAVERSECGRAPH, SCANFORWARD is executed first, then SCANBACKWARD is executed, and so on. Firstly, for any edge $(n, q) \rightarrow (n', q')$ in B, SCANFORWARD starts a traversal from (n, q) if $q = q_0$ (lines 6 and 9). For example, consider an edge $(n_{12}, q_0) \rightarrow (n_{10}, q_1)$ in Fig. 3(A). Since the state q_0 of (n_{12}, q_0) is the start state, we start a traversal from this edge. This edge is adjacent to an edge $(n_{10}, q_1) \rightarrow (n_2, q_2)$, which is outside B (in a backward direction). Thus we have to "suspend" the traversal and "restart" it from (n_{10}, q_1) later. To record this, we add $(n_{10}, q_1)_{(n_{12}, q_0)}$ to pq_bwd (lines 15 and 16, the subscript (n_{12}, q_0) is explained later). As an another example, consider an edge $(n_9, q_0) \rightarrow (n_8, q_1)$ in B (Fig. 3(A)). The edge is adjacent to $(n_8, q_1) \rightarrow (n_{11}, q_1)$, which is outside B (in a forward direction). Thus $(n_8, q_1)_{(n_9, q_0)}$ is added to pq_fwd (lines 17 and 18). We use pq_fwd and pq_bwd separately according to scanning directions. Since pq_fwd is an ascending priority queue, SCANFORWARD looks pq_fwd to obtain pairs at which traversals should be restarted in line 8 (i.e., the scanning direction coincides with the node ordering of pq_fwd). On the other hand, pq_bwd is a

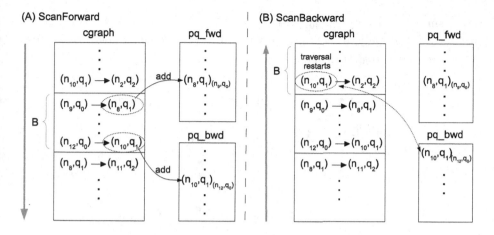

Fig. 3. SCANFORWARD and SCANBACKWARD

descending priority queue, and thus SCANBACKWARD looks pq_bwd to obtain pairs at which traversals should be restarted.

Let us explain the subscriptions. Suppose that a traversal is started from (n, q_0) and a pair (n', q') is added to pq_fwd/pq_bwd during this traversal. Then we have to record the fact that "the source of this traversal is (n, q_0)" since the node n of (n, q_0) is required when outputting an answer. Such a pair is added as the subscript of (n', q') when (n', q') is added to pq_fwd/pq_bwd. For example, for an edge $(n_{12}, q_0) \rightarrow (n_{10}, q_1)$ in Fig. 3(A), $(n_{10}, q_1)_{(n_{12}, q_0)}$ is added to pq_bwd by line 16 of SCANFORWARD.

After the first execution of SCANFORWARD, SCANBACKWARD is executed in which $cgraph$ is scanned in a backward direction (Fig. 3(B)). Here, for each edge $(n, q) \rightarrow (n', q')$ in B, if (n, q) is contained in pq_bwd, then we "restart" the traversal from (n, q). And if a pair (n'', q'') with $q'' \in F$ is found during the traversal, then this is outputted as an answer. For example, consider an edge $(n_{10}, q_1) \rightarrow (n_2, q_2)$ in Fig. 3(B). Then $(n_{10}, q_1)_{(n_{12}, q_0)}$ is contained in pq_bwd due to the previous SCANFORWARD execution, and thus the traversal is restarted. We repeat such scans until both pq_fwd and pq_bwd become empty.

4.3 Pruning Edges

In line 7 of MAIN, PRUNE(B,A) deletes unnecessary edges B w.r.t. NFA A from B. Let n be a node. This procedure determines which outgoing edges of n can be pruned according to the transition function of A and the incoming edges of n. Because of space limitation, we explain this procedure by an example. Consider pruning outgoing edges of n_1 in Fig. 4(b). It is clear that $n_1 \xrightarrow{f} n_7$ can be pruned since f does not appear in the NFA A in Fig. 4(a). On the other hand, $n_1 \xrightarrow{a} n_4$ is not pruned since $\delta(q_0, a) \neq \emptyset$ and thus this edge is required for a traversal starting from (n_1, q_0) (lines 2 and 3 of TRAVERSEBUF). $n_1 \xrightarrow{b} n_5$ is not pruned either since we have $q_2 \in \hat{\delta}(q_0, ab)$ and $n_2 \xrightarrow{a} n_1 \xrightarrow{b} n_5$, i.e., (n_5, q_2) is reachable from (n_2, q_0). Finally, $n_1 \xrightarrow{c} n_6$ is pruned since the incoming edge $n_2 \xrightarrow{a} n_1$ is labeled by a but $\hat{\delta}(q_0, ac) = \emptyset$.

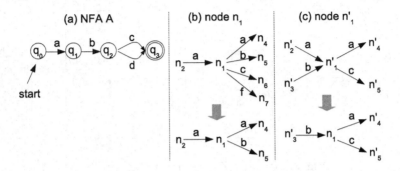

Fig. 4. Pruning outgoing/incoming edges

Incoming edges are pruned similarly. Consider the incoming edges of n_1' in Fig. 4(c). First, $n_3' \xrightarrow{b} n_1'$ is not pruned since n_1' has an outgoing edge $n_1' \xrightarrow{c} n_5'$ and $q_3 \in \hat{\delta}(q_1, bc)$, i.e., (n_5', q_3) is reachable from (n_3', q_1). Then consider $n_2' \xrightarrow{a} n_1'$. (n_1', q_1) is reachable from (n_2', q_0) and we have $\delta(q_1, b) = \{q_2\}$, but n_1' has no outgoing edge labeled by b (any traversal is stopped here). Thus $n_2' \xrightarrow{a} n_1'$ is redundant and it is pruned.

4.4 Correctness and I/O Cost

In this subsection, we show the correctness and I/O cost of the algorithm. The details are omitted because of space limitation.

We first show the correctness of the algorithm. Let $G = (V, E)$ be a graph, r be a regular path query, and $A = (Q, \Sigma, \delta, q_0, F)$ be an NFA such that $L(A) = L(r)$. Then the following theorem can be shown by induction on the number of boundary edges on p.

Theorem 1. *There exists a path $n \leadsto_p n'$ in G such that $q' \in \hat{\delta}(q_0, l(p))$ and that $q' \in F$ iff the algorithm outputs (n, n').* □

Second, we present some I/O costs. Following External Memory Model [1], we assume that data is transferred between external and internal memories in blocks of size B'. Thus, the parameters we are considering are as follows.

- Graph $G = (V, E)$, and node list N constructed from G
- Regular expression r, or NFA $A = (Q, \Sigma, \delta, q_0, F)$ such that $L(A) = L(r)$
- Area B allocated in main memory
- Block size B' mentioned above

Table 1. Summary of the dataset

	10 m	20 m	50 m	100 m
Resources	1,730,250	3,472,940	8,698,023	17,823,525
Literals	5,320,785	10,506,697	25,880,271	51,298,417
$|V|$ (resources+ literals)	7,051,035	13,979,637	34,578,294	69,121,942
$|E|$	10,000,451	20,000,686	50,000,863	100,000,374
Size of N (GBytes)	0.28	0.57	1.46	3.04

Then the I/O cost of constructing node list N from G is in

$$O(sort(|E|) + |V| \log_{B'} |V|),$$

where $sort(|E|) = \Theta \left(\frac{|E|}{B'} \log_{B/B'} \frac{|E|}{B'} \right)$. Thus N can be constructed from G efficiently. We next present the I/O cost of the algorithm. By the definition of N the size of N is in $O(|E|+|V|)$. If the size of *cgraph* does not exceed $|B|$, *cgraph* is scanned only once and the I/O cost of the algorithm is in

$$O((|E| + |V|)/B'). \tag{1}$$

Consider next the case where the size of *cgraph* exceeds $|B|$. Since MAIN reads N sequentially by using B, N is divided into $k = \Theta((|E| + |V|)/|B|)$ sublists. Assuming that the number of boundary edges in N is in $O(k)$, the total I/O cost of the algorithm is in

$$O((|E| + |V|)/B' + k^2|\delta|^2/B'),$$

where $|\delta| = |\{(q,q') \mid q' \in \delta(q,a), q,q' \in Q, a \in \Sigma\}|$ and "$k^2|\delta|^2/B'$" is the cost of TRAVERSECGRAPH. However, as shown in the next section, in most cases *cgraph* fits in B and the I/O cost of the algorithm follows (1).

5   Experimental Results

We implemented the algorithm in C++ and made evaluation experiments on I/O cost, execution time, and the size of *cgraph*. All the evaluations were executed on a machine with Intel Core i7 3.5GHz CPU, 8GB RAM, 2TB SATA HDD, and Linux OS (CentOS 7).

To construct the dataset for the experiments, we use SP^{2}Bench [11]. This is an RDF benchmark tool that can generate RDF files of arbitrary size. We generate four RDF files of different sizes by SP^{2}Bench, and construct node lists N from the RDF files. Table 1 presents a summary of the dataset.

To obtain a query set for the experiments, we made a program for generating regular path queries. In short, this program generates a sequence of labels and operators randomly, and constructs regular path queries by combining these elements. However, if a sequence of labels were generated completely randomly, the answers of most queries would be empty. Therefore, the program generates a sequence of labels so that adjacent labels are "connected", e.g., a sequence "$l_1 l_2$" is generated only if there exists at least one node n such that an incoming edge of n is labeled by l_1 and that an outgoing edge of n is labeled by l_2. By using this program we generate 15 regular path queries. The average size of the queries is 8.0 (5.6 labels and 2.4 operators), and 8 out of the 15 queries has non-empty answers.

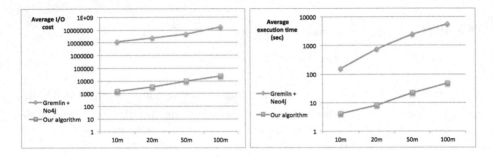

Fig. 5. Average I/O cost and execution time

With the dataset and queries, we examined the I/O cost and execution time of the algorithm. For comparison, we import the above dataset into Neo4j database by batch-import[7], and convert the 15 queries into equivalent Gremlin scripts[8]. Then we execute (1) the regular path queries by our algorithm ($|B|$ is set to 800 MB) and (2) the Gremlin scripts under Gremlin 2.6.0 environment with the imported Neo4j database[9]. Since Linux OS has a buffer cache to store previously accessed files, we clear the cache prior to each query execution. Figure 5 plots the average I/O cost (the sum of I/O reads and writes) and the average execution time of the 15 queries. As shown in this figure, both the I/O cost and execution time of our algorithm are considerably smaller than those of Gremlin+Neo4j. This is mainly because (i) our algorithm is based on a sequential access to N and does not require fetching nodes and edges repeatedly and (ii) by PRUNE (line 7 of MAIN) we can delete a significant portion of edges of N prior to graph traversal (note that the set of labels in a query is much smaller than the set of all labels of N).

Let us next show the results related to the size of *cgraph*. Figure 6(left) plots the sizes of *cgraph* for the same dataset as above. As shown in the figure, *cgraph* grows almost linearly to the size of input graph. We also examined how the size

---

[7] https://github.com/jexp/batch-import.

[8] Neo4j has a declarative query language Cypher but it does not fully support regular path query. Thus we chose Gremlin in this experiment.

[9] We also tried the same queries on Sparksee 5.1.0 and Apache Jena TDB, but obtained no result due to main memory exhaustion errors in both environments.

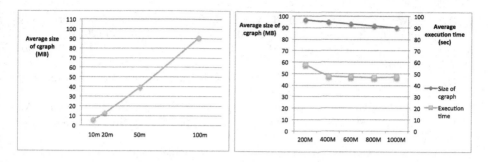

Fig. 6. Results on the sizes of *cgraph*

of B affects the size of *cgraph* and the execution time of the algorithm. In this examination, we use the same queries as above on the "100 m" node list in Table 1. We change the size of B from 200 to 1000 MBytes, and plot the average execution time of our algorithm and the average size of *cgraph* (Fig. 6(right)). The execution time in the case of $B = 200\,M$ is about 20 % larger than those of the other cases, which is due to the fact that in the case of $B = 200\,M$ three *cgraph*s of 15 exceed the size of B, while in the other cases no *cgraph* exceeds the size of B. These results suggest that (1) in most cases *cgraph* is enough small to fit in B and (2) the size of B hardly affects the execution time of the algorithm, unless B is extremely small.

6 Conclusion

In this paper, we proposed an external memory algorithm for the all-pairs regular path problem. Our algorithm finds the answers matching r by scanning the node list of G sequentially, which avoids random accesses to disk and thus makes regular path query processing efficient. The experiments suggest our approach is effective in solving the all-pairs regular path problem on large graphs.

However, this work has just started and we still have a lot things to do. First, we need to investigate the performance of our algorithm by using much more kinds of datasets as well as SP²Bench. Second, we have to compare our algorithm with more graph stores. Third, we have to consider node ordering of N. For example, [10] proposes a node ordering which achieves an efficient child and descendant traversal on tree structured data. We need to consider node ordering of N that makes the algorithm more efficient. Finally, we have to consider improving our algorithm to obtain better performance. For example, in this paper we assume that node list N is stored in a *single* disk. By partitioning N into several fragments and putting them into multiple disks, the algorithm may perform better.

References

1. Aggarwal, A., Vitter, J.S.: The input/output complexity of sorting and related problems. Commun. ACM **31**(9), 1116–1127 (1988)

2. Bai, Y., Wang, C., Ning, Y., Wu, H., Wang, H.: G-path: flexible path pattern query on large graphs. In: Proceedings of the WWW 2013, pp. 333–336 (2013)
3. Baeza, P.B.: Querying graph databases. In: Proceedings of the PODS 2013, pp. 175–188 (2013)
4. Hu, X., Tao, Y., Chung, C.-W.: Massive graph triangulation. In: Proceedings of the SIGMOD 2013, pp. 325–336 (2013)
5. Koschmieder, A., Leser, U.: Regular path queries on large graphs. In: Ailamaki, A., Bowers, S. (eds.) SSDBM 2012. LNCS, vol. 7338, pp. 177–194. Springer, Heidelberg (2012)
6. Losemann, K., Martens, W.: The complexity of regular expressions and property paths in SPARQL. ACM Trans. Database Syst. **38**(4), 24:1–24:39 (2013)
7. Luo, Y., Fletcher, G.H., Hidders, J., Wu, Y., De Bra, P.: External memory k-bisimulation reduction of big graphs. In: Proceedings of CIKM 2013, pp. 919–928 (2013)
8. Malewicz, G., Austern, M.H., Bik, A.J., Dehnert, J.C., Horn, I., Leiser, N., Czajkowski, G.: Pregel: A system for large-scale graph processing. In: Proceedings of the SIGMOD 2010, pp. 135–146 (2010)
9. Martínez-Bazan, N., Muntés-Mulero, V., Gómez-Villamor, S., Nin, J., Sánchez-Martínez, M.-A., Larriba-Pey, J.-L.: Dex: high-performance exploration on large graphs for information retrieval. In: Proceedings of the CIKM 2007, pp. 573–582 (2007)
10. Morishima, A., Tajima, K., Tadaishi, M.: Optimal tree node ordering for child/descendant navigations. In: Proceedings of the ICDE 2010, pp. 840–843 (2010)
11. Schmidt, M., Hornung, T., Lausen, G., Pinkel, C.: Sp2Bench: A SPARQL performance benchmark. In: Proceedings of the ICDE 2009, pp. 222–233 (2009)
12. Shoaran, M., Thomo, A.: Distributed multi-source regular path queries. In: Proceedings of the ISPA 2007, pp. 365–374 (2007)
13. Stefanescu, D.C., Thomo, A., Thomo, L.: Distributed evaluation of generalized path queries. In: Proceedings of the SAC 2005, pp. 610–616 (2005)
14. Tung, L.-D., Nguyen-Van, Q., Hu, Z.: Efficient query evaluation on distributed graphs with hadoop environment. In: SoICT 2013, pp. 311–319 (2013)
15. Wood, P.T.: Query languages for graph databases. SIGMOD Rec. **41**(1), 50–60 (2012)
16. Yildirim, H., Chaoji, V., Zaki, M.J.: Grail: a scalable index for reachability queries in very large graphs. VLDB J. **21**(4), 509–534 (2012)
17. Zhang, Z., Yu, J.X., Qin, L., Chang, L., Lin, X.: I/O efficient: computing SCCs in massive graphs. In: Proceedings of the SIGMOD 2013, pp. 181–192 (2013)
18. Zhang, Z., Yu, J.X., Qin, L., Zhu, Q., Zhou, X.: I/O cost minimization: reachability queries processing over massive graphs. In: Proceedings of the EDBT 2012, pp. 468–479 (2012)
19. Zou, L., Özsu, M.T., Chen, L., Shen, X., Huang, R., Zhao, D.: gStore: a graph-based SPARQL query engine. VLDB J. **23**(4), 565–590 (2014)

Special Section Globe 2015 – MapReduce Framework : Load Balancing, Optimization and Classification

An Efficient Solution for Processing Skewed MapReduce Jobs

Reza Akbarinia[1]([✉]), Miguel Liroz-Gistau[1], Divyakant Agrawal[2],
and Patrick Valduriez[1]

[1] INRIA and LIRMM, Montpellier, France
{Reza.Akbarinia,Miguel.Liroz_Gistau,Patrick.Valduriez}@inria.fr
[2] University of California, Santa Barbara, USA
agrawal@cs.ucsb.edu

Abstract. Although MapReduce has been praised for its high scalability and fault tolerance, it has been criticized in some points, in particular, its poor performance in the case of data skew. There are important cases where a high percentage of processing in the reduce side is done by a few nodes, or even one node, while the others remain idle. There have been some attempts to address the problem of data skew, but only for specific cases. In particular, there is no proposed solution for the cases where most of the intermediate values correspond to a single key, or when the number of keys is less than the number of reduce workers.

In this paper, we propose *FP-Hadoop*, a system that makes the reduce side of MapReduce more parallel, and efficiently deals with the problem of data skew in the reduce side. In FP-Hadoop, there is a new phase, called *intermediate reduce (IR)*, in which blocks of intermediate values, constructed dynamically, are processed by intermediate reduce workers in parallel, by using a scheduling strategy. By using the IR phase, even if all intermediate values belong to only one key, the main part of the reducing work can be done in parallel by using the computing resources of all available workers. We implemented a prototype of FP-Hadoop, and conducted extensive experiments over synthetic and real datasets. We achieved excellent performance gains compared to native Hadoop, e.g. *more than 10 times in reduce time and 5 times in total execution time*.

Keywords: MapReduce · Data skew · Load balancing

1 Introduction

MapReduce [3] is one of the most popular solutions for big data processing, in particular due to its automatic management of parallel execution in clusters of machines. Initially proposed by Google, its popularity has continued to grow over time, as shown by the tremendous success of Hadoop [1], an open-source implementation.

The idea behind MapReduce is simple and elegant. Given an input file, and two map and reduce functions, each MapReduce job is executed in two main

© Springer International Publishing Switzerland 2015
Q. Chen et al. (Eds.): DEXA 2015, Part II, LNCS 9262, pp. 417–429, 2015.
DOI: 10.1007/978-3-319-22852-5_35

Algorithm 1. Map and reduce functions for Example 1

map($id : \mathcal{K}_1$, $content : \mathcal{V}_1$ **)**

 foreach $line$ $\langle lang,\ page\_id,\ num\_visits,\ ...\rangle$ in $content$ **do**
  emit $(lang,\ page\_info = \langle num\_visits,\ page\_id\rangle)$

reduce($lang : \mathcal{K}_2$, $pages\_info : \mathrm{list}(\mathcal{V}_2)$ **)**

 Sort $pages\_info$ by num_visits
 foreach $page\_info$ in top $k\,\%$ **do**
  emit $(lang,\ page\_id)$

---

phases. In the first phase, called map, the input data is divided into a set of splits, and each split is processed by a map task in a given worker node. These tasks apply the map function on every key-value pair of their split and generate a set of intermediate pairs. In the second phase, called reduce, all the values of each intermediate key are grouped and assigned to a reduce task. Reduce tasks are also assigned to worker machines and apply the reduce function on the created groups to produce the final results.

Although MapReduce has been praised for its high scalability and fault tolerance, it has also been criticized in some points, particularly its poor performance in the case of data skew. There are important cases where a high percentage of processing in the reduce side ends up being done by only one node. Let us illustrate this by an example.

Example 1. Top accessed pages in Wikipedia. Suppose we want to analyze the statistics[1] that the free encyclopedia, Wikipedia, has published about the visits of its pages by users. In the statistics, for every hour, there is a file about the pages visited in that hour. More precisely, for each visited page there is a line containing some information about the page including its url, language and the number of visits. Given a file, we want to return the top-k % of pages accessed for each language (e.g. top 1 %). To answer this query, we can write a simple program as in the following Algorithm[2]:

In this example, there may be a high skew in the load of reduce workers. In particular, the worker that is responsible for reducing the English language will receive a lot of values. According to the statistics published by Wikipedia[3], the percentage of English pages over total was more than 70 % in 2002 and more than 25 % in 2007. This means for example that if we use the pages published up to 2007, when the number of reduce workers is more than 4, then we have no way for balancing the load because one of the nodes would receive more than 1/4 of the data. The situation is even worse when the number of reduce tasks is

---

[1] http://dumps.wikimedia.org/other/pagecounts-raw/.

[2] This program is just for illustration; actually, it is possible to write a more efficient code by leveraging the sorting mechanisms of MapReduce.

[3] http://en.wikipedia.org/wiki/Wikipedia:Size_of_Wikipedia.

high, e.g., 100, in which case after some time, all reduce workers but one would finish their assigned task, and the job has to wait for the responsible of English pages to finish. In this case, the execution time of the reduce phase is at least equal to the execution time of this task, no matter the size of the cluster.

There have been some proposals to deal with the problem of reduce side data skew. One of the main approaches is to try to uniformly distribute the intermediate values to the reduce tasks, e.g., by repartitioning the keys to the reduce workers [8]. However, this approach is not efficient in many cases, e.g. when there is only one intermediate key, or when most of the values correspond to one of the keys.

One solution for decreasing the reduce side skew is to filter the intermediate data as much as possible in the map side, e.g., by using a *combiner function*. But, the input of the combiner function is restricted to the data of one map task [12], thus its filtering power is very limited for many applications.

In this paper, we propose *FP-Hadoop*, a system that uses a new approach for dealing with the data skew in reduce side. In FP-Hadoop, there is a new phase, called *intermediate reduce (IR)*, whose objective is to make the reduce side of MapReduce more parallel. More specifically, the programmer replaces his reduce function by two functions: *intermediate reduce (IR)* and *final reduce (FR)* functions. Then, FP-Hadoop executes the job in three phases, each phase corresponding to one of the functions: map, intermediate reduce (IR) and final reduce (FR) phases. In the IR phase, even if all intermediate values belong to only one key (i.e., the extreme case of skew), the reducing work is done by using the computing power of available workers. Briefly, the data reducing in the *IR phase* has the following distinguishing features:

- **Parallel Reducing of Each Key:** The intermediate *values of each key* can be processed in parallel by using multiple intermediate reduce workers.
- **Distributed Intermediate Block Construction:** The input of each intermediate worker is a block composed of intermediate values *distributed over multiple nodes* of the system, and chosen using a *scheduling strategy*, e.g. locality-aware.
- **Hierarchical Execution:** The processing of intermediate values in the IR phase can be done in several levels (iterations). This permits to perform *hierarchical execution plans* for jobs such as top-k % queries, in order to decrease the size of the intermediate data more and more.
- **Non-overwhelming Reducing:** The size of the intermediate blocks is bounded by configurable maximum value that prevents the intermediate reducers to be overwhelmed by very large blocks of intermediate data.

We implemented a prototype of FP-Hadoop by modifying Hadoop's code. We conducted extensive experiments over synthetic and real datasets. The results show excellent performance gains of FP-Hadoop compared to native Hadoop. For example, in a cluster of 20 nodes with 120GB of input data, FP-Hadoop outperformed Hadoop by a *factor of about 10 in reduce time, and a factor of 5 in total execution time.*

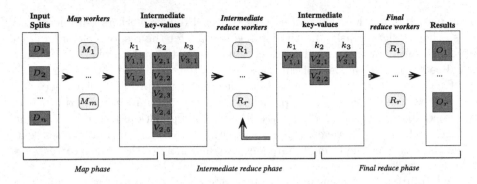

Fig. 1. FP-Hadoop job processing scheme

The rest of this paper is organized as follows. In Sect. 2, we propose FP-Hadoop. In Sect. 3, we report the results of our experiments done to evaluate the performance of FP-Hadoop. In Sect. 4, we discuss related work, and Sect. 5 concludes.

2 FP-Hadoop

In this section, we propose FP-Hadoop, a new Hadoop-based system designed for dealing with data skew in MapReduce jobs. We first introduce the programming model of FP-Hadoop, its main phases, and the functions that are necessary for executing the jobs. Then, Then, we provide a more detailed description of the FP-Hadoop design, such as our technique for constructing the working blocks of the IR phase, and the scheduling of intermediate workers.

2.1 Job Execution Model

In FP-Hadoop, the output of the map tasks is organized as a set of blocks (splits) which are consumed by the reduce workers (see Fig. 1). More specifically, the intermediate key-value pairs are dynamically grouped into splits, called *Intermediate Result Splits* (*IR splits* for short). The size of an IR split is bounded between two values, *minIRsize* and *maxIRsize*, that can be configured by the user. Formally, each IR split is a set of (k, V) pairs such that k is an intermediate key and V is a subset of the values generated for k by the map tasks.

In FP-Hadoop, the execution of a job is done in three phases: *map, intermediate reduce*, and *final reduce*. The map phase is almost the same as that of Hadoop in the sense that the map workers apply the map function on the input splits, and produce intermediate key-value pairs. The only difference is that in FP-Hadoop, the map output is managed as a set of *IR fragments* that are used for constructing IR splits (more details about the management of IR splits are given in Sect. 2.3).

There are two different reduce functions: *intermediate reduce* (*IR*) and *final reduce* (*FR*) functions.

In the *intermediate reduce phase*, the IR function is executed in parallel by reduce workers on the IR splits, which are constructed using a scheduling strategy from the intermediate values distributed over different nodes. More specifically, in this phase, each reduce worker takes an IR split as input, applies the IR function on it, and produces a set of key-value pairs which may be used for constructing future IR splits. When a reduce worker finishes its input split, it takes another split and so on until there is no more IR splits.

The intermediate reduce phase can be repeated in *several iterations*, to apply the IR function several times on the intermediate data, and incrementally decrease the size of the final splits which will be consumed by the FR function. The *maximum number of iterations* can be specified by the programmer, or be chosen adaptively, i.e., until the intermediate reduce tasks input/output size ratio is higher than a threshold (which can be configured by the user).

In the *final reduce phase*, the FR function is applied on the IR splits generated as the output of the intermediate reduce phase. The FR function is in charge of performing the final grouping and production of the results. Like in Hadoop, the keys are assigned to the reduce tasks according to a partitioning function. Each reduce worker pulls all IR splits corresponding to its keys, merges them, applies the FR function on the values of each key, and generates the final job results. Since in FP-Hadoop the final reduce workers receive the values on which the intermediate workers have worked, the load of the final reduce workers in FP-Hadoop is usually much lower than that of the reduce workers in Hadoop.

In the next subsection, we give more details about the IR and FR functions, and explain how they can be programmed.

2.2   IR and FR Functions

To take advantage of the intermediate reduce phase, the programmer should replace his/her reduce function by intermediate and final reduce functions. Formally, the input and output of map (M), intermediate reduce (IR) and final reduce (FR) functions are as follows:

$$M : (\mathcal{K}_1, \mathcal{V}_1) \to list(\mathcal{K}_2, \mathcal{V}_2)$$
$$IR : (\mathcal{K}_2, partial\_list(\mathcal{V}_2)) \to (\mathcal{K}_2, partial\_list(\mathcal{V}_2))$$
$$FR : (\mathcal{K}_2, list(\mathcal{V}_2)) \to list(\mathcal{K}_3, \mathcal{V}_3)$$

Notice that in IR function, any partial set of intermediate values can be received as input. However, in FR function, all values of an intermediate key are passed to the function.

Given a reduce function, to write the IR and FR functions, the programmer should separate the sections that can be processed in parallel and put them in IR function, and the rest in FR function. Formally, given a reduce function R, the programmer generates two functions IR and FR, such that for any intermediate key k and its list of values V, $R(k, V) = FR(k, \langle IR(k, V_1), ..., IR(k, V_n) \rangle)$ for every partition $V_1 \cup ... \cup V_n = V$.

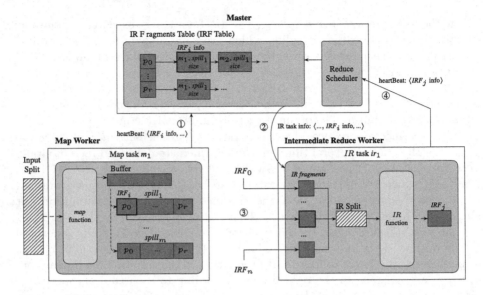

Fig. 2. Data flow between a map worker, the master and an intermediate reduce worker. The communicated messages are shown in their sent order.

The following example illustrates the IR and FR functions for a job that implements the average operation.

Example 2. **Avg.** Consider a job that computes the average of the numeric values that are in a big file. To implement this job in Hadoop, it is sufficient to emit(1, value) for each read value in the map function. Then, the reduce function computes the sum and count of all values and returns sum/count. In FP-Hadoop, in the IR function, we compute the (partial) sum and count of the values in the input IR split, and emit(1, {sum, count}). This allows to compute the partial counts and sums in parallel. Then, in the FR function, we compute the sum of partial sums and counts, and divide the total sum by the total count.

There are many functions for which we can use the original reduce function both in the intermediate and final reduce phases, i.e., we have $IR = FR = R$. Examples of such functions are Top-k, SkyLine, Union, SUM, MIN and MAX.

Please notice that if the programmer does not want to use the IR phase, he/she should specify no IR function. In this case, the final reduce phase starts just after the map phase completes, i.e., as in Hadoop.

2.3   Dynamic Construction of IR Splits

In this subsection, we describe our approach for constructing the IR splits that are the working blocks of the reducers in the intermediate reduce phase.

In Hadoop, the output of the map tasks is kept in the form of temporary files (called *spills*). Each spill contains a set of partitions, such that each partition involves a set of keys and their values. These spills are merged at the end of the map task, and the data of each partition is sent to one of the reducers.

In FP-Hadoop, the spills are not merged. Each partition of a spill generates an *IR fragment*, and the IR fragments are used for making IR splits. When a spill is produced by a map task of FP-Hadoop, the information about the spill's IR fragments, which we call *IRF metadata*, is sent to the master node by using the heartbeat message passing mechanism[4]. The data flow between FP-Haddop components is shown in Fig. 2.

For keeping IRF metadata, the master of FP-Hadoop uses a specific data structure called *IR fragment table* (*IRF Table*). Each partition has an entry in IRF Table that points to a list keeping the IR fragment metadata of the partition, e.g., size, spill and the ID of the map worker where the IR fragment has been produced. The master uses the information in IRF Table for constructing IR splits and assigning them to ready reduce workers. This is done mainly based on the scheduling strategies described in the next subsection.

2.4   Scheduling of Intermediate Tasks

For scheduling an intermediate reduce task, the most important issue is to choose the IR fragments that belong to the IR split that should be processed by the task. For this, the following strategies are actually implemented in FP-Hadoop:

- *Greedy.* In this strategy, the objective is to give priority to the IR fragments of the partition that has the maximum number of values. In our implementation of IRF Table, for each partition, we keep the total size of IR fragments. Thus, by a scan of IRF Table, we can find the partition that has the highest number of values until now. After finding the partition, we scan its list, and take from the head of the list the IR fragments until reaching the *maxIRsize* value, i.e., the upper bound size for an IR split. This strategy is the default strategy in FP-Hadoop.
- *Locality-Aware.* In this strategy, we try to choose for a worker w the IR fragments that are on its local disk or close to it. For this, the scheduler scans IRF Table, and finds the partitions whose total data size is at least *minIRsize* (the minimum defined size for IR split), and chooses among them the partition that has the maximum local data at w. After choosing the partition, say p, the scheduler chooses a combination of p's IR fragments at w with size between *minIRsize* and *maxIRsize*. If the total size of p's IR fragments at w is lower than *minIRsize*, then the scheduler completes the IR split by first choosing IR fragments from the same rack as that of w, and then if necessary from the same data center.

In FP-Haddop, the programmer can configure the system to execute the IR phase in several iterations, in such a way that the output of each iteration is

---

[4] This mechanism is used for communication between the master and workers.

consumed by the next iteration. There is a parameter *maxIter* that defines the maximum number of iterations. Notice that this parameter sets the maximum number, but in practice each partition may be processed in a different number of iterations, for instance depending on its size, input/output ratio or skew. By default, FP-Hadoop implements an approach in which an iteration is launched only if its input size is more than a given threshold, *minIterSize*. The default value for the threshold is the same as *minIRsize* (i.e., the minimum size of IR splits).

3   Performance Evaluation

We implemented a prototype of FP-Hadoop by modifying Hadoop's code. In this section, we report on the results of our experiments for evaluating the performance of FP-Hadoop. We first discuss the experimental setup such as the datasets, queries and the experimental platform. Then, we discuss the results of our tests done to study the performance of FP-Hadoop, particularly by varying parameters such as the number of nodes in the cluster, the size of input data, etc.

3.1   Setup

We have used the following combinations of MapReduce jobs and datasets to assess the performance of our prototype:

Top-k % (TK). This job, which is our default job in the experiments, corresponds to the query from the Wikipedia example described in the introduction of the paper. Our query consists of retrieving for each language the k % most visited articles. The default value of k is 1, i.e., by default the query returns 1 % of the input data. We have used real-world and synthetic datasets. The real-world dataset (TK-RD) is obtained from the Wikipedia page view statistics[5] stored in hourly log files. We also produced a synthetic dataset (TK-SK), where the number of articles per language follows a Zipfian distribution function with exponent $S = 1$ and N=10 (i.e. 10 languages). We have performed several tests varying the data size, among other parameters, up to 120GB. The query is implemented using a secondary sort [12], where intermediate keys are sorted first by language and then by the article's number of visits, but only grouped by language.

Inverted Index (II). This job consists of generating an inverted index with the words of the English Wikipedia articles[6], as in [8]. We used a RADIX partitioner to map letters of the alphabet to reduce tasks and produce a lexicographically ordered output. We have executed the job with a dataset containing 20GB of Wikipedia articles.

---

[5] http://dumps.wikimedia.org/other/pagecounts-raw/.
[6] http://dumps.wikimedia.org/enwiki/latest/.

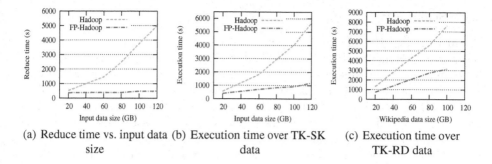

(a) Reduce time vs. input data size   (b) Execution time over TK-SK data   (c) Execution time over TK-RD data

Fig. 3. Scalability of FP-Hadoop

PageRank (PR). This query applies the PageRank algorithm to a graph in order to assign weights to the vertices. As in [8] we have used the implementation provided by Cloud9[7]. As dataset, we used the PLD graph from Web Data Commons[8] whose size is about 2.8GB.

Wordcount (WC). Finally, we have used the wordcount job provided in Apache Hadoop. We have applied it to a dataset generated with the RandomWriter job provided in the Hadoop distribution. We tested this job with a 100GB dataset.

The default values for the parameters which we used in our experiments are as follows. The default number of nodes which we used in our cluster was 20. Unless otherwise specified, the input data size in the experiments was 20 GB. In FP-Hadoop, the default value for *minIRsize* was set to 512 MB. The value of *maxIRsize* is always twice as that of *minIRsize*, and the maximum number of iterations is set to 1.

We compared FP-Hadoop with Hadoop and SkewTune [8] which is the closest related work to ours (see a brief description in Related Work Section).

In all our experiments, we used a *combiner function* (for Hadoop, FP-Hadoop and SkewTune) that is executed on the results of map tasks before sending them to the reduce tasks. This function is used to decrease the amount of data transferred from map to reduce workers, and so to decrease the load of reduce workers.

In our experiments, we have measured the *execution time*, which computes the time elapsed between the start and end of a job, and the *reduce time*, which only considers the time elapsed from the end of the last map task.

We run the experiments over a cluster of nodes (up to 50 nodes). The nodes are provided with Intel Quad-Core Xeon L5335 processors with 4 cores each, and 16GB of RAM. All the experiments were executed with a number of reduce workers equal to the number of machines. We have changed *io.sort.factor* to 100, as advised in [12], which actually favors Hadoop. For the rest of the parameters, we have used Hadoop's default values.

---

[7] http://www.umiacs.umd.edu/~jimmylin/Cloud9/docs/index.html.

[8] http://webdatacommons.org/hyperlinkgraph/.

3.2   Scalability

We investigated the effect of the input size on the performance of FP-Hadoop compared to Hadoop. Using TK-SK dataset, Fig. 3(a) and (b) shows the reduce time and execution time respectively, by varying the input size up to 120 GB, *minIRsize* set to 5 GB, and other parameters set as default values described in Sect. 3.1. Figure 3(c) shows the performance using TK-RD dataset with sizes up to 100 GB, while other parameters as default values described in Sect. 3.1. As expected, increasing the input size increases the execution time of both Hadoop and FP-Hadoop, because more data will be processed by map and reduce workers. But, the performance of FP-Hadoop is much better than Hadoop when we increase the size of input data. For example, in Fig. 3(b), the gain of FP-Hadoop vs Hadoop on execution time is around 1.4 for input size of 20GB, but this gain increases to around 5 when the input size is 120GB. For the latter data size, the reduce time of FP-Hadoop is more than 10 times lower than Hadoop. The reason for this significant performance gain is that in the intermediate reduce phase of FP-Hadoop the reduce workers collaborate on processing the values of the keys containing a high number of values.

3.3   Effect of Cluster Size

We studied the effect of the number of nodes of the cluster on performance. Figure 4(a) shows the execution time by varying the number of nodes, and other parameters set as default values described in Sect. 3.1. Increasing the number of nodes decreases the execution time of both Hadoop and FP-Hadoop. However, FP-Hadoop benefits more from the increasing number of nodes. In Fig. 4(a), with 5 nodes, FP-Hadoop outperforms Hadoop by a factor of around 1.75. But, when the number of nodes is equal to 50, the improvement factor is around 4. This increase in the gain can be explained by the fact that when there are more nodes in the system, more nodes can collaborate on the values of hot keys in FP-Hadoop. But, in Hadoop, although using higher number of nodes can decrease the execution time of the map phase, it cannot significantly decrease the reduce phase time, in particular if there are intermediate keys with high number of values.

3.4   Overhead of IR Phase for Balanced Jobs

Our performance evaluation results, reported until now, show that the IR phase in FP-Hadoop significantly improves the performance of skewed jobs. Let us now investigate its overhead in the case where there is no skew in the job. For this, we have chosen the word-count (WC) query over a uniform random dataset as described in Sect. 3.1. In this job, the key partitioner balances perfectly the reduce load among workers, thus there is no skew in the reduce side. Figure 4(b) shows the total execution time of FP-Hadoop and Hadoop for this job. The results show that the execution time of FP-Hadoop is a little (3 %) higher than Hadoop, and this increase corresponds to the overhead of the IR phase. Indeed,

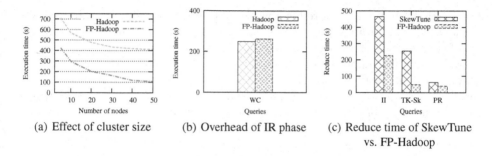

(a) Effect of cluster size   (b) Overhead of IR phase   (c) Reduce time of SkewTune
vs. FP-Hadoop

Fig. 4. Effect of different parameters, and comparison with SkewTune

FP-Hadoop spends some time to detect the lack of skew in the intermediate results, and then launches the final reduce phase. Thus, its execution time is slightly higher than Hadoop. We believe that this very slight overhead of IR phase in perfectly balanced jobs can usually be tolerated. If not, the programmer simply disables the IR phase, then FP-Haddop does not launch that phase.

3.5 Comparison with SkewTune Using Different Queries

We compared FP-Hadoop with SkewTune [8] using different queries. Figure 4(c) shows the reduce time and execution time of both approaches, using the data and parameters described in Sect. 3.1. For these experiments, we downloaded the SkewTune prototype[9]. The data which we used are the default data and sizes described in Sect. 3.1 (e.g. 20GB of data for TK-SK). As the results show, FP-Hadoop can outperform SkewTune with significant factors. The main reason is that SkewTune is unable to split the computation of the tuples assigned to the same intermediate key.

4 Related Work

In the literature, there have been many efforts to improve MapReduce [9]; these include supporting loops [2], adding index [4], caching intermediate data [5], balancing data skew [7,8,10]. Hereafter, we briefly present some of them that are the most related to our work.

The approach proposed in [10] tries to balance data skew in reduce tasks by subdividing keys with large value sets. It requires some user interaction or user knowledge of statistics or sampling, in order to estimate in advance the values size of each key, and then subdivide the keys with large values. Gufler et al. [7] propose an adaptive approach that collects statistics about intermediate key frequencies and assigns them to the reduce tasks dynamically at scheduling time. In a similar approach, Sailfish [11] collects some information about the intermediate keys, and uses them for optimizing the number of reduce tasks

---

[9] https://code.google.com/p/skewtune/

and partitioning the keys to reducer workers. However, these approaches are not efficient when all or a big part of the intermediate values belong to only one key or a few number of keys (i.e., fewer than the number of reduce workers).

SkewTune [8] adopts an on-the-fly approach that detects straggling reduce tasks and dynamically repartitions their input keys among the reduce workers that have completed their work. This approach can be efficient in the cases where the slow progress of a reduce task is due to inappropriate initial partitioning of the key-values to reduce tasks. But, it does not allow the collaboration of reduce workers on the same key.

Haloop [2] extends MapReduce to serve applications that need iterative programs. Although iterative programs in MapReduce can be done by executing a sequence of MapReduce jobs, they may suffer from big data transfer between reduce and map workers of successive iterations. Haloop offers a programming interface to express iterative programs and implements a task scheduling that enables data reuse across iterations. However, it does not allow hierarchical execution plans for reducing the intermediate values of one key, as in our intermediate reduce phase. SpongeFiles [6] is a system that uses the available memory of nodes in the cluster to construct a distributed-memory, for minimizing the disk spilling in MapReduce jobs, and thereby improving performance. Spark [13], an alternative to MapReduce, uses the concept of Resilient Distributed Datasets (RDDs) to transparently store data in memory and persist it to disk only when needed. The concept of intermediate reduce phase proposed in FP-Hadoop can be used as a complementary mechanism in the systems such as Haloop, SpongeFiles and Spark, to resolve the problem of data skew when reducing the intermediate data.

In general, none of the existing solutions in the literature can deal with data skew in the cases when most of the intermediate values correspond to a single key, or when the number of keys is less than the number of reduce workers. But, FP-Hadoop addresses this problem by enabling the reducers to work in the IR phase on dynamically generated blocks of intermediate values, which can belong to a single key.

5   Conclusion

In this paper, we presented FP-Hadoop, a system that brings more parallelism to the MapReduce job processing by allowing the reduce workers to collaborate on processing the intermediate values of a key. We added a new phase to the job processing, called intermediate reduce phase, in which the input of reduce workers is considered as a pool of IR Splits (blocks). The reduce workers collaborate on processing IR splits until finishing them, thus no reduce worker becomes idle in this phase. In the final reduce phase, we just group the results of the intermediate reduce phase. We evaluated the performance of FP-Hadoop through experiments over synthetic and real datasets. The results show excellent gains compared to Hadoop. For example, over a cluster of 20 nodes with 120GB of input data, FP-Hadoop can outperform Hadoop by a factor of about 10 in reduce time, and a factor of 5 in total execution time. The results show that the higher the number of nodes, the greater the potential gain from FP-Hadoop. They also show that the bigger the size of the input data, the larger the potential improvement from FP-Hadoop.

Acknowledgments. Experiments presented in this paper were carried out using the Grid'5000 experimental testbed, being developed under the INRIA ALADDIN development action with support from CNRS, RENATER and several universities as well as other funding bodies (see https://www.grid5000.fr).

References

1. Hadoop (2014). http://hadoop.apache.org
2. Bu, Y., Howe, B., Balazinska, M., Ernst, M.D.: The HaLoop approach to large-scale iterative data analysis. VLDB J. **21**(2), 169–190 (2012)
3. Dean, J., Ghemawat, S.: MapReduce: simplified data processing on large clusters. In: OSDI (2004)
4. Dittrich, J., Quiané-Ruiz, J.A., Jindal, A., Kargin, Y., Setty, V., Schad, J.: Hadoop++: making a yellow elephant run like a cheetah (without it even noticing). PVLDB **3**(1), 518–529 (2010)
5. Elghandour, I., Aboulnaga, A.: ReStore: reusing results of MapReduce jobs in pig. In: SIGMOD (2012)
6. Elmeleegy, K., Olston, C., Reed, B.: SpongeFiles: mitigating data skew in mapreduce using distributed memory. In: SIGMOD (2014)
7. Gufler, B., Augsten, N., Reiser, A., Kemper, A.: Load balancing in MapReduce based on scalable cardinality estimates. In: ICDE. IEEE, April 2012
8. Kwon, Y., Balazinska, M., Howe, B., Rolia, J.A.: SkewTune: mitigating skew in MapReduce applications. In: SIGMOD (2012)
9. Lee, K.H., Lee, Y.J., Choi, H., Chung, Y.D., Moon, B.: Parallel data processing with MapReduce: a survey. SIGMOD Rec. **40**(4), 11–20 (2011)
10. Ramakrishnan, S.R., Swart, G., Urmanov, A.: Balancing reducer skew in MapReduce workloads using progressive sampling. In: ACM Symposium on Cloud Computing, SoCC (2012)
11. Rao, S., Ramakrishnan, R., Silberstein, A., Ovsiannikov, M., Reeves, D.: Sailfish: a framework for large scale data processing. In: ACM Symposium on Cloud Computing, SoCC (2012)
12. White, T.: Hadoop - The Definitive Guide: Storage and Analysis at Internet Scale, 3rd edn. O'Reilly, Sebastopol (2012)
13. Zaharia, M., Chowdhury, M., Das, T., Dave, A., Ma, J., McCauly, M., Franklin, M.J., Shenker, S., Stoica, I.: Resilient distributed datasets: a fault-tolerant abstraction for in-memory cluster computing. In: NSDI (2012)

MapReduce-DBMS: An Integration Model for Big Data Management and Optimization

Dhouha Jemal[1(✉)], Rim Faiz[2], Ahcène Boukorca[3], and Ladjel Bellatreche[3]

[1] LARODEC, ISG Tunis, University of Tunis, Tunis, Tunisia
dh.jemal@gmail.com
[2] LARODEC, IHEC Carthage, University of Carthage, Carthage, Tunisia
rim.faiz@ihec.rnu.tn
[3] LIAS/ISAE-ENSMA Futuroscope, Chasseneuil-du-Poitou, France
{ahcene.boukorca,bellatreche}@ensma.fr

Abstract. The data volume and the multitude of sources have an exponential number of technical and application challenges. In the past, Big Data solutions have been presented as a replacement for the Parallel Database Management Systems. However, Big Data solutions can be seen as a complement to a RDBMS for analytical applications, because different problems require complex analysis capabilities provided by both technologies. The aim of his work is to integrate a Big Data solution and a classic DBMS, in a goal of queries optimization. We propose a model for OLAP queries process. Then, we valid the proposed optimized model through experiments showing the gain of the execution cost saved up.

Keywords: Big data · Mapreduce · RDBMS · Information retrieval · Integration · Optimization · Performance · Cost · OLAP

1 Introduction

The amount of data is increasing dramatically, with an increase use of multimedia, social media and the Internet across a range of devices, which creates an exponential growth in data. Data growth is estimated of 1.2 zettabytes per year 1. Nevertheless, the impact of data abundance extends well beyond volume: the multitude of sources leads to diversity of structures, 70 % of data are unstructured and will grow faster than structured data[1]. Moreover, beyond the storage, challenges will focus on the capacity to process the data and make it available to users. As described in [1], with the worldwide volume of data which does not stop growing, the classical tools for data management have become unsuitable for processing. New technologies are necessary to answer the data explosion, that it is a question of storing but also of making accessible and of analyzing.

---

[1] http://www.gartner.com.

© Springer International Publishing Switzerland 2015
Q. Chen et al. (Eds.): DEXA 2015, Part II, LNCS 9262, pp. 430–439, 2015.
DOI: 10.1007/978-3-319-22852-5_36

Hence, as described in [2], Big Data is a popular phenomenon used to identify the datasets that due to their large size and complexity, we can not manage them with our current methodologies. It proposes new solutions in order to provide an alternative to traditional solutions database and analysis. It is not just about storage of and access to data, but also to manage and analyze large amounts of data, which is getting increasingly larger because of evolution of digital data and information collection devices. Thus, MapReduce [3] is presented as one of the most efficient Big Data solutions mainly due to its salient features that include scalability, fault-tolerance, ease of programming, and flexibility. As described in [4], it is a programming model introduced by Google for processing very large data-sets.

The emergence of big data and its solutions contributes to rapidly transform traditional businesses as well. Data and information are becoming primary assets for many organizations, and the data they own and how they use it can make differentiate them than others. That's why, organizations prepare for these developments in the big data ecosystem to have efficient and effective decision-making processes with right data to make the decision the most adapted at a given moment. In 2013, 42 % state have invested in big data or are planning to do so within a year[2]; By 2015, 20 % of global 1000 organizations will have established a strategic focus on information infrastructure.[3] Today, most organizations try to collect and process as much data as possible, in order to rapidly respond to market needs and changes. The question is what technology to use for data analysis in such an environment.

Conscious of the need to a powerful and optimized tools to verify and analyze data in order to support the decision-making process, the aim of this work is to optimize queries to meet new organizations needs by determining the most efficient way to execute a given query. We propose to integrate two main categories of data management systems: classic Database management systems (DBMS) and NoSQL DBMSs. The idea is to integrate the ORDBMS [5] PostgreSQL [6] and MapReduce to perform OLAP queries in a goal of minimizing Input/Output costs in terms of the amount of data to manipulate, reading and writing throughout the execution process. The main idea of integration leans on the cost model to approve execution and selectivity of solutions based on the estimated cost of queries processing. We propose to compare the query execution costs by both paradigms with the aim of minimizing the Input/Output costs. We valid the proposed approach through experiments showing the significant gain of the cost saved up compared to executing queries independently on MapReduce and PostgreSQL.

The remainder of this paper is organized as follows. In Sect. 2, we give an overview of related work addressing data management systems integration search. In Sect. 3, we describe the MapReduce process. In Sect. 4, we present our proposed approach for optimizing the online analytical processing (OLAP) queries Input/Output execution cost, then in Sect. 5 we discuss the obtained results. Finally, Sect. 6 concludes this paper and outlines our future work.

---

[2] http://www.idc.com/.
[3] http://www.mckinsey.com/.

2   Related Work

A strong interest towards the term Big Data is arising in the literature actually. Indeed, in this field several studies focus on this actual research trends. A lot of work has been done to compare the MapReduce model with parallel relational databases. In addition, there exist some recent work on bringing together ideas from MapReduce and database systems.

In [7], McClean et al. consider broader themes of the paradigms rather than the specific implementations of MapReduce and Parallel DBMS. It discusses MapReduce and Parallel Database Management Systems as competing and complimentary paradigms with the aim of providing a high-level comparison between MapReduce and Parallel DBMS, in order to provide a selection of criteria, which can be used to choose between MapReduce and Parallel DBMS for a particular enterprise application. Nance et al. discuss the pros/cons of NoSQL in [8], and present NoSQL data modeling techniques. This work propose that the SQL and NoSQL models both have their own set of pros and cons that each business has to identify, and then decide which one is better for their company; or if they should use a combination of both SQL and NoSQL. Stonebraker et al. [9], discuss the differences in the architectural decisions of MapReduce systems and database systems in order to provide insight into how the systems should complement one another. This work argues that MapReduce is more like an extract-transform-load (ETL) system than a DBMS, as it quickly loads and processes large amounts of data in an ad hoc manner. As such, it complements DBMS technology rather than competes with it, since databases are not designed to be good at ETL tasks. Then, it describes what the ideal use of MapReduce technology is and highlights the different MapReduce and parallel DMBS markets. Several research works have been conducted between MapReduce and relational database management system (RDBMS) in the goal of integration. In [10], Gruska and Martin consider the two systems RDBMSs and MapReduce as complimentary and not competitors. In this work, taxonomy is provided to characterize several existing integration methods. It proposes a classification and characterization of current MapReduce and RDBMS integration technologies and argues the need for interoperability between a RDBMS and MapReduce system. Then, Yui and Kojima propose in [11] a database-Hadoop hybrid approach to scalable machine learning. In this approach, batch-learning is performed on the Hadoop platform, while incremental-learning is performed on PostgreSQL. The training speed is considered as the main metric for evaluating this work. The work presented by Abouzeid et al. [12], attempted to bridge the gap between the two technologies, that is, parallel databases and Map/Reduce model, suggesting a hybrid system that combines the best features from both. But this work evaluated the model having as a metric the computation time and efficiency. Pavlo et al. [13], conduct experiments to evaluate both parallel DBMS and the MapReduce model in terms of performance and development complexity. This work showed that the MapReduce model outperformed in scalability and fault tolerance, but at the same time underlined the performance limitations of the model, in terms of computation time. This lack was explained by the fact that the model was not originally designed to perform structured data analysis.

The presented research works which have been conducted between MapReduce and RDBMSs in order to provide an overview on each paradigm, discuss advantages and challenges and detailing main components, these works did not take into account many cases of complex analytical problems require the capabilities provided by both systems. Even several research studies have been conducted between MapReduce and RDBMSs in the goal of integration, but all these works evaluate the model having as a metric the computation time and efficiency. No attempt was carried out to analyse the I/O execution cost.

Our proposed work is at the level of the MapReduce and RDBMS integration technologies, and it aims to minimize the Input/Output cost for data processing. The next section will present an overview on the MapReduce framework involved in the integration model.

3   MapReduce Process

In this section we describe the MapReduce process. One who works in the world of databases and NoSQL, he certainly heard of MapReduce the powerful tool characterized by its performance for heavy processing to be performed on a large volume of data that it can be a solution to have the best performance hence makes it very popular with companies that have large data processing centers such as Amazon and Facebook. MapReduce, as presented in [4, 18] is a programming model for processing large data sets with a parallel, distributed algorithm on a cluster. It was created by Google in 2003, in order to simplify parallel processing and distributed data on a large number of machines with an abstraction that hides the details of the hardware layer to programmers.

The MapReduce model consists of two primitive functions: Map and Reduce. This involves dividing the data to be processed in independent partitions, treat the partitions in parallel and finally combining the results of these treatments. Two main components manage the Map/Reduce process: Job tracker (assigns tasks to the Tasktrackers to be performed), and Task tracker (accepts tasks map and reduce from the job tracker). MapReduce offciates the file system HDFS presented in [14] and proposed by Hadoop to perform processing on large data volumes.

The Map phase [15]: During each map phase, the mapper reads the input block and converts each record into a Key/Value pair. The user defined map function transforms each pair into a new Key/Value pair based on the user's implementation. The subsequent phase consists of partitioning/grouping and sorting the map function's outputs.

The Reduce phase [15]: The reduce tasks are the n run over the resulting data, to combine the outcome of the Map Results. During each reduce phase, the reducers retrieve the intermediate data from the mappers. In details, each reducer fetches the corresponding partition from each mapper. Based on the same key, each set of partitions will be merged in order to construct pairs of Key/List (Values). The new intermediate pairs of Key/List (Values) are combined based on the user's defined function to return a new key/value pair. The output pairs are stored on the HDFS in the output file.

4   MapReduce-PostgreSQL Integration Model

Organizations look to choose the technology to be used to analyze the huge quantity of data, and look for best practices to deal with data volume and diversity of structures for best performance and optimize costs. The question is about the selection criteria to be considered in processing the data while meeting the objectives of the organization.

Data processing needs are changing with the ever increasing amounts of both structured and unstructured data. While the processing of structured data typically relies on the well developed field of relational database management systems (RDBMSs), MapReduce is a programming model developed to cope with processing immense amounts of unstructured data.

For this purpose, we suggest a model to integrate the RDBMS PostgreSQL and The MapReduce framework as one of the Big Data solutions in order to optimize the OLAP queries Input/Output execution cost. We aim to integrate classic data management systems with Big Data solutions in order to give businesses the capability to better analyze data with a goal of transforming it into useful information as well as minimizing costs.

The basic idea behind our approach is based on the cost model to approve execution and selectivity of solutions based on the estimated cost of execution. To support the decision making process for analyzing data and extracting useful knowledge while minimizing costs, we propose to compare the estimates of the costs of running a query on Hadoop MapReduce compared to PostgreSQL to choose the least costly technology. For a better control, we will proceed to a thorough query analysis.

4.1   Pushed Analysis of a Query's Execution

The detailed analysis of the queries execution costs showed a gap mattering between both paradigms. Hence the idea of the thorough analysis of the execution process of each query and the implied cost. To better control the cost difference between costs of Hadoop MapReduce versus PostgreSQL on each step of the query's execution process, we propose to dissect each query for a set of operations that demonstrates the process of executing the query. In this way we can check the impact of the execution of each operation of a query on the overall cost and we can control the total cost of the query by controlling the partial cost of each operation in the information retrieval process.

In this context, we are inspired from a work done in [16] to provide a detailed execution plan for OLAP queries. This execution plan zooms on the steps sequence of the query execution process. It allows detailing the various operations of the process highlighting the order of succession and dependence.

In addition to dissect the implementation process, the execution plan details for each operation the amount of data by the accuracy of the number of records involved and the dependence implemented in the succession of phases. These parameters will be needed to calculate the cost involved in each operation. After distinguishing the different operations of the query, the next step is to calculate the unit cost of each operation. As part of our approach we aim to dissect each query and focus on each separate operation. That way we can control the different stages of the execution process for each query. The aim is to calculate the cost implied by each operation as well as its influence on the total cost of the query. Therefore we can control the cost of each query to support the decision making process and the selectivity of the proposed solutions based on the criterion of cost minimization.

4.2 Queries Cost Estimation

Having identified all operations performed during the query execution process the next step is then to calculate the cost implied in each operation independently, in both paradigms PostreSQL and MapReduce with the aim of controlling the estimated costs difference according to the operations as well as the total cost of query execution. At this stage, we consider each operation independently to calculate an estimate of its cost execution on PostreSQL on one hand then on MapReduce on the other hand.

Cost estimation on MapReduce. In a MapReduce system, a query star join between F (fact table) and n dimension tables, runs a number of phases, each phase corresponds to a MapReduce job. So, for MapReduce paradigm we have first to extract the number of jobs that will be run for executing the query. The MapReduce job number depends on the number of joint and the presence or absence of aggregation and sorting data. There are three cases of figure: The request contains only n successive joint operations between F and n dimension tables; Join operations are followed by a process of grouping and aggregation on the results of the joint; Sort is applied to the results. In this context, we propose to rely on the Eq. (1) presented below, and inspired from [17]. This equation allows determining the number of MapReduce jobs implied in the execution of a given OLAP query. This number can be estimated by the following formula:

$$Nbrjobq = n + x \tag{1}$$

The "n" refers to the number of dimension tables. The _x_ can be equal to: 0, if the query involves only join operations; 1, if the query contains grouping operations and aggregation; 2, if the results are sorted. After identifying all the jobs of query execution, the next step is to calculate the Input/Output cost implicated in each job. In this stage, we relied on the mentioned work presented in [16] to extract the amount of data and the number of records involved in each operation. These two parameters will be used in a MapReduce cost model that we implemented to approve our proposed approach, in order to calculate the Input/Output cost of each job.

Cost estimation on PostgreSQL. PostgreSQl provides the possibility of itemize each operation by an incremental value of the Input/Output cost implied in each step.

In PostgreSQL platform, the command "explain" shows the execution plan of a statement. This command displays the execution plan that the PostgreSQL planner generates for the supplied statement. Besides the succession of the executed operations, the most critical part of the display is the estimated statement execution cost (measured in cost units that are arbitrary, but conventionally mean disk page fetches). It includes information on the estimated start-up and total cost of each plan no de, as well as the estimated number of rows. Actually two numbers are shown: the start- up cost before the first row can be returned, and the total cost to return all the rows. Therefore for each operation, the cost should be the difference between these two values. This way, using "explain" statement, we extract the cost implicated in each operation of the query execution process.

4.3 Analysis: OLAP Queries Processing Smart Model

The estimated costs results analysis independently for each operation showed the high cost of the first join operation executed on PostgreSQL, and a noticeable difference for the Hadoop MapReduce paradigm. This can be explained by the fact that in the case of data warehouses, the fact table is still the largest table in terms of number of tuple, which explains the high cost of its analysis. Hence our idea of performing the first joint operation that integrates fact table on MapReduce framework, which proves competence for heavy processing to be performed on a large volume of data. In this way we try to minimize the cost of the query execution by minimizing the cost of the most expensive operation. According to the observation, other operations required by the query such as aggregation, sorting, in addition to other join operations are most often less costly on PostgreSQL. Therefore, we propose to generalize this case and pass these operations on PostgreSQL.

5 Experimental Results

In order to validate our proposed approach, we present in this section the results of experiments conducted to evaluate the proposed model performance as well as gain cost compared with the cost required by each platform independently.

The experiments involve two DBMSs: an ORBMS PostgreSQL and a NoSQL DBMS Hadoop MapReduce. To test and compare our theoretical expectations with the real values, we set up a cluster consisting of one node. For all the experiments, we use the version 9.3 of PostgreSQL. For Hadoop, we used the version 2.0.0 with a single node as worker node hosting DataNode, and as the master node hosting NameNode.

We worked on a workload of 30 queries OLAP. The training data consisted of a data warehouse of 100 GB of data with a fact table and 4 dimension tables.

Our approach proposes an hybrid model between ORDBMS and Hadoop MapReduce, based on the comparison of Input/Output costs on the both paradigms. The results of the application of the proposed approach are presented in the Table 1. It shows the total Input/Output cost of running the workload on Hadoop MapReduce (Tot_cost_MR), the total Input/Output cost of running the workload on ORDBMS PostgreSQL (Tot_cost_PG), the total Input/Output cost of the workload by applying our proposed approach (Tot_cost_intg).

Table 1. Proposed approach results

Tot_cost_MR	Tot_cost_PG	Tot_cost_intg
505579305,9	786297800,8	369473436,7

Observing the values presented in Table 1 illustrate the difference of the cost saved up by the application of the proposed approach. The cost is presented in page unit (4096 bytes). The operations required by the query such as aggregation, sorting, in addition to join operations except the first join operation are most often less costly on PostgreSQL. In our smart model, we proposed to generalize this case and pass these operations on PostgreSQL.

The histogram presented in the Fig. 1 shows the gain of the Input/Output cost by running the workload independently on Hadoop MapReduce and on postgreSQL. In addition, it compare the Input/Output cost of the workload obtained by applying our proposed approach to the Input/Output the cost obtained if we choose the lower cost for each operation contained in the execution plan of each query.

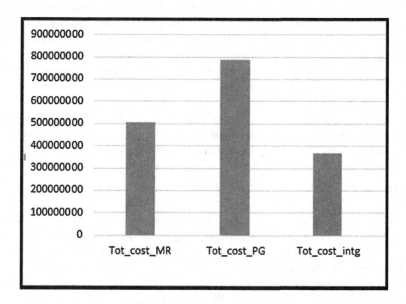

Fig. 1. Proposed approach results

The Fig. 1 illustrates the notable difference of the Input/Output cost of running the workload by applying the proposed approach compared to the Input/Output running cost on the two paradigms PostgreSQL and Hadoop MapReduce separately.

The gain estimated of the Input/Output cost saved up by running the workload tanks to applying the proposed approach is of 27 % compared to Hadoop MapReduce and 53 % compared to PostgreSQL. This percentage gain proves the performance of our proposed approach.

6   Conclusion

There is no question that we are living in an era of data and information explosion. For instance, there is the challenge of managing large amounts of data, which is getting increasingly larger. But, the classical tools for data management have become unsuitable for processing. Thus, the emergence of Big Data, which requires a revolutionary step forward from traditional data analysis.

Therefore, it is important to benefit the diversity of the solutions proposed for data analysis. It is important to pick the right database technology for the task at hand. Depending on what problem the organization is trying to solve, the classic database model and Big Data solutions are each good for specific applications. But, many complex analytical problems require the capabilities provided by both systems. This requirement motivates the need for interfaces between MapReduce system and DBMSs that allow each system to do what it is good at. The result is a much more efficient overall system than if one tries to do the entire application in either system.

We have proposed in this paper a new OLAP queries process smart model integrating an ORDBMS with the MapReduce framework in a goal of minimizing Input/Output costs in terms of the amount of data to manipulate, reading and writing throughout the execution process. Our work intended to find useful information within massive data stores, while keeping the lower cost. The experimental evaluation we conducted on a workload of OLAP queries shows the highly performance of our search results confirming our model's effectiveness.

Looking ahead, we plan to conduct experiments on a greater workload. We also need to highlight the influence of the node's number in the cluster. We plan to consider other criteria in the strategy of the integration model as the execution time.

References

1. Ordonez, C.: Can we analyze big data inside a DBMS?. In: Proceedings of the Sixteenth International Workshop on Data Warehousing and OLAP, ACM, pp. 85–92 (2013)
2. Fan, W., Bifet, A.: Mining big data: current status, and forecast to the future. In: SIGKDD Explorations, vol. 14, issue 2 (2011)
3. Doulkeridis, C., Nørvåg, K.: A survey of large-scale analytical query processing in MapReduce. VLDB J. **23**(3), 355–380 (2014)
4. Dean, J., Ghemawats, S.: Mapreduce: Simplified data processing on large clusters. Commun. ACM **51**(1), 107–113 (2008)
5. Brown, P.G.: Object-Relational Database Development: A Plumber's Guide. Prentice Hall PTR, Upper Saddle River (2000)
6. Douglas, K., Douglas, S.: PostgreSQL: A Comprehensive Guide to Building, Programming, and Administring PostreSQL Databases, 1st edn. Sams Publishing, (2003)
7. McClean, A., Conceicao, R., O'halloran, M.: A comparison of MapReduce and parallel database management systems. In: ICONS 2013 The Eighth International Conference on Systems, pp. 64–68 (2013)
8. Nance, C., Losser, T., Iype, R., Harmon, G.: NOSQL VS RDBMS – why there is room for both. In: Proceedings of the Southern Association for Information Systems Conference Savannah USA, pp. 111–116 (2013)
9. Stonebraker, M., Abadi, D., Dewitt, D., Madden, S., Paulsone, E., Pavlo, A., Rasin, A.: Mapreduce and parallel dbmss: friends or foes? Commun. ACM **53**(1), 64–71 (2010)
10. Gruska, N., Martin, P.: Integrating MapReduce and RDBMSs. In: Proceedings of the 2010 Conference of the Center for Advanced Studies on Collaborative Research, IBM Corp, pp. 212–223 (2010)

11. Yui, M., Kojima, I.: A database-hadoop hybrid approach to scalable machine learning. In: Big Data 2013 IEEE International Congress on IEEE, pp. 1–8 (2013)
12. Abouzeid, A., Pawlikowski, K., Abadi, D., Silberschatz, A., Rasin, A.: Hadoopdb: an architectural hybrid of mapreduce and dbms technologies for analytical workloads. In: Proceedings of the VLDB Endowment, pp. 922–933 (2009)
13. Pavlo, A., Rasin, A., Madden, S., Stonebraker, M., Dewitt, D., Paulson, E., Shrinivas, L., Abadi, D.: A comparison of approaches to large scale data analysis. In: Proceedings of the 2009 ACM SIGMOD International Conference on Management of data, pp. 165–178 (2009)
14. Chandrasekar, S., Dakshinamurthy, R., Seshakumar, P.G., Prabavathy, B., Babu, C.: A novel indexing scheme for efficient handling of small files in hadoop distributed file system. In: Computer Communication and Informatics (ICCCI), 2013 International Conference, pp. 1–8 (2013)
15. Pavlo, A., Paulson, E., Rasin, A., Abadi, D.J., DeWitt, D.J., Madden, S., Stonebraker, M.: A comparison of approaches to large-scale data analysis. In: Proceedings of the 2009 ACM SIGMOD International Conference on Management of data ACM, pp. 165–178 (2009)
16. Boukorca, A., Faget, Z., Bellatreche, L.: What-if physical design for multiple query plan generation. In: Decker, H., Lhotská, L., Link, S., Spies, M., Wagner, R.R. (eds.) DEXA 2014, Part I. LNCS, vol. 8644, pp. 492–506. Springer, Heidelberg (2014)
17. Brighen, A.: Conception de bases de données volumineuses sur le cloud. In: Doctoral dissertation, Université Abderrahmane Mira de Béjaia (2012)
18. Fabrizio, M., Domenico, T., Paolo, T.: P2P-MapReduce: parallel data processing in dynamic cloud environments. J. Comput. Syst. Sci. **78**(5), 1382–1402 (2012)

A Travel Time Prediction Algorithm Using Rule-Based Classification on MapReduce

HyunJo Lee, Seungtae Hong, Hyung Jin Kim, and Jae-Woo Chang[(✉)]

Department of Computer Engineering, Chonbuk National University,
Jeonju, Jeollabuk-Do, South Korea
{o2near,dantehst,yeon_hui4,jwchang}@jbnu.ac.kr

Abstract. Recently, the amount of trajectory data has been rapidly increasing with the popularity of LBS and the development of mobile technology. Thus, the analysis of trajectory patterns for large amounts of trajectory data has attracted much interest. To improve the quality of trajectory-based services, it is essential to predict an exact travel time for a given query on road networks. One of the typical schemes for travel time prediction is a rule-based classification method which can ensure high accuracy. However, the existing scheme is inadequate for the processing of massive data because it is designed without the consideration of distributed computing environments. To solve this problem, this paper proposes a travel time prediction algorithm using rule-based classification on MapReduce for a large amount of trajectory data. First, our algorithm generates classification rules based on the actual traffic statistics and measures adequate velocity classes for each road segment. Second, our algorithm generates a distributed index by using the grid-based map partitioning method. Our algorithm can reduces the query processing cost because it only retrieves the grid cells which contain a query region, instead of the entire road network. Furthermore, it can reduce the query processing time by estimating the travel time for each segment of a given query in a parallel way. Finally, we show from our performance analysis that our scheme performs more accurate travel time prediction than the existing algorithms.

Keywords: Travel time prediction · Rule-based classification · Mapreduce

1 Introduction

Recently, the amount of trajectory data has been rapidly increasing with the popularity of LBS and the development of mobile technology. The travel time prediction on road networks is essential for ATIS (Advanced Travelers Information Systems) and ITS (Intelligent Transport Systems) [1]. Existing travel time prediction methods can be categorized into link-based prediction and path-based prediction. The path-based method predicts travel time by searching for similar paths to a given query. The link-based method predicts travel time by measuring the average time and the speed of moving objects for a road segment contained in the query path. To enhance the accuracy of travel time prediction on road networks, various factors affecting traffic condition

© Springer International Publishing Switzerland 2015
Q. Chen et al. (Eds.): DEXA 2015, Part II, LNCS 9262, pp. 440–452, 2015.
DOI: 10.1007/978-3-319-22852-5_37

must be considered. For instance, a time variable such as morning, evening and rush hour is critical for accurate travel time prediction because vehicle speeds differ greatly throughout the day. However, existing methods tend to have poor accuracy because they do not reflect various traffic conditions such as time, day, month, season, weather and road type. Moreover, the existing methods are inadequate for the processing of massive data because they predict the travel time of a query path by sequentially estimating the travel time of each road segment in the path.

To resolve the problems, we propose a new travel time prediction algorithm using rule-based classification on Hadoop's MapReduce. First, we propose a rule generation algorithm on MapReduce for supporting the efficient processing of spatial big data. By defining rules using actual traffic statistics, adequate velocity classes can be determined for each road section. Second, we propose a grid-based distributed indexing scheme to efficiently find velocity classes for road sections. Based on the proposed indexing scheme, our query processing algorithm can estimate the accurate travel time of the road sections being contained in a query path in a parallel way.

The rest of the paper is organized as follows. In Sect. 2, we introduce the existing methods for travel time prediction. In Sect. 3, we describe the proposed travel time prediction algorithm using rule-based classification on MapReduce. In Sect. 4, we do performance analysis of our scheme. Finally, we conclude this paper with future work in Sect. 5.

2 Related Work

2.1 Travel Time Prediction Method on Road Network

Existing travel time prediction methods can be categorized into path-based prediction algorithm and link-based prediction one. The path-based prediction methods [2–9] are suitable for predicting the travel time of long distance driving such as highway driving. By computing the travel time of the similar paths with a given path, they focus on reducing the errors of travel time prediction. Therefore, when the number of alternative paths is relatively small (e.g. highway), they have a high accuracy of travel time prediction because the daily/weekly traffic volume is consistent. On the other hand, in the case of an urban area, the number of alternative paths is very large, so the accuracy is decreased. To solve this problem, link-based prediction methods [10–12] have been proposed. The link-based prediction methods predict the travel time for all road segments which constitute the given query. For this, they measure the average time and the speed of moving objects per road segment included in the query path. However, they have a low accuracy since they do not consider various hourly/daily traffic conditions except the average travel time.

2.2 Hadoop MapReduce

Hadoop MapReduce is a framework for parallel processing of massive data sets. A job to be performed using the MapReduce framework is composed of two functions: map function and reduce function. The map function produces intermediate results in the form of key/value pairs. The reduce function merges the intermediate results based on the key and returns the final results to users. For example, when counting the frequency of words in a document by using MapReduce, the document is tokenized into words in the map phase. The map function generates the key-value pair {key = word, value = 1 (frequency)}. In the reduce phase, <word, 1> pairs are grouped by their keys. The reduce function measures the frequency of each word by aggregating frequencies for key groups.

3   Travel Time Prediction Algorithm on MapReduce

The procedure of our travel time prediction algorithm on MapReduce is as follows. First, in order to accurately estimate the travel time, our algorithm generates classification rules by using traffic data from traffic monitoring system (TMS) [13] in the pre-processing step. Secondly, our algorithm measures adequate velocity classes for each road segment on MapReduce in the velocity class measurement step. Thirdly, our algorithm generates a distributed index by using the grid-based map partitioning method in the grid-based indexing step. Finally, in the query processing step, our algorithm estimates the travel time for each segment of a given query on MapReduce in a parallel way

3.1   TMS-Based Classification Rule Generation

To accurately estimate the travel time, we define classification rules by using traffic data from TMS. For this, we use six classification groups; departure time, day of the week, month, weather, road-type and season group.

(1) Departure time group
    Figure 1 shows an hourly traffic count from TMS. The traffic volume increases dramatically at 7:00 AM and decreases rapidly at 6:00 PM. Especially, the traffic volume grows rapidly from 3:00 PM to 6:00 PM. Table 1 shows the Departure Time Group by considering the hourly permanent traffic count. The rush hour (16:00~20:00) is divided into two groups: late afternoon and quitting time. The late afternoon has a highest traffic volume, while the traffic volume decreases dramatically in the quitting time.

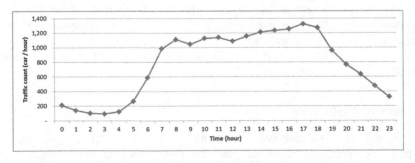

Fig. 1. Hourly permanent traffic count

Table 1. Departure time group

Departure time	Group	Description	Departure time	Group	Description
00:00–04:30	1	Midnight	14:00–16:00	7	Afternoon
04:30–07:00	2	Dawn	16:00–18:00	8	Late afternoon
07:00–09:00	3	Rush hour	18:00–20:00	9	Quitting time
09:00–12:00	4	Morning	20:00–22:00	10	Night
12:00–13:00	5	Lunch	22:00–24:00	11	Late night
13:00–14:00	6	Late lunch			

(2) Day of week group

Figure 2 shows a day of the week traffic count. Sunday has a lowest traffic volume, while Saturday has a highest one. Table 2 shows the day of week group by considering the day of week traffic count.

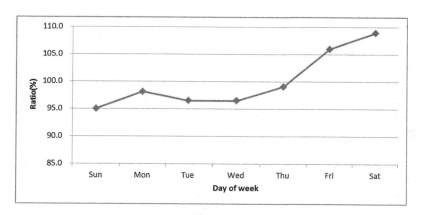

Fig. 2. Day of week permanent traffic count

Table 2. Day of week group

Departure time	Group	Description
Monday	M	Beginning of the week1
Tuesday, Wednesday	W	Beginning of the week2
Thursday	T	Midweek
Friday, Saturday	F	Weekend
Sunday	S	–

(3) Monthly group

Figure 3 shows a monthly traffic count. The traffic volume of August is the largest among 12 months. Table 3 shows the month group by considering the monthly traffic count.

Table 3. Month group

Month	Group	Description	Month	Group	Description
January	JA	Beginning of the year	July	JUL	Summer2 + vacation
February	FE	Opening day of the school year	August	AU	Summer3 + vacation
March	MA	New semester	September	SE	Autumn1
April	AP	Spring1	October	OC	Autumn2
May	MAY	Spring2	November	NO	Winter1
June	JUN	Summer1	December	DE	Winter2 + end of the year

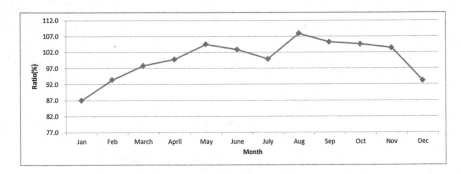

Fig. 3. Monthly permanent traffic count

(4) Weather, road-type and seasonal group

Table 4 shows the weather group by using open DB from the weather center. Table 5 shows the road-type group by considering the speed limit. Meanwhile, the seasonal group consists of four seasons; spring, summer, autumn and winter.

Table 4. Weather group

Weather	Group	Weather	Group
Clean	C	Fog	F
Cloudy	D	Hail	S1
Rain	R	Etc.	E
Snow	S		

Table 5. Road-type group

Speed limit	Road-type	Description
High	CA	Driveway
Low	CH	Child protection
Medium	CV	General

3.2 Velocity Class Measurement

In order to accurately measure the velocity class of moving objects per road segment, we devise a velocity class measurement algorithm on MapReduce. We define three velocity classes: VB, B and F which mean Very Busy, Busy and Free, respectively. When the speed of a vehicle is less than 35 km/h, the traffic jam is assumed to occur. Table 6 shows the velocity class for a road Section.

Table 6. Velocity class

Speed range (km/h)	Rank
0–40	VB
41–80	B
>81	F

(1) Velocity class measurement using rule-based classification

By considering attributes such as road section, time zone and day of the week, our algorithm analyzes a road condition for classifying the speed of the road section. Then, our algorithm measures the velocity class by computing the average speed of the vehicle under the generated road condition. We make a decision rule as the combination of road condition and velocity class.

IF (R(E) = Road section ID)
AND (R(T_G) = Departure time group)
AND (R(D_G) = Day of week group)
THEN Velocity class = (VB, B, F)

Here, R(E) means the road segments which constitute the path between departure point and arrival point. R(T_G) and R(D_G) means the Departure time group and Day of week group, respectively. The IF-clause of the decision rule is an antecedent which combines one or more attributes such as R(E), R(T_G), R(D_G). The THEN-clause is a consequent which is one of three velocity classes, like VB, B and F. If the condition in a rule antecedent holds true for a given tuple, we can say that rule antecedent is satisfied and that the rule covers the tuple. Accuracy is a measure to assess any rule R. The accuracy of R is calculated as the percentage of tuples covered by the same antecedent.

(2) Velocity class measurement algorithm

Figure 4 shows a velocity class measurement algorithm. First, a master node divides the input data into equal-sized splits by considering the number of slave nodes (line 1). Then, the master node allocates splits to each mapper (line 2). Second, in the Map phase, each mapper measures the velocity class by analyzing the assigned trajectory data. Then, each mapper outputs the intermediate data in the form of < EdgeID, (road condition, velocity class) > (line 3~7). The intermediate data is allocated to each reducer based on their key. Finally, in the Reduce phase, each reducer merges the intermediate data and measures the velocity class (line 8~9). Also, the final results are returned to the user (line 10).

```
//Input : Trajectory Dataset
//         <Trajectory ID, List of (edge, timestamp),
/              year, month, day, OtherConditions>
//Output : Rules for each edge
   i)Master
1. SplitData(# of Mappers);
2. Set(SplitData(i), Mapper(i));
   ii)Mapper
3. For each trajectory data{
4.       Rule = GradeConditions();
5.       Key = EdgeID;
6.       Value = Rule;
7.       ContextWrite(EdgeID, Rule); }
   iii)Reducer
8. MergedData = Merge(List of (Edge, Rule));
9. Result = StatisticalAnalysis(MergedData);
10. Return Result;
```

Fig. 4. 1st MapReduce: Velocity class measurement algorithm

3.3   Grid-Based Rule Indexing

In order to efficiently find velocity classes for road sections in parallel way, our algorithm generates a distributed index by using the grid-based map partitioning method. For this, our algorithm divides the whole road network into several grid cells. A grid cell stores a set of velocity classes and the traffic conditions of overlapping road segments.

Figure 5 shows a grid-based indexing algorithm. First, a master node divides the input data into equal-sized splits based on the number of slave nodes (line 1). Then, the master node allocates splits to each mapper (line 2). Secondly, in the Map phase, each mapper measures grid CellIDs based on the location of start/end point of a trajectory data. By using grid CellIDs, each mapper outputs the intermediate data in the form of <CellID, EdgeID> (line 3~7). The intermediate data is allocated to each reducer based on their key. Finally, in the Reduce phase, each reducer merges the intermediate data and generates a grid-based distributed index (line 8~9).

```
//Input : Road Network
//        <EdgeID, Two Points for edge,
//              List of (neighbor edges, timestamp)>
//Output : Grid-Edge Table
   i)Master
1. SplitData(# of Mappers);
2. Set(SplitData(i), Mapper(i));
   ii)Mapper
3. For each point in the edge{
4.      GridID = CalculateGID(Point(i));
5.      Key = GridID;
6.      Value = EdgeID;
7.      ContextWrite(GridID, EdgeID); }
   iii)Reducer
8. MergedData = Merge(List of (GridID, EdgeID));
9. GenerateTable(MergedData);
```

Fig. 5. 2nd MapReduce: Grid-based indexing algorithm

Figure 6 shows an example of grid-based indexing. Each grid cell stores <RoadID, Classification group, Velocity class, Frequency>. When additional data are generated, our algorithm only updates the corresponding cells. As a result, our algorithm can reduce the update cost, compared to the existing tree-based indexing scheme.

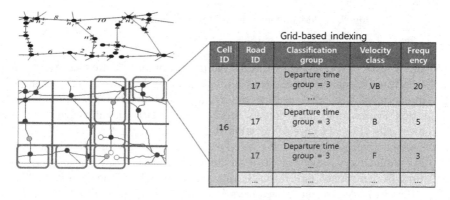

Fig. 6. Example of grid-based indexing

3.4   Query Processing for Travel Time Prediction

Because our algorithm estimates the travel time by using a grid-based index on MapReduce, it can reduce the query processing time because it estimates the travel time for each segment of a given query path in a parallel way, instead of the sequential searching of a whole query path. Figure 7 shows a travel time prediction algorithm. First, a master node receives from a user a query which consists of query path, departure time and day of week. Based on the grid index, the master node selects the grid cell IDs for estimating the travel time for a given path (line 1~2). Then, the master node allocates each grid cell to mappers (line 3). Second, each mapper loads the data located in the assigned grid cell (line 4~5). Then, each mapper chooses a decision rule based on the given query condition (line 6). Thirdly, each mapper calculates the travel time by using decision rule and estimates the arrival time of given section (line 7). Then, each mapper sends both the estimated travel time and edge ID to a reducer (line 8). At last, in the Reduce phase, each reducer merges the travel times for all sections and returns the final results to users (line 9~10).

```
// Input : Query <Query ID, List of edges, Start_time(year, month, day, timestamp),
//         OtherConditions>
// Output : Predicted travel time
   i)Master
1. For each edge in Query{
2.      GridID(i) = Calculate(Edge, Grid-Edge Table);
3.      Set(GridID(i), Mapper(i));}
   ii)Mapper
4. GridData = Load(GridID_file);
5. Rules(Edge) = Extract(GridID, Edge);
6. SelectRule(Query, Rules(Edge));
7. pTime(Edge) = CalculateTravelTime(Edge, Query);
8. GenerateMapResult(Edge, pTime);
   iii)Reducer
9. FinalpTime = Merge(List of (Edge, pTime));
10. Return FinalpTime;
```

Fig. 7.   3rd MapReduce: Travel time prediction algorithm

Figure 8 shows the example of travel time prediction on MapReduce. We assume that a vehicle moves on the pre-defined path, like Link1 → Link2 → Link3 → Link4. In the Map phase, each mapper measures the average speed and the travel time for each road section. For example, the average speed of Link1 in Grid1 is calculated by mapper #1. The velocity class of Link1 is VB at the start time. By using the records 1, 2 and 3, the mapper can calculate the average speed as 28 km/h. As a result, the travel time of Link1 is estimated as 21 min (Fig. 8(a)). In the Reduce phase, each reducer merges the travel times from mapper1 (Grid1, Link1), mapper2 (Grid2, Link2) and mapper3 (Grid3, Link3) (Fig. 8(b)). As a result, the reducers estimate the total travel time as 65 min and the final arrival time as AM 11:10.

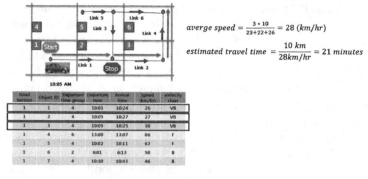

$$averge\ speed = \frac{3 * 10}{23 + 22 + 26} = 28\ (km/hr)$$

$$estimated\ travel\ time = \frac{10\ km}{28km/hr} = 21\ minutes$$

Road Section	Object ID	Departure time group	Departure time	Arrival time	Speed (km/hr)	Velocity class
1	1	4	10:01	10:24	26	VB
1	2	4	10:05	10:27	27	VB
1	3	4	10:05	10:25	30	VB
1	4	6	13:00	13:07	86	F
1	5	4	10:02	10:11	67	F
1	6	2	6:01	6:13	50	B
1	7	4	10:30	10:43	46	B

(a) Map phase

1. 10:05(AM) → Departure time group 4 → 21 minutes → 10:26(AM)
2. 10:26(AM) → Departure time group 4 → 22 minutes → 10:48(AM)
3. 10:48(AM) → Departure time group 4 → 22 minutes → 11:10(AM) → Arrival

Mapper#1(Grid 1, Link1)		Mapper#2(Grid 2, Link2)		Mapper#3(Grid 3, Link4)	
Departure time group	Estimated travel time	Departure time group	Estimated travel time	Departure time group	Estimated travel time
1	-	1	15 minutes	1	14 minutes
2	-	2	11 minutes	2	9 minutes
3	-	3	37 minutes	3	36 minutes
4	26 minutes	4	22 minutes	4	22 minutes
5	-	5	26 minutes	5	25 minutes
6	-	6	31 minutes	6	29 minutes
7	-	7	37 minutes	7	34 minutes
8	-	8	41 minutes	8	36 minutes
9	-	9	30 minutes	9	26 minutes
10	-	10	18 minutes	10	17 minutes
11	-	11	15 minutes	11	15 minutes

(b) Reduce phase

Fig. 8. Example of travel time prediction process on MapReduce

4   Performance Analysis

We compare the performance of our travel time prediction algorithm with that of the existing algorithms. For this, we measure the Mean Absolute Relative Error (MARE) [1] as shown in Eq. (1). Here, t is a time stamp, x(t) is the observation value at t, x*(t) is a predicted value at t, and N is the number of samples. The MARE measures the magnitude of the relative error over the desired time range.

$$MARE = \frac{1}{N} \sum_t \frac{|x(t) - X * (t)|}{x(t)} \tag{1}$$

We use a synthetic dataset containing 4 million trajectories for one year by using a trajectory data generator which was developed based on the actual traffic condition of Busan city, South Korea [14]. Here, the average length of trajectory is 10. This means a trajectory contains 10 road sections on average. For the query processing performance, we create query sets consisting of 1,000 trajectories, which are randomly selected from the synthetic dataset. In our experiment, the time window is ranged from AM 8:00 to PM 6:00 as people moves actively. We used two time lags, 30 and 60 min, from the departure time of a vehicle, t. For example, when a given query time t is AM 8:00 and a time lag is set as 30 min, the departure time of vehicle in query path become AM 8:30. On the other hand, we compare our algorithm with the existing link-based algorithm [1] and Micro T* algorithm [10].

Figures 9 and 10 shows the MARE of travel time prediction algorithms when time lags are set as 30 min and 60 min, respectively. As a result, our algorithm shows the best performance in all cases, compared with the existing algorithms. When a time lag is 30 min, our algorithm shows about 17 % better performance during PM 3:00–6:00 than the existing work. When a time lag is 60 min, our algorithm shows about 10 % better performance than the existing work. This is because the prediction error per each road section is larger than that of time lag = 30 min.

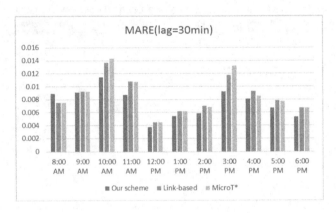

Fig. 9. MARE of travel time prediction algorithms (time lag = 30 min)

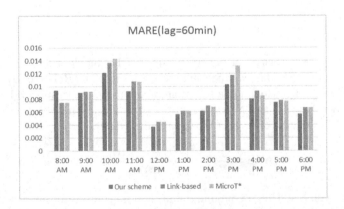

Fig. 10. MARE of travel time prediction algorithms (time lag = 60 min)

Figure 11 shows the average MARE of travel time prediction algorithms. When the time lag is 30 min, the prediction performance (MARE) of our algorithm is improved about 12 % and 13 %, compared to those of the link-based algorithm and Micro T* algorithm, respectively. When a time lag is 60 min, the prediction performance (MARE) of our algorithm is improved about 8 % and 9 %, compared to those of the link-based algorithm and Micro T* algorithm, respectively. This is because our algorithm predicts the travel time of a vehicle accurately by classifying the velocity classes of road sections based on actual traffic statistics.

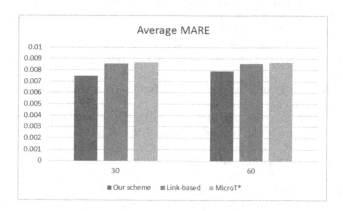

Fig. 11. Average MARE of travel time prediction algorithms (time lag = 30 min)

5   Conclusion

In this paper, we propose a new travel time prediction algorithm using a rule-based classification on MapReduce. Our algorithm can analyze the large-scale spatial data in a parallel way and can accurately predict the travel time of the road sections in a query path. Because our algorithm utilizes the rule-based classification method using actual traffic statistics, it shows better performance on the prediction accuracy than the existing algorithms. As a future work, we plan to apply our algorithm to a real large-scale trajectory data of moving objects.

Acknowledgement. This research was supported by Basic Science Research Program through the National Research Foundation of Korea (NRF) funded by the Ministry of Education (2014065816).

References

1. Chen, M., Chien, S., Dynamic freeway travel time prediction using probe vehicle data: Link-based vs. Path-based. J. Transp. Res. Rec. TRB Paper No. 01-2887 (2001)
2. Chun-Hsin, W., Chia-Chen, W., Da-Chun, S., Ming-Hua, C., Jan-Ming, H.: Travel time prediction with support vector regression. In: IEEE Intelligent Transportation Systems Conference (2003)
3. Park, D., Rilett, L.: Forecasting multiple-period freeway link travel times using modular neural networks. J. Transp. Res. Rec. **1617**, 163–170 (1998)
4. Park, D., Rilett, L.: Spectral basis neural networks for real-time travel time forecasting. J. Transp. Eng. **125**(6), 515–523 (1999)
5. Kwon, J., Coifman, B., Bickel, P.J.: Day-to-day travel time trends and travel time prediction from loop detector data. J. Transp. Res. Rec. No. 1717, TRB, National Research Council, pp. 120–129 (2000)
6. Zhang, X., Rice, J.: Short-term travel time prediction. Transportation Research Part C, vol. 11, pp. 187–210 (2003)
7. Van der Voort, M., Dougherty, M., Watson, S.: Combining KOHONEN maps with ARIMA time series models to forecast traffic flow. Transp. Res. Part C **4**, 307–318 (1996)
8. Rice, J., Van Zwet, E.: A simple and effective method for predicting travel times on freeways. IEEE Trans. Intell. Trans. Syst. **5**(3), 200–207 (2004)
9. Schmitt Erick, J., Jula, H.: On the limitations of linear models in predicting travel times. In: IEEE Intelligent Transportation Systems Conference (2007)
10. Kwon, J., Petty, K.: A travel time prediction algorithm scalable to freeway networks with many nodes with arbitrary travel routes. In: Transportation Research Board 84th Annual Meeting (2005)
11. Lee, C.W., Park, J.Y, Kho, S.Y.: Prediction of path travel time using Kalman filter. J. Korean Soc. Civil Eng. **22**(5), 871–880 (2002)
12. Lee, Y.W.: Establish for link travel time distribution estimation model using fuzzy. J. Korean Soc. Civil Eng. **26**(2D), 233–239 (2006)
13. Traffic Monitoring System. http://www.road.re.kr/main/main.asp
14. Kim, B.R, Lee, S.H., Li, K.J.: Generating trajectory of road network-based moving objects. In: KIPS Fall Conference, vol. 12, No. 2, pp. 1–4 (2005)

Special Section Globe 2015 – Security, Data Privacy and Consistency

Numerical SQL Value Expressions Over Encrypted Cloud Databases

Sushil Jajodia[1], Witold Litwin[2(✉)], and Thomas Schwarz[3]

[1] George Mason University, Fairfax, VA, USA
jajodia@gmu.edu
[2] Université Paris Dauphine, Paris, France
witold.litwin@dauphine.fr
[3] Universidad Centroamericana, El Salvador, Germany
tjschwarz@scu.edu

Abstract. Cloud databases often need client-side encryption. Encryption however impairs queries, especially with numerical SQL value expressions. Fully homomorphic encryption scheme could suffice, but known schemes remain impractical. Partially homomorphic encryption suffices for specific expressions only. The additively homomorphic Paillier scheme appears the most practical. We propose the homomorphic encryption for standard SQL expressions over a practical domain of positive values. The scheme uses a version of Paillier's formulae and auxiliary tables at the cloud that are conceptually the traditional mathematical tables. They tabulate encrypted log and antilog functions and some others over the domain. The choice of functions is extensible. We rewrite the expressions with any number of SQL operators '*', '/' '^' and of standard aggregate functions so they compute over encrypted data using the tables and Paillier's formulae only. All calculations occur at the cloud. We present our scheme, show its security, variants and practicality.

Keywords: Cloud SQL database · Outsourcing · Table-based homomorphic encryption · Value expressions

1 Introduction

It is the common knowledge that client-side encryption is desirable for privacy of data outsourced to a cloud database. However, the capabilities of SQL queries on encrypted data are limited, [3, 7, 14, 17]. If we use a popular encryption such as AES, an SQL query with a numerical value expression becomes impossible to execute over encrypted data. We recall that such an expression may contain the standard operators '+', '-', '*', '/' or '^' (exponentiation). It also may contain the standard aggregate functions COUNT, SUM, AVG, VAR, or STD. SQL dialects make available other aggregate functions and different collections of scalar functions. *Homomorphic* cryptography was therefore proposed for the use of the above operators (directly) over encrypted data. A *fully* homomorphic scheme should allow for expressions with unlimited number of above operators over any encrypted numerical values. Current proposals turned however unsafe or impractical for databases, [14]. Computing with Gentry's scheme and its

© Springer International Publishing Switzerland 2015
Q. Chen et al. (Eds.): DEXA 2015, Part II, LNCS 9262, pp. 455–478, 2015.
DOI: 10.1007/978-3-319-22852-5_38

variations is yet billions of times slower than the same plaintext calculations [13]. An 8-bit multiplication with Gentry's scheme takes 15 min and a calculation that takes a second in plaintext would last three centuries, [13]. Another implementation, specifically for a cloud DB, claims 23 min per multiplication of two encrypted 16-bit integers [6]. Consequently, it took seven days to find a given row in 10-record DB. To make the homomorphic encryption more practical, the *somewhat* homomorphic scheme analyzed in [9] restricts the calculations to expressions with a single '*'. The authors claim execution times of 1 ms per '+', and 60 ms per '*'. Yet another direction are the *semi*-homomorphic a.k.a. *partially* homomorphic systems. These are notably faster, but each only deals with selected operators. Especially, the *additively* semi-homomorphic systems support basically only '+' and '-', but are orders of magnitude faster than other homomorphic systems. Among these, the Paillier cryptosystem seems the most popular [4, 10, 14, 15, 17]. 15 μs may suffice per addition or subtraction over ciphertexts, [13]. Thus, e.g., 1.5 s may suffice to sum up 100 K encrypted values, what seems acceptable for the database world.

Database applications nevertheless need every available SQL operator and function, with possibly several invocations of each in an expression. One approach is to decompose the query so that the client post-processes the decrypted (plaintext) data for the operations impossible over the encrypted ones on the cloud, under the partially or somewhat homomorphic scheme used. Another approach uses per-row pre-computation of a plaintext expression over attributes of a table, e.g., a*b, for encryption and later upload of the computed (dynamic, virtual...) attribute [17]. Yet another approach is to encrypt only the data which are not destined for value expressions [3]. The future of all these proposals remains to be seen.

Efficiency with respect to value expressions must further preserve usual requirements on any relational DB. The prime one are Select-Project-Join (SPJ) queries. Paillier cryptosystem is a *probabilistic* one that is the same plaintext is about never encrypted to the same ciphertext. This precludes the equality comparison blocking the frequency analysis what is good for the security. Without that capability, SPJ-queries are impossible over outsourced encrypted columns. These queries are nevertheless the must for a relational DB as widely known. A naïve approach could be to split the query so that the cloud DBMS leaves all the SPJ operations to the client, for the post-processing after the decryption of the adequately retrieved encrypted ones. Even at the glance, one sees that such an attempt would made the outsourcing usually a non-sense. SPJ-queries require at present that the same plaintext maps always to the same ciphertext, i.e., they need a *deterministic* encryption. This one may be vulnerable to the frequency analysis, but can also resist to, [1]. The client may help, e.g., through decoy values, obfuscating power laws possibly serving disclosures, [2]. To get best of both worlds an approach perhaps practical is to use Paillier's scheme and a deterministic encryption like AES on every column subject to both: value expressions and SPJ-queries, [17]. The cloud supports then the SPJ-queries directly. In turn, the resistance to frequency analysis becomes that of the client data under the deterministic scheme used. There is also an evident storage and processing overhead.

Next in-line practical requirement for many plaintext DBs is the exploration of the total order. Related queries involve Θ-joins or GROUP BY, or TOP K etc. To perform these operations on ciphertexts, one approach is an order-preserving deterministic encryption. Any such encryption by definition discloses the order. This somewhat harms the security by itself. A safer alternative solution is again to split a query so that these operations are left for the post-processing on the client after the decryption, [17]. Order-dependent operations are less frequent than SPJ-queries. The strategy may perhaps thus be useful. In turn, it is obviously inefficient with respect to the plaintext DB, whenever the order-dependent operation reveals highly selective over a large encrypted dataset. We are not aware of any other proposal.

The final practical requirement is that of selected scalar or even aggregate functions beyond the SQL standard ones. Each SQL dialect offers some such functions out-of-the-box. The major DBMSs support also user-defined (UDF) functions. The client of an outsourced encrypted DB has to be entitled to similar capabilities.

Below, we propose a novel scheme for an encrypted cloud DB. It enables the homomorphic encryption sufficient for the above outlined standard numerical SQL expressions over a domain of positive values. While this is our primary goal, the scheme also conforms to some extent the other discussed constraints. SPJ-queries with the value expressions evaluate entirely at the cloud. The order-depend operation may occur there, mainly at additional storage cost. Perhaps surprisingly this capability does not disclose the order. Next, the client can add any scalar function, also at some storage cost. Finally, the cloud DBMS may entirely evaluate any additional aggregate functions defined by the value expressions combining the standard aggregate ones and the available scalar functions. For instance, the covariance aggregation defines that way.

Our encryption is a novel deterministic version of Paillier's scheme. The cloud DB stores the encrypted data in the tables defined by the conceptual schema. The DB contains further two auxiliary tables called each a *Scalar Function Table* (SFT). An SFT tabulates encrypted values of selected scalar functions. The client creates each SFT and uploads it to the cloud, before the DB becomes operational. Figure 1 shows the concept with related components of our scheme.

Conceptually, an SFT is an old-fashioned mathematical table, e.g., a printed table of logarithms. Actually however, an SFT usually should have many times more rows, e.g., hundreds of millions. Our scheme uses SFTs primarily to evaluate SQL expressions with operators other than '+' and '-' over encrypted data or with the aggregate functions, as people used the classical tables. Namely, the client (site) rewrites the query before sending it out, so that only the above operators and values retrieved from SFTs are used at the cloud. For example, to compute $x*y$ over encrypted x and y, the rewriting applies the high-school identity (rule) $x*y$ = antilog (log x + log y). The rewritten query looks up the tabulated encrypted logarithms that is the log function, then calculates the '+' over these values, finally, looks up the tabulated encrypted results of exp function (that is the antilog). The '+' evaluation applies Paillier's formulae we reuse within our kernel additively homomorphic scheme, Fig. 1. The kernel scheme uses also our variant of Paillier's encryption and the original decryption that remains applicable to. The whole construction provides the application with a homomorphic encryption providing *in fine* more operators than effectively used. This is the Gentry's "blueprint" of constructing a fuller homomorphism through the internal use of a less powerful kernel.

Our core scheme has as the domain of input and output values of the value expressions the monetary type values 0.01, 0.02...1,000,000.00. Below, we often note

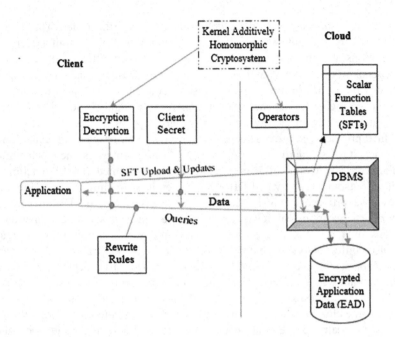

Fig. 1. High-level architecture of THE scheme

the upper bound as 1M. This domain should suffice for many, perhaps most, of practical DBs. It becomes a subset of the key column of an SFT, termed SFT1, whose rows tabulate, in a column, the encrypted logarithm for each domain value. For reasons that will appear, we tabulate the natural logarithm. SFT1 contains also other rows, useful for the aggregate functions, and a column for encrypted '∧' calculus. The core scheme uses further a second SFT, termed SFT2. This one tabulates antilogarithms (exponentials). It acts conceptually as a printed table, but with a practical difference compensating for the impossibility of interpolation over encrypted values.

The core SFTs suffice for standard SQL value expressions over encrypted data in the domain, including the '∧' and the standard aggregate functions. Notice that we could not find how the above discussed fully or somewhat homomorphic schemes calculate the exponentiation. We could not determine their efficiency for the aggregate functions neither. As we mentioned, our scheme also always provides for SPJ-queries over the encrypted data at the cloud, unlike Paillier's scheme. We show further that SFTs can get additional columns, providing for more scalar and aggregate functions and the Θ – joins. It should be also possible, although remains future work, to have the negative values in a domain.

The storage for SFTs depends in general on the range and precision of the encrypted data. It also depends on the number of supported functions. For the core domain, 50 GB should suffice. SFTs enters then easily the RAM of a modern cloud server. An operator over ciphertexts executes then in 15 μs or less. In other words, 1.5 s suffice for 100 K operations. This speed should satisfy many, if not most current applications. Experimental performance evaluation remains nevertheless the further work. At present, apparently, our scheme is the only to provide the described capabilities.

Our scheme is secure for a cloud DB under the popular honest-but-curious model. Its security is due in particular to a new cloud security paradigm. The traditional one is that all the cloud node is potentially insecure, i.e., the intruder can eventually disclose any data there. Any sensitive client (meta)data stays at the client. Our paradigm, restricts insecurity to the stored (permanent, long-term…) cloud data. The transient (volatile, short-lived…) run-time internal cloud DBMS variables are secure. The client may safely send sensitive data. The paradigm proves attractive for our scheme. The data that the client sends are called below *client secret*.

Despite more than three decades of high-level research, the traditional paradigm did not lead to usable schemes. The new one proves a way out through our scheme at least. High-level rationale is that having safely some security related data at the cloud, eases the apparent hardness of getting efficiency under the former paradigm, precluding that possibility. In turn, an effort is necessary to effectively secure a cloud DBMS. Such hope was probably unreasonable in the past. The progress in secure software engineering, e.g., through the moving target defenses, [11], makes it henceforward rational.

The next section introduces our scheme. We call it *Table-based Homomorphic Encryption*, *THE* scheme (crypto system) in short. We first overview the high-level architecture of THE scheme. Next, we recall Paillier's cryptosystem and present our kernel crypto system. Then, we define the SFTs, the rewrite rules, the value representation and storage structures for SFTs. Afterwards, we discuss the secure storage of the ciphertexts in SFTs, as shares of the client defined secret-sharing. Next, we define THE security model that fits the popular honest-but-curious one. We show that cloud DBs under THE scheme are secure under our model. Section 2 determines the processing times and storage space for SFTs. Section 3 discusses variants of THE schema. We conclude in Sect. 4, where we also point towards further work.

2   THE Scheme

2.1   High-Level Architecture

Figure 1 depicts the high-level architecture of THE scheme. The client and the cloud sides communicate as usual through query and data flows. At the cloud, the secure cloud DBMS manages the (cloud) DB, consisting of potentially insecure SFTs and of Encrypted Application Data (EAD). EAD component stores all the data defined by the conceptual schema. These would be in plaintext in the client DB relations if the client hasn't outsourced them.

The additively homomorphic Kernel Cryptosystem serves both the client and the cloud side. At the client, the component handles the encryption of all the outsourced plaintexts, i.e., in EAD. It serves similarly the upload of ciphertexts for SFTs and updates to. It eventually also encrypts query constants. Finally, it decrypts the retrieved data. At the cloud, the kernel scheme component provides the operators over ciphertexts in EAD and SFTs. These serve the addition and subtraction of plaintexts encrypted by the ciphertexts at the cloud, as well as the multiplication of a ciphertext by a plaintext. The kernel operators use a version of Paillier's formulae.

The Client Secret component serves the upload of SFTs and secure their use at the cloud. Queries with value expressions also use it. The secret remains in transient cloud DBMS internal variables, as already mentioned. The Rewrite Rules, finally, serve the numerical expressions within an application query, using operators other than these provided by the kernel or aggregate functions. These are rewritten into equivalent expressions using, over ciphertexts in EAD, only the kernel operators and scalar functions tabulated in SFTs. The equivalent expressions compute entirely at the cloud.

2.2 Paillier Cryptosystem

The Paillier Cryptosystem defines an additively semi-homomorphic encryption [10]. We recall only the properties most relevant to our work. Formally, plaintexts and ciphertexts are integers in Z_n for a specific n that is the product of two large primes. We use italics to denote a value x in plaintext and, sometimes, a bold font x to represent its encryption (ciphertext). Alternatively, we may denote the encrypted x as $E(x)$. We denote $D(x)$ the decryption of x.

Paillier cryptography uses (g,n) as the public key with g being a random number. It encrypts a plaintext x mod n after choosing some random number r in Z_n^* as:

$$x = g^x * r^n \bmod n^2. \tag{1}$$

The scheme is additively homomorphic since:

$$x + y \bmod n = D(x * y \bmod n^2). \tag{2}$$

We can also add a ciphertext x and a plaintext k:

$$x + k \bmod n = D(x * g^k \bmod n^2). \tag{3}$$

Finally, we can multiply a ciphertext x with a plaintext k via

$$D(x^k \bmod n^2) = D(k^x \bmod n^2) = kx \bmod n. \tag{4}$$

The decryption of the ciphertext uses the formulae presented in [10]. The random r, makes Paillier scheme probabilistic, making occurrences of the same plaintext rarely encoded into the same ciphertext. Notice that (2) means that if we encrypt x with r_1 and y with r_2 then $x + y$ is encrypted with r_1*r_2. We can also subtract ciphertexts since:

$$D(g^x * r_1^n / g^y * r_2^n \bmod n^2) = D(g^{x-y} * (r_1/r_2)^n \bmod n^2). \tag{5}$$

The right side decrypts $x - y$ mod n encrypted with r_1/r_2.

Paillier ciphertext at least doubles the storage necessary for the plaintext. The ciphertext of a typical 64 bit numerical value requires at least 128 b, i.e., 16 B per encrypted value. It believed however necessary that for adequate level of security the plaintext is 1024b at least. Smaller actual plaintexts may then be concatenated into a sufficiently large single plaintext vector, e.g., [17]. As we will show, such grouping is not necessary for our scheme.

2.3   THE Kernel Cryptosystem

By default we choose $r = 1$ for all clients. Every client gets (g, n) from DBA as *client* (encryption) *key*. To encrypt a plaintext x, the clients apply formula (1) for $r = 1$, i.e., $x = g^x \bmod n^2$. For decryption, the client reuses the Paillier formulae in [10]. The cloud DB server(s) get from DBA the *cloud key* that is n^2. This enables the cloud to manipulate the ciphertext. The cloud reuses the Paillier's '+' calculation through formula (2). Formula (3) is of no use since cloud key does not include g. A query constant involved in '+' operation sent to the cloud must be consequently encrypted. THE scheme reuses also Formula (4). A query constant involved in '*' operation may thus remain in plaintext when processed at the cloud.

In what follows, we will call *encrypted* addition (or '+') of x and y the operation over the ciphertext x and y at the right side of formula (2). Likewise we talk about encrypted subtraction, multiplication etc., referring implicitly to the manipulations over the ciphertexts.

Paillier's formulae apply to integers. Database values are usually reals. THE kernel scheme maps these reals into integers at the client, using column scale factors. Details are in Sect. 2.5. If real x is scaled to integer m, we encrypt x as $g^m \bmod n^2$. The encrypted operations concern then only ciphertexts of reals with the same scale factor. The scheme decrypts a ciphertext at the client accordingly to the representation. First thus to m, through Paillier's formulae. Then to x, through the scale factor. As shortcut, below we still denote $E(x)$ or simply x the ciphertext $g^m \bmod n^2$ encrypting x.

While random r makes Paillier scheme probabilistic, fixed r makes ours deterministic. As already mentioned, the equality of plaintexts holds then over the ciphertexts, what equality comparison is required for the select-project-join (SPJ) queries at the cloud. We come back to the issue specifically for THE scheme in Sect. 2.9. General comparison of vices and virtues of both classes of schemes is beyond our work as well-known, [1, 7]. All things considered, a deterministic scheme appears at present the only general choice for a homomorphic encryption for relational DB.

2.4   Scalar Function Tables

We aim on the homomorphic encryption supporting standard numerical SQL (value) expressions. These may contain '+', '-', '*', '/' or '^' operators. They also may contain the standard aggregate functions that are COUNT, SUM, AVG (the average), VAR (the variance), or STD (the standard deviation). As usual for a major DBMS, we further aim on the extensibility of the core pack with optional dialect-dependent functions, e.g., the popular INT scalar function. We wish all these capabilities available for possibly vast majority, of 'practical' DBs. The basic need seems the money manipulation. The values are then mainly all positive integers and reals with the 2-digit typical monetary precision, all within some 'practical' range. THE scheme is conceptually independent on this range. We need however some to evaluate the storage needs. Our 'practical' range, apparently sufficient for a vast majority of DBs, is from 0.01 to 1 million (1 M). Thus amounts of money, e.g., a price, should be 0.01, 0.02…1.00 M\$ (or €…).

We call V' the set of values in the range. V' is our *domain*. The domain values are the only allowed as input for a value expression using an operator other than '+' or computing a standard SQL aggregate function at the cloud DB over EAD, Fig. 1. For the (plaintext) value expression output values they should be in V' or in set V'' containing every value x^2 where $x \in V'$, while $x^2 \notin V'$. We may use V'' for intermediate values during VAR and STD computation at the cloud, Sect. 2.8. We basically constraint the result c of any operator or aggregate function to $c \in V = V' \cup V''$. A value returned by a query should furthermore fit V'. A ciphertext *overflowing* V' otherwise, may nevertheless get stored in EAD. It is not expected however to become input value again. Next, for our domain, any results returned by an expression have 2-digit precision, e.g., $10/3 = 3.33$. Finally, any intermediate result of a computation at the cloud overflowing V constitutes an *exception*. By name these are not supposed to happen, but if ever, they are the only values triggering the post-processing of an expression at the client.

As already mentioned, THE (core) scheme has two SFTs, called SFT1, Fig. 2 and SFT2 at Fig. 3. These tables are conceptually the old-fashioned mathematical tables. The difference is that every 'classical' plaintext x becomes the ciphertext $E(x)$ in SFTs and that there can be many times more values in these tables than classically. Actually, the ciphertexts, shown at the figures are encrypted further (obfuscated) for storage at the cloud, Sect. 2.6. We ignore the obfuscation till then.

VAL	LOG	LG2
$g\chi((0.01^\wedge 2)E5)$		
$g\chi((0.02^\wedge 2)E5)$		
…		
$g\chi((0.09^\wedge 2)E5)$		
$g\chi(0.01E5)$	$g\chi((\ln 0.01)E8)$	
…	…	
$g\chi(1.00E5)$	$g\chi((\ln 1.00)E8)$	
$g\chi(1.01E5)$	$g\chi((\ln 1.01)E8)$	$g\chi((\ln(\ln 1.01))E8)$
…	…	….
$g\chi(1,000,000.00E5)$	$g\chi(\ln(1,000,000.00)E5)$	$g\chi(\ln(\ln(1,000,000.00))E8)$
$g\chi((1000.01^\wedge 2)E5)$		
…		
$g\chi((1,000,000.00^\wedge 2)E5)$		

Fig. 2. Table SFT1. Ciphertexts are integer powers of g. Plaintexts are reals, scaled up to become integers (Sect. 2.5). Scaling factors 5 and 8 are in E notation. Ciphertexts are stored obfuscated through the client secret s.

SFT1 scheme is SFT1 (VAL, LOG, LG2). Informally, VAL is the key column with the ciphertext $E(x)$ for every x in V. LOG column tabulates for every such $E(x)$, the ciphertext $E(Log(x))$, where $Log(x)$ is the result of the usual SQL scalar Log function returning the natural algorithm, i.e., ln (x). Likewise, LG2 tabulates the ciphertexts E (Log (Log (x))). We use these for '^' operator. More formally, the plaintexts corresponding to both columns have some p-digit precision after the digital dot. Actually, we have $p = 8$, Sect. 2.5. Since Log function can return a value with more than eight digits, the functions tabulated in SFT1 are thus formally in SQL vocabulary: E (Round (Log (x), p)) and E (Round (Log (Log (x)), p)).

More in detail, SFT1 columns are as follows.

VAL. This is the key column. For our domain, Fig. 2, it contains $\{E\ (0.01^2)$, $E\ (0.02^2)...E\ (0.09^2)$, $E\ (0.01)$, $E\ (0.02)...E\ (10^6)$, $E\ (1000.01^2)...\ E\ (10^{12})\}$. The figure reminds that for every $x \in V$, $E\ (x) = g^{m\ (x)}$, where m is the integer representing x using the scale factor of the column.

LOG. It tabulates with 8-digit precision E (ln (x)), i.e. $g^{m(\ln\ (x))}$ at the figure, for every $x \in V$. Stated differently, we have LOG $(x) = E$ (log $_B$ (x)), where the logarithm base B is $B = e$ and $p = 8$. The decrypted (plaintext) LOG values range from $-4.60517019 = \ln\ (0.01)$ to $13.81551056 = \ln\ (1,000,000)$. The rationale for our choice of B and p are experiments with various bases and precisions. They have shown that, for the bijection between VAL and LOG columns, obviously necessary, the choice of ln provides for the smallest precision that is $p = 8$. These settings minimize accordingly the size of SFT2 table, Sect. 3.2.

LG2. The column tabulates, also with 8-digit precision, E (ln (ln (x))), for every $x \in V'$ such that ln $(x) > 0$. In other words and as again at the figure, for every x, LG2 contains $g^{m(\ln(\ln\ (x)))}$. We use LG2 for x^y operations.

Our 2nd core table is SFT2 (DLG, EXP, EX2), Fig. 3. This table serves as the old-fashioned antilog table. With again the difference of many more values and of their encryption. That one makes impossible the traditional interpolation of log values that may result the additions/subtractions of plaintexts D (LOG). Such values may fall into the "gaps" between two successive D (LOG) values, as well-known. The bounds of the gaps are unknown to the cloud since encrypted. Unlike for the printed tables, SFT2 has therefore conceptually a row with the antilog not only for every value in LOG, but also for every values within each gap.

DLG has in this way by far more rows than LOG. The number of rows clearly increases with p chosen for LOG column. As already hinted, we come back to this issue in Sect. 3.2. Some SFT2 rows tabulate also the antilog of antilog, for '^' calculation we explain soon. All antilog values are interpolated so to fit V.

The SFT2 columns are in detail as follow:

DLG. The name stands for *dense* logarithm. As hinted, it means that DLG tabulates E (x) for every p-digit precise value x such that (i) $x \geq$ LOG (min (V')) and $x \leq$ LOG (max (V')) or (ii) $x =$ LOG y with $y \in V''$. For our 'practical' V, DLG contains thus, in ascending order of plaintexts, round-up to $p = 8$ digit precision, (a) E (ln (0.01^2)), E (ln (0.02^2))... E (ln (0.99^2)). Next, DLG contains E (ln $(0.01)) = E\ (-4.60517019)$. We recall for sake of example that this stands for the value $g^{-4.60517019}$. This is followed by $E\ (-4.60517020)$, then by $E\ (-4.60517021)...E\ (0 = \ln\ 1)$, $E\ (0.000,000,01)$, until $E\ (13.81551056 = \ln\ (1,000,000))$. At the end, we have, (c), E (ln 1000.01^2), E (ln 1000.02^2)...$E\ (10^{12})$.

EXP. This column tabulates the encrypted natural antilogarithms and linearly interpolated natural antilogarithms of the plaintexts of DLG values. The plaintexts of

DLG	EXP	EX2
g˘(-921034037)	g˘((0.01^2)E5)	
g˘(-782404601)	g˘((0.02^2)E5)	
...	...	
g˘(-481589121)	g˘((0.09^2)E5)	
g˘(-460517019)	g˘(0.01E5)	g˘(1.01E5)
g˘(-460517018)	g˘(0.01E5)	g˘(1.01E5)
...	...	...
g˘(0)	g˘(1.00E5)	g˘(2.72E5)
g˘(1)	g˘(1.00E5)	g˘(2.72E5)
...	...	...
g˘(262611682)	g˘(13.82E5)	g˘(1,000,000.00E5)
...	...	
g˘(1381551056)	g˘(1,000,000.00E5)	
g˘(1381553056)	g˘((1000.01^2)E5)	
...	...	
g˘(2763102111)	g˘((1,000,000.00^2)E5)	

Fig. 3. Table SFT2 with its ciphertexts. These are actually stored as secret shares in DLG and as pointers to their VAL values displayed here otherwise. Plaintexts are again reals, represented as integers. Scale factor is implicitly 8 in DLG column and explicitly 5 in EXP and EX2.

EXP values are all in V. Thus their encryptions are in VAL. More precisely, given a row (x, y, z) in SFT2, there are two possibilities. (1) x is in LOG column. Then, $y = \text{EXP} (x)$, where EXP means here the SQL function rounded up to precision p of V. (2) x is not in LOG. Then y is interpolated, as it would be for printed log/antilog tables. Namely, let d_1 be the maximal plaintext D (LOG) such that $d_1 < x$ and let d_2 be the minimal plaintext D (LOG) such that $d_2 > x$. These are the bounds on the gap where x is. Next, let it be $d_3 = (d_1 + d_2)/2$, round up to 8-digit precision. Then, $x \le d_3$ implies $y = \text{VAL} (v)$ with v in row $(v, d_1...)$ of SFT1. Otherwise $y = \text{VAL} (v)$ with v in $(v, d_2...)$. Since DLG has many times more values than VAL, as we said, most EXP column values are duplicates.

EX2. The column tabulates the function $f = \text{EXP} (\text{EXP} (x))$, rounded up to the precision p of V, for every x in DLG and such that $f \in V'$. In our case, we recall, we have thus $f = 0.01...1$ M. Like for LG2, we use EX2 for $x^\wedge y$ clauses with both values encrypted.

Example. Consider two plaintexts 23.43 and 23.44 successive in V. In SFT1, they give rise to rows (**23.43, 3.15401725, 1.14867696**) and (**23.44, 3.15444397, 1.14881224**). These rows are also successive in SFT1 with respect to plaintexts of their VAL values. The plaintexts of their LOG values are successive in this sense as well. There is thus the gap between these, constituted by the successive values 3.15401726, 3.15401727...3.15444396. We have $d_1 = 3.15401725$, $d_2 = 3.15444397$ and $d_3 = 3,15423061$ for this gap. SFT2 has then for our plaintexts 23.43 and 23.44 successive rows that are (**3.15401725, 23.43**), (**3.15401726, 23.43**)... (**3.15423060, 23.43**), (**3.15423061, 23.44**)... (**3.15444397, 23.44**). The antilog of any encrypted logarithm value falling within the gap will be interpolated accordingly to **23.43** or **23.44**.

Consider now the query Select... Price*VAT... finding the encrypted plaintexts Price = 19.53\$ and VAT = 1.2, reflecting the 20 % tax on every price. As the calculus of $x*y$ over x and y in previous sections mentioned, the cloud applies the high-school rewrite rule, formalized in Sect. 2.F, that yields EXP (LOG (**19.53**) + LOG (**1.20**)). Through lookups into SFT1 the cloud finds two rows t and t' with respectively VAL (t) = **19.53** and VAL (t') = **1.20.** It finds further LOG (t) = **2.97195175**, equal thus to E (LOG (19.53)). Likewise, the cloud finds E (LOG (1.20)) that is E (LOG (t')) = 0.18232156. It then performs the addition of the encrypted values using the above discussed THE kernel formula for it. The encrypted sum is **3,15427331**. The cloud looks up then SFT2 for this value in DLG column, to ultimately find the encrypted antilog-arithm in EXP column. The sum is within the above gap and is above d_3. The cloud gets finally (**3,15427331**) = **23.44.** This is the correct result, as one may easily verify.

2.5 Real Number Representation

As said in Sect. 2.3, Paillier encryption requires integers, while DB plaintexts are basically reals. THE scheme represents therefore every real x to be encrypted in SFTs and EAD as the couple $x = (m,i)$, where m is an integer and i is a scale factor. Scale factors are known at the client only. We use m as power of g to produce the ciphertext of x that we note x by convenience as already discussed. Figures 2 and 3 show the ciphertexts in SFTs with E notation for scale factors. The actual plaintext is $m * 10^{-i}$, i.e., mE-i. For instance, 0.01 may be represented as (1000, 5), being then encrypted as g^{1000} mod n^2. This ciphertext is denoted, we recall, as E (0.01) or **0.01**. Every encrypted operation involving **0.01** will use g^{1000} mod n^2 as the operand. The decryption of the ciphertext first use the Paillier' formulae to produce 1000, finally made 0,01 again as $1000*10^{-5}$.

For SFTs, as well as for each table in EAD of course, there is single i per column. With our core V, the client should choose for VAL column at least $i = 5$. One reason is that the smallest plaintext to be encrypted in VAL is 0.0001. Using $i = 4$ could suffice, but higher i benefits the security, Sect. 2.8. For LOG, LG2 and DLG columns we set $i = 8$. Thus, e.g., for the above mentioned VAL (t) = **19.53**, i.e., with the ciphertext $g^{1953000}$ mod n^2, since $i = 5$, we have LOG (t) = $g^{2.97195175}$. A negative number $-m$ for these columns is, as usual, represented as the (positive) complement $n - m$ in Z_n. Thus, e.g., we represent the plaintext ln (0.01) = -4.60517019 as (positive) complement

$c = n$ -460517019 mod n and encrypt it as g^c mod n^2.

2.6 Storage of SFTs and of EAD

We store every ciphertext in SFTs as one of the shares of the basic two-share secret sharing, for some secret. The obfuscation of the ciphertexts by secret-sharing serves the security of THE scheme as we show in Sect. 2.9. The secret is client-defined as we spoke about and illustrated at Fig. 1. It is a pseudorandom number $s \in Z_n'$ where Z_n is also the space of our plaintexts and ciphertexts, we recall. The ciphertext c serves as the 1st share. Accordingly to the secret-sharing principles, we then form the 2nd share as

$c' = c$ XOR s and use c' as storage representation of c. In practice for our large n, we always have $c' \neq c$. The cloud DBMS extracts c for calculations from its stored representation as $c = c'$ XOR s. The s value comes to the cloud with every query using SFTs and only in such cases. The SFT lookups serving the query use then s and the latter formula. The DBMS discards s, at the end of the query at latest.

For both SFTs, key values, i.e., VAL for SFT1 and DLG for SFT2, are pseudo-random. Both tables are also basically static. We store therefore each SFT in a static hash file. Each row is represented by one record. The record (hash) key is the share c' of the table key ciphertext c, i.e., $c' = c$ XOR s with $c \in$ VAL for SFT1 file and $c' = c$ XOR s with $c \in$ DLG for SFT2 file. There are no duplicated record keys. Each record contains the key and the non-key values in the row, as in Figs. 2 and 3.

As usual each key c' hashes to some bucket, $h(c')$ where h is the hash function used. For instance, h is the popular hashing by division over N-bucket file, i.e., $h: c' -> c'$ mod N. The buckets have all the same capacity of a dozen of records. The records in a bucket share the hash result over different keys, we recall. The in-bucket search for the record with the given key is basically sequential. As well-known, fixed buckets with the above capacity provide then for fast overall search. Especially in RAM that should be the usual location for SFTs, Sect. 3.

We actually represent EXP and EX2 values as 4B pointers to their actual 16B long VAL values. The obvious rationale is smaller storage for SFT2. We revisit this issue in Sect. 3.2. We discuss the actual creation of SFT files in Sect. 3.3.

The file structures in EAD are client-defined. Records may contain rows or columns.

2.7   Rewrite Rules for Operators

A rewrite rule modifies the original value expression to an *equivalent* one, i.e., with valid SQL syntax and yielding the same result, but executable over the encrypted data. Incidentally, the rewriting includes the attribute and column names. These are also client-side encrypted for cloud security, Sect. 2.9. The equivalent query has every operator other than '+' or '_' and every SQL scalar or aggregate function invoked replaced by look-ups into SFTs and calculations using only encrypted '+' or '-'. As already mentioned, the client encrypts eventually also every constant subject to '+' operation.

The clients sends the equivalent query to the cloud. The result sent back consists normally of data to be decrypted by the client. Exceptionally, data that the cloud could not process come back as well. E.g., when an operation produces the data not in SFTs, i.e., with a plaintext above or below values in V. The client is in charge of post-processing then. In-depth analysis of the exceptions remains future work.

The query format somehow distinguishes numerical values representing plaintext from numerical values representing ciphertext. When the cloud gets the query, it produces an execution plan. The plan is a transaction formed by SQL queries searching SFTs and, perhaps, by a host language statements.

In what follows, to simplify notation, the operands and the signs of operations are overloaded depending on the context. We do not expressively distinguish between operations over ciphertexts and over plaintexts, unless necessary. We consider first only numerical expressions that do not involve aggregate functions. They consist of column values or of the results of operations only.

(a) Addition/Subtraction: The cloud performs any addition or subtraction in cipher-text, as discussed in Sects. 2.2–2.3. No rewrite of clause $x \pm y$ occurs, except, perhaps, for above mentioned plaintext constant encryption for x or y.

(b) Multiplication: if both operands are in ciphertext, then, as already mentioned, we use the well-known identity: $x*y =$ antilog $(\log(x) + \log(y))$. Using our tabulated scalar functions this translates to the expression

$$E\,(x * y) = \mathrm{EXP}\,(\mathrm{LOG}\,(x) + \mathrm{LOG}\,(y)).$$

The expression has valid SQL syntax. The cloud DBMS knows nevertheless that '+' is the encrypted addition. Next, it knows that the scalar functions are the tabulated ones. Also, since EXP has DLG as the domain, the sum must be within DLG as well, to have the result in V, as wished. An exception to post-process by the client occurs otherwise.

(c) Multiplication by a plaintext constant. The cloud uses rule (3). No rewrite.

(d) Exponentiation $x^\wedge y$, supposing, x encrypted and y either (i) plaintext constant or (ii) encrypted as well. The rewrite uses basically the high-school identity

$x^\wedge y = \exp\,(y * \log(x))$. For case (i), using SFTs and encrypted '*' defined through formula (4) the identity leads to the rule:

$$E\,(x^\wedge y) = \mathrm{EXP}\,(y * \mathrm{LOG}\,(x)).$$

For case (ii), first, observe that the basic identity expands to the following, also well-known identity $x^\wedge y = \exp\,(\exp\,(\log(x) + \log(\log(y))))$. This leads to the following rule, using once more an SQL expression referring to our tabulated functions:

$$E\,(x^\wedge y) = \mathrm{EX2}\,(\mathrm{LOG}\,(y) + \mathrm{LG2}\,(x)),$$

Again, the rule refers to the encrypted '+' and the sum must be in DLG. In addition, EX2 must evaluate to a value in VAL, hence in V.

The rewrite rules apply recursively. In particular, - when an expression x with the rewrite $R\,(x)$, gets equated in a query to a plaintext constant C in the clause $x = C$. The rewrite rule is then simply $R\,(x) = \mathbf{C}$ where $\mathbf{C} = E\,(C)$, we recall. Notice that this rule applies even if x is only an attribute name.

Example. (a) The popular expression K*(1 + R)^Y calculates the value of capital K placed for Y years at fixed annual rate R. To be calculated over the encrypted values, our rules yield the following equivalent one:

$$\mathrm{EXP}\,(\mathrm{LOG}\,(K) + \mathrm{LOG}\,(\mathrm{EX2}\,(\mathrm{LOG}\,(Y) + \mathrm{LG2}\,(1 + R)))).$$

The expression has valid SQL syntax. When the client sends it out, the column names are replaced with meaningless cloud identifiers, say 123 for C, etc. The cloud evaluates the scalar functions looking up the SFTs and calculating the encrypted '+'.

(b) Consider the clause qty * price = 123. Its rewrite is:

$$\text{EXP } (\text{LOG}\,(\text{qty}) + \text{LOG}\,(\text{price})) = \mathbf{123}$$

2.8   Rewrite Rules for Aggregate Functions

The result of COUNT function is a plaintext that obviously does not need arithmetic calculations. The functions SUM, AVG, VAR, and STD do. The calculation of SUM uses the additions as discussed. No need for a rewrite. For (encrypted) AVG calculation, the basic rewrite rule is to replace it with the SUM, assumed then in V', divided by COUNT. The final result is computed through Paillier formula (4). If, presumably rarely, SUM is not in V', while the overall AVG result is, the rewrite is more complex, Section IV.

For encrypted variance, i.e., VAR, the rewrite rule applies König-Huygens formula (or Steiner translation) $\text{VAR}(A) = \text{AVG}(A^2) - \text{AVG}(A)^2$. The full rewrite goes in fact recursively further, given the rewrite of AVG itself and of '^'. The 1st iteration is:

$$\text{VAR } (A) = \text{SUM } (\text{EXP } (2 * \text{LOG } (A)))/\text{COUNT } (A)$$
$$+ \text{EXP } (2 * \text{LOG } (\text{AVG } (A))).$$

We assume again the result of AVG (A) in V'. Then, E (AVG (A)) is in VAL column and 2 * LOG... in the second clause is in DLG. The cloud finds EXP value in this row. This value is necessarily in V' or in V''. The squaring was the rationale for V''', we recall. Similarly for each LOG (A) and EXP summed up in the first clause. The cloud can now terminate evaluating the expression using the kernel scheme operators only.

The rewrite of STD (standard deviation) starts with the well-known property STD (A) = VAR (A)(1/2). Hence, we have STD (A) = EXP (1/2 * LOG (VAR (A)). Applied at the cloud, the rule means first that E (VAR (A)) must be in SFT1. The (encrypted) result of LOG (tabulated) function is in SFT1 as well and in DLG. The result of EXP (tabulated) function is logically in SFT2, although actually in VAL column in SFT1. The encrypted multiplication by ½ uses Paillier's rule (4).

2.9   Security of THE Scheme

Security Model. We consider the encrypted data within a cloud DB using our scheme, i.e., stored in SFTs and EAD, Fig. 1. Our security model is the popular honest-but-curious one. The cloud intruder may succeed accessing any data constituting our cloud DB, i.e., data in SFTs and EAD. S/he may use, e.g., some storage dump, perhaps easily available to a cloud management insider. The intruder knows our method. S/he can do off-line any calculations over the ciphertexts. S/he is ultimately "curious" to disclose the application data, i.e., to decrypt ciphertexts in EAD. But, behaves "honestly", i.e. only reads EAD. To succeed, the intruder may also 'honestly' read SFTs.

EAD respects furthermore all the well-known conditions necessary for the security of any deterministically encrypted cloud DB, [1]. These are beyond our scheme, depending solely on the application. In practice, they mean first no prior knowledge of any client plaintexts. Next, they presume high min-entropy of every column of EAD, i.e., many equally likely values. All this prevents the frequency analysis. The client may help, as already mentioned.

Next, if the client calculates at the cloud an SQL scalar function beyond the core ones, e.g., the LOG10 mentioned in Sect. 4, then all the tabulating columns are within SFTs. Finally, the intruder cannot access the run-time (temporary) variables within the DBMS code, instantiated for a query processing and cleared afterwards. Perhaps, - just since they are short-lived or at locations known only to the DBMS. Or, since, wisely in these troubled times, the code, with its run-time variables, is protected, e.g. by a moving target defense like [11]. We recall that these recent techniques aim on secure cloud software through secret randomization of the code and variables, blocking injections and reverse engineering. Among the variables concerned in our specific case are these with the shared secrets s and s'. They are instantiated each time a query using SFTs brings the secrets in, being all discarded at most by the query end. Are similarly concerned the variables instantiated by values retrieved from EAD or SFT files. Same for the variables that are set during the query time for search in SFT or EAD files, e.g., are instantiated with a record key.

As said, our paradigm of the secure cloud DBMS internals is new. Up to now, the research on encrypted outsourced DBs considered simply that the client never releases its sensitive data, e.g., encryption keys. As already stressed as well, this somehow natural paradigm unfortunately did not produce any practical solutions for more than three decades. In ours, the cloud DBMS internals constitute a secure space for transient metadata. Just like a bank vault for transient money. The hope is are practical algorithms that could not exist otherwise. Somehow like vaults lead to operations unthinkable without.

In turn, we displace part of the difficulty of creating a practical cloud DB encryption scheme towards that of creating an infrastructure of a secure cloud DBMS. We believe that on-going generic cloud software security research, especially the one we mentioned, makes the goal already feasible.

Security Proof. The data in any cloud DB using THE scheme are secure (safe) under this model. In other words, whatever is ciphertext c within EAD, the intruder cannot disclose the plaintext D (c) in practical time. In particular every ciphertext c stored obfuscated in an SFT as some c', is secure in this sense as well. Otherwise, if the intruder could disclose a plaintext d for some E (d) there, then one could apply the kernel operators to calculate E (d + d + ...+d) or E (k*d), for various plaintext constants k. By matching the results with EAD, a disclosure of all outsourced plaintexts could follow.

As stressed already, we consider EAD immune against any attacks possibly disclosing values in a DB, independently of the deterministic encryption used. As we will show, EAD alone is then secure since the intruder cannot disclose g from any contents it may have. Reading SFTs is of no help to the intruder. The overall rationale the missing knowledge of the client secrets s and s'. The intruder can't get them under our model neither from stored nor from run-time values. *A fortiori* the intruder can't get knowledge of any couple (c, c') with c in EAD stored obfuscated as the share c' in VAL, calculated in Sect. 2.6. This blocks any use of a property of c in SFTs to disclose the plaintext. The security of EAD follows.

Now, in detail, first, to disclose g, the intruder could start with guessing likely scale factors. Then, could pick up a ciphertext v expected to be g and try plausible values of m. This until equality of v^m mod n^2 possibly occurs with some ciphertext in EAD. Trying out accordingly perhaps even every c in EAD. This process should be obviously already typically rather tedious. In addition, under our core scheme there are only values g^m mod n^2 with $m > 1$ in the cloud DB. This is due to the choice of the scale factors ($s \geq 5$ we recall). The appearance of g value for any visited v is extremely unlikely. The discussed attack is hopeless in practice.

Same conclusion applies to the guess of any two values v, V' as some g^m and g^{m-1}, (modulo n^2), yielding g value as $g = v/V'$. Likewise, the intruder cannot calculate g from some ciphertext g^m mod n^2 with $m > 1$, either in SFT or EAD. Here, m is the discrete logarithm of g private and large. While no algorithm calculating the discrete logarithm for large g in practical time is known, the 128b wide ciphertext is also believed not large enough for sufficient security. But, this belief concerns g used as a public key, hence known to the intruder. This is not our case, what basically multiplies the complexity of any algorithm proposed for the discrete logarithm computation by a factor up to 2^{64}. On the average the complexity increases 2^{63} times as g is randomly picked up in Z_n, we recall. The result appears high enough for many years of number crunching by any supercomputers at present.

Next, with respect to the secret sharing, observe first that the intruder cannot disclose the secret from SFTs alone. The reason is that whatever is ciphertext c present in SFTs, the intruder cannot determine c from c' obfuscating it. The ciphertexts are in all the columns of SFT1 and DLG column in SFT2, we recall. *A fortiori;* the intruder cannot determine D (c).

The two-share secret sharing is indeed known to be secure for its secret assumed pseudorandom, as in our case, provided the 1st share that is here c, is also pseudorandom enough to resist any frequency analysis. This is our case as well. First, our encryption formula g^m mod n^2 acts as the hashing by division. This one usually randomize the hashed values well. Next, in every column with the shares, every (hash) value is unique, keeping the distribution uniform. We have a single secret hidden into two shares as many times as there are obfuscated values c' in SFTs. Since each c is pseudo random, c' values cannot disclose the secret. The intruder cannot thus calculate any c from its c' by analyzing only the related columns, i.e., cannot disclose any c.

In contrast, the values in EXP and EX2 are not the shares. The columns have also duplicates. It is easy to see that for every $v_1, v_2 \in V$ with $v_1 > v_2$, more rows point to v_2 than to v_1. The intruder may accordingly order the shares in VAL. Moreover, the intruder may luckily guess the scale factors and logarithm base used. E.g., if the client simply applies the core scheme. A count of duplicates may then identify a plaintext. For instance, the count of 42680 identifies 23.43 from our example in Sect. 2.4.

However, for any record in SFT1 file, this knowledge still does not disclose any ciphertext c that c' obfuscates. In other words, for any above D (c), the value of c obfuscated in VAL column remains unknown if the secret is unknown. We consider the latter to be the case of the intruder from now on. We will show progressively that it is actually so. For instance therefore, c with D (c) = 23.43 above remains unknown to the intruder. The intruder cannot disclose therefore the plaintext D (c) for any c in EAD using SFTs alone. Natural way to progress is then to explore both SFTs and EAD. Basically, - to identify for some c' in VAL 'its' c in EAD. Guessing pairs (c', c) by brute force however, i.e., by trying for given c' every c there, as possibly sharing D (c), is clearly a nonsense. One must restrict number of the trials. Possibly towards a single pair.

One hypothetical way to follow then is the access path from some c in EAD to the (unique) bucket in SFT1 file containing a single record. The plaintext found for c' there would be D (c), hence the disclosure would succeed. Such bucket may exist with a reasonable probability in SFT1 file, as known for hash files. However, to access a bucket within SFT1 file one needs to hash c'. Whatever is c, the intruder cannot determine c', as discussed. To disclose D (c) is then impossible as well. Same holds for the other way around. That is, - the access to the bucket with unique c in EAD, starting from some c' in VAL. Clearly, without knowing s, no exploration of our physical data structures for SFTs and EAD may relate c' in SFTs to 'its' c in EAD and vice versa.

The same rationale blocks the only remaining way. That one is the sequential (brute force) search through EAD alone. Since g is unknown, the "Holy Grail" ciphertext should have the plaintext being a unique solution to some algebraic identity or equation. It thus should be the only equalizing two selected different expressions. For instance, 2 is the only to solve $x + x = x*x$. Likewise 3 uniquely solves the identity $x + x + x = x*x$ etc. If one may calculate an identity over EAD, e.g., may test E $(x + x) = ?$ E $(x*x)$, the disclosure obviously follows. The performance analysis in Sect. 3 shows that this could happen then in less than an hour for an EAD with even dozens of millions of values. The result would be similar for the search through VAL if the stored values there were not obfuscated. A rapid disclosure of entire EAD could follow the wise use of kernel operators above discussed.

However, the intruder cannot even calculate the '+' above, because of s' obfuscation of EAD ciphertexts. Besides, '*' unfortunately (for the intruder) requires in addition again SFTs. Not knowing the secrets, the intruder can't thus test the identity. Same applies to above mentioned additional SQL scalar functions, e.g., LOG10n that could serve '*' alternatively. Their tabulations are in SFTs hence are secret for the intruder as well. Identities using operators other than '*', e.g., $x + x = x^\wedge x$ solved by $x = 2$, obviously can't help neither. Algebraic identities always use at least one non-kernel operator. This way of proceeding thus still cannot bring the disclosure under our model.

Finally, the intruder could consider getting knowledge of "useful" values while they are processed for a query. As already mentioned, they are then and only then, in some run-time variables within the DBMS. The ciphertext c could come from EAD. Or could be brought by the value expression, e.g., in the clause $2^\wedge x$, with $c = E$ (2). In both cases, some run-time variables are instantiated then also with the secrets s and s', c' and h (c'), to read the relevant bucket, e.g., with LG2 (c). Reading s or c and c' or c and h (c'), would again trigger a disclosure. However, as we discussed, the run-time variables are out of reach for the intruder in our model. This strategy does work neither thus. In particular, the intruder is definitively unable to disclose the secrets. As we only supposed till now, we recall, since proving this disclosure impossible from SFTs alone. Altogether, THE scheme is thus secure under our model.

3   Performance Analysis

3.1   Processing Time

Processing a rewritten query requires a few lookups (searches for records), XORing with the secrets s and s', and time for one or a few computations of the kernel operators. The latter times clearly dominate. The overall time to process an operator other than the kernel one, should be thus only a few times longer than for a single encrypted addition. The result is practical as long as that time is. Authors of [14] report the actual speed of about 12 μs per Paillier's addition. This is the slowdown of eighty times with respect to the plaintext addition. Nevertheless, this is still, e.g., only 1.2 s for 100 K additions. Processing speed of that order remains practical for a DB. The same number of plaintext additions needs 14 ms, [14]. In comparison, the Gentry scheme would need over four years. Likewise, the somewhat homomorphic scheme we cited, would still require 100 s. This would clearly be usually impractical, being also about eighty times more than with our scheme.

The multiplication operation with our scheme requires the time of an addition and of four lookups: two for LOG values, one for DLG and for EXP pointer in the same record, and one follow up into VAL. As it appears below, SFTs may and therefore preferably should, reside in RAM of a modern (database) server. The lookup time through a bucket of a hash RAM file should be up to 1 μs. Time to get ciphertexts from their storage representations is negligible since we use XOR. The result is 16 μs per multiplication, i.e., 1.6 s for our 100 K multiplications. This should be obviously usually practical. The result holds obviously for the division. We recall that the somewhat homomorphic scheme that appears the fastest supporting both '+' and '*' known till now, requires 60 ms per '*'. In other words, it appears 4000 slower, notwithstanding its inherent limitation on '*'.

If SFTs better do not reside in RAM of the cloud server, the flash storage is the next candidate on the memory hierarchy. The lookup time is essentially the flash read time, i.e., about 1 ms. The lookups then dominate the time for the multiplication and brings the response time for a single '*' to 3 ms, i.e., to five minutes per 100 K. The latter timing might be already long for some users. The location of SFTs in flash storage instead of RAM should be therefore possibly avoided. Nevertheless, even so, the processing is still twenty times faster than for the somewhat homomorphic scheme. Finally, with their 5–10 ms per access on the average, the next in the memory hierarchy magnetic disks, are clearly too slow to appear as practical for SFTs.

The evaluation of '^' shows that for RAM resident SFTs, 100 K such operations should take also less than 2 s. We skip fastidious details and evaluation for the flash. The calculus of an aggregate function depends mainly on the number of selected values. In practice the latter should be usually under a few thousands. The result should be under a second for SFTs in RAM. Again we skip the details.

3.2   Storage for SFTs

We consider our core domain V. We evaluate the resulting storage first for SFT1 values, then for SFT2 values. Next, we determine the total storage for the hash files with these values. Then, we discuss formulae for the storage, applicable to variants of core choices. We determine finally the expected storage amount per core domain value as the thumb rule for comparing variants.

As already said, Paillier's ciphertext requires at least 128b, i.e., 16B. For SFT1, the key column VAL has $|V| = 200{,}000{,}000$ values. These values require thus $16 * |V| / 1024^3 \cong 3$ GB of storage. The non-key column LOG has only $|V'| = 100{,}000{,}000$ values. Those need 1.5 GB. LG2 column has a value $E (\ln (\ln (x)))$ only for $x \in V'$ and only if $\ln (x) > 0$. The difference to $|\text{LOG}|$ is negligible. LG2 values require thus practically 1.5 GB as well. SFT1 values need therefore 6 GB total.

For SFT2, DLG column has, first, the ciphertexts for the logarithms of the values forming our domain V'. These are the 100,000,100 ciphertexts in LOG. DLG has also all the values within each gap among logarithm values encrypted in LOG. The natural logarithms range for V', we recall, from 4.60517018 to 13.81551055, with the bounds provided with our precision $p = 8$. Altogether, for our V' and p, we need thus in DLG 1,842,068,074 values. Notice that this is eighteen times more than in LOG.

DLG has furthermore the ciphertexts for V''' values. We recall that these values are necessary for the aggregate functions. There is one such value first for each $x \in V'$ such that $x > 1000$. We have then $x^2 > 1$ M and so $x^2 \in V'''$. We have 99,900,000 such values in V'''. We also have in V''', 9 values for $v < 0.1$, hence $v^2 < 0.01$, our lower bound on V'. The grand total is 1,941,968,083. This leads to 29 GB for DLG values.

Notice that for our V, the choice of the logarithm base $B \neq e$, would lead to precision $p > 8$. Hence, it would lead to even more values in DLG. Experiment, e.g., with the popular bases $B = 2$ or $B = 10$.

EXP and EX2 columns contain only the already mentioned 4B pointers. Such pointers suffice for up to $2^{32} - 1$ values, hence largely enough for our "practical" V. The storage for EXP values is thus 7.3 GB. EX2 column only has the values that fit V'. The maximal possible DLG value is $\ln (\ln (10^6) = 2.62579191$. We have therefore $460{,}517{,}019 + 100 + 262{,}579{,}192 = 723{,}096{,}311$ values in EX2. They need 2.7 GB. Altogether, SFT1 and SFT2 values require thus $6 + 29 + 7.3 + 2.7 = 45$ GB of storage.

As we said our storage structure for an SFT column is a hash file. The load factor can then exceed 90 % with almost no access performance deterioration by collisions. 50 GB total should suffice thus for our core SFTs.

Mass-produced workstations and cloud servers, offer now routinely 64 GB and easily up to 512 GB of RAM. As already mentioned, our SFTs may therefore reside in RAM storage and thus typically should. Alternatively, there are cheap flash cards and SSDs for those, as mentioned already as well.

More generally, the storage amount for SFT1, say S_1, calculated as above, is linear with the number of values in VAL. Thus let it be $x = |\text{SFT1}|$. Next, let w be the ciphertext width, $w = 16$ B in our case. We have at least $S_1 = O (3*w*x)$. Each optional column we talk about in Sect. 4 costs for its values $w*x$ bytes. This, except for the plaintext although also secret RANK costing $8*x$ bytes for modern 64 bit arithmetic. For SFT2, the cost is $S_2 = O ((w + 4 + 4) * x * o (x, b))$. Here, for the first sum, we consider again 4B pointers for EXP and EX2, sufficing for up to $2^{32} - 1$ pointed values, we recall. For the last expression, o denotes the overhead of dense log, $o = |DLG|/x$. This one depends on x and B. For our V, $B = e$ appears best as we already pointed out, with still almost 2000 % overhead. For other values of x another B can perhaps lower o. Finally, since we needed 50 GB for $|V| = 100$ M, the thumb rule for our (core) scheme is 0.5 KB per domain value.

3.3   SFT Upload and Update

To create the SFTs at the cloud, the client has to create SFTs ciphertexts elsewhere than at the cloud DB server(s) and upload them there. Suppose first that the client creates all the ciphertexts at its own site. We call this scheme *centralized*. To evaluate the time for SFTs creation at the cloud then, say T, let w is the total number of encrypted columns in an SFT, and x the average cardinal of a column. Then T is proportional to $w*x$. T includes the encryption time at the client, say $T_C = O (w*x*c)$, where c is here the time to encrypt a single value. It also includes the transfer time and the time to create the tables at the server. The client-side encryption and the other operations can be parallel. It suffices to send group encrypted values, while the encryption goes-on. T_C becomes then the dominant component of T.

Our basic hardware assumption for the centralized scheme is that the client site is a PC with usual high or low speed connection to the cloud. This was the basis for the experiments reported in [5, 12]. These papers are short and lack important details. Nevertheless, it appears that T_C for 100 K values is $3 \div 15$ min. Notice that the upper bound is ten-year old, while the lower is only two-year old. We have the total of about 2,241 M values to encrypt: the LG2 column of SFT1 and the DLG column of SFT2. The latter reuses the 100 M ciphertexts of LOG column. Even the best case, this leads to 47 days. This is hardly practical, although remains subject to caution, requiring experiments on current hardware. To this time adds up the creation of the pointers in EXP and EX2 columns of SFT2 at the client that must follow that of SFT1. That time is however relatively negligible. One may expect four hours, assuming reasonably at most 5μs per record creation, for our almost 2,700 M records to produce for EXP and EX2 columns.

While the centralized scheme is the classical configuration, a modern way towards a practical T_C is clearly the use by the client of some popular MapReduce tool distributing the calculations over a *trusted* cloud. That cloud is other than the one with the cloud DB, being chosen by the client as safe. For instance, if the DB is outsourced to Google cloud, Amazon EC2 could be the trusted one. One way is that the client creates then locally a file with all the plaintexts for VAL, say PVL, and, similarly a file PDL of plaintexts for DLG column. The tool dynamically partitions then each file over the trusted cloud during the Map phase. For each value of PVL and of PDL, each node calculates in parallel the rows of each SFT and forms the records. The Reduce phase sorts the records by bucket addresses in the cloud DB files. The results are uploaded to the cloud DB as inserts to the buckets, initially empty.

Provided the trusted nodes have the same throughput, the partitioning by Map/Reduce over N nodes should reduce T_C of the centralized scheme on the average

N times. The client requests thus enough nodes for an acceptable value. If this one is, e.g., on the average 1 h, 1100 nodes may do. The total time should be somehow longer. First, there is time spent to create PVL and PDL files, possibly in RAM. Next, there are transfers of GBytes of plaintexts towards the trusted nodes, during the Map phase. There are also transfers towards the cloud DB from the reducers. There should be usually high parallelism between theses transfers, as well as with the encryption calculations. The dominating transfer is the one of 45G of SFT values for our core domain from the reducers to the cloud DB. For the Internet fiber optics transfer speed of 200 + MBs, the overhead to T_C should be altogether a couple of minutes.

An update to SFTs should merely consist of appending optional scalar functions we describe below. Using the basic scheme, this operation has the time complexity of $O(|V'|)$ per function. This could still appear impractical for our core V', lasting mainly because of the encryption time perhaps 2 days. Again, one can make it almost N times faster through the N-node trusted cloud.

3.4   Variants

We now outline some variants tailoring the core scheme to specific need. First, the storage for SFTs can be reduced. Results in [16, 17] point to impressive almost 90 % savings in ciphertexts representation. One should study the application of these techniques to SFTs. On the simple side, consider, e.g., the secret decision to represent plaintext 2.34 as 1.22 + 1.12. We may represent **2.34** then as two pointers. Instead of 16 B we use 8 B only. The application of the strategy on massive scale to SFTs leads to almost 50 % saving. The obvious price to pay is longer processing time. For our **2.34** and a 16 B ciphertext, we about double the '+' and '*' times.

Next, the domain V' of our core scheme has $|V'| \cong 100$ M. The algorithmic is nevertheless independent on $|V'|$, provided V' contains positive numbers only. An obvious direction is thus to use a smaller or larger V', perhaps fitting better a DB. The storage space for SFTs is affected linearly, as we have shown. A logarithm base B other than our e one may then perhaps provide to a smaller DLG for a given domain. Providing therefore a smaller SFT2 storage than our base would do.

Yet another direction is more rewrite rules. For instance the expression $d = a * b/c$ evaluated as discussed till now may result in an overflow of V' after $a * b$ calculation, while d may still be in V'. An alternative rule could rewrite d as $(a/c) * b$ perhaps avoiding the overflow. Likewise, the straight calculus $a*b - a*c$ could fail because of $a*b$ overflow. But, perhaps the rewrite $a * (b - c)$ would do. The high-school books seem a pond for more rules.

The issue occurs also for the aggregate functions. For AVG, in particular, the following recursive formula may avoid an overflow:

$$AVG\,(A_1,\ A_2 \ldots A_{n+1}) = n/(n + 1) * AVG\,(A_1,\ A_2 \ldots A_n) + 1/(n + 1)\ A_{n+1}.$$

The formula extends to VAR computation.

An important direction is furthermore the expansion of SFT1 with *optional* scalar functions. These are functions other than the core ones we have discussed. Each optional function give rise to a dedicated column, with its value entering the core row and record. For instance, a useful add-up may be the column INT tabulating in each row $(v,...)$, the ciphertext INT (v) of the popular function. This could lead e.g., to two rows with (VAL...INT) being **(23.43...23)** and **(23.44...23)**. Likewise one may add the RANK column tabulating in plaintext the descending or ascending order of plaintexts of ciphertexts in VAL. We recall that RANK plaintexts would be nevertheless stored obfuscated as the client secret shares. Under our security model, they would not disclose the order, what the order preserving encryption schemes naturally do, [17]. The function allows to compute at the cloud the ORDER BY clause on value expressions, and thus TOP k, MIN and MAX aggregate functions. Likewise it allows for Θ-joins. The RAM sizes up to 512 GB discussed in Sect. 3, easily allows for many more optional functions.

Another goal is a domain with zero and negative numbers for operators other than '+' and '-'. One way towards it is, first, an additional column, say MVL, in SFT1, tabulating –**a** for each **a** in VAL, assuming the representation of the negative numbers as complements mod n in $0...n$. Next, the rewriting rules have to be revised and optional functions added, to deal with the sign of a value expression where an operand may be negative. E.g., the rule for a * b where perhaps $a,b < 0$, may become SIGN (EXP (LOG (ABS (a)) + LOG (ABS(b)))), where SIGN depends trivially on the operands. Some rules have to deal then with forbidden cases, e.g., LOG (0) or $a^{1/2}$ for any $a < 0$.

Finally, with respect to SFT1 and SFT2 upload, an application of a scheme in [8], instead of MapReduce, would first bring the advantage of the guaranteed time limit over T_c at each node, e.g., $T_C \leq 1$ h. Unlike for MapReduce tools, the limit would hold even if some trusted nodes had smaller throughput, even heavily, as often in practice. Next, the client would comparatively save the overhead time and storage related to the creation and transfer to the nodes of PVL and PDL files. These would exist only as virtual files, partitioned over the trusted cloud nodes. For our SFTs, 1100-node cloud could then again suffice for the previously discussed 1 h limit on T_C, if the trusted nodes have the same throughput. More than 1100 nodes would get involved dynamically for heterogeneous throughput. Known MapReduce tools lead then to processing times perhaps even much longer than the expected T_C/N, executing over statically allotted $N = 1100$. However, while there are numerous MapReduce tools, no application of schemes in [8] appears known as yet.

4   Conclusion

Despite thirty five years quest, a practical fully homomorphic encryption remains yet the Holy Grail. THE scheme offers the homomorphic capabilities appearing practical for many, perhaps most, DBs. It evaluates the numerical SQL expressions over encrypted data with any number of operators other than '+' and '-', and with standard SQL aggregate functions. It uses additively homomorphic Paillier-based kernel cryptosystem, auxiliary tables tabulating selected scalar functions and rewrite rules. It also uses a novel paradigm of the cloud DBMS secure for the transient client metadata. The research until now presumed to the contrary that the client never sends out such data.

Our paradigm appears practical, through the progress towards secure cloud software, using the moving target defenses especially.

THE scheme processes the expressions normally entirely at the cloud. It is secure under our security model against honest-but-curious intrusions. THE scheme is finally deterministic, supporting thus SPJ queries over ciphertexts. Efficiency of SPJ operations is not subject of our work. Processing such queries entirely at the cloud is nevertheless well-known to be the key to the success of relational cloud DBs. To our best knowledge, THE scheme is the first with this capability to the extent shown.

Further work should address the prototyping. As for other proposals, this is necessary to evaluate THE scheme in depth. We outlined several variants. Future efforts should aim at these as well.

Acknowledgements. Discussions with Ken Smith at the 2013 Cloud Security Meeting at CSIS (GMU), after the presentation [13] laid the basis of this work.

References

1. Boldyreva, A., Fehr, S., O'Neill, A.: On notions of security for deterministic encryption, and efficient constructions without random oracles. In: Wagner, D. (ed.) CRYPTO 2008. LNCS, vol. 5157, pp. 335–359. Springer, Heidelberg (2008)
2. Blank, A., Solomon, S.: Power laws in cities population, financial markets and internet sites (scaling in systems with a variable number of components). Physica A **287**(1), 279–288 (2000)
3. De Capitani, S., Foresti, S., Samarati, P.: Selective and fine-grained access to data in the cloud. In: Jajodia, S., et al. (eds.) Secure Cloud Computing. LNCS, vol. 5735, pp. 123–148. Springer, New York (2014)
4. Encounter software library. http://plaintext.crypto.lo.gy/article/658/encounter
5. Farah, S., Javed, Y., Shamim, A., Navaz, T.: An experimental study on performance evaluation of asymmetric encryption algorithms. In: Recent Advances in Information Science, Proceeding of the 3rd European Conference of Computer Science, (EECS-2012), Paris, pp. 121–124. WSEAS Press (2012)
6. Gahi, Y., et al.: A secure database system using homomorphic encryption schemes. In: Proceedings of 3rd International Conference on Advances in Databases, Knowledge, and Data Applications (DBKDA 2011), St. Maarten, Netherlands Antilles, pp. 54–58. International Academy, Research, and Industry Association (IARIA) (2011)
7. Hacigumus, H., et al.: Search on encrypted data. In: Yu, T., Jajodia, S. (eds.) Secure Data Management in Decentralized Systems. Advances in Information Security, vol. 33, pp. 383–427. Springer, New York (2007)
8. Jajodia, S., Litwin, W., Schwarz, Th.: Scalable distributed virtual data structures. In: Proceedings of Second ASE International Conference on Big Data Science and Computing, Stanford (2014)
9. Lauter, K., et al.: Can homomorphic encryption be practical? In: Proceedings of 3rd ACM Workshop on Cloud Computing Security (CCSW 2011), Chicago, pp. 113–124 (2011)
10. Paillier, P.: Public-key cryptosystems based on composite degree residuosity classes. In: Stern, J. (ed.) EUROCRYPT 1999. LNCS, vol. 1592, pp. 223–238. Springer, Heidelberg (1999)

11. Portokalidis, G., Keromytis, A.D.: Global ISR: towards a comprehensive defense against unauthorized code execution. In: Jajodia, S., Ghosh, A.K., Swarup, V., Wang, C., Wang, X. S. (eds.) Moving Target Defense. Advances in Information Security, vol. 54, pp. 49–76. Springer, New York (2011)

12. Subramaniam, H., Wright, R.N., Yang, Z.: Experimental analysis of privacy-preserving statistics computation. In: Jonker, W., Petković, M. (eds.) SDM 2004. LNCS, vol. 3178, pp. 55–66. Springer, Heidelberg (2004)

13. Smith, K., Allen, D., Sillers, A., Lan, H., Kini, A.: How practical is computable encryption? http://csis.gmu.edu/albanese/events/march-2013-cloud-security-meeting/04-Ken-Smith.pdf

14. Smith, K., Allen, M.D., Lan, H., Sillers, A.: Making query execution over encrypted data practical. In: Jajodia, S., Kant, K., Samarati, P., Singhal, A., Swarup, V., Wang, C. (eds.) Secure Cloud Computing, pp. 173–190. Springer, New York (2014)

15. Thep. The homomorphic encryption project. https://code.google.com/p/thep/

16. Tingjian, G., Zdonik, S.: Answering aggregation queries in a secure system model. In: Proceedings of 33rd International Conference on Very Large Databases (VLDB 2007), Vienna, Austria, VLDB Endowment, pp. 519–530, 23–28 Sept 2007

17. Tu, S., Kaashoek, M.F., Madden, S., Zeldovich, N.: Processing analytical queries over encrypted data. Proc. VLDB Endow. **6**(5), 289–300 (2013)

A Privacy-Aware Framework for Decentralized Online Social Networks

Andrea De Salve[1,2]([⊠]), Paolo Mori[2], and Laura Ricci[1]

[1] Department of Computer Science, Largo B. Pontecorvo, 56127 Pisa, Italy
{desalve,ricci}@di.unipi.it
[2] IIT-CNR, via G. Moruzzi 1, 56124 Pisa, Italy
paolo.mori@iit.cnr.it

Abstract. Online social networks based on a single service provider suffer several drawbacks, first of all the privacy issues arising from the delegation of user data to a single entity. Distributed online social networks (DOSN) have been recently proposed as an alternative solution allowing users to keep control of their private data. However, the lack of a centralized entity introduces new problems, like the need of defining proper privacy policies for data access and of guaranteeing the availability of user's data when the user disconnects from the social network. This paper introduces a privacy-aware support for DOSN enabling users to define a set of privacy policies which describe who is entitled to access the data in their social profile. These policies are exploited by the DOSN support to decide the re-allocation of the profile when the user disconnects from the social network. The proposed approach is validated through a set of simulations performed on real traces logged from Facebook.

Keywords: Decentralized online social network · Privacy · Data availability

1 Introduction

In the last few years, Online Social Networks (OSNs) have become one of the most popular Internet services and they have changed the way of how people interact with each other. The most popular OSNs are based on a centralized architecture where the service provider takes control over users' information. Centralized OSN architectures present several problems that include both technical and social issues that emerge as a consequence of the centralized management of the services [8]. If not properly protected, data of the OSNs can be used by malicious users to infer personal information or to perform other harmful activities [1]. Recent events have shown that, in addition to malicious users (internal or external to the OSN), also the centralized service provider [10] and the third-party applications [17] introduce new security and privacy risks.

A current trend for developing OSN services is towards the decentralization of the OSN infrastructure. A DOSN [8] is an OSN implemented in a distributed

© Springer International Publishing Switzerland 2015
Q. Chen et al. (Eds.): DEXA 2015, Part II, LNCS 9262, pp. 479–490, 2015.
DOI: 10.1007/978-3-319-22852-5_39

and decentralized way, such as a P2P systems or opportunistic networks. By decentralizing the OSN service, the concept of service provider changes because there is no central control authority which manages the system, stores the data and decides the term of the service. Instead, DOSNs are based on a set of peers that take on and share the tasks needed to keep on the system. Therefore, DOSNs shift the storage and the control over data to the end user and thus solving some, but introducing new security and privacy issues. Among them, the strategy for the allocation and replication of those data to the nodes of the DOSN which guarantee the protection of the their privacy, as described in Sect. 3.

In this paper, we focus on DOSNs where each user is associated with a *profile* which is a digital representation of the user including her contents (i.e. text, snippets, pictures, videos and music, etc.), also referred interchangeably as information or data. The privacy of the users' profile data in DOSNs presents new interesting challenges which involve two different aspects: *(i)* the access control on the contents of the user's profile, and *(ii)* the storage (allocation and replication) of the profile data on the nodes which builds up the DOSN system. Indeed, the user data must be kept private according to the preferences expressed by its owner but, at the same time, they should be available to authorized users even when the owner is not online. Most of the existing DOSNs implement very basic access control mechanisms to protect users' privacy, simply making a user able to decide which information is accessible by other members. Moreover, these access control mechanisms exploit only a limited set of the information available in OSNs. As a result, the privacy support of DOSNs would benefit from the adoption of an advanced language to express privacy policies which allow DOSNs users to exploit social network-related information to define the access rights to their contents. Moreover, to the best of our knowledge, existing DOSNs protect the privacy of the profile data allocated on users' nodes adopting encryption techniques. Hence, none of existing DOSNs take into account users' privacy policies to drive the tasks of the underlying support, e.g. to perform a smart allocation of the profile data to the users' nodes in order to avoid encryption when possible. Instead, we believe that making the underlying infrastructure aware of the privacy preference of the users may further improve the design of the current DOSNs in terms of privacy. In particular, in this paper we focus on a main challenge in DOSN which is guaranteeing availability of the data without compromising the privacy of the data owner. To this aim, we propose a framework which enhances the privacy support of DOSNs by:

- allowing users to define flexible privacy policies to regulate the accesses to the content they have shared by means of a proper Privacy Policy Language.
- exploiting users' privacy policies for making decisions about the allocation and replication of the users' profile data on the peers, in order to preserve as much as possible the expected users' privacy.

The remaining of the paper is organized as follows. In Sect. 2 we discuss the access control methods provided by the current DOSNs. Sections 3 and 4 describes the approach used to allocate users' data in DOSN and the framework's architecture. Section 5 presents the evaluation of our proposal by using privacy policies. Finally, Sect. 6 reports the conclusions and discusses future works.

2   Related Work

This section provides an overview of current approaches used by DOSNs in order to enforce privacy control over users' data. In order to help users to protect their personal content, current DOSNs adopt simple privacy policies by coupling distributed approaches with encryption techniques. Typically, the users are able to decide which personal information is accessible by other members by settings a given content as public, private, or accessible to their direct contacts, or by providing simple variants to this basic setting.

Table 1 summarizes the privacy options of main current DOSNs. In Diaspora[1] users organize their contacts into "aspects" (i.e. groups of contacts) where members can define access policies for each content by selecting the aspects that can access it. In Safebook [7] user's data can be private, protected, or public. In the first case data is not published, in the second case it is published and encrypted, and in the third case it is published without encryption. Personal information is organized into atomic attributes that allow the user to define privacy policies. PeerSoN [5] allows its members to define simple access policies based on individual users where the user's data are encrypted with the public keys of the users who have access to it. In LotusNet [2] users are able to define privacy policy based on the identities and regular expression, that is a compressed list of all the allowed content types. SuperNova [16] allows users to define the access policy of a content as public, protected for limited access to a subset of friends or private and inaccessible to anyone. LifeSocial.KOM [9] does not allow users to define complex access policies but provides a security layer where users are able to define Access Control Lists (ACLs). Vis-a-Vis [15] assumes that users, or preferably providers of the storage services, must properly configure their access-control policies on their platforms. Members of My3 [12] leverage their mutual trust relationships to enforce access control on the access requests but the system does not allow its members to define privacy policies on data. In Cachet [13] users' data are protected by a lower-level cryptographic hybrid structure that allows users to define their privacy policies based on attributes. Members can use friends' identities and relationships as attributes to define two kinds of policies: identity-based policies that define user-specific access and attribute-based policies that define access for a group of social contacts sharing some content. In Persona [4] the access control to system's resources is enforced by ACL. It supports group permission policy by using either public-key with symmetric cryptography or attribute-based encryption (ABE). Private user data is always encrypted with a symmetric key. The symmetric key is encrypted with an ABE key or with a traditional public key.

3   Privacy-Aware Allocation of Users' Profiles on DOSN

In DOSNs, users are free from centralized service provider since every participating node can be a server and a client depending on context. The data representing

---

[1] http://joindiaspora.com/.

Table 1. Privacy options of the most popular DOSNs.

DOSN	Protection options
Diaspora	Public, private, selected contacts
Safebook	Private, protected (group), public on profile attributes
PeerSoN	Not designed
LotusNet	Selected contacts and regular expression on content type
SuperNova	Public, private, selected contacts
LifeSocial.KOM	Public, private, selected contacts
Via-a-Vis	Not designed
My3	Trusted contacts, 1st degree contacts
Cachet	Identity or lower level attribute-based policy
Persona	Public, private, selected contacts

users' profiles are stored on users' nodes, and they are available as long as users are connected to the DOSN (i.e. until the user decides to log out the system by switching offline their peer), after that, users' data can not longer be obtained. To enhance availability, users data are replicated on distinct nodes of the system and reallocated to other nodes when one or more of them go offline.

This paper proposes an approach to preserve the privacy of DOSN's users based on a proper allocation of the data representing their profiles on the DOSN's nodes. This approach goes beyond traditional DOSN information allocation mechanisms [6], which typically replicate users' profiles on nodes that are chosen randomly or among their friends, and which need to employ encryption tools, that are inefficient in terms of storage overhead, to preserve users' privacy. In the following we will suppose that each user of the social network is paired with a different node of the distributed system, thus the term user and node will be used in an interchangeable way.

The main novelty introduced by the proposed approach is that it exploits the users' privacy policies to choose on which nodes the profile of a user U, say P_U, should be allocated. Our approach is based on the trust that the user U has in the other users, which is defined by the privacy policy of U. Let us suppose that U's privacy policy gives to user V the right to access to all the contents in P_U. This implies that user U believes that user V, once accessed P_U, will not disclose his contents to other users (using the DOSN itself or other tools) who may not be allowed to access them according to the privacy policy of U. In this case we say that user U trusts user V for storing his profile in clear because, if the files representing P_U are stored on the V's node, V cannot collect additional information by inspecting these files, with respect to the ones he can access exploiting the DOSN interface. In other words, the proposed allocation strategy tries to store the files representing the profile of U on the nodes of those users who are already allowed to access the data included in those files by exploiting the DOSN interface. In particular, the proposed allocation procedure works as follows. Let us suppose that, initially, P_U is stored on the node of U. To create a set of profile replicas, and when U disconnects from the DOSN, an election procedure selects a set of online DOSN nodes that could host P_U (e.g., U's online friends), and U's privacy policy is exploited to choose the nodes where P_U can be copied in order guarantee the profile availability while preserving user's privacy. Hence, U's profile is moved on the node V if the privacy policy of U states that V is allowed to access U's profile. In this way, there is no need of deploying any (encryption) support to protect the confidentiality of U's data when stored on V's node because V is entitled to access these data according to U's privacy policy. Eventually, if none of the selected nodes are eligible to host the profile of a user in clear, traditional encryption based approaches are used to store the profile on some of these nodes.

Although the integrity of the user profiles is another important security feature that the DOSN should guarantee, this paper does not cover it, and it will be addressed in a future work. Furthermore, we do not discuss the re-election problem, i.e., the case when a user V, whose node also hosts the profile of user U, disconnects from the network and the election procedure must be executed to move both U's and V's profiles.

3.1   Privacy Policy and Reference Examples

The proposed approach exploits a Privacy Policy Language [11] to enable users to define flexible privacy policies. These policies describe access rights by taking into account dynamic features of users and contents derived from the knowledge which can be extracted from the DOSN. The policy language that best matches the requirements of our scenario is the eXtensible Access Control Markup Language (XACML) [14]: an OASIS standard language based on XML, for defining access control policy on resources. XACML has gained the most attention because of its standardization and of its ability of expressing access control constraints on the basis of dynamic properties (attributes) of the owner, or resources, or requesters, rather than on fixed values. The full details of the language are discussed in [14].

We define a set of reference policies that will be used in the rest of the paper both to illustrate the capabilities of our approach and to provide a clear evaluation of the system under discussion. Differently from access control mechanisms proposed by current DOSNs, our approach is able to both model complex privacy policies that leverage the knowledge provided by the DOSN and exploits such policies to produce smart contents allocation that meets the privacy preferences defined by users. In order to capture multiple privacy aspects of real-life social networks, the reference policies are based on the the friendship types and on the common relationships with other users. Moreover, we use an attribute to model another key property of the relationship between users A and B: the tie strength which is a numerical representation of the intensity of the relationship between two users and can be approximated by using the number of interactions occurred between them [3]. Although the enforceable privacy policies are expressed in XACML, for the sake of clarity, in the following, we express the policy examples in natural language. Consider the user Alice. By using our framework, Alice may define privacy policies like the following ones:

Policy 1. Only users who have a friendship relation with Alice can see Alice's profile.

Policy 2. Only users who have a friendship relation with Alice and at least k common friends with Alice can see Alice's profile.

Policy 3. Users who have a friendship relationship with Alice can see Alice's profile provided that they do not have a friendship relationship with Bob.

Policy 4. Only users who have a friendship relationship with Alice and tie strength greater than a threshold t can see Alice's profile information.

4   The Framework Architecture

This section describes the architecture of the proposed framework. We consider a general DOSN reference architecture, without limiting ourselves to a specific implementation. A DOSN is defined as a service that allows users to articulate connections with the other users and share information with them. Each user of a DOSN is uniquely identified and corresponds to a node of the distributed network. The architecture of each node of the DOSN consists of two layers where each layer interacts with a module, the authorization system, to enforce the privacy policy.

- The *distributed online social network service* layer provides higher-level functionalities useful for DOSNs' users to manage their identity, profiles or contents and to access other users' ones. We assume that this layer implements all basic functionalities and features that are provided by contemporary centralized social networking services.
- The *distributed online social network infrastructure* layer provides the core functionalities to support the services used at the DOSN services layer, such as overlay network management, storage management, bootstrap, information diffusion, etc.

Figure 1 shows the layered DOSN platform described above (boxes on the left) together with the proposed privacy support (box on the right). The privacy support we propose is based on an authorization system which acts on both layers of the general DOSN platform. At service level it evaluates users' privacy policies to regulate access to the contents shared by users in their profiles, while at infrastructure level it evaluates the same privacy policies to allow the users' profiles allocation mechanism of the DOSN to perform decisions that preserve the expected users' privacy. From the architectural point of view, the authorization system is invoked by the Policy Enforcement Points (PEPs), which are software components that have been embedded, respectively, in the service layer and in the infrastructure layer of the DOSN. In particular, supposing that the node in Fig. 1 host U's profile, the PEPs at service layer invokes the authorization system when a user V wants to access the contents in the profile of user U,

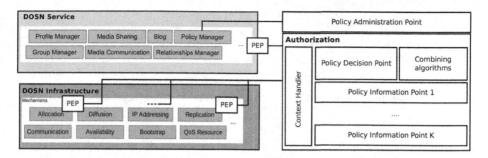

Fig. 1. The framework architecture along with the general DOSN architectural model.

while the PEP at infrastructure layer invokes the authorization system when the allocation component tests whether the node of V can be selected to store the profile of U.

The authorization system follows the XACML reference architecture, where the main components are: the Context Handler, which receives the access request from the PEPs and manages the interactions among the other components of the authorization system, the Policy Decision Point (PDP), which is the system entity that evaluates applicable policy based on request, and generates an authorization decision, the Policy Information Point (PIP) that acts as a source of attribute values about subjects, or resources, or environment to the PDP, which are required to evaluate policies, and the Policy Administration Point (PAP) which allows to edit privacy policies. The PAP is invoked by the Policy Manager in the service level layer when the user wants to set up or update his privacy policy. A detailed description of the components' functionalities as well as the data flow between them is presented in [14].

5 Evaluation

This section provides a quantitative evaluation of the proposed framework. We focus on the delay introduced by the authorization system in the user's profile re-allocation phase, since it is critical with respect to the system performance. In fact, to provide the privacy-aware ranked list of the nodes where the profile of user U could be copied, the privacy policy must be evaluated on each user V which is online and has a friendship relation with U, thus simulating an access of the user V to the profile of U.

With the aim of evaluating the efficiency of such approach, we have developed a set of simulations of our system using the Peersim[2] simulator. We have implemented a Facebook application, called *SocialCircles!*[3] able to retrieve the following sets of information from registered users:

---

[2] Available at http://peersim.sourceforge.net/.
[3] Available at http://www.socialcircles.eu/.

Table 2. Statistics of the dataset

# Users	144,481	Avg. clustering coefficient	0.63
# Friendships	3,683,458	Avg. session number	5
Avg. degree	486.9	Avg. session length (min)	6
Avg. modularity	0.46	Avg. inter-arrival time (min)	90

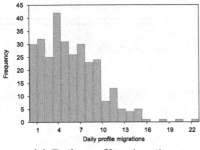   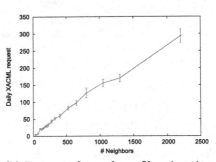

(a) Daily profile migrations   (b) Requests for each profile migration

Fig. 2. Number of profile migrations and requests

- Friendships and Profile information of registered users;
- Interaction Information between users registered to the application and their friends, such as posts, comments, likes, tags and photos. Due to technical reasons, we restrict the interaction information retrieved up to 6 months prior to user application registration;
- By using the Facebook chat status we track the online status of Facebook's users by sampling all users every 8 min for 10 consecutive days (from Tuesday 3 June to Friday 13 June 2014);

The dataset obtained contains 328 registered users, for a total of 144.481 users (registered users and their friends). The resulting Facebook population has the advantage of representing a very heterogeneous population: 213 males and 115 females, with age range of 15–79 with different education, background and geographic location. Table 2 shows the characteristics of our dataset. In our experiments we consider the users who are going to disconnect from the DOSN and therefore their profile must be copied on other node(s). The simulation duration is the same as the crawling period, so that each node exactly simulates the behaviour of the corresponding user during the crawling, in particular his connections and disconnections from the social network. For each disconnecting user U, an election mechanism selects as neighbours the online users having a friendship relation with U. Then, the authorization procedure is executed on each user V in the set of neighbours, i.e., the framework checks whether V is authorized to access the profile of U. As we model the profile of a user as a unique object, access requests may only return permit (i.e. access level equals to 1) or deny (access level equals to 0). It is important to note that the policies

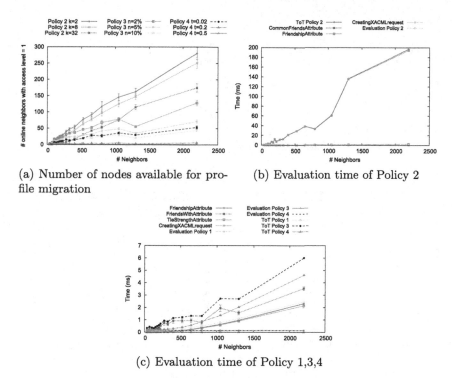

(a) Number of nodes available for profile migration

(b) Evaluation time of Policy 2

(c) Evaluation time of Policy 1,3,4

Fig. 3. Evaluation time of the reference policies

must be re-evaluated every time the user U goes offline because attributes change
over time. We have evaluated the authorization component for different scenarios
based on the policies described in Sect. 3, namely: *(i) Policy 1, (ii) Policy 2* with
a number of common friends (k) equal to 2, 8 and 32, *(iii) Policy 3* with a fraction
of randomly chosen users who are friend of Bob (n) equal to 2 %, 5 % and 10 %
of the user's neighbours and *(iv) Policy 4* with tie strength (t) equal to 0.5, 0.2
and 0.02. Figure 2(a) shows the distribution of the elections triggered in a day,
i.e., the number of times a node disconnect from the social network and triggers
the election mechanism. The number of authorization requests performed by a
node in each execution is equal to the number of active friends and they are
distributed according to the graph shown in Fig. 2(b) (with 95 % C.I.). Since
the number of executions and the requests' number depend on the availability
pattern of both the users and their friends, they remain the same for each policy.
Specifically, the average number of online friends of each registered user ranges
between 4 and 287.

Figure 3(a) shows the average number of online neighbours who fulfill the
policies, as a function of the users' neighbourhood size. Indeed, on the basis of
the policies, these neighbours have the rights of access the user's profile and they
are good candidates to be chosen as replicas for the user's profile. As shown by
the figure, the selected privacy policies allow the user to exploit several levels of

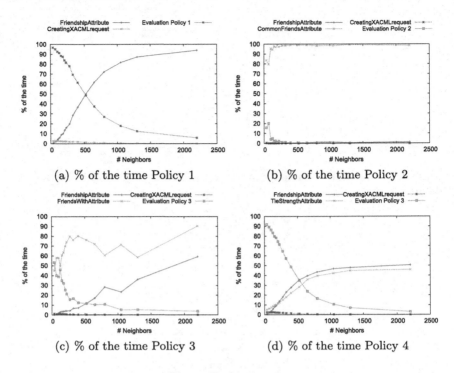

Fig. 4. Evaluation time

privacy preferences. Figure 3(c) and (b) show the average time (in ms) needed
to create the XACML request (denoted as CreatingXACMLrequest in 3(b)), to
compute the attributes required by the policies, to evaluate the single policy
(denoted as Evaluation Policy 1, 2, 3 and 4) and the overall evaluation time
of each policy (denoted as ToT Policy1, 2, 3 and 4). The attributes used by the
policies are:

FriendshipAttribute: It models the friendship relation between two users as
a boolean value. Required by every policy.
CommonFriendsAttribute: It models the number of common friendship relations between two users as an integer value. Required by policy 2.
FriendsWithAttribute: It models the friendship relation between a user and
a set of selected users as a boolean value. Required by policy 3.
TieStrengthAttribute: It models the min-max normalization of the interactions' number that occur towards a specific user as an float value. Required
by policy 4.

The attributes' values are computed on the fly, each time they are required by
the authorization module. The evaluation time required to compute attributes
increases as the number of the users' neighbours. The computation of the *CommonFriendsAttribute* has the highest execution time compared to the other
attributes (see Fig. 3(b)) since it needs to check all the possible friendship

relations between two users' neighborhoods while the others attributes require only to check for a friendship relation (*FriendshipAttribute* and *FriendsWithAttribute*) or for a value (*TieStrengthAttribute*). Both the policies evaluation time and the creation time of the XACML request (CreatingXACMLrequest) remain almost the same for all users. The other evaluation times are negligible and they consume only a few milliseconds.

In order to assess the impact of the phases that build up the policy enforcement, we compared the percentage of time consumed by each phase as a function of the number of users' neighbours. Figure 4(a), (b), (c) and (d) show the measurements obtained by the policies 1, 2, 3 and 4, respectively. In the case of Policy 1, most part of the time is spent for the policy evaluation phase when the number of neighbours is less than 500. As the number of users' neighbours is greater than 500, the attribute evaluation phase takes more time. The same thing happens in the case of the policy 4, where the impact of the attributes evaluation phase grows as the number of user's neighbours increases. For Policy 3 the computation of the *FriendsWithAttribute* attribute take most of the total time. Instead, for Policy 2, the computation of the *CommonFriendsAttribute* takes more than 80 % of the total time while the other phases remain negligible as compared to it. Finally, the *creatingXACMLrequest* phase is negligible compared to the others.

6   Conclusion and Future Works

This paper presented a framework for DOSNs which allows users to define flexible privacy policies based on attributes modelling the DOSN knowledge. These policies, besides regulating the access to the contents shared on the users' profiles, are also exploited to support the underlying data allocation mechanism of the DOSN in order to preserve users' privacy. In particular, we have proposed a privacy preserving strategy for the allocation of users' profiles, defined the related architecture, described a reference implementation based on a simulator, and presented some performance results on a real dataset, which showed that the overhead introduced by our allocation strategy is quite low. In fact, in our experiments, the total policy evaluation time ranges from 2 ms (for Policy 1) to 195 ms (for Policy 2), although it depends on the complexity of the attributes exploited in the policies. Moreover, in our experiments, the policy-driven mechanism provided by our framework always succeeded in performing data allocation decisions that preserve users' privacy policies.

The proposed privacy-preserving framework raises different challenges that we plan to investigate as future works. We will extend the set of attributes used in policies in order to ensure high degree of expressiveness demanded by users. At the same time, a fast authorization evaluation requires caching techniques able to speed up attributes' retrieval. A further extension of our framework concerns the definition of a support ensuring the integrity of users' profiles. Finally, we plan to enhance the allocation strategy in order ensure that the profiles are always maintained available on trusted nodes based on the variation in the users' privacy policies.

References

1. Acquisti, A., Gross, R.: Predicting social security numbers from public data. Proc. Natl. Acad. Sci. **106**(27), 10975–10980 (2009)
2. Aiello, L.M., Ruffo, G.: Lotusnet: tunable privacy for distributed online social network services. Comput. Commun. **35**(1), 75–88 (2012)
3. Arnaboldi, V., Guazzini, A., Passarella, A.: Egocentric online social networks: analysis of key features and prediction of tie strength in facebook. Comput. Commun. **36**(10), 1130–1144 (2013)
4. Baden, R., Bender, A., Spring, N., Bhattacharjee, B., Starin, D.: Persona: an online social network with user-defined privacy. ACM SIGCOMM Comput. Commun. Rev. **39**(4), 135–146 (2009)
5. Buchegger, S., Schiöberg, D., Vu, L.H., Datta, A.: Peerson: P2p social networking: early experiences and insights. In: Proceedings of the Second ACM EuroSys Workshop on Social Network Systems, pp. 46–52, ACM (2009)
6. Conti, M., De Salve, A., Guidi, B., Pitto, F., Ricci, L.: Trusted dynamic storage for dunbar-based P2P online social networks. In: Meersman, R., Panetto, H., Dillon, T., Missikoff, M., Liu, L., Pastor, O., Cuzzocrea, A., Sellis, T. (eds.) OTM 2014. LNCS, vol. 8841, pp. 400–417. Springer, Heidelberg (2014)
7. Cutillo, L.A., Molva, R., Strufe, T.: Safebook: a privacy-preserving online social network leveraging on real-life trust. Commun. Mag. IEEE **47**(12), 94–101 (2009)
8. Datta, A., Buchegger, S., Vu, L.H., Strufe, T., Rzadca, K.: Decentralized online social networks. In: Furht, B. (ed.) Handbook of Social Network Technologies and Applications, pp. 349–378. Springer, New York (2010)
9. Graffi, K., Gross, C., Stingl, D., Hartung, D., Kovacevic, A., Steinmetz, R.: Lifesocial. kom: a secure and p2p-based solution for online social networks. In: 2011 IEEE Consumer Communications and Networking Conference, pp. 554–558, IEEE (2011)
10. Greenwald, G., MacAskill, E.: Nsa prism program taps in to user data of apple, google and others. The Guardian **7**(6), 1–43 (2013)
11. Kumaraguru, P., Cranor, L., Lobo, J., Calo, S.: A survey of privacy policy languages. In: Workshop on Usable IT Security Management: Proceedings of the 3rd Symposium on Usable Privacy and Security, ACM, Citeseer (2007)
12. Narendula, R., Papaioannou, T.G., Aberer, K.: My3: a highly-available p2p-based online social network. In: 2011 IEEE International Conference on Peer-to-Peer Computing, pp. 166–167, IEEE (2011)
13. Nilizadeh, S., Jahid, S., Mittal, P., Borisov, N., Kapadia, A.: Cachet: a decentralized architecture for privacy preserving social networking with caching. In: Proceedings of the 8th International Conference on Emerging Networking Experiments and Technologies, pp. 337–348, ACM (2012)
14. OASIS: Xacml version 3.0 (2013). http://docs.oasis-open.org/xacml/3.0/xacml-3.0-core-spec-os-en.html
15. Shakimov, A., Lim, H., Cáceres, R., Cox, L.P., Li, K., Liu, D., Varshavsky, A.: Vis-a-vis: privacy-preserving online social networking via virtual individual servers. In: Third International Conference on Communication Systems and Networks, pp. 1–10, IEEE (2011)
16. Sharma, R., Datta, A.: Supernova: super-peers based architecture for decentralized online social networks. In: Fourth International Conference on Communication Systems and Networks, pp. 1–10, IEEE (2012)
17. Steel, E., Fowler, G.: Facebook in privacy breach. The Wall Street Journal, 18 Oct 2010

CaLibRe: A Better Consistency-Latency Tradeoff for Quorum Based Replication Systems

Sathiya Prabhu Kumar[1,2]($\boxtimes$), Sylvain Lefebvre[1], Raja Chiky[1],
and Eric Gressier-Soudan[2]

[1] LISITE Laboratory, ISEP Paris, Paris, France
`sathiya-prabhu.kumar@isep.fr`
[2] CEDRIC Laboratory, CNAM Paris, Paris, France

Abstract. In Multi-writer, Multi-reader systems, data consistency is ensured by the number of replica nodes contacted during read and write operations. Contacting a sufficient number of nodes in order to ensure data consistency comes with a communication cost and a risk to data availability. In this paper, we describe an enhancement of a consistency protocol called LibRe, which ensures consistency by contacting a minimum number of replica nodes. Porting the idea of achieving consistent reads with the help of a registry information from the original protocol, the enhancement integrate and distribute the registry inside the storage system in order to achieve better performance.

We propose an initial implementation of the model inside the Cassandra distributed data store and the performance of LibRe incarnation is benchmarked against Cassandra's native consistency options ONE, ALL and QUORUM. The test results prove that using LibRe protocol, an application would experience a similar number of stale reads compared to strong consistency options offered by Cassandra, while achieving lower latency and similar availability.

Keywords: Distributed storage systems · Eventual consistency · Quorum systems

1 Introduction

In distributed data storage systems, data is replicated to improve the performance and availability of the system. However, ensuring data consistency with higher availability and minimum request latency is notoriously challenging [1,9]. In order to efficiently handle these challenges, the Dynamo system [7] designed by Amazon uses a quorum-based voting technique that facilitates configurable tradeoffs between Consistency, Latency and Availability. This technique inspired subsequent distributed data storage systems such as Cassandra [14], Voldemort [22] and Riak [11]. The quorum-based voting technique ensures consistency based on the math behind the intersection property of the quorum systems [18,23]. This intersection property can be expressed by the formula $R + W > N$. This formula symbolizes that the system can ensure consistency if the sum of the

© Springer International Publishing Switzerland 2015
Q. Chen et al. (Eds.): DEXA 2015, Part II, LNCS 9262, pp. 491–503, 2015.
DOI: 10.1007/978-3-319-22852-5_40

nodes acknowledging the Read (R) and the nodes acknowledging the Write (W) is greater than the total number of replicas (N). Since more nodes have to be contacted, ensuring consistency comes with a communication cost and a threat to the system availability. Hence, in order to provide fast response time, these storage systems rely on eventual consistency and do not satisfy the intersection property by default. Therefore the user can configure the number of quorum members to contact during read and write time in order to ensure consistency on demand.

If the intersection property cannot be satisfied, the system will reject the operation. The three popular ways of satisfying the intersection property to ensure strong consistency are as follows: write to all and read from one node, write to one and read from all nodes, write and read to/from a majority of nodes.

Most of these storage systems are optimized for write intensive workloads, which requires the system to acknowledge writes as fast as possible and reconcile conflicts at read time. For these systems to guarantee the reads with minimum latency, the default eventual consistency option (non-overlapping quorum) is desirable. But, if a data item is written with the minimum write quorum (one node), the only mode to preserve consistency for this data during read time is reading from all the replicas. However, if one of the nodes is temporarily down or can not be contacted, the read will fail. The same risk applies when the system writes to all nodes and reads from one node. When reading and writing from/to a majority of nodes, failure of one or few replica nodes is tolerable. In that case, availability guarantees will still be affected if the system is not able to communicate to a majority of replica nodes. Besides, the latency for both reads and writes will be affected as well.

Performance of these modern storage systems rely on caching most of the recent data in order to handle the read requests faster. When a request is forwarded to all or majority of replicas to retrieve a data item, the possibility that all replicas have the right data in their cache would be improbable. Hence, the slowest replica node responding to the request will increase the request latency and the advantage of cache memory is lost.

The existing "strong" consistency options offered by these storage systems are strong enough to ensure data consistency when no partition occurs, but add some extra communication cost and a risk to data availability. To our knowledge, there is no softer option that can ensure consistency guarantees with availability and latency guarantees similar to the default level of eventual consistency.

In this paper, we discuss improvements on a consistency protocol called LibRe [12], which acts as an in-between consistency strategy between the default eventual consistency and the strong consistency options derived from the intersection property.

The original LibRe protocol used a registry, which records the list of replica nodes containing the most recent version of the data items. Hence, referring to the registry during read time helps to forward the read requests to a replica node holding the most recent version of the needed data item.

Instead of relying on a synchronization service such as Zookeeper [10], as initially proposed in [12], the enhanced protocol distributes the registry over each node in the cluster and manages the data items entries in the registry only until all the replicas converge to a consistent state. These mechanisms are detailed in Sect. 2.4, along with the protocol description. Since the improvements are inclined towards higher availability and minimum request latency, its consistency guarantee is slightly relaxed compared to the original LibRe design [12]. For the sake of simplicity, in the following sections, the enhanced LibRe protocol that the paper intended to describe is termed as LibRe and its initial version proposed in [12] is termed as original LibRe or simply original protocol.

The following section provides a description of the LibRe protocol. The implementation inside the Cassandra distributed data storage system [14] and its performance evaluations are discussed in Sect. 3. We describe some of the related works at Sect. 4 before presenting our conclusions and future works in the last section.

2   LibRe

The LibRe acronym stands for "Library for Replication". The protocol collects information about writes until they are fully propagated to all the replica nodes. If an update is not propagated to all the replica nodes, then an entry for this data item is added to an in-memory data structure called the LibRe Registry, along with a *version-id* and the list of replica nodes holding this latest version. The *version-id* is a monotonically increasing value representing the recent version of the data item, for instance it can be the timestamp of the operation or a version vector. Consulting the registry at read time helps to avoid forwarding the read requests to a replica node where the recent update is not effected.

2.1   Targeted System

Key-Value data stores that store an opaque value for a given Key seen enough popularity in the modern distributed storage systems. Some of these systems ensure strong consistency, whereas most of the systems such as Dynamo [7], Riak [11] and Voldemort [22] rely on tunable consistency. In order to tune the consistency level of the system on a per-query or per-table basis, these storage systems follow a quorum-based voting technique such as described in Sect. 1. These systems mostly use a Distributed Hash Table (DHT) [24] for identifying the replica nodes of a data item. LibRe protocol targets the Key-Value data stores that offer tunable consistency based on quorum based voting technique using DHT. In building LibRe we assume that the underlying system provides failure detection and tolerance mechanisms, which are building blocks for the reliability of our Registry. Currently, we do not consider the use of LibRe for inter-cluster replication.

2.2   LibRe Registry

The core of the LibRe protocol is its Registry. The registry is an in-memory key-value data structure that takes the data identifier as the Key and the list of replica nodes id (IP addresses) holding the most recent *version-id* of the data item as the Value. During write operation, each replica node tries to add its id in the list. If the number of ids in the list reaches the total number of replica nodes for the data item, then confirming the convergence of all replicas, the entry for the data item in the registry can be safely removed.

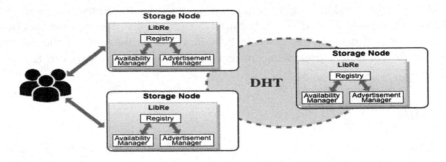

Fig. 1. LibRe architecture diagram

The registry is distributed over all the nodes in the cluster, but at any point in time only one copy of the entry for a data item exists. An entry for a data item d_i will be stored on only one of the available replica nodes that is responsible for storing the data item d_i.

Figure 1 shows the position of LibRe in the system architecture and the components of the LibRe protocol. Let R_i be a replica set for a data item d_i, such that $R_i = \{r_1, r_2, ..r_n\}$, where r_x is a node identifier, and n is the number of nodes in the replica set. So, one of the available replica nodes in R_i (say the first one: r_1) will hold the registry and the other two supporting components of LibRe: the Availability-Manager and the Advertisement-Manager, as shown in Fig. 1. The node that holds the registry and the supporting components is called the Registry Node for the particular data item. For any data item d_i, the id of the first replica node obtained via consistent hashing function is the registry node id. In other words, the replica node that has the lowest token id is considered as the registry node. The registry is distributed over each node in the cluster, so each node plays the role of replica node as well as registry node.

2.3   LibRe Messages

The LibRe protocol is based on two types of messages, namely: the Advertisement Message (Fig. 2a) and Availability Message (Fig. 2b), corresponding respectively to the Advertisement-Manager and Availability-Manager of the LibRe

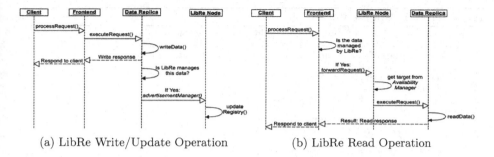

(a) LibRe Write/Update Operation   (b) LibRe Read Operation

Fig. 2. LibRe sequence diagram

components shown in Fig. 1. The Fig. 2 and the corresponding Algorithms 1 and 2 discussed at the Sect. 2.4 show the improvements of the LibRe protocol over the original algorithm discussed in the paper [12].

Advertisement Message: From Fig. 1, a client can connect to any node in the system. Some storage systems call this node the Coordinator node [14]. In the usual system behavior, a write request will be forwarded to all the replica nodes that are available. If the coordinator node receives back the required number of acknowledgements (or votes) from the replica nodes for the write, the coordinator issues a success response to the client. If the sufficient number of vote is not received within a timeout period, the coordinator issues a failure response to the client. LibRe protocol follows this default system behavior, but in addition, after a successful write operation, the replica node sends an *advertisement message* to the registry node asynchronously. The replica node sends advertisement message only for the data item that are configured to use LibRe. The advertisement message consists of the data key, version-id and originating node id.

Availability Message: When the coordinator node receives a read request that is configured to use LibRe protocol, the coordinator sends an *availability message* to the Registry Node of this particular data item. The availability message contains the original *read message* received from the client and the data key of the needed data item. When the registry node receives an *availability message*, it finds a replica node from the registry and forwards the original *read message* to that replica node. The replica node sends the read response directly to the coordinator node and the coordinator forwards it to the client. If an entry for a data key is not found in the registry, then the *read message* will be forwarded to one of the available replica nodes.

2.4 LibRe Protocol

Algorithm 1 describes the role of LibRe's *Advertisement Manager* during update operation. The update for a data item can be issued when its replicas are in converged state or in diverged state.

Algorithm 1. Update Operation

$r_k = \{n_i, n_j, \ldots\}$: set of replica nodes holding recent version of data item k.
$e_k = <v_k, r_k>$: record where v is a version-id and r is a replica set.
$R : k \rightarrow e_k$: Map of data item keys k to corresponding entry record.
N: Number of replicas
1: **function** ADVERTISEMENTMANAGER(k, v, n)
2:     $entry \leftarrow R.k$
3:     **if** $entry \neq \emptyset$ and $entry.v = v$ **then**
4:         $entry.r \leftarrow entry.r \cup \{n\}$
5:         **if** $|entry.r| = N$ **then**
6:             $R \leftarrow R \setminus \{k\}$
7:         **end if**
8:     **else if** $entry = \emptyset$ or $v > entry.v$ **then**
9:         $entry.v \leftarrow v$
10:        $entry.r \leftarrow \{n\}$
11:    **end if**
12:    $R.k \leftarrow entry$
13: **end function**

When a replica node sends an advertisement message regarding an update, the *Availability Manager* of the LibRe node takes on the following actions. First the protocol checks whether the *data-key* already exists in the *Registry*: line 3. If the *data-key* exists in the registry (replicas are in diverged state), line 3: the *version-id* logged in the registry for the respective data-key is compared to the one sent with the update message. If the *version-id* logged in the registry matches with the *version-id* of the operation (replica convergence), then line 4: the node id (IP-address) will be appended to the existing replica list. Lines 5–6: If the number of replicas in the list is the same as the total number of replicas for data item k, then replicas are in converged state, and the entry is deleted. If the entry does not exist in the registry or the *version-id* of the operation is greater than the existing *version-id* in the registry (line 8): the entry is created or reinitialized with the operation's *version-id* and the sender node id (lines 9 and 10). Finally the entry is recorded in the registry (line 12). This setup helps to achieve Last Writer Wins policy [17,19].

Read Operation: Algorithm 2 describes the LibRe policy during the read operation. According to the algorithm, since the Registry keeps information about the replica nodes holding the recent version of data item k, the nodes information will be retrieved from the registry (line 2). Line 3: If an entry for the data-key exists in the registry, line 4: one of the replica nodes from the entry will be chosen as the target node. The method *first()* (lines 4 and 6) returns the closest replica node sorted via proximity. Line 5: if the Registry does not contain an entry for the needed data-key, then, line 6: one of the replica nodes that are responsible for storing the data item will be retrieved locally via DHT lookup. Finally, the read message will be forwarded to the chosen target node: line 8.

2.5   LibRe Reliability

As mentioned in Sect. 2.1, the reliability and fault tolerance of the LibRe protocol relies on the guarantees of the targeted system. The systems that use DHT

Algorithm 2. Read Operation

$r_k = \{n_i, n_j, \ldots\}$: set of replica nodes holding recent version of data item k.
$e_k = <v_k, r_k>$: record where v is a version-id and r is a replica set.
$R : k \to e_k$: Map of data item keys k to corresponding entry record.
$D_k = <n_i, n_j, \ldots>$: replica nodes for data item k that are obtained via default method.

```
1: function GETTARGETNODE(k, M_read)
2:      replicaNodes ← R.k.r
3:      if replicaNodes ≠ ∅ then
4:          n ← replicaNodes.first()
5:      else
6:          n ← D_k.first()
7:      end if
8:      forward(M_read, n)
9: end function
```

---

for quorum based voting actively take care of the ring membership and failure detection [7]. The underlying DHT helps to find the first available replica node. In the event of node joining and/or leaving the cluster, the Consistent Hashing technique supports minimal redistribution of the nodes keys. In such case, there will be a change in the first available replica node (registry node) for a few data items and the registry information for these data-keys would not be available. In that case, the registry information will be rebuilt on the new registry node by sending successive *advertisement messages* to this node. This may lead to small and time limited inconsistencies in the system. Therefore, LibRe sacrifices Consistency in favor of Availability: cf. Algorithm 2. If a registry node that has been unavailable joins back the cluster, the stale registry information has to be flushed during a handshaking phase. Besides, periodic local garbage collection is needed to keep the registry information clean between replica nodes.

2.6 LibRe Cost

The tradeoff provided by the LibRe protocol comes at the expense of some additional cost on message transfers and memory consumption.

Extra Message Transfers: In LibRe, a lookup in the registry is required during a read operation on contacting the *availability manager* of the registry node to read from a right replica node. However, this operation represent constant cost, as the number of messages sent for achieving the consistent read does not depend on the number of replicas involved, as mentioned in Sect. 2.3. Besides, the latency spent during this lookup can be gained back via managing the cache memory efficiently. During write operations, notifying the *advertisement manager* about an update is asynchronous and does not affect the write latency. Although these messages are an additional effort when compared to the default eventual consistency option, it is better than the strong consistency options that communicate to a majority or all replica nodes during reads and/or writes.

Registry In-memory Data Structure: LibRe manages the registry information in-memory. This information is distributed among all the nodes in the cluster and is maintained only for the data items whose recent update is not effected on all replica nodes. Moreover, eventual consistency guarantees of the targeted system and the periodic local garbage collection of the LibRe protocol helps to reduce the amount of information to be kept in-memory.

3   CaLibRe: Cassandra with LibRe

Cassandra [14] is one of the most popular open-source NoSql systems that satisfies the system model specified in Sect. 2.1. Hence, we decided to implement the LibRe protocol inside the Cassandra workflow and evaluate its performance against Cassandra's native consistency options: ONE, QUORUM and ALL. Although Cassandra is a column family store, we used it as a Key-Value store during test setup: refer Sect. 3.1. LibRe protocol was implemented inside Cassandra release version 2.0.0. In the native workflow, while querying a data, the endpoints (replica nodes) addresses are retrieved locally via matching the token number of the data item over the nodes token numbers. The IP-address of the first alive endpoint (without sorting the endpoints by proximity), will be chosen as the registry node for the replica sets it is responsible. A separate thread pool for the LibRe messaging service is designed for handling the LibRe messages effectively. On system initialisation, all LibRe registries are empty until a write request is executed, which will trigger the protocol and start filling up the registries. CaLibRe can be configured to work either by passing a list of data items in a configuration file, or by specifying directly the name of the table(s) to monitor. Currently, the *version-id* used in CaLibRe is the hash of the modifying value. However, it will be replaced by a timestamp in the future.

3.1   CaLibRe Performance Evaluation Using YCSB

Test Setup: The experiment was conducted on a cluster of 19 Cassandra and CaLibRe instances that includes 4 medium, 4 small and 11 micro instances of Amazon EC2[1] cloud service and 1 large instance for the YCSB test suite [6]. All instances were running Ubuntu Server 14.04 LTS - 64 bit. The workload pattern used for the test suite was the "Update-Heavy" workload (*workload-a*), with a record count of 100000, operation count of 100000, thread count of 10 and the Replication-Factor as 3. YCSB by default stores 10 columns per RowKey. We used RowKey as the data key, for which, an entry will be managed in the LibRe Registry. Using RowKey as the data key could leads to a situation like, if one or few columns of a RowKey is updated on a replica node r_n, the registry would assume r_n contains the recent version for all the columns of the RowKey. In order to avoid this situation, we update all the 10 columns during each update. The test case evaluate the performance and consistency of the 19 Cassandra instances with different consistency options (ONE, QUORUM and ALL) against 19 CaLibRe instances with a consistency option ONE. Performance is evaluated

---

[1] http://aws.amazon.com.

by measuring read and write latencies and consistency is evaluated for each
level by counting the number of stale reads. In order to simulate a significative
number of stale reads, a partial update propagation mechanism was injected into
the Cassandra and CaLibRe cluster to account for the system performance under
this scenario [13]. Hence, during update operations, instead of propagating the
update to all 3 replica, the update will be propagated to only 2 of the replicas.

Test Evaluation: Figure 3a, b and c respectively show the evaluation of Read
Latency, Write Latency and the number of Stale Reads of Cassandra with different consistency options against CaLibRe: Cassandra with LibRe protocol.
In Fig. 3a, the entity ONE represents the read and write operations with consistency option ONE. The read and write operations with consistency option
QUORUM is indicated by the entity QUORUM. The entity ONE-ALL represents the operations with write consistency option ONE and read consistency
option ALL. The entity CALIBRE represents our implementation of the LibRe
protocol developed inside Cassandra. Due to the injection of the partial update
propagation, ROWA (Read One, Write All) principle could not be tested, as
writes would always fail.

The read latency graph in Fig. 3a, shows that the 95th Percentile Latency of
CALIBRE is similar to the other consistency options of Cassandra. The 99th Per-

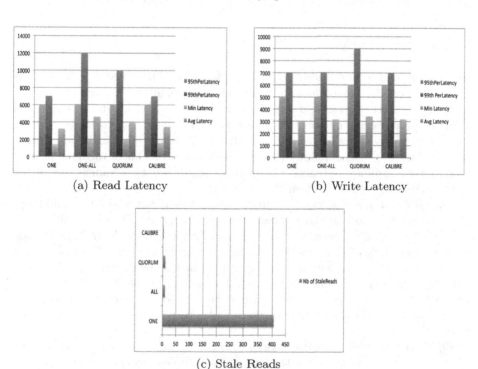

(a) Read Latency   (b) Write Latency

(c) Stale Reads

Fig. 3. CaLibRe performance evaluation (Color figure online)

centile Latency of CALIBRE and cassandra with consistency level ONE remains same and better than the other options ONE-ALL and QUORUM. The minimum and average latencies of CALIBRE are slightly higher when compared to Cassandra with consistency level ONE but better than the consistency options QUORUM and ONE-ALL. This is due to the fact that LibRe protocol imposes an additional call to the registry for all requests.

The write latency graph in Fig. 3b shows that the 95th percentile write latency of CALIBRE is the same as the 95th percentile latency of QUORUM, and that CALIBRE is faster in other metrics: 99th Percentile, Minimum and Average latencies of QUORUM. However, while comparing to the entities ONE and ONE-ALL, some of the write latency metrics of CALIBRE are slightly higher (but are not significant). This is due to the fact that both in ONE and ONE-ALL, writes need only one acknowledgement from a replica node. In CALIBRE also writes need only one acknowledgement but there is an extra messaging service in the background.

Graph 3c shows the number of stale reads for each level of consistency. Cassandra with consistency level ONE shows the highest number of stale reads. There were a few stale reads in the other consistency options, but these numbers are negligible when compared to the total number of requests. From these results, it is possible to conclude that CaLibRe offers a level of consistency similar to one provided by the QUORUM and ONE-ALL levels with better latency.

4   Related Works

Quorum systems are well studied in the literature. There are multiple works aiming at improving the performance and reliability of quorum systems [2, 15, 16]. However, in all these works, a sufficient amount of nodes has to be contacted in order to satisfy the intersection property. Apart from the works on quorum systems, there are also a few works in the literature whose approaches are similar to some of the approaches followed in the LibRe protocol. One of the most famous work that has similar approach of the LibRe protocol is the 'NameNode' of the Hadoop Distributed File System (HDFS) [20].

The HDFS NameNode manages metadata of files in the file system and helps to locate needed data in the cluster. But, on the contrary to the LibRe registry, which maintains metadata about small data items, the HDFS NameNode manages metadata of large file blocks. In addition to this, the NameNode is a centralized registry that stores information about the whole cluster, which can make the whole system unavailable in case of failure of the NameNode. In our approach, the LibRe registry only stores the location of partially propagated writes, and in case of failure of a registry node, availability of the system is not affected.

BigTable [4], which is a data store designed by Google, uses a two level lookup before contacting the actual data node for accomplishing reads and writes. In BigTable, the UserTable that needs to be contacted for accomplishing reads and writes is found by looking at a ROOT tablet followed by a METADATA tablet.

This enables to have a scalable and fast lookup. The earlier version of HBase [8], which is an open source implementation of BigTable, used a similar two-level lookup for finding data in the system. In the later version, the two-level lookup is reduced to a single lookup in the METADATA table. However, BigTable and HBase ensure strong consistency, so there is no context of stale replicas in these data stores. In LibRe, we use a single lookup to identify a fresh replica node only for reading some data items that are configured to use LibRe protocol.

In [21], Tlili et al. designed a reconciliation protocol for collaborative text editing over a peer to peer network using a Distributed Hash Table (DHT). According to the protocol, for each document, a master peer is assigned via the lookup service of the DHT. The master peer holds the last modification timestamp of the documents in order to identify missing updates of a replica peer in order to avoid update conflicts. This master-peer assignment is similar to the *Registry Node* assignment in the LibRe protocol. However, in LibRe, the registry node holds the version-id of the recent update and the replica nodes holding it in order to avoid reading from a stale replica.

The Global sequence protocol designed by Burckhardt et al. in [3] uses two states for an update: known sequence and a pending updates sequence. When an update is issued by a client, the update is kept in the pending updates list, and broadcasts its origin and a sequence number to all the replicas. Once an echo is received confirming all the needed copies received the update, the particular update from the pending updates list is removed. Similarly, in LibRe, when an update is applied on a replica copy, the version-id of the update along with the replica id is kept in a *Registry*. Once a confirmation is received from all the replica nodes, the entry for the corresponding data item will be removed from the registry. However, GSP focusses on ordering the write operations, whereas LibRe focusses on reading the value of the recent write.

The PNUTS Database [5] from Yahoo uses a per-record mastership over per-table or per-tablet mastership and forwards all updates of the record to this master in order to provide timeline consistency during read operations. In contrast, LibRe allows any replica to process an update and chooses a registry node per data item in order to identify the most recent version of the data item.

5   Conclusion and Future Works

The work described in this document aims at enhancing the tradeoffs between Consistency, Latency and Availability of an eventually consistent Key-Value store. Our protocol: LibRe prevents the system from forwarding read requests to the replica nodes that contain stale replica for the needed data item. In order to identify replica nodes that contain stale replicas, LibRe uses a monotonically increasing version-id for each data item. The initial implementation of LibRe protocol was developed inside Cassandra NoSql data store. This so-called 'CaLibRe' implementation offers LibRe protocol as an additional consistency option for Cassandra storage system.

The performance of the CaLibRe implementation was benchmarked against the native consistency options of Cassandra using YCSB on a 19 nodes CaLibRe and Cassandra cluster. The performance results prove that CaLibRe provides lower request latency compared to the strong consistency levels offered by Cassandra, combined to a similar number of stale reads. Hence we can safely conclude that using the LibRe protocol gives a new tradeoff between consistency, latency and availability. However, the performance results were not tested under nodes joining or leaving the clusters. During such events, LibRe protocol would experience temporary inconsistency, which has to be studied in the future works.

Additional works are required to optimize the performance of the CaLibRe implementation. Another perspective to this work is to study the influence of the nature of the version-id (timestamp, version vector, vector clocks, ...). Also, evaluating the LibRe performance under a real world use case is considered.

References

1. Abadi, D.J.: Consistency tradeoffs in modern distributed database system design: cap is only part of the story. Computer **45**(2), 37–42 (2012)
2. Agrawal, D., El Abbadi, A.: The generalized tree quorum protocol: an efficient approach for managing replicated data. ACM Trans. Database Syst. **17**(4), 689–717 (1992). http://doi.acm.org/10.1145/146931.146935
3. Burckhardt, S., Leijen, D., Protzenko, J., Fähndrich, M.: Global sequence protocol: a robust abstraction for replicated shared state. Technical report, Microsoft Research (2015). http://research.microsoft.com/apps/pubs/default.aspx?id=240462
4. Chang, F., Dean, J., Ghemawat, S., Hsieh, W.C., Wallach, D.A., Burrows, M., Chandra, T., Fikes, A., Gruber, R.E.: Bigtable: a distributed storage system for structured data. ACM Trans. Comput. Syst. (TOCS) **26**(2), 4 (2008)
5. Cooper, B.F., Ramakrishnan, R., Srivastava, U., Silberstein, A., Bohannon, P., Jacobsen, H.A., Puz, N., Weaver, D., Yerneni, R.: Pnuts: Yahoo!'s hosted data serving platform. Proc. VLDB Endow. **1**(2), 1277–1288 (2008). http://dx.doi.org/10.14778/1454159.1454167
6. Cooper, B.F., Silberstein, A., Tam, E., Ramakrishnan, R., Sears, R.: Benchmarking cloud serving systems with YCSB. In: Proceedings of the 1st ACM Symposium on Cloud Computing, SoCC 2010, pp. 143–154. ACM, New York (2010). http://doi.acm.org/10.1145/1807128.1807152
7. DeCandia, G., Hastorun, D., Jampani, M., Kakulapati, G., Lakshman, A., Pilchin, A., Sivasubramanian, S., Vosshall, P., Vogels, W.: Dynamo: amazon's highly available key-value store. SIGOPS Oper. Syst. Rev. **41**(6), 205–220 (2007). http://doi.acm.org/10.1145/1323293.1294281
8. George, L.: HBase: The Definitive Guide, 1st edn. O'Reilly Media, Sebastopol (2011)
9. Gilbert, S., Lynch, N.: Brewer's conjecture and the feasibility of consistent, available, partition-tolerant web services. SIGACT News **33**(2), 51–59 (2002). http://doi.acm.org/10.1145/564585.564601
10. Hunt, P., Konar, M., Junqueira, F.P., Reed, B.: Zookeeper: wait-free coordination for internet-scale systems. In: Proceedings of the 2010 USENIX Conference on USENIX Annual Technical Conference, USENIXATC 2010, p. 11. USENIX Association, Berkeley (2010). http://dl.acm.org/citation.cfm?id=1855840.1855851

11. Klophaus, R.: Riak core: building distributed applications without shared state. In: ACM SIGPLAN Commercial Users of Functional Programming, CUFP 2010, p. 14:1. ACM, New York (2010). http://doi.acm.org/10.1145/1900160.1900176
12. Kumar, S.P., Chiky, R., Lefebvre, S., Soudan, E.G.: Libre: a consistency protocol for modern storage systems. In: Proceedings of the 6th ACM India Computing Convention, Compute 2013, pp. 8:1–8:9. ACM, New York (2013). http://doi.acm.org/10.1145/2522548.2522605
13. Kumar, S., Lefebvre, S., Chiky, R., Soudan, E.: Evaluating consistency on the fly using YCSB. IWCIM **2014**, 1–6 (2014)
14. Lakshman, A., Malik, P.: Cassandra: a decentralized structured storage system. SIGOPS Oper. Syst. Rev. **44**(2), 35–40 (2010). http://doi.acm.org/10.1145/1773912.1773922
15. Malkhi, D., Reiter, M.: Byzantine quorum systems. In: STOC 1997, pp. 569–578. ACM (1997). http://doi.acm.org/10.1145/258533.258650
16. Malkhi, D., Reiter, M., Wright, R.: Probabilistic quorum systems. In: PODC 1997, pp. 267–273. ACM (1997). http://doi.acm.org/10.1145/259380.259458
17. Shapiro, M., Preguica, N., Baquero, C., Zawirski, M.: A comprehensive study of convergent and commutative replicated data types. RR-7506, INRIA (2011)
18. Naor, M., Wool, A.: The load, capacity, and availability of quorum systems. SIAM J. Comput. **27**(2), 423–447 (1998). http://dx.doi.org/10.1137/S0097539795281232
19. Saito, Y., Shapiro, M.: Optimistic replication. ACM Comput. Surv. **37**(1), 42–81 (2005). http://doi.acm.org/10.1145/1057977.1057980
20. Shvachko, K., Kuang, H., Radia, S., Chansler, R.: The hadoop distributed file system. In: 2010 IEEE 26th Symposium on Mass Storage Systems and Technologies (MSST), pp. 1–10, May 2010
21. Tlili, M., Akbarinia, R., Pacitti, E., Valduriez, P.: Scalable P2P reconciliation infrastructure for collaborative text editing. In: 2010 Second International Conference on Advances in Databases Knowledge and Data Applications (DBKDA), pp. 155–164, April 2010
22. Voldemort, P.: Physical architecture options, April 2015. http://www.project-voldemort.com/voldemort/design.html
23. Vukolic, M.: Remarks: the origin of quorum systems. Bull. EATCS **102**, 109–110 (2010). http://dblp.uni-trier.de/db/journals/eatcs/eatcs102.html
24. Zhang, H., Wen, Y., Xie, H., Yu, N.: Distributed Hash Table - Theory, Platforms and Applications. Springer Briefs in Computer Science. Springer, New York (2013). http://dx.doi.org/10.1007/978-1-4614-9008-1

Special Section Globe 2015 – Query Rewriting and Streaming

QTor: A Flexible Publish/Subscribe Peer-to-Peer Organization Based on Query Rewriting

Sébastien Dufromentel$^{(\boxtimes)}$, Sylvie Cazalens, François Lesueur,
and Philippe Lamarre

CNRS INSA-Lyon, LIRIS, UMR5205, Université de Lyon, 69621 Lyon, France
{sebastien.dufromentel,sylvie.cazalens,
francois.lesueur,philippe.lamarre}@liris.cnrs.fr

Abstract. Peer-to-peer publish/subscribe architectures are an interesting support for scalable distributed data stream applications. Most approaches, often based on brokers, have a static organization which is not much adaptive to different configurations of the participants' capacities. We present *QTor* (*Query Torrent*) a generic organization that enables dynamic adaptation providing a continuum from centralized to fully decentralized solutions. Based on query rewriting and equivalence, *QTor* proposes a definition of *communities* and their relations that decouples the logical and physical aspects of the problem, while efficiently reducing organizational and functional costs.

1 Introduction

Peer-to-peer publish/subscribe organizations are an essential support for scalable distributed data stream applications. A lot of them [5,8] rely on brokers to benefit from efficient local processing optimizations. More recent propositions [6,12] are fully distributed, asking the users to contribute by sharing their results, which requires to take care of their limited resources.

Among several possible overlays to organize the system, query-oriented organizations take advantage of already computed results [2,4,6]. In this field, an approach based on query rewriting [1,9] provides the most general solution that is both generic and language independent and that eases the decoupling of logical and physical layers. In addition, we consider the case of popular queries which, even if their expressions differ, are *equivalent*, which means they give the same results when executed on the same dataset.

We present QTor, an organization driven by the queries in which participants are grouped into *communities*, regarding query equivalence relation. Those communities are independent from each other and autonomous for their local organization. Connections between communities come from query rewritings combined with multiple criteria (computing and networking costs, latency…). This results in a system with low organizational and functional costs that is adaptive to the incoming participants' capacities while preserving a low latency.

© Springer International Publishing Switzerland 2015
Q. Chen et al. (Eds.): DEXA 2015, Part II, LNCS 9262, pp. 507–519, 2015.
DOI: 10.1007/978-3-319-22852-5_41

In Sect. 2, we present some preliminaries and formally define the problem. The main concepts of our approach and the resulting system are described in Sect. 3. Section 4 discusses QTor flexibility. Experimental results of Sect. 5 show its good performances, and draws a comparison with two other approaches. Finally, Sect. 6 presents related work while Sect. 7 concludes and exposes some perspectives.

2   Background and Problem Statement

In this section, we formally define the problem as finding a system on which, for any submitted query, a rewriting graph can be mapped, while complying with the participants' limitations.

2.1   System Organization Based on Query Rewriting

A distributed system is composed of software *participants* which originate some streams, submit users' queries, and bring the system some resources (memory, storage and computational capacities).

Definition 1 (System). *A system is a labelled graph $\langle P, PA \rangle$ such that:*

- *P is a set of participants. For each participant $p \in P$, p.sources is the set of queries representing the data streams p originates and p.queries is the set of queries it submits to the system.*
- *PA is a set of triples $\langle p_i, p_j, Q_{i,j} \rangle$ with $(p_i, p_j) \in P^2$ and with $Q_{i,j}$ non empty finite set of queries representing the data streams p_j gets from p_i.*

The bottom part of Fig. 1 is an example of such system.

In order to organize such a system, relations between queries have to be studied. Two queries q_1 and q_2 are equivalent, noted $q_1 \equiv q_2$, if and only if they provide the same results regardless of the sources' data streams contents. Rewriting query q means finding a query q' equivalent to q but expressed over a set of queries q_i (representing data streams or queries) different from the one used to express q. This is noted $q' = (q \longleftarrow \{q_1, q_2, \ldots, q_n\})$, or simply $q \longleftarrow \{q_1, q_2, \ldots, q_n\}$, when q' is of no need.

Definition 2 (Rewriting schema). *A rewriting schema of a query q_r, is a cycle free, totally connected graph $\langle Q, E \rangle$, with Q a set of queries and E a set of arcs, characterized by $q_r \in Q$ with:*

- *$\forall \{q, q_1, q_2 \ldots q_n\} \in Q^{n+1}$, if $(\{q_1, q_2 \ldots q_n\}, q) \in E$ then there exists a rewriting $q \longleftarrow \{q_1, q_2, \ldots, q_n\}$, and*
- *the in-degree of each node is at most one, and*
- *q_r is the only node with an out-degree equal to zero.*

Despite the number of possible rewritings between queries, a rewriting schema contains at most one possible rewriting for each query and avoids cycles, as the configuration graph used in [6]. An example of rewriting schema is illustrated on the upper part of Fig. 1.

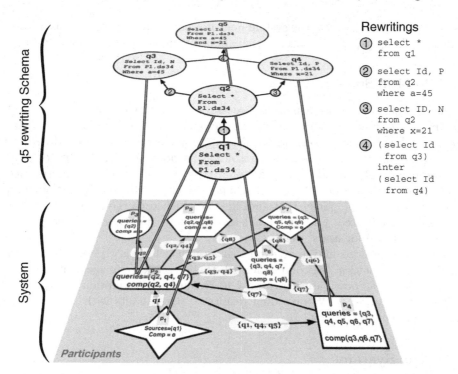

Fig. 1. The system implements a rewriting schema for $q5$.

System: Participant $p1$ is the unique source participant that originates a data stream ($p1.ds34$, represented by $q1$). Participant $p2$ submits queries $q2$, $q4$ and $q7$ to the system. It gets $q1$ from $p1$ and $q7$ from $p4$. It computes $q2$ and $q4$. It serves $q1$ to $p4$; $q2$ to $p3$ and $p5$; $q3$ to $p6$ and $p7$; $q4$ to $p4$, $p5$ and $p6$; $q5$ to $p4$ and $p7$. And so on for other participants.

Rewriting schema for $q5$: $q2$ (`select * from p1.ds34`) is rewritten from $q1$ (`select * from p1.ds34`) using identity (noted rewriting ① on the figure) ; $q3$ and $q4$, are rewritten using $q2$ with a selection and a projection (`select Id, P from q2 where x=21`) ; $q5$ is rewritten as the intersection of $q3$ and $q4$.

Definition 3 (Well founded system). *A system* $\langle P, PA \rangle$ *is well founded if and only if for any query* q *issued by any participant* p *there exists a rewriting schema* $\langle q_r, Q_r, E_r \rangle$ *such that:*

- $q_r = q$ *(syntactic equality), and*
- *there is a mapping* m *from* Q_r *to* P *such that*
  - $m(q_r) = p$, *and*
  - $\forall(\{q_1, q_2 \ldots q_n\}, q_g) \in E_r,\ \forall q_i \in \{q_1, q_2 \ldots q_n\},$
    $(m(q_i) = m(q_g)) \lor (\langle m(q_i), m(q_g), Q_{i,g}\rangle \in PA \land q_i \in Q_{i,g})$, *and*
  - $\forall q_i \in Q_r,$ *if* q_i *is a leaf then* $q_i \in m(q_i).sources.$

The use of rewriting schemas ensures each participant can compute the results it is interested in using the streams it gets. Furthermore, because the rewriting schema is acyclic, the system does not embed any cycle. This is important to avoid loops, which make participants waiting for each others' results.

2.2 Problem Statement

To further define the problem it is necessary to pay attention to real world constraints. *Physical constraints:* the participants have limited computational and network capacities. Overload of each capacity has to be avoided. Moreover, the solution should consume as few resources as possible considering not only its functional cost, but also organizational costs (setting up and maintenance costs). *Social constraints:* the solution has to comply with each participant's resource management policy that specifies to which uses the resources that it provides are dedicated. Last but not least, in many applications, latency is of prime importance for users' welfare.

The problem is to find a *well founded system*, which is (i) consistent with participants capacities, (ii) compliant with their resource management policies, (iii) as efficient as possible with respect to resources consumption (computing, network, energy), and (iv) providing results to participants with a latency as small as possible.

3 QTor System

We present *QTor* (*Query Torrent*), a generic well founded system which addresses the problem. We focus on its organization principles, which ensure the system flexibility avoiding details of the participants' inner management. Our approach is based on the notion of community, which is in charge of the computation of a set of equivalent queries and of the resulting data stream. A graph of those communities provides the overall organization.

3.1 Graph of Communities

We define a **computing unit** as a software component which is created by a participant p to declare some interest in a query q, formally noted $u = \langle q, p \rangle$.

A participant creates a unit for each expressed or computed query. In addition, each unit requires sufficient network ressources to get its data streams in and to transmit its data stream output to at least another unit.

Definition 4 (Community). *A community C is a couple $C = \langle q, U \rangle$ where q is a query and U is a set of computing units such that $\forall u \in U$, $u.q \equiv q$.*

Notice that we do not require all units with equivalent queries to be part of the same community, due to pragmatic reasons. Indeed, all equivalences are not always provable in a reasonable time.

As a community C is defined by a query $C.q$, the semantic overlay can be structured taking advantage of rewriting technics.

Definition 5 (Graph of Communities). *A Graph of Communities is a cycle free hyper-graph* $\phi = \langle \mathscr{C}, RR \rangle$, *with*

- $\mathscr{C}$ *is a finite set of communities where*
  $\forall (C_i, C_j) \in \mathscr{C}^2$, *if* $i \neq j$ *then* $C_i \cap C_j = \emptyset$.
- *RR is a set of hyper-arcs such that:*
  - *if* $(\{C_1, C_2 \ldots, C_n\}, C) \in RR$ *then*
    there exists a rewriting $C.q \longleftarrow \{C_1.q, C_2.q, \ldots, C_n.q\}$, *and*
  - *the in-degree of any source-community is equal to zero; the in-degree of any other community is equal to one.*

As each community is dedicated to a query (or a set of equivalent queries), the graph of communities describes and implements a rewriting schema.

The use of communities has three major effects. First, there are often much less communities than units, due to the popularity of queries. This reduces the graph size. Second, it avoids wasting time and energy to compute several times the same rewritings for each equivalent query (and it also avoids cycles between those equivalent queries). Third, it splits the physical organization problem into several smaller, independent tasks, while enabling units to collaborate when they compute complex queries which would be out of reach of a single participant.

Materializing this theoretical model into a functional system requires to answer some concrete questions like: "how does a community get the data streams it needs?"; "how to disseminate the data stream corresponding to the result of a query?"; and "how to compute a query within a community?". Without loss of generality, we simplify this latter point by assuming that the computation is ensured by a single unit with enough resources. Indeed, introducing distributed computing within a community is an orthogonal problem which does not impact the other points.

3.2   Building the Whole System

Reorganizing the whole system at each insertion, without considering the existing system, may be too expensive to implement. A periodic reorganization as in Delta [6] is possible, but it is difficult to define this periodicity and how to handle insertion of new participants during the period. This is why we choose to rely on an incremental approach. For example, when a participant expresses a query, it searches for an existing community working on an equivalent query, through a tracker that knows all the communities and their organization. If one is found, the participant creates a unit which joins it. Otherwise, a new community is created: a cost model, considering processing, network loads and latency is used to select the best rewriting schema among those computed using the existing communities already organized according to rewriting schemas. Then, the new community has to be connected with the others. Notice that, contrary to a data integration problem, in this case, it is not necessary to compute all the possible rewritings [9].

3.3   Participants' Spreading Over Communities

Given a community C, associated with query q and taking its data stream input from another community C', we define two mechanisms enabling a participant to create a new unit to contribute to an additional community. Notice that these mechanisms keep the communities disjoint and that they can be generalized to a rewriting involving several parent communities.

SPC: participant's Spreading to a Parent Community. A participant p_1 having a unit u_1 in community C creates and introduces a new unit u'_1 into community C' which enables p_1 to obtain the data stream from C'. From this, assuming u_1 (i.e. p_1) is powerful enough, it computes the rewriting of q defined by the rewriting schema and spreads the resulting data stream to other units of community C.

SCC: participant's Spreading to a Child Community. A participant p_2 having a unit u'_2 in the community C' creates and introduces a new unit u_2 into community C. u_2 has to be powerful enough to compute the rewriting query of q defined by the rewriting schema. Then, it spreads the resulting data stream to other units of community C.

In both cases, the new unit participates to its community tasks as any unit does. Depending on the situation, one solution may be better than the other. For example, due to a high selectivity of query q, the data stream of community C can have a lower throughput than the one of C'. If in addition, the rewriting of query q is easy to compute, the SCC mechanism becomes particularly interesting.

Choosing among these two mechanisms has to be done when a participant joins the system with a query that requires to create a new community. This choice is questioned at the departure of units concerned by one of these mechanisms or at the arrival of a new unit that is more powerful than those implied by the mechanism. This choice depends on specific situations, so this decision is left to participants, who will decide to use one or the other mechanism. Notice that communication between two communities is thus ensured by a same participant, involving a unit in each of them. Hence, the stream exchange is implemented without consuming ressources theoretically allocated to the community C', which would create dependencies between communities. As a consequence, each community is autonomous to organize itself.

3.4   Inner Organization of a Community

Due to the previous choices, each community can organize itself autonomously. The SCC and SPC mechanisms and the assumption that a query computation is done by a single unit define the way both data acquisition and query computation are ensured. The computing unit is the one belonging to the participant in charge of the incoming communication mechanism.

Results dissemination within a community is formalized as a diffusion tree with minimum latency. Under the simplifying assumption that the network is homogeneous we search for a minimal depth spanning tree rooted by the computing unit.

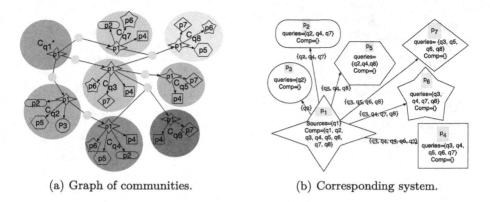

(a) Graph of communities.   (b) Corresponding system.

Fig. 2. Powerful data stream producer that accepts to compute for others.

4   QTor Flexibility

An interesting property of our proposal is its ability to adapt to different situations. We illustrate this point with two scenarios involving several subscribers and a single source $p1$ that provides a data stream represented by query $q1$ (same elements as in Fig. 1). In our figures, a participant is represented by a specific form (e.g. circle, star...) which is also used for its units, with a smaller size.

In Fig. 2, $p1$ is powerful enough to compute all queries and to disseminate results. By using several times the SCC mechanism, it creates a unit in each community leading to Fig. 2(a). Thus $p1$ computes all queries and broadcasts results. The other participants get their results directly from the data producer without any effort. Figure 2(b) describes the resulting system which turns out to be a classical publish/subscribe system.

In Fig. 3, the producer can send its data stream to many units but has no computing capability. Hopefully, some participants are powerful enough: $p2$ and $p6$ use SPC to get data stream from $p1$. Participant $p6$ uses also SCC to create units in community C_{q5}. Participant $p2$ already has units in C_{q2} and C_{q4} so using SPC or SCC does not change anything. As shown on Fig. 3(b) participants $p2$ and $p6$ bear all the queries computation and most of the diffusion.

The limitation of the number of connections provided by data sources is a problem that dissemination systems are facing. A benefit of the SPC mechanism is to enable participants to contribute to the diffusion of the source streams, by creating units in the source community (as $p2$ and $p6$ do in the previous scenario).

Based on fairly simple concepts such as graph of communities and the creation of units to provide data streams to communities, the formalism captures systems previously considered as very different. As a consequence, our approach is very flexible providing a continuum from centralized to fully decentralized solutions.

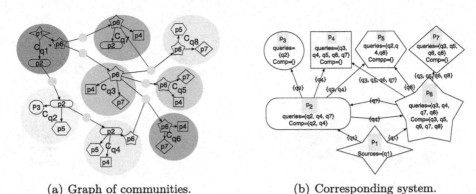

(a) Graph of communities.   (b) Corresponding system.

As in Figure 1, $p1$ originates the data stream represented $q1$

Fig. 3. A data stream producer with no computation capacity, with altruist consumers.

5   Experiments

We perform several experimentations on a java simulator using PeerSim [7]. We focus on the efficiency and the adaptability of the QTor organization. Thus without loss of generality, we consider participants expressing boolean queries (using both conjunctive and disjunctive operators) over a single data source, which publishes some randomly generated tuples. Participants capacities follow a Poisson law around 30. We use [4] and [6] as baselines.

Semantic peer-to-peer overlays [4] (called later "SPO") is a containment-based approach that places participants in a common spanning tree. Query equivalence is used to guide the organization (participants sharing equivalent queries are linked together, as this is the best way to reduce costs), but without any explicit notion of community: placements are computed considering the whole graph of participants. The original approach does not take care of the participants' capacities, considering a single node can send data to all the others. We developed an algorithm to fix this problem, and did not compare the aspects that were mainly related to this algorithm (Figs. 4(a), 5, and 7(b)).

Delta [6] is a recent proposition in which each participant is used as a view in a rewriting system. The system organization is obtained in five steps: an embedding graph of queries is computed, from which cycles are removed using a generic algorithm. It is used to build a rewriting graph. An ILP solver is then called to find a minimal cost solution under capacity constraints. Then, a last algorithm, called LOGA, reduces the distance between participants and the source to decrease latency (which is a non linear problem) avoiding to connect them directly to the source. However, it may choose rewritings that need more processing than the one chosen by the ILP solver.

Figure 4 (a) illustrates QTor scalability. We consider up to 15.000 participants with popular queries following a Zipf distribution. Neither Delta nor SPO are shown here. Indeed, this kind of setup results in huge communities grouping more

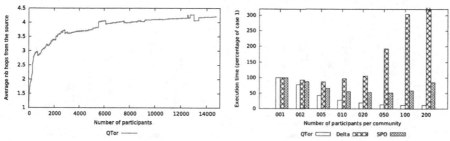

(a) Latency evolution in QTor, size of communities following a Zipf distribution

(b) Time spent to organize all the 2.000 participants, same size for all communities

Fig. 4. Scalability of the organizations

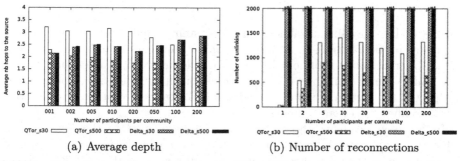

(a) Average depth

(b) Number of reconnections

2.000 participants, variable source capacity (30 or 500), same size for all communities.

Fig. 5. System organisation

than 500 participants. Delta (especially the LOGA part) becomes very slow with such communities and it is too long to obtain results. SPO is not relevant.

In the following, to conduct comparisons, we consider smaller systems with up to 2.000 participants and a variable query popularity, i.e. community size.

In Fig. 5 we evaluate the ability of the systems to adapt to different source capacities. We consider two cases: One source with capacities similar to other participants' with 30 output connexions. Another one with 500 output connexions. Figure 5(a) shows the average depth of participants in the final system which is related to the latency of obtaining results for participants. Delta does not distinguish between those two situations. Indeed, participants are linked to the source if and only if they have no other possibility. This means that Delta may overload or under use the source. In QTor, the source is never overloaded (other participants take extra charge using SPC mechanism) and, as the other participants, a source can deploy units and participate to many communities (up to its capacity limit, using SCC mechanism) which is a way to reduce latency. In Fig. 5(b) we evaluate the stability of the organization and so the cost of the maintenance of the system. In Delta, it is recommended to run the reorganization algorithm periodically. In its current form, Delta does not consider the existing

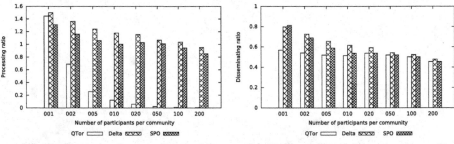

(a) Comparison of processing ratio   (b) Comparison of disseminating ratio

2.000 participants, same size for all communities.

Fig. 6. Evaluation of functional costs

system, and it may impact almost all the system graph. So, even with a low frequency, participants' re-connexions are more numerous than in our solution, and a powerful source is of no help. In QTor, the insertion of a new participant may lead to some community reorganizations, but its impacts are circumscribed to a part of the system graph. Furthermore, QTor takes advantage of a powerful source to limit the organization cost.

In Fig. 6 we evaluate the functional costs associated to the organization (which is independent from the participants' fan-out limitation). (a) shows the processing ratio (which means, the average number of analysis performed per produced tuple), while (b) shows the disseminating ratio (the average number of tuples delivered by a participant per number of tuples produced by the source). SPO has good results, but their limitation to containment relations (which means one view per rewriting) makes them loose some opportunities. Delta's results in low redundancy are higher than QTor's one for the same reasons its latency is lower: algorithm LOGA makes the functional costs increase. In QTor, communities avoid this fact.

In Fig. 7 we study two limit cases: (i) the queries of the communities are separate, i.e. they cannot be rewritten one for the others, and (ii) all the participants are grouped into a sigle community, i.e. they all query for same results. In the first case, QTor is the only proposition providing a mechanism that avoids source's overwhelming. In the second case, the QTor organization has an efficient inner organization, while Delta's one is not really adapted.

6   Related Work

Some distributed Publish/Subscribe systems, with brokers such as SemCast [8], or deployed over all the participants in a full peer-to-peer organization such as DPS [2], deploy complex organizations having several diffusion tree for the different predicates or several indexing levels. DHTrie [12] uses an extended Distributed Hashtable to register subscriptions, which, as often, highly depends on the chosen structure (query language, data schema). This gives organizations

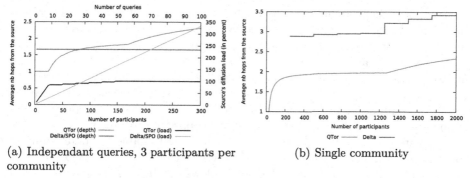

(a) Independant queries, 3 participants per community

(b) Single community

Source capacity: 30 in both cases.

Fig. 7. Limit cases for the organization

that are hard to adapt, while the use of rewritings allows a generic organization that can efficiently be adapted.

Some other distributed systems, like CDNs [10] or Multicast-inspired systems (for instance, SplitStream [3]), are based on the possibility for some participants to obtain all the data produced by the source, and to re-send them unchanged to the others. Such approaches lead to huge, redundant processing, as each participant has to work by its own, but may be used to design efficient diffusion scheme inside a QTor community.

A lot of propositions deal with local multi-query optimizations. For instance, in NiagaraCQ [5], grouping the query by their "signatures" based on the physical operators allow to limit the I/O. In RoSeS [11], a local query graph is optimized by factorization. Those kind of propositions alone do not cope with huge number of participants, which require a network-wide organization, but may, in a QTor system, efficiently be used to decrease the processing load of participants that compute for several communities.

To the best of our knowledge, the closest propositions to ours are [4] and Delta [6]. SPO [4] is a very dynamic and incremental solution, but it is limited to containment and does not address the participants' capacity limitations. Delta [6] is based on query rewriting but proposes a global, non incremental, system reorganization with higher organizational and functional costs without decisive advantage on latency. Furthermore, it is not adapted at all to popular queries. These systems don't take full advantage of powerful sources/participants and may overwhelm sources which is a problem our solution avoids. Finally, the notion of query-based community introduces an intermediate abstraction which favors scalability in presence of popular queries.

7   Conclusion and Perspectives

This paper shows that the QTor approach provides an efficient organization, taking care of several aspects to result in a reliable system: capacity limitations are never overwhelmed and participants having high capacity can efficiently help the others. The use of communities enables to highly reduce redundant works in case of popular queries, while simplifying the organization.

Those results are even more interesting considering there still are possible improvements: in this paper, collaboration inside a community is limited to sharing results. When the chosen rewritten query is still unreachable for a single participant, our model enables distributed processing, shared by participants having similar interests. Moreover, it may be possible to create communities that do not correspond to any expressed queries, but enable to share common works for several existing queries, as in the way locally used in RoSeS [11].

Decoupling logical and physical aspects allows to efficiently take care of both of them, without concurrent objectives, in a flexible, reliable organization. Some aspects still have to be considered, but the already obtained effects allow to consider that the QTor approach promises interesting surroundings.

Acknowledgements. This work has been partially funded by the French ANR Socio-Plug project under grant No. ANR-13-INFR-0003.

References

1. Abiteboul, S., Duschka, O.M.: Complexity of answering queries using materialized views. In: Proceedings of the Seventeenth ACM SIGACT-SIGMOD-SIGART Symposium on Principles of Database Systems, PODS 1998, pp. 254–263. ACM, New York (1998)
2. Anceaume, E., Gradinariu, M., Datta, A.K., Simon, G., Virgillito, A.: A semantic overlay for self- peer-to-peer publish/subscribe. In: 26th IEEE International Conference on Distributed Computing Systems, 2006, ICDCS 2006, pp. 22–22 (2006)
3. Castro, M., Druschel, P., Kermarrec, A.-M., Nandi, A., Rowstron, A., Singh, A.: Splitstream: high-bandwidth multicast in cooperative. In: SOSP 2003: Proceedings of the Nineteenth ACM Symposium on Operating Systems Principles. ACM Request Permissions, December 2003
4. Chand, R., Felber, P.: Semantic peer-to-peer overlays for publish/subscribe networks. In: Cunha, J.C., Medeiros, P.D. (eds.) Euro-Par 2005. LNCS, vol. 3648, pp. 1194–1204. Springer, Heidelberg (2005)
5. Chen, J., DeWitt, D.J., Tian, F., Wang, Y.: NiagaraCQ: a scalable continuous query system for internet databases. In: Proceedings of the 2000 ACM SIGMOD International Conference on Management of Data, SIGMOD 2000, pp. 379–390. ACM, New York (2000)
6. Karanasos, K., Katsifodimos, A., Manolescu, I.: Delta: scalable data dissemination under capacity constraints. PVLDB 7(4), 217–228 (2013)
7. Montresor, A., Jelasity, M.: PeerSim: a scalable P2P simulator. In: Proceedings of the 9th International Conference on Peer-to-Peer (P2P 2009), Seattle, WA, pp. 99–100, September 2009
8. Papaemmanouil, O., Cetintemel, U.: Semcast: semantic multicast for content-based data dissemination. In: Proceedings of the 21st International Conference on Data Engineering, 2005, ICDE 2005, pp. 242–253 (2005)
9. Pottinger, R., Halevy, A.: Minicon: a scalable algorithm for answering queries using views. VLDB J. 10(2–3), 182–198 (2001)

10. Ouveysib, I., Bektasa, T., Oguza, O.: Designing cost-effective content distribution networks. Comput. Oper. Res. **34**, 2436–2449 (2005)
11. Creus Tomàs, J., Amann, B., Travers, N., Vodislav, D.: RoSeS: a continuous content-based query engine for RSS feeds. In: Hameurlain, A., Liddle, S.W., Schewe, K.-D., Zhou, X. (eds.) DEXA 2011, Part II. LNCS, vol. 6861, pp. 203–218. Springer, Heidelberg (2011)
12. Tryfonopoulos, C., Idreos, S., Koubarakis, M., Raftopoulou, P.: Distributed large-scale information filtering. In: Hameurlain, A., Küng, J., Wagner, R. (eds.) TLDKS XIII 2014. LNCS, vol. 8420, pp. 87–116. Springer, Heidelberg (2014)

Model for Performance Analysis of Distributed Stream Processing Applications

Filip Nalepa[✉], Michal Batko, and Pavel Zezula

Faculty of Informatics, Masaryk University, Brno, Czech Republic
f.nalepa@gmail.com

Abstract. Nowadays, a lot of data is produced every second and it needs to be processed immediately. Processing such unbounded streams of data is often applied in a distributed environment in order to achieve high throughput. There is a challenge to predict the performance-related characteristics of such applications. Knowledge of these properties is essential for decisions about the amount of needed computational resources, how the computations should be spread in the distributed environment, etc.

In this paper, we propose a model to represent such streaming applications with the respect to their performance related properties. We present a conversion of the model to Colored Petri Nets (CPNs) which is used for performance analysis of the original application. The behavior of the proposed model and its conversion to the CPNs is validated through experiments. Our prediction was able to achieve nearly 100% precise maximum delays of real stream processing applications.

Keywords: Stream processing · Performance analysis · Data stream model

1 Introduction

1.1 Motivation

Nowadays, a lot of data is produced every second. Also such a huge amount of data often needs to be reprocessed later on. There are two basic ways of doing so. The data can be processed in batches, i.e., at first the data is stored, and subsequently the whole dataset is processed. However, there are scenarios when the data needs to be processed immediately as soon as it is acquired, e.g., analyzing surveillance video footage. For such types of applications, so called stream processing is appropriate.

The base of a stream processing application is a set of tasks (atomic computation units) that are linked by precedence constraints. This is often called workflow [2]. The tasks are used to process data streams (potentially infinite sequences of data items) which enter the application. In order to achieve high throughput, the applications can be deployed in a distributed environment. In such cases, the tasks are placed on individual computational resources, and the

© Springer International Publishing Switzerland 2015
Q. Chen et al. (Eds.): DEXA 2015, Part II, LNCS 9262, pp. 520–533, 2015.
DOI: 10.1007/978-3-319-22852-5_42

data streams are sent between the resources so that the required operations can be evaluated.

For example, let there be a stream of images uploaded to a social network. Suppose we need to extract visual features of the images, classify the images, annotate them, or detect faces in them. Some of these tasks can be performed independently of each other, but for some of them, precedence constraints have to be employed. E.g., features of an image have to be extracted before the image is annotated.

When building streaming applications, performance metrics (such as a delay or a throughput) of the final system are often a big concern. For instance, it may be required that each data item can be fully processed in five seconds, i.e., the maximum acceptable delay of the whole process is five seconds. Whether or not the system is able to meet such criteria is not always obvious for complex applications. The performance metrics are heavily dependent on the number of allocated resources and on the way how the tasks are spread throughout the network.

It may be too late to start measuring the performance characteristics once the application has been deployed in a distributed environment since it can be difficult or expensive to deal with the load balancing then. A more appropriate time to be interested in the performance is before the deployment when decisions about needed resources and placement of the tasks on them are made.

Therefore, it is important to be able to carry out performance analysis of streaming applications without the need to deploy them. The performance analysis can serve to derive characteristics of different settings of the system, e.g., different numbers of used resources, different task placements, or different arrival rates of data items in the streams. Considering all these aspects is a key to successful planning of distributed stream processing applications.

1.2 Objectives

In this paper, we propose a model of stream processing applications which captures important properties useful for performance analysis of the applications. We focus mainly on multimedia data streams and on detecting various events in them.

Deriving performance metrics of systems that work with multimedia data brings challenges that do not appear when working with textual data. Among specific characteristics of multimedia data are largely variable sizes of data items (e.g., users upload images of different sizes) and variable lengths of processing (e.g., higher processing cost of large images).

Once an application is represented by the proposed model, we proceed with the performance analysis. For this purpose, we describe a mechanism how to transform our model into Colored Petri Nets [8]. After that, simulations of the Petri Net are run to derive performance qualities of the application.

1.3   Related Work

A lot of work has been devoted to models and performance analysis of distributed computing systems.

Targeted at streaming applications in embedded systems, several formalisms have been proposed [3–5]. They all work with irregular arrival patterns of data streams which cannot be described using standard periodic or sporadic event models. They make use of the arrival function for a description of a variable number of data items that can arrive to a particular component of the system each time unit. This approach allows to define a rich collection of arrival sequences.

In [3], authors extend Real-Time Calculus-based functional modeling which avoids explicit state modeling, and hence the state-space explosion problem which other approaches may suffer from. They also provide techniques for analysis of the modeled systems.

A new task structure is proposed in [5] to accurately model the software structures of stream processing applications such as conditional branches and different end-to-end deadlines for different types of input data items.

In [4], a new model called Event Count Automata is presented which captures the timing properties of irregular streams. The automata are able to model a wide variety of protocols and scheduling policies. A global model of a complex network of automata leads to the state explosion problem.

The previous approaches are intended for multiprocessor architectures. Therefore they do not consider network operations such as data transfers between resources. The analytical methods focus on predicting maximal/minimal bounds of the systems whereas we want to be able to analyze also expected (most probable) behavior of the applications. The approaches do not provide automatic support for features typical for distributed systems, e.g. task replication. In addition, each task is supposed to emit a new data item for each processed data item. This is a limiting factor since we work with event detecting tasks which emit new data only if an event is detected.

In [2], an extensive survey of models and algorithms for workflow scheduling is given. They organize various characteristics of the applications into three components: the workflow model, the system model, and the performance model.

Another survey of workflow models is carried out in [10]. They propose several novel taxonomies of the workflow scheduling problem, considering five facets: workflow model, scheduling criteria, scheduling process, resource model, and task model.

Both surveys provide a general structured look on the known results in the area of workflow modelling and scheduling. They do not mention any approaches to handle variable processing costs of a single task nor variable data sizes output by a single task. Also, the tasks are assumed to emit a new data item for each processed data item. In our research, we focus on a specific area of workflow systems (event detection in multimedia streams) whose characteristics are not completely dealt with by the current approaches.

Petri Nets and their extensions have been widely used as tools for analysis of distributed systems. In [9], Queueing Petri Nets are used to estimate throughput in computer networks of modern data centers. Colored Petri Nets are utilised for analysis of workflow models in [4, 7].

In summary, there are both general and specific models and methods for performance analysis of stream processing systems. However, we see a lack of approaches depicting specific features of distributed streaming applications which deal with multimedia data (e.g., variable processing costs).

2 Model

In this section, we present a model of stream processing applications adopted to systems working with multimedia data. The model is focused on those aspects of tasks (atomic computation units), data, and underlying network of resources which enable us to derive performance qualities of the applications.

The whole model consists of three different perspectives on the applications (partly following [2]). The workflow model describes streams of data and tasks which process the streams. The system model is used to characterize the infrastructure of the network of available computational resources. The deployment model puts both models together and represents mapping of individual tasks to computational resources.

Here is a list of the model features:

- Workflow model
  - Workflow graph (task dependencies)
  - Processing cost (the cost to process data items)
  - Output frequency (how frequently new data items are output)
  - Data size (the size of generated data items)
- System model
  - Topology of the underlying network (resource connectivity)
  - Resources (computational power)
  - Connections (bandwidth)
- Deployment model
  - Deployment graph (task to resource mapping).

2.1 Workflow Model

The workflow model is based on a directed acyclic graph where nodes represent tasks, and edges show flow of data streams between the tasks. Nodes with no income edges represent entry points of the streams to the application. An example of such a graph is in Fig. 1a.

We consider the following aspects of the tasks and the streams which are crucial for performance analysis. We need to know the processing costs of individual data items at each task, i.e., how much computational resource is required. Other important aspects are sizes of data items transferred between the tasks and also how frequently the data is transferred (i.e., how often the tasks output data to their successors). Let us have a deeper look at all these properties.

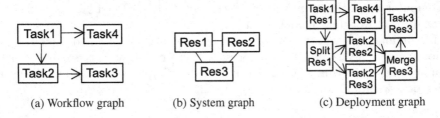

(a) Workflow graph   (b) System graph   (c) Deployment graph

Fig. 1. Graph examples of workflow and system models, and a corresponding deployment model

Processing Cost. In workflow models, the cost to process a single data item is quite frequently represented using a single value [2]. It is very convenient to compute with a single value, and its sufficient in some scenarios. However, it does not reflect the reality correctly when the processing costs can vary a lot. If just average processing costs are considered, it is not possible to model the worst case delays. If just the highest costs are taken into consideration, the results of analysis are too pessimistic since it does not take such a long predicted time to process every data item.

For the reasons above, we decided to adapt the arrival function for irregular streams presented in [3]. To capture the variability in processing costs, we define the highest bound of processing cost cumulatively for individual lengths of data sequences. Formally, $cost(\Delta) = x$ where x is the highest possible number of processing units (e.g., CPU cycles) which are needed to process any sequence of data items of length Δ.

For instance, let $cost(1) = 10$, $cost(2) = 20$, $cost(3) = 30$, $cost(4) = 35$. It means that it costs up to 10 units to process a single data item; it costs up to 20 units to process any sequence of two data items, etc. In this way, we can express that there can be a sequence of three data items which are all "hard to process" (cost of 10 units). But if we consider any data sequence of four data items, there always has to be an item whose cost is less than 10 units.

Analogically, the cost function can be defined to set the minimal limits.

In addition to the maximal and minimal limits defined above, also the probability density cost function [6] is specified so that it is possible to work with the distribution of possible costs during performance analysis.

To sum it up, the maximal and minimal cumulative bounds are useful for deriving maximal and minimal performance qualities. When the bounds are combined with the probability density function, we can also predict the expected (the most probable) performance characteristics.

Data Size. When working with data of large sizes (such as multimedia), it is important to consider how much data is transferred through the network since it influences the metrics of the performance analysis.

For the data size modelling, the same principle can be used as for the processing cost. I.e., $size(\Delta) = x$ where x is the highest possible overall size in bytes of any sequence of data items of length Δ transferred between two given tasks.

Again, this can be defined analogically for minimal size limits. In addition, the cumulative functions are complemented by the probability density function so that the distribution of data sizes can be considered during performance analysis.

Output Frequency. In the previous, we have defined processing costs of data items and sizes of transferred data. Finally, we need to express how frequently the data is emitted from a task and thus transferred between two tasks.

The output frequency is specified for each output stream of a task. There are two possibilities how the output frequency can be specified. One is applied when the rate of output data items is based on the number of processed data items, e.g., a new data item is emitted for each processed data item. The other option is used to specify time dependent output streams. This one is especially important for tasks representing entry points to the application; e.g., one could specify that a new data item is generated every two seconds.

Let us look in detail at the option based on the number of processed data items. For some tasks, the output rate is not fixed, e.g., a face detection task may output new data items only if some face is detected. To specify such variability, we can make use of the same principle as we do for processing cost and data size: $output(\Delta) = x$ where x is the highest possible number of data items output per any sequence of Δ processed data items. The minimal limits are defined analogically. In addition, we define the probability density function of a number of data items which are output per one processed data item.

The output rates of time dependent streams, similarly to the previous option, are not always fixed. For instance, images uploaded to a social network enter the system in irregular intervals. Therefore we define maximal (minimal) numbers of output data items: $output(\Delta) = x$ where x is the highest (lowest) possible number of data items output per any time interval of Δ time units. To work with the probabilities of output data items, the maximal and minimal limits are complemented by the probability density function of a number of data items output per a single time unit.

2.2 System Model

The system model represents the infrastructure of the network of available computational resources. Only two components of this model are differentiated: computational resources and connections between them. The topology of the network is modeled by an undirected graph where nodes represent computational resources, and edges show their connections via the computer network (see Fig. 1b). For the sake of simplicity the resources are defined just by the amount of processing units they provide per time unit (e.g., CPU cycles per second), and the connections are described just by their bandwidth (i.e., bytes per second).

2.3 Deployment Model

Finally, we have to specify which task is placed on which resource. This can be represented as a directed acyclic graph. Each node is of the form (t, r) where t

is a task and r is a resource where the task is placed. The edges represent flow of the data. Each task has to be placed on at least one resource. For the cases when a task is put on more than one resource, we define two special types of nodes: $(split, r)$ and $(merge, r)$.

$(split, r)$ splits one stream into two or more partial streams. Every data item which arrives at the split node is sent to exactly one partial stream. Each partial stream S is assigned a cumulative function $split(\Delta) = x$ where x is the maximal number of data items sent to the stream S per any sequence of data items of length Δ output by the split node. This function enables to specify occasional bursts of data items sent to a single partial stream.

$(merge, r)$ merges two or more streams output by different instances of the same task. That is the results from the multiple input streams are sequentially multiplexed into the single output stream.

See an example of the deployment graph in Fig. 1c.

3   Performance Analysis

As soon as an application is described using the proposed model, we may proceed to the performance analysis.

One of the possible ways to carry out the performance analysis would be to develop a new analytical method suited directly for the presented model. However, we decided to go in a different direction. The model is converted to an existing well-defined model for which analytical methods already exist. Specifically, all the aspects of the model are represented as a Colored Petri Net [8] which is an extension of the standard Petri Nets.

Once the Petri Net is created, it can be analyzed using state space exploration techniques. If a technique for a full space exploration is used, the state space explosion problem is likely to emerge [4]. Therefore, we focus on simulation based techniques in our experiments for the sake of efficiency.

3.1   Colored Petri Nets

Colored Petri Nets (CPNs) [8] is a discrete-event modelling language combining Petri nets with the functional programming language Standard ML. Petri nets provide basic primitives for modelling concurrency, communication, and synchronisation; Standard ML provides primitives for the definition of data types and describing data manipulation. CPNs allow to model a system as a set of modules; it also has a support for time representation in the modelled systems.

3.2   Model to CPN

In this part of the paper, we show how an application represented by the proposed model can be described as a Colored Petri Net.

The final CPN is built gradually by creating small components and joining them into bigger ones. The structure of the components is depicted in Fig. 2a.

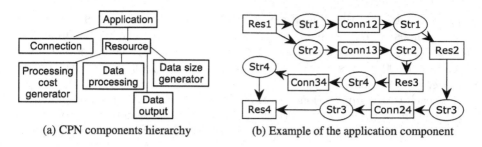

(a) CPN components hierarchy   (b) Example of the application component

Fig. 2. Model to CPN

At the top of the hierarchy, there is a view of the whole application. On this level, it can be observed how the streams are sent through connections between resources (see Fig. 2b). The places are used to keep data items of the streams; the transitions represent resource and connection components.

The connection component is used to simulate transfer of data items between two resources. Its behaviour is based on the properties defined in the system model.

The resource components are made of several smaller components covering the properties defined for the workflow, system and deployment models. There is a component generating processing costs which are based on the cost functions defined earlier. The data processing component simulates the processing of individual data items with respect to the computational power defined in the system model. The data output component decides when new data items should be output based on the frequency functions defined in the workflow model. Finally, the component for data size generation is used to assign data size for newly created data items. The size is based on the size functions defined in the workflow model.

The final CPN captures all the aspects of the model presented in Sect. 2. As of now, the conversion of the model to CPN is done manually. In future, we intend to make the conversion automatic.

4   Experiments

To validate the proposed model and its representation in CPNs, two experiments were conducted. For each of them, a stream processing application was created and deployed in a distributed environment. During the run of the applications, performance related characteristics were gathered. Also the applications were modelled using the proposed approach and subsequently converted to a CPN. After that, simulations of the CPNs were run, and performance related properties were derived. Finally, the results were compared.

Both the applications consume an image stream, detect faces in the images, and search for similar images in the Profiset data collection[1]. The applications were implemented using the Apache Storm technology [1] and deployed as a

---

[1] http://mufin.fi.muni.cz/profiset/.

Storm topology in a distributed environment. The Storm cluster is run on virtual machines with 4 available CPUs and 4 GB RAM. The virtual machines are managed by OpenNebula[2] cloud manager. To generate the image stream, a dataset of 1,000 images is used which are repeated in cycles to get an infinite sequence of data items.

For each of the applications, we show its workflow, system and deployment model. Then we present the performance characteristics derived through the simulations and the statistics measured during the run of the Storm topology.

4.1   Experiment #1

First, we describe the workflow model of the application. In Fig. 4a, we can see the workflow graph. The *image generator* serves as the entry point of the stream to the application. The *face detector* task detects faces in each image and sends the result to the *face event detector* if any faces are found. The *feature extractor* task is used to extract features of the images which are then sent to the *find similar* task to search for similar images in the Profiset database. If any similar images are found, the information is sent to the *similar event detector*. The *face* and *similar event detectors* save the information about the detected events to a local disk.

The processing cost, data size, and output frequency functions were derived based on our sample image stream.

In Fig. 3, there are shown parts of the minimal and maximal cumulative cost functions for the *face detector*. It can be observed that the processing cost can vary a lot. Similar variability was observed also for the *feature extractor* and the *find similar* task. Since the stream consists of images of different sizes, the data size bounds are similar too.

Let us have a look at the output frequencies of the tasks. The *image generator* outputs a new image every 1,850 ms. The *face detector* outputs a new data item only if any faces are detected. For a sample of the image stream, it was measured there are 57 to 78 images containing faces in every sequence of 100 images. The *feature extractor* outputs one new data item for every processed image. The *find*

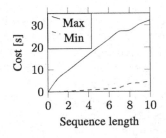

Fig. 3. Face detector processing cost bounds

---

[2] http://opennebula.org/.

similar task outputs a data item only if any similar images are found. Using a sample of the image stream, similar images are found for up to 9 in every sequence of 100 data items. All these output patterns can be represented in the terms of the output frequency functions as defined in Sect. 2.

Now, we can proceed to the system model. In Fig. 4b, there is the infrastructure of the computational resources and connections between them. The resources *Header*, *Worker1* and *Worker2* form the Storm cluster. There is also the *SimEngine* resource where the "find similar" service resides.

Figure 5 shows the deployment graph of the application. The *image generator* is placed on *Header* where the dataset is present. The *face detector* is duplicated and placed on *Header* and *Worker2* so that the images can be processed in parallel. The split node distributes the images to both instances of the detector equally. The *face event detector* receives merged face stream on *Worker2*. The *feature extractor* resides on *Worker1* and communicates with the *find similar* task on *SimEngine*. The *similar event detector* stores the results of the events on *Worker1*.

During the run of the application in Storm and also during simulations of the CPN, delays at the individual tasks were measured, i.e., the time since an image is output by the *image generator* until it is processed by a particular task.

In Table 1, we can see the results concerning the maximum (minimum) measured delays. In the CPN rows, there are the maximum (minimum) delays retrieved during the simulation of the CPN. In the Storm rows, there are the percentages of data items whose delay was equal or less (greater) than the delay

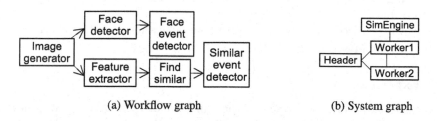

(a) Workflow graph   (b) System graph

Fig. 4. Experiment #1 model

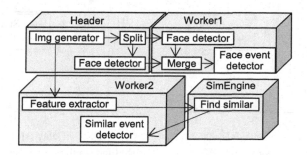

Fig. 5. Experiment #1 deployment graph

Table 1. Experiment #1 maximum delays at CPN and Storm comparison

	Face det. Header	Face det. Worker2	Feat. ext. Worker1	Find sim. SimEngine
CPN: max delay [ms]	7718	7778	3782	7623
Storm: % of delays ≤ CPN max	99.66	99.75	99.96	99.97
CPN: min delay [ms]	26	29	1369	1683
Storm: % of delays ≥ CPN min	99.74	99.75	99.99	99.42
CPN: avg delay [ms]	2082	2125	1793	2826
Storm: avg delay [ms]	2174	2174	1775	2384

predicted by the CPN. It can be observed that nearly 100 % of the data items processed by Storm were processed within the limits set by the CPN. Two last rows of the table show average delays obtained during the CPN simulation and the Storm run.

In Fig. 6, there are relative frequencies of the delays measured in Storm and retrieved during the CPN simulations. Only two graphs are shown due to space constraints.

The values for the relative frequencies are computed as follows. The timeline is separated into time intervals (mostly 1 s long), and for each of the intervals, number of data items having the corresponding delay is counted. Finally, the number is divided by the overall amount of the data items to obtain the relative frequency. The frequencies are represented as a continuous line connecting the discrete values.

It can be noticed that the most probable delays measured in Storm correspond to those predicted by the CPN.

4.2   Experiment #2

The second experiment uses almost the same set of tasks but the workflow is different (see Fig. 7a). At first, an image is sent to the *face detector*. If any faces

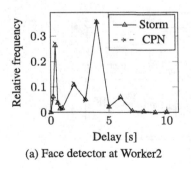

(a) Face detector at Worker2

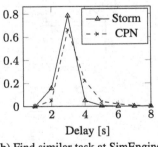

(b) Find similar task at SimEngine

Fig. 6. Experiment #1 delays

are detected, the image is passed on to the *feature extractor* and subsequently to the *find similar* task. A new image is generated every 1,200 ms. The rest of the workflow model is the same as for the first experiment.

The system model is exactly the same as for the first experiment. The deployment graph is slightly different and is shown in Fig. 7b. The *face event detector* is removed; the other tasks are placed on the same resources as for the Experiment #1; the edges are adjusted so that they match the workflow model.

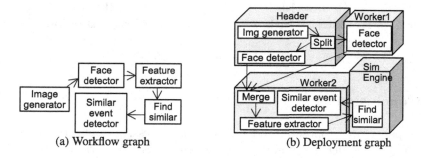

(a) Workflow graph   (b) Deployment graph

Fig. 7. Experiment #2 model

In Table 2, there is a comparison of maximum, minimum and average delays measured in Storm and those which were derived during CPN simulations. We can notice higher delays than we saw at the Experiment #1. This is due to the shorter output frequency of the *image generator* (1,850 ms vs. 1,200 ms) which results in temporary queues at individual tasks. Nearly 100 % of the data items processed by Storm were processed within the limits predicted by the CPN.

Table 2. Experiment #2 maximum delays at CPN and Storm comparison

	Face det. Header	Face det. Worker2	Feat. ext. Worker1	Find sim. SimEngine
CPN: max delay [ms]	14685	15203	34063	34619
Storm: % of delays ≤ CPN max	99.59	99.30	98.70	98.65
CPN: min delay [ms]	27	63	1580	2015
Storm: % of delays ≥ CPN min	99.88	99.67	100	99.96
CPN: avg delay [ms]	3377	3462	11583	12631
Storm: avg delay [ms]	3952	4093	13499	14194

The graphs in Fig. 8 show relative frequencies of the delays measured in Storm and computed during CPN simulations. It can be observed that the distributions of the delays retrieved through the simulations can be used to reliably predict expected delays in Storm.

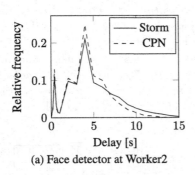 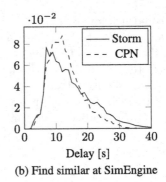

(a) Face detector at Worker2   (b) Find similar at SimEngine

Fig. 8. Experiment #2 delays

5   Conclusion

In this paper, we propose a model of stream processing applications which is focused on depicting performance related properties. For performance analysis of the application, the model is converted to a Colored Petri Net which is subsequently used to derive performance characteristics. The accuracy of the model was validated through a couple of experiments dealing with multimedia data. CPN simulations results were compared to performance qualities of applications running in a Storm cluster. It was able to predict nearly 100 % of the maximum delays precisely. Moreover, it is possible to reliably predict the distribution of the delays using the proposed methods.

The applicability of our proposed model is not restricted to just multimedia domain. It will serve well for other kinds of streaming applications which consist of tasks with variable processing costs, and variable amounts and frequencies of produced data. In the future work, we intend to apply the model and the performance analysis to security applications such as surveillance video analysis. In addition, we plan to extend the model to reflect the reality more precisely; for instance, scheduling strategies would be useful to specify whether any of the tasks placed on the same resource should be prioritized. Finally, we intend to make the conversion of the model to CPN automatic.

Acknowledgements. This work was supported by the Czech national research project GBP103/12/G084. The hardware infrastructure was provided by the META-Centrum under the programme LM 2010005.

References

1. Apache Storm. https://storm.apache.org/
2. Benoit, A., Çatalyürek, Ü.V., Robert, Y., Saule, E.: A survey of pipelined workflow scheduling: models and algorithms. ACM Comput. Surv. (CSUR) **45**(4), 50 (2013)

3. Bouillard, A., Phan, L.T., Chakraborty, S.: Lightweight modeling of complex state dependencies in stream processing systems. In: 15th IEEE Real-Time and Embedded Technology and Applications Symposium, 2009, RTAS 2009, pp. 195–204. IEEE (2009)
4. Chakraborty, S., Phan, L.T., Thiagarajan, P.: Event count automata: a state-based model for stream processing systems. In: 26th IEEE International Real-Time Systems Symposium, 2005, RTSS 2005, pp. 87–98. IEEE (2005)
5. Chakraborty, S., Thiele, L.: A new task model for streaming applications and its schedulability analysis. In: Proceedings of Design, Automation and Test in Europe, 2005, pp. 486–491. IEEE (2005)
6. Evans, M., Hastings, N., Peacock, B.: Probability density function and probability function. In: Hastings, N. (ed.) Statistical Distributions, pp. 9–11. Wiley, New York (2000)
7. Gottumukkala, R.N., Shepherd, M.D., Sun, T.: Validation and analysis of JDF workflows using colored petri nets. US Patent 7,734,492, 8 June 2010
8. Jensen, K., Kristensen, L.M.: Coloured Petri Nets: Modelling and Validation of Concurrent Systems. Springer, Heidelberg (2009)
9. Rygielski, P., Kounev, S.: Data center network throughput analysis using queueing petri nets. In: 2014 IEEE 34th International Conference on Distributed Computing Systems Workshops (ICDCSW), pp. 100–105. IEEE (2014)
10. Wieczorek, M., Hoheisel, A., Prodan, R.: Towards a general model of the multi-criteria workflow scheduling on the grid. Future Gener. Comput. Syst. **25**(3), 237–256 (2009)

Author Index